Computing the Brain

Computing the Brain

A Guide to Neuroinformatics

Edited by

Michael A. Arbib
Center for Neural Engineering
University of Southern California
Los Angeles, California

Jeffrey S. Grethe
Center for Cognitive Neuroscience
Dartmouth College
Hanover, New Hampshire

ACADEMIC PRESS
A Harcourt Science and Technology Company

San Diego   New York   Boston   London   Sydney   Tokyo   Toronto

Front cover photograph: Adaptation of a screenshot of NeuARt's Display Manager. (For more details, see Chapter 4.3, Figure 7.)

This book is printed on acid-free paper.

Copyright © 2001 by ACADEMIC PRESS

All Rights Reserved.
No part of this publication may be reproduced or transmitted in any form or by any means, electronic or mechanical, including photocopy, recording, or any information storage and retrieval system, without permission in writing from the publisher.

Requests for permission to make copies of any part of the work should be mailed to: Permissions Department, Harcourt Inc., 6277 Sea Harbor Drive, Orlando, Florida 32887-6777

Academic Press
A Harcourt Science and Technology Company
525 B Street, Suite 1900, San Diego, California 92101-4495, USA
http://www.academicpress.com

Academic Press
Harcourt Place, 32 Jamestown Road, London NW1 7BY, UK
http://www.academicpress.com

Library of Congress Catalog Card Number: 00-109535

International Standard Book Number: 0-12-059781-0

PRINTED IN THE UNITED STATES OF AMERICA
01 02 03 04 05 06 EB 9 8 7 6 5 4 3 2 1

Contents

Contributors ix
Preface xi

PART 1
INTRODUCTION

CHAPTER 1.1
NeuroInformatics: The Issues 3
Michael A. Arbib

1.1.1 Overview 3
1.1.2 Modeling and Simulation 8
1.1.3 Databases for Neuroscience Time Series 14
1.1.4 Visualization and Atlas-Based Databases 17
1.1.5 Data Management and Summary Databases 19
1.1.6 The NeuroInformatics Workbench 26
References 27

CHAPTER 1.2
Introduction to Databases 29
Wen-Hsiang Kevin Liao and Dennis McLeod

Abstract 29
1.2.1 An Overview of Database Management 29
1.2.2 Historical View of Key Database Developments 30
1.2.3 Relational Database Model 31
1.2.4 SQL 32
1.2.5 Object-Based Database Models 33
1.2.6 Object-Relational Database Model 37
1.2.7 An Overview of Federated Database Systems 38
References 39

PART 2
MODELING AND SIMULATION

CHAPTER 2.1
Modeling the Brain 43
Michael A. Arbib

Abstract 43
2.1.1 Modeling Issues 43
2.1.2 Parietal-Premotor Interactions in the Control of Grasping 46
2.1.3 Basal Ganglia 48
2.1.4 Cerebellum 53
2.1.5 Hippocampus, Parietal Cortex, and Navigation 61
2.1.6 Discussion 67
References 67

CHAPTER 2.2
NSL Neural Simulation Language 71
Michael A. Arbib, Amanda Alexander, and Alfredo Weitzenfeld

2.2.1 Modeling and Simulation of Neural Networks 72
2.2.2 NSL Modules and Simulation 74
2.2.3 The NSL System 83
2.2.4 Simulating a Model—The Maximum Selector Model 85
2.2.5 Maximum Selector Model 85
2.2.6 Available Resources 89
References 89

CHAPTER 2.3
EONS: A Multi-Level Modeling System and Its Applications 91
Jim-Shih Liaw, Ying Shu, Taraneh Ghaffar, Xiaping Xie, and Theodore W. Berger

2.3.1 Introduction 91
2.3.2 EONS Object Library 92
2.3.3 Protocol-Based Simulation 98
2.3.4 Conclusion 100
References 100

CHAPTER 2.4
Brain Imaging and Synthetic PET 103
Amanda Bischoff-Grethe and Michael A. Arbib

Abstract 103
2.4.1 PET Imaging and Neurophysiology 103
2.4.2 Defining Synthetic PET 104
2.4.3 Example: A Model of Saccadic Eye Movements 105

2.4.4 Synthetic PET for Grasp Control   107
2.4.5 Discussion   111
References   112

Part 3
DATABASES FOR NEUROSCIENCE TIME SERIES

Chapter 3.1
Repositories for the Storage of Experimental Neuroscience Data   117
Richard F. Thompson, Jeffrey S. Grethe, Ted Berger, and X. Xie

3.1.1 Introduction   117
3.1.2 Protocols: A Data Model To Address Schema Complexity in Neuroscience   118
3.1.3 Considerations of the User Community   119
3.1.4 Building a Time-Series Database for *In Vivo* Neurophysiology: Cerebellum and Classical Conditioning   120
3.1.5 Building a Time-Series Database for *In Vitro* Neurophysiology: Long-Term Potentiation (LTP) in the Hippocampal Slice   125
3.1.6 Building a Database for Human Neuroimaging Data   129
3.1.7 Discussion   131
References   132

Chapter 3.2
Design Concepts for NeuroCore and NeuroScience Databases   135
Jeffrey S. Grethe, Jonas Mureika, and Edriss N. Merchant

3.2.1 Design Concepts for Neuroscience Databases   135
3.2.2 The Three Main Components of the NeuroCore Database   135
3.2.3 Detailed Description of the NeuroCore Database   137
Conclusion   150
References   150

Chapter 3.3
User Interaction with NeuroCore   151
Edriss N. Merchant, David A.T. King, and Jeffrey S. Grethe

3.3.1 Introduction   151
3.3.2 NeuroCore Schema Browser   153
3.3.3 Java Applet for Data Entry (JADE)   156
3.3.4 Database Browser   157
3.3.5 DataMunch   160
3.3.6 Discussion   161
References   163

Part 4
ATLAS-BASED DATABASES

Chapter 4.1
Interactive Brain Maps and Atlases   167
Larry W. Swanson

Abstract   167
4.1.1 General Features of Maps   167
4.1.2 Overall Structure of the Brain   168
4.1.3 Experimental Circuit-Tracing Methods   169
4.1.4 Atlases: Slice-Based Sampling and Standard Brains   169
4.1.5 Transferring Data from Experimental Brain to Standard (Atlas Reference) Brain   173
4.1.6 Toward Textual and Graphical Databases on the Web   174
4.1.7 Three-Dimensional Computer Graphics Models of the Brain   174
4.1.8 Two-Dimensional Flatmaps: Schematic Circuit Diagrams and Distribution Patterns   176
4.1.9 The Future: Atlases as Expandable Databases and Models   176
References   177

Chapter 4.2
Perspective: Geographical Information Systems   179
Cyrus Shahabi and Shuping Jia

Abstract   179
4.2.1 Introduction   179
4.2.2 Overview of GIS   180
4.2.3 Atlas-Based Neuroscientific Data   184
4.2.4 Raster Data   185
4.2.5 Conclusion and Web Resources   186
References   187

Chapter 4.3
The Neuroanatomical Rat Brain Viewer (NeuARt)   189
Ali Esmail Dashti, Gully A.P.C. Burns, Donna M. Simmons, Larry Swanson, Shahram Ghandeharizadeh, Cyrus Shahabi, James Stone, and Shuping Jia

4.3.1 Introduction   189
4.3.2 The NeuARt System   190
4.3.3 Discussion   200
References   200

Chapter 4.4
Neuro Slicer: A Tool for Registering 2–D Slice Data to 3–D Surface Atlases   203
Bijan Timsari, Richard M. Leahy, Jean-Marie Bouteiller, and Michael Baudry

Abstract   203
4.4.1 Introduction   203
4.4.2 Classification Criteria   204
4.4.3 Intrasubject Image Matching   205
4.4.4 Intersubject Image Matching   206
4.4.5 Neuro Slicer: USCBP Histological Registration Tool   208
References   213

Chapter 4.5
An Atlas-Based Database of Neurochemical Data   217
Rabi Simantov, Jean-Marie Bouteiller, and Michael Baudry

Abstract   217
4.5.1 Synaptic Neurotransmission: Molecular and Functional Aspects   217
4.5.2 Roles of Glutamatergic Synapses in LTP and LTD   218

4.5.3 Glutamate Receptor Regulation and
Synaptic Plasticity   218
4.5.4 How To Build a Useful Neurochemical Database   219
4.5.5 Incorporating Neurochemical Data into the NeuroCore Repository of Empirical Data   222
4.5.6 Available Resources   225
4.5.7 Conclusion   227
References   227

Part 5
DATA MANAGEMENT

Chapter 5.1
Federating Neuroscience Databases   231
Wen-Hsiang Kevin Liao and Dennis McLeod

Abstract   231
5.1.1 Introduction   231
5.1.2 Information Discovery   232
5.1.3 Semantic Heterogeneity Resolution   233
5.1.4 System-Level Interconnection   233
5.1.5 Characteristics of Sharing Patterns   234
5.1.6 System Architecture for Sharing Primitives/Tools   236
5.1.7 Federating Neuroscience Databases   237
References   238

Chapter 5.2
Dynamic Classification Ontologies   241
Jonghyun Kahng and Dennis McLeod

Abstract   241
5.2.1 Introduction   241
5.2.2 Heterogeneity   242
5.2.3 Common Ontology   244
5.2.4 Classification   246
5.2.5 Dynamic Classification Ontology   247
5.2.6 Mediators for Information Sharing   250
5.2.7 Conclusions   253
References   254

Chapter 5.3
Annotator: Annotation Technology for the WWW   255
Ilia Ovsiannikov and Michael Arbib

Abstract   255
5.3.1 Introduction   255
5.3.2 Overview of Existing Annotation Software   256
5.3.3 Annotation Technology: An Integrative Approach   256
5.3.4 Annotator   261
References   263

Chapter 5.4
Management of Space in Hierarchical Storage Systems   265
Shahram Ghandeharizadeh, Douglas J. Ierardi, and Roger Zimmermann

Abstract   265
5.4.1 Introduction   265
5.4.2 Target Environment   268
5.4.3 Four Alternative Space Management Techniques   269

5.4.4 Performance Evaluation   276
5.4.5 Analytical Models   280
5.4.6 Conclusions   283
References   283

Part 6
SUMMARY DATABASES AND MODEL REPOSITORIES

Chapter 6.1
Summary Databases and Model Repositories   287
Michael A. Arbib and Amanda Bischoff-Grethe

Abstract   287
6.1.1 The Database Typology and the NeuroInformatics Workbench   287
6.1.2 An Overall Perspective   288
6.1.3 General Considerations on Model Repositories   291
6.1.4 Brain Models on the Web   291
6.1.5 NeuroScholar   292
6.1.6 NeuroHomology   294
6.1.7 Future Plans   296

Chapter 6.2
Brain Models on the Web and the Need for Summary Data   297
Amanda Bischoff-Grethe, Jacob Spoelstra, and Michael A. Arbib

Abstract   297
6.2.1 Storing Brain Models   297
6.2.2 The BMW-SDB Relationship   298
6.2.3 Reviewing a Model: The Dart Model of Prism Adaptation   299
6.2.4 Database Design   300
6.2.5 Accessing the Database   308
6.2.6 Future Plans   316
References   317

Chapter 6.3
Knowledge Mechanics and the Neuroscholar Project: A New Approach to Neuroscientific Theory   319
Gully A.P.C. Burns

6.3.1 An Introduction to Knowledge Mechanics   319
6.3.2 Concept of "Theory" in Neuroscience   320
6.3.3 High-Level Software Requirements and Fundamental Design Concepts of the NeuroScholar System   321
6.3.4 Neuroscholar in Detail   327
6.3.5 The Significance of Knowledge Mechanics   333
References   334

Chapter 6.4
The NeuroHomology Database   337
Mihail Bota and Michael A. Arbib

6.4.1 Introduction: The Definition of the Concept of Homology in Neurobiology   337
6.4.2 Theory of Degrees of Homology   339

6.4.3 The NeuroHomology Database: Description 340
6.4.4 The NeuroHomology Database: Brain Structures 341
6.4.5 NeuroHomology Database: Connectivity Issues 343
6.4.6 The NeuroHomology Database: Homologies 346
6.4.7 Conclusion and Future Development 348
References 350

APPPENDICES

APPENDIX A1
Introduction to Informix 355
Jonas Mureika and Edriss N. Merchant

Informix SQL Tutorial: A Practical Example 355

APPENDIX A2
NeuroCore TimeSeries Datablade 359
Jonas Mureika and Edriss N. Merchant

Introduction 359
Internal Structure and Description 359
Support Functions 360
Additional SQL-Invoked Routines 360
Comparing Data 360

Discussion 361

APPENDIX A3
USCBP Development Team 363

APPENDIX B1
Informix SQL Quick Reference 365
Jonas Mureika and Edriss N. Merchant

Introduction 365
SQL Statements 365

APPENDIX C1
USC Brain Project Research Personnel 367

APPENDIX C2
Doctoral Theses from the USC Brain Project (May 1997–August 2) 369

Index 371

Contributors

The numbers in parentheses indicate the pages on which the authors' contributions begin.

Amanda Alexander (71)
University of Southern California Brain Project, University of Southern California, Los Angeles, California 90089-2520

Michael A. Arbib (3, 43, 71, 103, 255, 287, 297, 337)
University of Southern California Brain Project, University of Southern California, Los Angeles, California 90089-2520

Michel Baudry (203, 217)
University of Southern California Brain Project, Biological Sciences Department, University of Southern California, Los Angeles, California 90089-2520

Theodore W. Berger (91, 117)
Biomedical Engineering Department, University of Southern California, Los Angeles, California 90089-2520

Amanda Bischoff-Grethe (103, 287, 297)
Center for Cognitive Neuroscience, Dartmouth College, Hanover, New Hampshire 02755

Mihail Bota (337)
University of Southern California Brain Project, University of Southern California, Los Angeles, California 90089-2520

Jean-Marie Bouteiller (203, 217)
University of Southern California Brain Project and Neuroscience Program, University of Southern California, Los Angeles, California 90089-2520

Gully A. P. C. Burns (189, 319)
University of Southern California Brain Project and Neuroscience Program, University of Southern California, Los Angeles, California 90089-2520

Ali Esmail Dashti (189)
Computer Engineering Department, College of Engineering and Petroleum, Kuwait University, Al-Khaldya, Kuwait

Taraneh Ghaffar (91)
Biomedical Engineering Department, University of Southern California, Los Angeles, California 90089-2520

Shahram Ghandeharizadeh (189, 265)
University of Southern California Brain Project and Computer Science Department, University of Southern California, Los Angeles, California 90089-0781

Jeffrey S. Grethe (117, 135, 151)
Center for Cognitive Neuroscience, Dartmouth College, Hanover, New Hampshire

Douglas J. Ierardi (265)
Computer Science Department, University of Southern California, Los Angeles, California 90089-2520

Shuping Jia (179, 189)
Intuit Inc., San Diego, California 92122

Jonghyun Kahng (241)
Live365.com, Foster City, California 94404

David A. T. King (151)
Neuroscience Program, University of Southern California, Los Angeles, California 90089-2520

Richard M. Leahy (203)
Department of Electrical Engineering-Systems, University of Southern California, Los Angeles, California 90089-2520

Wen-Hsiang Keven Liao (29, 231)
Live365.com, Foster City, California 94404

Jim-Shih Liaw (91)
University of Southern California Brain Project and Biomedical Engineering Department, University of Southern California, Los Angeles, California 90089-2520

Dennis McLeod (29, 231, 241)
University of Southern California Brain Project and Computer Science Department, University of Southern California, Los Angeles 90089-2520

Edriss N. Merchant (135, 151, 355, 359, 365)
University of Southern California Brain Project, University of Southern California, Los Angeles 90089–2520

Jonas Mureika (135, 355, 359, 365)
Physics Department, University of Toronto, Toronto, Ontario Canada M5S 1A7

Ilia Ovsiannikov (255)
University of Southern California Brain Project, University of Southern California, Los Angeles 90089–2520

Ying Shu (91)
Computer Science Department, University of Southern California, Los Angeles, California 90089–0781

Cyrus Shahabi (179, 189)
University of Southern California Brain Project and the Computer Science Department, University of Southern California, Los Angeles, California 90089–0781

Ying Shu (91)
Seibel Systems, Inc., San Mateo, California 94404

Rabi Simantov (217)
Molecular Genetics Department, Weizmann Institute of Science, Rehovot, 76100, Israel

Donna M. Simmons (189)
Biological Sciences Department, University of Southern California, Los Angeles, California 90089–2520

Jacob Spoelstra (297)
HNC Software, Inc., San Diego, California 92120

James Stone (189)
Neuroscience Program, University of California, Davis, California 95616

Larry W. Swanson (167, 189)
The Neuroscience Program and University of Southern California Brain Project, University of Southern California, Los Angeles, California 90089–2520

Bijan Timsari (203)
Netergy Microelectronics, Inc., Santa Clara California 95054

Richard F. Thompson (117)
The Neuroscience Program and Departments of Psychology and Biological Sciences, University of Southern California, Los Angeles, California 90089–2520

Alfredo Weitzenfeld (71)
Instituto Tecnologico Autonomo de Mexico, Departmento Academico de Computación, San Angel Tizapan, CP01000, Mexico DF, Mexico

Xiaping Xie (91, 117)
Biomedical Engineering Department, University of Southern California, Los Angeles, California 90089–2520

Roger Zimmermann (265)
Integrated Media Systems Center and Computer Science Department, University of Southern California, Los Angeles, California 90089–2561

Preface

For many workers in the field, *Neuroinformatics* is the use of databases, the World Wide Web, and visualization in the storage and analysis of neuroscience data. However, in this book we see the structuring of masses of data by a variety of computational models as essential to the future of neuroscience, and thus broaden this definition to include *Computational Neuroscience*, the use of computational techniques and metaphors to investigate relations between neural structure and function.

In recent years, the Human Genome Project has become widely known for its sequencing of the complete human genome, and its placing of the results in comprehensive databases such as GenBank. This has been made possible by advances in gene sequencing machinery that have transformed the sequencing of a gene from being a major research contribution publishable in *Science* to an automated process costing a few cents per base pair. The resultant data are of immense importance but are rather simple, since the key data are in the form of annotated base-pair sequences of DNA. By contrast, the Human Brain Project (HBP)—a consortium of U.S. federal agencies funding work in neuroinformatics—has the problem of building databases for immensely heterogeneous sets of data. The brain is studied at multiple levels, from the behavior of the overall organism through the diversity of brain regions down through specific neural circuits and beyond. The human brain contains on the order of 10^{11} neurons, such neurons may have tens of thousands of synapses (connections) from other neurons, and these synapses are themselves complex neurochemical structures containing many macromolecular channels or receptors. Not only do we have to contend with the many orders of magnitude linking the finest details of neurochemistry to the overall behavior of the organism, but we also have to integrate data gathered by many different specialists. Neuroanatomists characterize the brain's connectivity patterns. Neurophysiologists characterize neural activity and the "learning rules" which summarize the conditions for, and dynamics of, change. Neurochemists seek the molecular mechanisms which yield these "rules", while computational neuroscientists seek to place all these within a systems perspective.

The first "map" of neuroinformatics was provided in the edited volume *Neuroinformatics: An Overview of the Human Brain Project* (Koslow and Huerta, 1997). The present volume is both broader than its predecessor—it gives a much fuller view of the computational neuroscience components of neuroinformatics and of underlying issues in research on databases—and narrower in that it has rather little to say on human brain imaging, which is a major thrust of the HBP consortium. Indeed, the book focuses on the work of the University of Southern California Brain Project (USCBP), funded in part by a Program Project (P20) grant from the Human Brain Project (P01MH52194), with contributions from NIMH, NIDA, and NASA. At first this focus might seem a weakness. However, what has distinguished USCBP from other HBP efforts is its emphasis on *integration,* and we are thus able to offer an integrated overview of neuroinformatics which was missing in the previous volume, which gathered contributions from a number of laboratories with very different foci.

We do not claim that our work subsumes the many contributions made by other laboratories engaged in neuroinformatics. Much research has been conducted on neuroinformatics, with HBP funding and under other auspices, both in the U.S. and elsewhere. To get a sense of what is being done beyond the material presented in this book, the reader should start with the HBP Website (http://www.nimh.nih.gov/neuroinformatics/index.cfm), and follow the links from there.

What we do claim is that the present volume offers a unified perspective that is available nowhere else, a perspective in which the diverse contributions of many laboratories can be better appreciated and evaluated than would otherwise be possible. Indeed, the material in this book grows not only from our own research but also from our experience in teaching, three times in five years, a graduate course in neuroinformatics to a total of 80 students in biomedical engineering, computer science, neuroscience, and other departments. We have thus kept the needs of graduate students coming to neuroinformatics research from diverse disciplines, as well as the needs of neuroscientists seeking a comprehensive introduction to neuroinformatics, very much in mind. In this spirit, this book aims to show how to approach "Computing the Brain," integrating database, visualization, and simulation technology to gain a deeper, more integrated view of the data of neuroscience, assisting the conversion of data into knowledge.

The book is divided into 6 parts:

Part 1. Introduction: The first chapter, "Neuroinformatics: The Issues," both sets the stage for the study of neuroinformatics in general, and also introduces the feature that makes USCBP unique among all other HBP projects, namely that we have created the **NeuroInformatics Workbench**, a *unified architecture for neuroinformatics*. This is a suite of tools to aid the neuroscientist in constructing and using databases, and in visualizing and linking models and data. At present, the Workbench contains three main components: **NSLJ**, a modular, Java-based language and environment for neural simulation; **NeuroCore**, a system for constructing and using neuroscience databases; and **NeuARt**, a viewer for atlas-based neural data (the **Neu**ro**A**natomical **R**egistration Viewer). The second chapter, "Introduction to Databases," provides the expository role of its title. Our approach to databases exploits object-relational database management and is adaptable to any database management of this kind. The specific implementation of our database uses the Informix Universal Server which provides the ability to construct new data types as Datablades (a new base type along with its associated functions) which can be "plugged in" to the Informix architecture. These facilities are described in the appendices.

Part 2. Modeling and Simulation: We start with a chapter, "Modeling the Brain" which provides an overview of work in computational neuroscience, providing both a general perspective and a brief sampling of models constructed at USCBP. For a variety of behaviors, we seek to understand what must be added to the available databases on neural responsiveness and connectivity to explain the time course of cellular activity, and the way in which such activity mediates between sensory data, the animal's intention, and the animal's movement. The attention paid by neuroscience experimentalists to computational models is increasing, as modeling occurs at many levels, such as (i) the systems analysis of circuits using the NSL Neural Simulation Language developed at USC; (ii) the use of the GENESIS language developed at Caltech and the NEURON language from the University of North Carolina and Yale to relate the detailed morphology of single cells to their response to patterns of input stimulation; and (iii) the EONS library of "Essential Objects of Nervous Systems" developed at USC to model activity in individual synapses in great detail. The next two chapters introduce the USC contributions, "NSL Neural Simulation Language" and "EONS: A Multi-Level Modeling System and Its Applications." Since the neuroinformatics of human brain imaging is so well covered by many research groups with and without HBP funding, this has not been a focus of USCBP research. However, we have been concerned with the following question: "How can the data from animal neurophysiology be integrated with data from human imaging studies?" The chapter "Brain Imaging and Synthetic PET" presents our answer.

Part 3. Databases For Neuroscience Time Series: The first chapter provides our general view of how to build "Repositories for the Storage of Experimental Neuroscience Data." We see the key to be the notion of the *experimental protocol* which defines a class of experiments by specifying a set of experimental manipulations and observations. When linking empirical data to models, we translate such a protocol into a simulation interface for "stimulating" a simulation of the empirical system under analysis and displaying the results in a form which eases comparison with the results of biological experiments conducted using the given protocol. The chapter "Design Concepts for NeuroCore and NeuroScience Databases" introduces NeuroCore, a novel extendible object-relational database schema implemented in Informix. The schema (structure of data tables, etc.) for each NeuroCore database is an extension of our core database schema which is readily adaptable to meet the needs of a wide variety of neuroscience databases. In particular, we have constructed a new Datablade which allows neurophysiological data to be stored and manipulated readily in the database. (See the appendix "NeuroCore TimeSeries Datablade.") The final chapter of Part 3, "User Interaction with NeuroCore," describes the various components we have developed of an *on-line notebook* that provides a laboratory independent "standard" for viewing, storing, and retrieving data across the Internet. We also present our view that the article will continue to be a basic unit of scientific communication, but envision ways in which articles can be enriched by manifold links to the federated databases of neuroscience.

Part 4. Atlas-Based Databases: How are data from diverse experiments on the brains of a given species to be integrated? Our answer is to register the data—whether the locations of cells recorded neurophysiologically, the tract tracings of an anatomical experiment, or the receptor densities revealed on a slice of brain in a neurochemical study—against a standard brain atlas for the given species. The chapter "Interactive Brain Maps and Atlases" provides a general view of such atlases, while "Perspective: Geographical Information Systems" notes the similarities and differences between maps of Earth and brain. The key chapter of Part 4 is "The Neuroanatomical Rat Brain Viewer (NeuARt)" The chapter "Neuro Slicer: A Tool for Registering 2–D Slice Data to 3–D Surface Atlases" addresses the problem of registering data against an atlas when the plane of section for the data is different from that of a plate in the atlas. The key is to reconstitute a 3–D atlas from a set of 2–D plates, and then reslice this representation to find a plane of section against which the empirical data can be registered with minimal distortion. Part 4 closes with the presentation of "An Atlas-Based Database of Neurochemical Data."

Part 5. Data Management: "Federating Neuroscience Databases" addresses the important issue that there will not be a single monolithic database which will store all neuroscience data. Rather, there will be a federation of databases throughout the neuroscience community. Each database has its own "ontology," the set of objects which create the "universe of discourse" for the database. However, different databases may use different ontologies to describe related material, and the chapter on "Dynamic Classification Ontologies" discusses strategies for dynamically linking the ontologies of the databases of a database federation. We then present "Annotator: Annotation Technology for the WWW" as a means to expand scientific (and other) collaboration by constructing databases of annotations linked to documents on the Web, whether they be for personal use, or for the shared use of a community. Part 5 closes with "Management of Space in Hierarchical Storage Systems," an example of our database research addressing the issue of how to support a user community that needs timely access to increasingly massive datasets.

Part 6. Summary Databases: "Summary Databases and Model Repositories" describes the essential role of databases which summarize key hypotheses gleaned from a wide variety of empirical and modeling studies in attempting to maintain a coherent view of a nearly overwhelming body of data, and how such summary database may be linked with model repositories both to ground model assumptions and to test model predictions. "Brain Models on the Web and the Need for Summary Data" describes the construction of a database which not only provides access to a wide range of neural models but also supports links to empirical databases, and tools for model revision. "Knowledge Mechanics and the NeuroScholar Project: A New Approach to Neuroscientific Theory" offers both a general philosophy of the construction of summary databases, and a specific database for analyzing connections of the rat brain exemplifying this philosophy. Finally, "The NeuroHomology Database" presents a database design which supports the analysis of homologies between the brain regions of different species, returning us to the issue of how best to integrate the findings of animal studies into our increasing understanding of the human brain.

The majority of chapters end with a section on "Available Resources" which describes the availability of our software and databases as this book goes to press. Much of the material is available for downloading; in other cases the prototypes are not yet robust enough for export, but in many cases may nonetheless be viewed online through demonstrations. The USCBP Website may be found at http:\\www-hbp.usc.edu, and will be continually updated to give the reader expanding access to currently available materials.

Michael A. Arbib
Los Angeles, California
Jeffrey S. Grethe
Hanover, New Hampshire

PART 1

Introduction

CHAPTER 1.1

NeuroInformatics: The Issues

Michael A. Arbib
*University of Southern California Brain Project and Computer Science Department,
University of Southern California, Los Angeles, California*

1.1.1 Overview

We see the structuring of masses of data by a variety of computational models as essential to the future of neuroscience; thus, we define *neuroinformatics* as the integration of: (1) the use of databases, the World Wide Web, and visualization in the storage and analysis of neuroscience data with (2) computational neuroscience, using computational techniques and metaphors to investigate relations between neural structure and function. The challenge to be met is that of going back and forth between model data (i.e., synthetic data obtained from running a model) and research data obtained empirically from studying the animal or human. Research will pursue a theory-experiment cycle as model predictions suggest new experiments and models improve as they are adapted to encompass more and more of these data.

We view it as crucially important to develop computational models at all levels, from molecules to compartments and physical properties of neurons up to neural networks in real systems constrained by real connections and real physiological properties. These can then be tested against the empirical data and that is why it is so valuable to maintain an architecture for a federation of empirical databases in which the results from diverse laboratories can be integrated, and to provide an environment in which we can develop computational modeling to the point where we can make quantitative verifiable or disprovable predictions from the model to the database.

The University of Southern California Brain Project (USCBP) approach to neuroinformatics is thus distinguished not only by its concern with the development of models to summarize and yield insight into data, but also in that we are developing general *architectures* for the support of neuroinformatics. To focus our work in software development, we are building *The NeuroInformatics Workbench*™, a collection of neuroinformatics tools which we summarize below. But, it is important to realize that many other groups will be developing neuroinformatics tools, so part of our work addresses the key issue—for databases, simulators, and all the other tools discussed in this volume—of *interoperability*, ensuring that tools and databases developed by different subcommunities can communicate with each other despite their idiosyncrasies.

Simulation, Databases, and the World Wide Web

Our approach to neuroinformatics is shaped by three technologies: (1) The classical use of computers for executing programs for numerical manipulation (this lies at the heart of our work in modeling and simulation); (2) the development of database management systems (DBMSs), which make it easy to generate a wide variety of databases (organized collections of structured facts) stored in a computer for rapid storage and retrieval of data; and (3) the World Wide Web, which has been transformed with startling rapidity from a tool for computer researchers into a household utility allowing resources appropriately stored on one computer hooked to the Internet (the "server") to be accessed from any other computer on the Internet (the "client") provided one has the URL (universal resource locator) for the resource of interest.

THE WORLD WIDE WEB

The World Wide Web has indeed become so familiar that we will assume that every reader of this book knows how to use it, and we will repeatedly provide URLs to the databases and tools described in the pages that follow.

DATABASES

On the other hand, we will not assume that the reader has any deep familiarity with databases. Chapter 1.2 introduces the basic concepts. *Relational databases* (introduced in the 1970s) provide a very structured way of describing information using "tables." The current standard for a data manipulation language for querying and modifying a database is SQL (a contraction of SEQUEL, the Structured English QUEry Language introduced by IBM). *Object-based databases* (introduced in the 1980s) organize as "objects"—a rich variety of formal structures. Key structures then include objects, classes (collections of objects with something in common), and inter-relationships which structure the semantic connections between objects and classes. *Object-relational databases* (introduced in the 1990s) combine the "best of both worlds" of relational databases and object-based databases. Our approach to databases in this volume is adaptable to any object-relational DBMS; the implementations available on our Website employ a specific object-relational DBMS, namely the Informix Universal Server. Chapter 5.1 will take up the theme of "Federating Databases."

PROGRAMS FOR NUMERICAL MANIPULATION

We shall not assume that the reader has mastery of a specific programming language such as Java or C++ but will rather provide a view of the simulation environments we have built atop these languages (Chapters 2.2 and 2.3). The expert programmer can follow the URLs provided to see all the "gory details." Here we simply note that Java is an object-oriented programming language (see Chapter 2.2 for the explanation of "object-oriented") that runs on the Web. Most Web browsers today are "Java enabled," meaning that the browser provides the "virtual machine" that Java needs to run its programs on the client's machine. *Applets* are programs that run under a browser on the client machine but as a security measure do not write to the disk on the client machine. *Applications* are programs that do not run under a browser (unless as a plug-in) but can write to the user's disk. Our work on the NSLJ simulation environment (Chapter 2.2) emphasizes the use of Java.

The Challenge of Heterogeneous Data

The brain is to be studied at multiple levels, from the behavior of the overall organism through the diversity of brain regions or functional schemas down through specific neural circuits to neurons, synapses, and macromolecular structures. Consider some of the diverse data that neuroscientists use. For example, the study of animals integrates anatomy, behavior, and physiology. In studying the monkey we may note that there are hundreds of brain regions and seek to provide for each such region the criteria by which it is discriminated from other regions—whether it be gross anatomy, the cytoarchitectonics, the input and output connections of the region, or the physiological characterization, or some combination, that drives this discrimination. Then, for a variety of behaviors of interest—whether it be eye movements, various aspects of motor control, performance on memory tasks, etc.—we may seek to characterize those regions of the brain that are most active or most correlated with such behaviors and then characterize the firing of particular populations of neurons in temporal correlation with different aspects of the task. For example, we have studied the role of the intraparietal sulcus in the control of eye movements (modeling data on the lateral intraparietal sulcus, LIP) and in the control of hand movements (modeling data on the role of the anterior intraparietal sulcus, AIP, and area F5 of the premotor cortex). In such modeling studies, we seek to understand what must be added to the available database on neural responsiveness and connectivity to explain the time course of cellular activity and the way in which they mediate between sensory data, the animal's intention, and the animal's movement.

Increasingly, our studies of animals can be related to the many insights we are now gaining from new methods of human brain imaging, such as those afforded by position emission tomography (PET) and functional magnetic resonance imaging (fMRI). Such methods are based on characterization of very subtle differences in the regional blood flow within particular subregions of the brain during one task as compared to another. As such, it is difficult to determine whether the fact of *lowered* significance in a particular region implies *non*-significance for a task. Moreover, the resolution of human brain imaging is very coarse in both space and time compared to the millisecond-by-millisecond study of individual cell activity in the animal. It is thus a great challenge for data analysis and for modeling to find ways, such as the Synthetic PET method for synthesizing predictions of PET activity developed at USC (Chapter 2.4), to relate the results of the observations of individual neural activity in the animal to the overall pattern of comparative regional activity seen in humans.

All this reinforces our point that the comparison of models and experiments is a crucial and continuing challenge, even though much neuroscience to date has paid relatively little attention to the role of explicit computational modeling of brain function. However, this inattention to explicit models is diminishing, as modeling occurs at many levels, such as: (1) the systems analysis of circuits using, for example, the NSL (Neural Simulation Language) developed at USC to compare such things as the effects of different hypotheses in bringing the activity of model circuitry of the cerebellum and related areas in accordance with observations in the Thompson laboratory during classical conditioning experiments; (2) the use of the GENESIS language developed at Caltech and the NEURON language from the University of North Carolina and Yale to relate the detailed morphology of single cells to their response to patterns of input stimulation; and (3) the EONS library of "essential objects of the nervous system" developed at USC to model activity in individual synapses in explicit detail. A challenge for future research is to better integrate the tools developed for the different levels into an integrated suite of multi-level modeling tools.

A crucial challenge, then, is to provide a powerful set of methods for comparing the predictions made by a model with relevant data mined from empirical databases developed under the Human Brain Project and related initiatives in neuroinformatics. We see the key, both for the construction of databases of empirical data and for the comparison of empirical data with simulation results, to be the notion of the *experimental protocol*. Such a protocol defines a class of experiments by specifying a set of experimental manipulations and observations. As a basis for further comparisons, we translate such a protocol into a simulation interface for driving a simulation of the empirical system under analysis and displaying the results in a form which eases comparison with the results of biological experiments conducted using the given protocol.

Federating a Variety of Databases

We here offer two typologies of databases to indicate the different ways in which we will organize the data and the related models and articles, but first we present the notion of federated databases.

FEDERATION

We do not envision there being a single repository of all the data of neuroscience. The way the Web is going, even a single field such as neuroscience may see hundreds, possibly thousands, of databases. There were over 20,000 presentations at the last meeting at the Society for Neuroscience. We expect there to be both personal or laboratory databases and public databases maintained by particular research communities say, people working on cerebellum or on cerebellum for classical conditioning, etc. Each subcommunity may have a shared public database linked to their private databases, thus, workers in neuroinformatics have to understand how to build a *federation* of databases such that it is easy to link data from these databases to gain answers to complex problems. The challenge is to set up databases so that they can be connected in a way that gives the user the illusion of having one wonderful big database at his or her disposal. When a query is made for data from a database federation, the data required may not be in any one of those databases but will be collated from a set of these databases. Users then have the choice of whether to keep the data in their computers as part of their own personal databases or to post the data on one of the existing databases as new information for others to share.

The classic idea of a database federation is to link databases so that each may be used as an extension of the other. We envision a federation linking a multitude of databases accessed through the Web. Our primary strategy has been to design NeuroCore, a database construction system based on an extendable schema (information structure) for neuroscience databases (Chapters 3.1 and 3.2), which makes it easy to link databases that share this common structure. More generally, we envision a "cooperative database federation" linking the neuroscience community. In this approach, the import schema of a given database specifies what data you want to bring in from other databases, and the export schema says what data you are prepared to share and how you will format them for that sharing purpose (Chapter 5.1). In order to be able to connect and access other databases, certain "hooks" have been included in the core database schema to foster such communication. This allows the database to reference and access other databases concerned with published literature as well as on-line electronic atlases. Another possible avenue for database federation in the future is with other neurophysiological database systems using platform-independent transfer protocols such as the TSDP (Time Series Data Protocol) developed by Gabriel and colleagues (Payne *et al.*, 1995). *Databases may be virtual*, integrating partial views gleaned from multiple databases. For example, a database on the neurochemistry of synaptic plasticity might actually be a federation of databases for different brain regions. Moreover, *databases must be linked*: our NeuARt technology (Chapter 4.3) enables an atlas of brain regions to be used to structure data both on the location of single cells (a link to a neurophysiology time series database) and for standardizing slice-based data (such as stains of receptor activity in a brain slice recorded in a neurochemistry database).

TYPOLOGY 1: THE TYPES OF DATA STORED

Article Repositories Many publishers are now going on-line with their journals. There are going to be many such *Article Repositories*, including preprint repositories, technical report repositories, and so on. Article Repositories provide an important class of databases—repositories for articles in electronic form, whether they are journal articles, chapters, or technical reports. Even if articles migrate from linear text to hypertext, such narratives about the data—"This is the recent experiment that I did," "Here is my review," etc.—are going to be very important and will often provide the way for humans to get started in understanding what is going on in some domain, even if they will eventually search specific datasets of the kind described below.

Repositories of Empirical Data What most often comes to mind when one talks about databases for neuroscience is what we call a *Repository of Empirical Data*. This is where we get data from different laboratories and make them available either to laboratory members or more generally. Our approach to Repositories of Empirical Data emphasizes the notion of a protocol. In your own laboratory, you can have a bunch of data and place the electronic recordings of what happened at a particular time on disks or tapes and find them in some drawer as needed. But, if you want other people to look at your data, you need to provide a *protocol:* information on the hypotheses being tested, the experimental methods used, etc. It is this protocol that will allow people to search for and find your experimental data even if they did not conduct the experiment. We have developed NeuroCore™ as our basic design for such databases. If your laboratory has special data structures, you can extend this core in a way that makes it simple for other users to understand the structure of your data. One analogy is with the Macintosh desktop, which is designed to meet certain standards in such a way that if you encounter a new application, you can figure out how to use key elements of the application even without reading the manuals. The idea of NeuroCore™ is to provide a general *data schema* (i.e., a basic structure for the tables of data in the database) which other people can extend readily to provide a tailored data structure that is still easy to understand. We have also invested some energy into the MOP prototype Model for On-line Publishing (Chapter 3.3), which increases the utility of on-line journals, etc. by offering new ways to link them to repositories of empirical data and personal databases.

Summary Databases A *Summary Database* is the place where you go for high-level data, such as assertions, summaries, hypotheses, tables, and figures that encapsulate the "state of knowledge" in a particular domain. A Summary Database is like a review article but is structured as entries in a database rather than as one narrative. If you want to know what is true in a field, you may start with a Summary Database and either accept a summary that it presents to you and work with it to test models or design experiments, or you may follow the links or otherwise search the database federation for data that support or attempt to refute the particular summary. In Summary Databases, assertions can be linked not only to primary literature but also to models or empirical data. One of the issues to be faced below is that, in many fields, there is no consensus as to just which hypotheses have been firmly established. Once you leave the safe world of airline reservations and look at databases for the state of research in any domain of science, you go from a situation where you can just say true or false to the situation where there is controversy, with evidence offered for and against a particular position. Different reviewers may thus assign different "confidence levels" to different primary data, and these will affect the confidence level of assertions in the Summary Database. One contribution of USCBP is the development of Annotation Technology (Chapter 5.4) for building a database of annotations on documents and databases scattered throughout the Web. This may be a personal database for private use or may be a database of annotations to be shared—whether between the members of a collaboratory or with a larger public. In particular, a Summary Database can be seen as a form of annotation database, with each summary serving as an annotation on all the clumps (selected items) that it summarizes. Once annotations are gathered within a database, rather than being embedded in the text of widely scattered documents, it becomes easy to efficiently search the annotations to bring related information together from these many documents. The key idea of Annotation Technology is to provide an extended URL for any "clump" (i.e., any material selected from a document for its interest) which tags the start and endpoint of the clump as well as the URL of the document that contains it. The extended URL methodology then makes it simple to jump to documents, whose relevance can then be determined.

Model Repositories Finally, very important to our concern to catalyze the integration of theory and experiment, is the idea of a *Model Repository*, which is a database that not only provides access to computational models but also links each model to the Empirical and Summary Databases to provide evidence for hypotheses in the model or data to test predictions from simulation runs made with the model. When we design an experiment or make a model of brain function, we have various assertions that summarize what we know for example, the key data from particular laboratories, a table that summarizes key connections, a view of which cells tend to be active during this type of behavior, etc. We have viewed the protocol as a way of understanding what an experiment is all about. When we design a model, we will often give an interface which mimics the protocol so that operations on the model capture the manipulations the experimenter might have made on the nervous system. This will allow the experimenter to make corresponding manipulations through the computer interface to see if the model replicates the results. This makes it easy for somebody not expert in detailed modeling to nonetheless evaluate a model by seeing how it runs in a variety of situations.

In particular, we will emphasize USCBP's model repository, Brain Models on the Web (BMW; see Chapter 6.2). BMW will serve as a framework for electronic publication of computational models of neural systems, as a database that links model assumptions and predictions to databases of empirical data, and as an environment for the development and testing of new models of greater validity. Current work focuses on four types of structures to be stored in the database:

1. *Models:* High-level views of a model linked to the more detailed elements that follow.
2. *Modules:* These are hierarchically structured components of a model.
3. *Simulations:* For each "useful" run of a model, we need to record the parameters and input values used and annotate key points concerning the results.
4. *Interfaces:* To aid non-experts using a model, interfaces must be available to provide a natural way to emulate a number of basic classes of experiments.

With the above typology of databases, we can already see many opportunities for database federation: Entries in a Summary Database may be supported by links to articles in an Article Repository as well as directly to data in a Repository of Empirical Data; articles may come with explicit links from summaries in the articles (figures, tables, assertions in the text) to more detailed supporting of data in the Repositories of Empirical Data; and hypotheses may be supported by models as well as data, thus assertions in a Summary Database may also be linked to predictions in BMW.

TYPOLOGY 2: ACCESS TO DATA

In our view, the database federation will include both lightweight personal databases corresponding to personal and collaboratory databases, as well as integrated public databases that serve a whole community. The issue is to foster both development of these individual databases and federation between them which gives each user the most powerful access to relevant data.

Our next typology is based on considerations of *security*. Every item in a database can be tagged for access by specific individuals or groups and *refereeing* items can be tagged for whether they have been posted by "just anybody" or by a member of some qualified accredited group, or whether an editorial board has looked at and passed an item and said, "Yes, that meets our standards." This provides the benefits of immediate access to results that are not guaranteed to be of high quality and delayed access to results that have been refereed. This is a useful model for using the Web to disseminate results. Thus, not only will databases differ in their type and in the particular scientific data on which they focus, they will also differ in their levels of access and refereeing. We see this as containing at least four levels:

1. *Personal laboratory databases:* These contain all the data needed by an individual or a particular laboratory: both data generated within the laboratory (some of which are too preliminary for publication) and data imported from other sources which are needed for the conduct of experimentation or modeling in that laboratory.

2. *Collaboratory databases:* Such databases will be shared by a group of collaborators working on a common problem. This will include all or part of the data in the personal laboratory databases of the collaborators, but because these collaborators may be scattered in different parts of the country or different countries of the world, these various subsets of the shared data must be linked through the Internet.

3. *Public "refereed" databases:* Whereas the above two kinds of databases are the personal property of an individual or a small group which accepts responsibility for the quality of the data they themselves use, there will also be public databases whose relation to the private data is similar to the relation of a published article to preliminary drafts and notes. Just as journals are now published by scientific societies and publishers, so do we expect that public scientific databases will be maintained by scientific societies and commercial publishers. A governing body for each database will thus take responsibility for some form of refereeing as well as ensuring the archival integrity of the database. Given the large size of datasets, we do not envision that in general such a dataset will be reviewed in detail. Rather, we envision two tracks of publication—for articles and datasets—in which there may be a many-to-many relationship between articles and datasets. The articles will be refereed in the usual fashion. A dataset will be endorsed to the extent that it can be linked to articles that have been refereed and support the data; however, there will also be a role for "posters" that have not been refereed but are supported by membership in an established scientific community.

4. In addition, of course, there will be the *World Wide Web*, in which material can be freely published by individuals irrespective of their expertise or integrity. It will be a case of "*caveat emptor*" (buyer beware), as not all scientists are reliable, while lay persons will often come up with interesting perspectives on scientific questions. Clearly, then, there will be many databases of many kinds in the federation that serves neuroscience, in particular, and science more generally.

One of the primary concerns that people have in contemplating the formation of a database federation such as that we envisage for neuroscience is the issue of what is to be done with old data. In the case of private databases, the data can simply "wither away" when the owner of the database no longer maintains the computer on which the data have been stored and provides no alternative means of access to the relevant databases. On the other hand, once a public database has been established, and once a proper form of references has been set up so that people will come to rely on the data that are referred to, then the data "cannot" be deleted. Yet, as time goes by, the way in which such archived data are treated can indeed reflect their changing status in light of new information. Published data can be *annotated* with personal annotations, refereed annotations, and links to subsequent supporting, competing, and completing material. Data that have proved of less and less current relevance, or whose subsequently questionable status makes them less likely to be referred to, can be demoted to low-cost, slow-access, tertiary storage, thus reducing the cost while the increase in retrieval time becomes of only marginal concern. This is an example of the importance of database research addressing the issue of how to support a user community that needs timely access to increasingly massive datasets (*cf.* Chapter 5.5, Management of Space in Hierarchical Storage Systems).

More generally, within the context of scientific databases, a crucial feature of the USCBP strategy is the linkage of empirical data to models and hypotheses so that the currently dominant ones can help provide and maintain coherent views of increasingly massive datasets. Datasets can then be demoted either because they have become completely subsumed by models that make it far easier to calculate values than to look them up or because the success of the models to fit a wide body of data has made the anomalous data seem suspect—whether because they have been superseded by data gathered with newer experimental techniques or because they no longer seem relevant as challenges useful for the restructuring of theory.

1.1.2 Modeling and Simulation

The term "neural networks" has been used to describe both the networks of biological neurons that constitute the nervous systems of animals and a technology of adaptive parallel computation in which the computing elements are "artificial neurons" loosely modeled after simple properties of biological neurons (Arbib, 1995). Modeling work for USCBP addresses the former use, focusing on computational techniques to model biological neural networks but also including attempts to understand the brain and its function in terms of structural and functional "networks" whose units are at scales both coarser and finer than that of the neuron. While much work on artificial neural networks focuses on networks of simple discrete-time neurons whose connections obey various learning rules, most work in brain theory now uses continuous-time models that represent either the variation in average firing rate of each neuron or the time course of membrane potentials. The models also address detailed anatomy and physiology as well as behavioral data to feed back to biological experiments.

Levels of Detail in Neural Modeling

Hodgkin and Huxley (1952) demonstrated how much can be learned from analysis of membrane properties and ion channels about the propagation of electrical activity along the axon; Rall (see Rall, 1995, for an overview) led the way in showing that the study of a variety of connected "compartments" of membrane in dendrite, soma, and axon can help us understand the detailed properties of individual neurons. Nonetheless, in many cases, the complexity of compartmental analysis makes it more insightful to use a more lumped representation of the individual neuron if we are to analyze large networks. To this end, detailed models of single neurons can be used to fine-tune the more economical models of neurons which serve as the units in models of large networks.

The simplest "realistic" model of the neuron is the *leaky integrator* model, in which the internal state of the neuron is described by a single variable, the *membrane potential* m(t) at the spike initiation zone. The time evolution of m(t) is given by the differential equation:

$$\tau dm(t)/dt = -m(t) + \sum_i w_i X_i(t) + h$$

with resting level, h; time constant, τ, $X_i(t)$, the firing rate at the ith input; and w_i, the corresponding synaptic weight. A simple model of a spiking cell, the integrate and fire model, was introduced by Lapicque (1907) and that coupled the above model of membrane potential to a threshold; a spike would be generated each time the neuron reached threshold. Hill (1936) used two coupled leaky integrators, one of them representing membrane potential and the other representing the fluctuating threshold. What I shall call the leaky integrator model *per se* does not compute spikes on an individual basis, firing when the membrane potential reaches threshold, but rather defines the firing rate as a continuously varying measure of the cell's activity. The *firing rate* is approximated by a simple, sigmoid function of the membrane potential, $M(t) = \sigma(m(t))$.

It should be noted that, even at this simple level of modeling, there are alternative models (e.g., using shunting inhibition or introducing appropriate delay terms on certain connections); there is no modeling approach that is automatically appropriate. Rather, we seek to find the simplest model adequate to address the complexity of a given range of problems. In general, biological neurons are far more subtle than can be captured in the leaky integrator model, which thus takes the form of a useful first-order approximation. An appreciation of neural complexity is necessary for the computational neuroscientist wishing to address the increasingly detailed database of experimental neuroscience, but it should also prove important for the technologist looking ahead to the incorporation of new capabilities into the next generation of artificial neural networks. (For an introduction to subtleties of function of biological neurons, the reader may wish to consult the articles "Axonal Modeling" (Koch and Bernander, 1995), "Dendritic Processing" (Segev, 1995), "Ion Channels: Keys to

Neuronal Specialization" (Bargas and Galarraga, 1995), and "Neuromodulation in Nervous Systems" (Dickinson, 1995).)

We may thus distinguish multiple levels of modeling, which include at least the following:

1. System models simulate many regions, with many neurons per region; neuron models such as the leaky integrator model permit economical modeling of many thousands of neurons and are supported by simulation systems such as the Neural Simulation Language (NSL; Chapter 2.2).
2. Compartmental models permit the modeling of far fewer neurons, unless unusually massive computing resources are available, and are supported by simulation systems such as GENESIS (Bower and Beeman, 1998) or NEURON (Hines and Carnevale, 1997).
3. Even more detailed models may concentrate on, for example, the diffusion of calcium in a single dendritic spine or the detailed interactions of neurotransmitters and receptors underlying synaptic plasticity, long-term potentiation (LTP), etc., as will be seen in the EONS Library (Chapter 2.3).

A Range of Models

Among the foci for USCBP modeling have been

1. *Basal ganglia:* The role of the basal ganglia in saccade control and arm control, as well as sequential behavior, and the effects on these behaviors of Parkinson's disease have been examined.
2. *Cerebellum:* Both empirical and modeling studies of classical conditioning, as well as modeling studies of the role of cerebellum in motor skills, have been conducted.
3. *Hippocampus:* Neurochemical and neurophysiological investigations of LTP have been related to fine-scale modeling of the synapse; we have also conducted systems-level modeling of the role of rat hippocampus in navigation, exploring its interaction with the parietal cortex.
4. *Parietal-premotor interactions:* We have worked with empirical data from other laboratories on the monkey, and designed and analyzed PET experiments on the human, to explore interactions between parietal cortex and premotor cortex in the control of reaching and grasping in the monkey, and, via our Synthetic PET methodology, have linked the analysis of the monkey visuomotor system to observations on human behavior.
5. *Motivational systems:* Swanson has conducted extensive anatomical studies to show the fine division of the hypothalamus into different motor pattern generators and to show their linkage to many other parts of the brain. Our work on modeling mechanisms of navigation also includes a motivational component related to this work.

The essential results for a number of these models will be summarized in Chapter 2.1.

Hierarchies, Models, and Modules

A great deal of knowledge of available neural data goes into the construction of a comprehensive model. In Chapter 2.1 we will present a model of interaction of multiple brain regions involved in the control of saccadic eye movements. Here, we simply want to preview some of the methodological issues involved.

For each brain region, a survey of the neurophysiological data calls attention to a few basic cell types with firing characteristics strongly correlated with some aspect of saccade control. For example, some cells fire most strongly near the onset of the target stimulus, others seem to be active during a delay period, and others are more active near the time of the saccade itself. The modeler using USCBP's NSL Neural Simulation Language then creates one array of cells for each such cell type. The data tell the modeler what the activity of the cells should be in a variety of situations, but in many cases experimenters do not know in any quantitative detail the way in which the cell responds to its synaptic inputs, nor do they know the action of the synapses in great detail.

In short, the available empirical data are not rich enough to define a model that would actually compute. Thus, the modeler has to make a number of hypotheses about some of the unknown connections, weights, time constants, and so on to get the model to run. The modeler may even have to postulate cell types that experimenters have not yet looked for and show by computer simulation that the resulting network will indeed perform in the observed way when known experiments are simulated, in which case: (1) it must match the external behavior; and (2) internally, for those populations that were based on cell populations with measured physiological responses, it must match those responses at some level of detail. What raises the ante is that (1) the modeler's hypotheses suggest new experiments on neural dynamics and connectivity, and (2) the model can be used to simulate experiments that have never been conducted with real nervous systems. The models considered in Chapter 2.1 are fairly complex, yet a few years from now we will consider these models simple, for the new models will both examine the interactions of a larger number of brain regions and analyze cells within each region in increasing detail. There is no way we would be able to keep cognitive track of these models if we had to look at everything at once. Our approach is to represent complex models in an object-oriented way, using a hierarchy of interconnected modules (Chapter 2.2 presents the particular formal approach to modules employed in

NSL). A module might be an interconnected set of brain regions; each region in turn might itself be a module composed of yet smaller modules that represent arrays of neurons sharing some common anatomical or physiological property (Fig. 1). In any case, a module is either decomposable, in which case this "parent module" is decomposed into submodules known as its children modules, or the module is a "leaf module" which is not decomposed further but is directly implemented in the chosen programming language such as Java or C++. In many NSL models, the neuron array provides the leaf modules for a model. In other models, decomposition can proceed further. There are basically two ways to proceed for a complex model. One is to focus on some particular subsystem, some module, and carry out studies of that. The other is to step back and look at higher levels of organization in which the details of particular modules are hidden. We can get both a hierarchical view of the model, where we can step back and analyze the whole model in terms of its overall relationship, or zoom in on subsystems and study them in detail.

NSL Neural Simulation Language

The NSL Neural Simulation Language developed at USC is especially designed for systems analysis of interacting circuits and brain regions. Chapter 2.1 focuses especially on NSLJ, written in Java. The main advantages with Java, of course, are (1) portability: you write it once and "it runs everywhere"; (2) maintainability: you only have to maintain one version of the software; (3) it runs on the client side of the Web; and (4) Java has parallel processing capabilities and a plan for future work is to develop a parallel version of our software.

Schematic Capture

NSL offers module composition to create hierarchical models. It provides layers of leaky integrator neurons connected by masks of weights as the base module for large scale simulations, but finer neuron models may be substituted. Currently, it emulates parallel execution mode. Essentially it has a fairly simple *scheduler* that will take each module in turn to execute the modules sequentially, but because the modules are all double

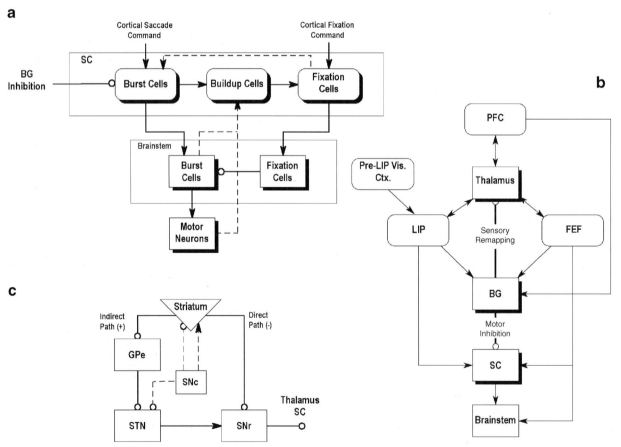

Figure 1 (a) A basic model of reflex control of saccades involves two main modules, one for superior colliculus (SC) and one for brainstem. Each of these is decomposed into submodules, with each submodule defining an array of physiologically defined neurons. (b) The model of (a) is embedded into a far larger model which embraces various regions of cerebral cortex (represented by the modules Pre-LIP Vis. Ctx., LIP, PFC, and FEF), thalamus, and basal ganglia (BG). While the model may indeed be analyzed at this top level of modular decomposition, we need to further decompose BG, as shown in (c), if we are to tease apart the role of dopamine in differentially modulating (the two arrows shown arising from SNc) the direct and indirect pathways within the basal ganglia.

buffered, it appears as though they are all firing simultaneously.

Given a rich library of modules, users will be able to fashion a rich variety of new models from existing modules, connecting them together and running a simulation without having to write code beyond tweaking a few parameters. A useful new aid to this is the development of a graphical user interface called the Schematic Capture System (SCS), which lets the user do much of the programming at the level of diagrams rather than having to type in every aspect of the model as line after line of code. The SCS lets modelers just draw boxes and label them. When one draws a box, one has to specify what its inputs are, what its outputs are, and what the data types are for each of them. The system will either fill in the information automatically or leave blanks for the modeler to fill in. A drawing tool lets one position copies of the boxes and click to form connections. Again, the SCS will automatically create NSL code for connecting those modules. In the same vein, one can specify, for example, that "basal ganglia" is a unitary module, BG, at the start of model design. Later on, one can click on the BG icon to create a new window in which one can decompose it graphically—with NSL code being generated automatically—until finally reaching the level where one either calls on preprogrammed modules for neural arrays or neurons or writes out the NSLJ code for the leaf modules oneself.

This approach to modular, graphical programming will be made easier by access to libraries containing modules that model portions of cerebral cortex, cerebellum, hippocampus, and so on. These can be plugged together using SCS to build novel models. The SCS is, in a sense, a "whiteboard" that makes it easy to connect different modules out of the library to make a new model and then run it.

Current work at USCBP will provide ways to interface diagrams generated by the SCS with various other databases to link assumptions made in constructing the model to empirical data. Correspondingly, other work on Brain Models on the Web (BMW, Chapter 6.2) will link simulation results to the data which test, whether supporting or calling into question, predictions made with the model.

The SCS style of programming has (at least) two advantages:

1. It makes it easy to program. It is a tool that lets the user place on the screen icons which represent modules already available or yet to be coded and then allows the user to make further copies of these modules and connect them to provide a high level view of a neural model. Any particular module may then be refined or modified to be replaced by a new module within the context of an overall system design.
2. When one views an existing model, the schematics make the relationship between modules much easier to understand. Using the SCS, an experimentalist who does not know how to program would still be able to sketch out at least a high-level view of the model, thus making it easier for the experimentalist and the modeler to interact with each other. A related virtue of the SCS approach is that it encourages collaboration between modelers and experimentalists who can examine an SCS representation of the model and analyze the various connections so displayed and the assumptions on which they rest.

We return to the key notion of the *experimental protocol*, which defines a class of experiments by specifying a set of experimental manipulations and observations. Another tool to aid comparison of experiment and model is the use of simulation interfaces which represent an experimental protocol in a very accessible way, thus making it easy for the non-modeler to carry out experiments on a given model. For example, the interface (Fig. 2) designed for the double saccade experiment described in Chapter 2.1 allows the user to simply click on points of a rectangle representing the visual field to determine the location of the fixation point as well as of targets 1 and 2. Similarly, sliding various bars on the display allows the user to specify the time periods of activation of the fixation and target points. Once this is done, the user has simply to press a "start" button to initiate the simulation and to see various panels representing the activity of different arrays of neurons. Various tools are available to change the chosen set of displays and the graphing conventions used for them. Tools are also available for the recording of particular activity patterns and their printing.

BRAIN MODELS ON THE WEB

A major goal of our work is to model the brain in a way that is tightly integrated with experimentation. We are interested in both function and learning. In a sense the whole brain is involved in every task, but holism is not very helpful when one wants to do science. Our modeling strategy, then, for a particular range of behaviors is to start with a data survey to determine a list of brain regions that are involved. Modeling may then concentrate initially on just a few regions to explore what range of behavior is involved, while other models may emphasize other regions. The driving idea is that if all details are modeled initially then it will be almost impossible to understand the effect of any one detail, but if models are built incrementally—both by adding regions and by adding details to the model of a particular region—one will better understand the implications of each part of the model and, it is hoped, the features so represented in the actual brain. It is in this spirit that we have developed NSLJ and SCS to ease the construction and "versioning" of models. These design considerations also motivate our design for Brain Models on the Web (BMW), a database of models with links to Summary

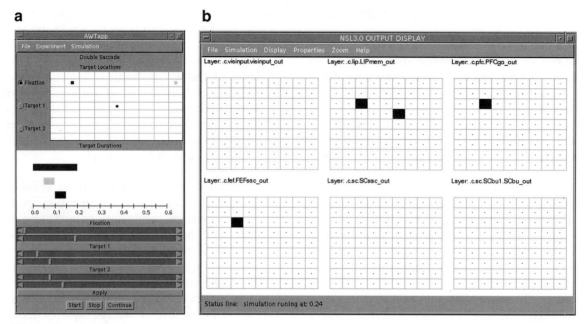

Figure 2 A simulation interface for the double saccade protocol in which a monkey fixates a fixation point during which time two targets are briefly flashed. After the fixation point is removed, the monkey saccades to the remembered position of the two targets in turn. **(a)** The position of the three targets for the simulated experiment is fixed by clicking on the display in the upper panel, and the duration of each stimulus is determined by the sliders in the lower panel. **(b)** This display presents the changing activity during the simulation in six of the arrays of the model shown in Fig. 1c. The menu at the top of the display lets one control the display and change what aspects of the simulated activity are displayed and the type of graphics used to display them.

Databases and Repositories on Empirical Data to support hypotheses and test predictions (Chapter 6.2). The results of analyzing a specific model in relation to the empirical data will in many cases establish a wide range of validity for the model, making confident predictions that can then be checked against empirical data or can be used to design new experiments. In other cases, comparison of predictions with empirical data will enable us to isolate defects in a given model which will lead us to develop new models. It is thus a crucial feature of BMW that it supports both modular structure and the versioning tools which allow one not only to build new models by combining or altering modules from existing models but also to document the efficacy thus gained in explaining a broader set of data, or using fewer assumptions, or gaining greater computational efficiency.

EONS: A Multi-Level Modeling System and Its Applications

The GENESIS and NEURON modeling systems have already been mentioned briefly. Each is designed most explicitly to address the issue of detailed modeling of neurons when the form-function relation of those neurons are to be explained by charting the pattern of currents and membrane potentials over diverse compartments of the structured neuron. At USC, we have addressed an even finer level of analysis, looking at how neural compartments can be further decomposed even down to the level of individual channels placed in spatial relationship across the cell membrane, with diffusion of calcium and other substances in the synaptic cleft defined by these membranes. The idea is, again, to adopt an object-oriented approach, with these "Elementary Objects of the Nervous System" (EONS) being placed together by a composition methodology like that offered by NSL. In fact, in some EONS models (Chapter 2.3), the top module is very small indeed, being a synapse which is then represented by a connection of objects for membranes and the synaptic cleft, and each of these can be further refined in turn.

A major concern in the development of EONS (and it is certainly a consideration for all groups seriously concerned about linking simulation to the data of neuroscience) has been to formalize this process of interaction between modeler and experimentalist. One side of the story, described in later sections of this chapter and volume, is to structure the experimental databases such that a modeler can easily find relevant data by constructing a search based on protocols. The other side of the story is to develop a model that will stimulate the experimenter to test various hypotheses. Whether involving the large-scale study of neural mechanisms of cognitive behavior or the fine scale of spatio-temporal patterns of synaptic transmission, one of the major paths to understanding is by studying the underlying mechanism by way of decomposing an existing model to include lower level features. Another point is matching model parameters with an external protocol so that the experimentalist can look at the protocol and transfer the parameters and then manipulate the model in novel ways. If a model fails to match the experimentalist's needs, then one needs ways for experimentalists to contribute to the design of new models. Doing so benefits from tools to facilitate sharing and exchange of available models. In this spirit, EONS (following the modular approach of NSL) enables models to be made up from self-contained objects that are described with the neurobiological terms that experimenters use and can form a library of neural objects. A synapse to a biologist is a synapse. It does not matter whether its model is just a number as in most artificial networks, or is an alpha function, or includes the presynaptic release mechanism and the kinetics of the receptors. With this system, we can construct varied models and then ask the question of what would happen if one manipulates them at the molecular level, by emulating the application of certain agonists or antagonists to determine what would happen at a synapse or in network dynamics.

From our modeling point of view, various experimental databases provide different experimental data for constraining and testing the model. On the other hand, the modeler will provide ways for experimentalists to test their hypotheses. In relating to the issues of database management and data mining, the models will also be part of the database search; therefore we can do intelligent searches and provide links for the search. A future goal is to develop a taxonomy of protocols to enable the database system to provide an intelligent search to find and query relevant data more easily.

At present, the EONS library of objects and methods includes numerical methods for the study of molecular kinetics, including diffusion, boundary conditions, and meshing, and provides objects describing axon terminals, the synaptic cleft, the postsynaptic spine, and their further subdivision down to the level of ion channels and receptor channels. One set of simulations has looked in detail at a two-dimensional slice across the synaptic cleft, representing the way in which vesicles release neurotransmitters into the cleft and how calcium diffusion influences the way in which neurotransmitters affect the receptors in the postsynaptic membrane. It has been shown that not only can the position of the vesicle relative to the receptors be important, but the very geometry (as revealed by EM) of the membranes can have a dramatic effect on synaptic efficacy.

MULTI-LEVEL SIMULATION: COMPLEXITY VS. EFFICACY

We close with the interesting fact that a careful simulation of several seconds of activity in a single synapse at this level of resolution requires 24 hours of computation by a moderately powerful workstation of 1998 vintage. Jim Bower (personal communication) reports that, in 2000, the world's fastest supercomputer can only handle six of his GENESIS simulations of the Purkinje cells of the cerebellum. Recall that there may be of the order of 10,000 synapses on a "typical" neuron, millions of neurons in a single region, and hundreds of regions in a brain. Clearly, any simulation methodology which simply required one to simulate every synapse or every neuron in such detail would be doomed to failure. No short-term increase in computer power will allow us to reduce the simulation of a system with 10^{15} synapses from 10^{15} days (even ignoring all the overhead of connectivity and non-synaptic membranes) down to a single second.

A major challenge for our work in multilevel simulation is thus to understand how to use detailed simulation at one level to validate a (possibly context-dependent) approximation that can be used in far more efficient large-scale simulations at the next level. For example, a NSL model might employ a neuron module that is far simpler than a corresponding compartmental model developed in NEURON but that has been validated by careful studies to yield an economical but effective approximation to it. Or a GENESIS modeler might want to check that a model of a compartment provides a satisfactory approximation to a far more detailed EONS model. All this raises two important challenges for the neural simulation community. One is to increase the range of tools currently available for comparing model to model, as well as model to data (with the parameter search methods that this implies). The other is to develop "wrapping" technology, so that modules developed using one simulator can indeed be used to replace objects (whether to simplify them or attend to crucial new details) in an existing model developed using another simulator. For example, if we had a large network model in NSL using leaky integrator neurons, we might like to plug in a more subtle neuron model of the individual neurons. It would then be more efficient to wrap a GENESIS or NEURON model of each neuron to serve as a module in a new version of the overall NSL model than to reprogram these complex neuron models in Java to fit them into the NSLJ environment directly. This topic is one of the USCBP goals for outreach to the broader neuroinformatics community, creating a set of standards for modularity, versioning, and data linkage so that wrapping technology will enable BMW to document and provide tools for linking models built using multiple neural simulators, not just the NSL system.

Brain Imaging and Synthetic PET

Since the neuroinformatics of human brain imaging is so well covered by many research groups with and without Human Brain Project funding, this has not been a major focus of USCBP research. However, we have been concerned with the question: "How can the data from animal neurophysiology be integrated with data from human imaging studies?" Our answer is *Synthetic PET Imaging* (Chapter 2.4), a technique for using computational models derived from primate neurophysiological data to predict and analyze the results of human

PET studies. This technique makes use of the hypothesis that regional cerebral blood flow (rCBF) is correlated with the integrated synaptic activity in a localized brain region. We first design NSL models of a key set of brain regions in the monkey, specifying the simulation of visual input and motor output in relation to the neural networks of the model. Synthetic PET measures are then computed for a set of simulated experiments on visually guided behavior in monkeys and then compared to the results of a similar human PET study. The human PET results may be used to further constrain the computational model. Moreover, the method is general and can potentially accommodate other hypotheses on single-cell correlates of imaged activity; it can thus be applied to other imaging techniques, such as functional MRI, as they emerge. Thus, although the present study uses Synthetic PET, we emphasize that this is but one case of the broader potential for systems neuroscience of synthetic brain imaging (SBI) in general.

1.1.3 Databases for Neuroscience Time Series

Neuroscience provides many examples of time series data (Chapter 3.1). Fig. 3 shows the well-known Pavlovian paradigm of "classical conditioning" as used in the Thompson laboratory at USC, with a blinking rabbit rather than a salivating dog. Puffing some air at the eye (the *unconditioned stimulus*) yields a blink (the *unconditioned response*; actually a closure of the nictitating

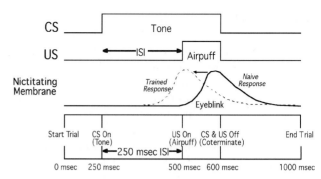

Figure 3 The protocol for using a tone to condition the eyeblink response of the rabbit.

membrane, the "third eyelid") a little later. Thompson precedes the airpuff with a tone as the *conditioned stimulus*, and eventually the animal learns this relationship and will eyeblink at each tone in anticipation of the airpuff, thus avoiding the noxious stimulus. The issue for Thompson's laboratory for many years now has been to go beyond Pavlov's behavioral studies to track down the neural mechanisms underlying that phenomenon, looking for cells in the brain responding in relation to these different effects and changing as conditioning proceeds. In fact, such changes are crucially observed in portions of the cerebellar cortex and the interpositus nucleus which lies beneath it.

Consider the time series data for such a study of classical conditioning shown in Fig. 4a. The top panel presents separate traces from separate trials of the same

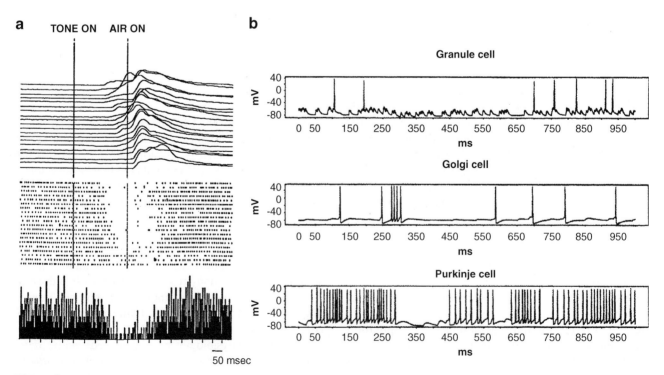

Figure 4 (a) Data from a Purkinje cell in temporal relation to the eyeblink behavior. (b) Result of a model of cellular interactions. These displays indicate the importance of linking empirical data and synthetic data (simulation results) to a common protocol for the comparison of data and model.

experiment, showing the movement of the eyelid. Each line in the middle panel is a "raster display," a series of dots corresponding to the firing of an action potential along the axon of a single neuron. At the bottom is the histogram produced by adding the firings of the neuron over the trials of the second panel which emphasizes the pause in firing that precedes the movement of the eyelid. This example brings up the database issues: How do we store time series? How do we register data with time stamps to facilitate interesting processing of sets of data? We need to store data with a *protocol* making explicit what hypotheses were being tested and what experimental methods were used. This must be supplemented with explicit data on what conditions were required to elicit each data set. To address these issues, we have developed NeuroCore, a general structure for the design of neuroscience databases (Chapter 3.2). Fig. 4b provides results of simulation with a detailed model, stressing the need for the use of a common protocol to structure comparison of model and data. We want to be able to take real neural data and compare them with the results of elaborate simulations of various brain regions. In this case, we use results from a model developed at USC by Gabor Bartha. He developed a network model, with biophysical properties built into the neurons, and then predicted patterns of firing of cells in cerebellar cortex and interpositus in the untrained and trained animal. Fig. 4 compares simulations predicting a shutdown in Purkinje cell activity with real data showing just this effect. We relate real data to computational predictions.

Fig. 5a shows the NeuroCore database architecture we have developed at USCBP for database management. The left-hand side of the figure shows the software required to keep track of queries in the standard query language (SQL; see Chapter 1.2) and to structure, enter, and retrieve data. The Informix architecture allows one to plug in a set of "Datablades." Consider how a Swiss Army knife has ordinary blades for cutting and then a set of additional devices for removing stones from horses' hooves and other important operations. In the same way, if one has a database functionality to add to the basic relational structure (the standard SQL processes), one can design or purchase a Datablade to structure and process data appropriately. The Web Datablade makes it easy to use a friendly Web interface to post queries and get the results. There are various Datablades available for two- and three-dimensional pictures and images. The previously available time series Datablade for financial applications was not suitable for neuroscience applications, so we developed a new neuroscience time series Datablade (Appendix 2) to handle the sort of spike data and behavioral data shown in Fig. 4(a).

NeuroCore provides a "core schema," a novel extendible object-relational database schema implemented in Informix. The schema (structure of data tables, etc.) for each NeuroCore database is an extension of our core database schema adapted to meet the needs of some group of neuroscience laboratories. Fig. 5b shows how the core database schema can be extended to accommodate Thompson's data. As shown in Fig. 6, the Core Experimental Framework, which we can link to the neuroanatomical and neurochemical concepts, provides an extendible specification of items needed in most experimental records, such as research subject, experimental manipulation, structure of the research data, and the statistics performed on the data. We see a slot for research data and a standard extension for handling time series data. This is then extended for the needs of this particular laboratory to provide fields for eyeblink (nictitating membrane response) data as well as unit data from the cells, whether from one unit or many units at a time. These are the sort of data we saw in Fig. 4a. Chapter 3.1 has more to say on this example and also discusses in some detail a protocol for intracellular recordings from hippocampal slices as an example of the flexibility of NeuroCore in developing Repositories of Empirical Data for neuroscience. NeuroCore comes with a Java applet called the Schema Browser, which allows one to learn the structure of a particular laboratory's database by showing, for each familiar core table, the extensions particular to that laboratory. Thus, the database structure becomes easy to understand.

Fig. 5a also indicates the use of Netscape or other browsers to go beyond whatever standard interfaces are given by the Web driver to develop an *on-line notebook* interface which make it easy both for the experimenter to enter comments and ideas during an experiment and for anybody to analyze the data (Chapter 3.3). The aim is to replace the situation where uninterpreted data are stored on disks or reels of tape with experimenter's comments in a separate handwritten notebook by a format that allows the experimenter to enter easily everything that would have been entered in the written notebook, and moreover, to have it time stamped and locked to the electronic data which themselves are coupled with that protocol information so the nature of the experiment, the fine data, and the comments are all electronically linked together. For the database and on-line notebook to foster inter- as well as intra-laboratory collaborations and communication, there need to be security protocols that allow researchers to "publish" and/or share their data in a secure fashion. Currently a simple security scheme has been implemented that allows us to track usage of the database through the on-line notebook. Built-in Informix security features allow a researcher to store his own data securely in the database; however, we are currently implementing a more complete security scheme to allow researchers to share their data in a secure fashion as well. Another way of extending the core, shown in Fig. 6, is by federating the given database with other databases, providing interfaces with, for example, BMW (our Model Repository) and other databases of neural data and literature (Article Repositories) and Summary Databases.

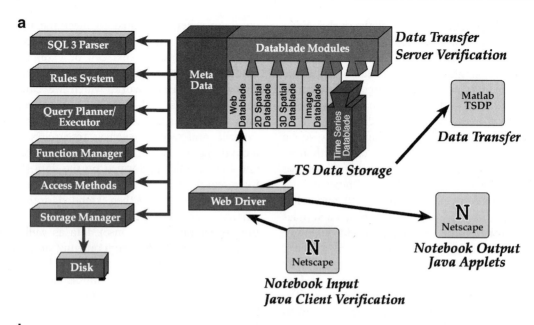

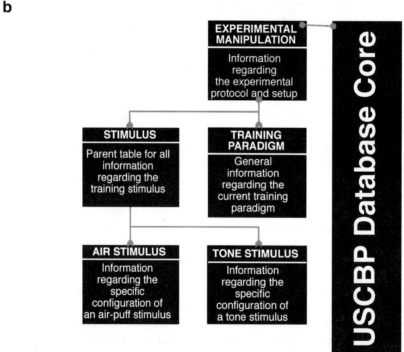

Figure 5 (a) The NeuroCore system, the general structure for the design of neuroscience databases developed by the USC Brain Project, as implemented in Informix, with linkage to the Web. (b) How to embed the Classical Conditioning Protocol in NeuroCore, the extensible structure of NeuroCore.

At USC, the protocols used by various laboratories are different because the research is fairly different. But, as we build a database protocol, we can converge with laboratories doing similar work at other institutions. For example, people doing classical conditioning on the cerebellum might have a shared extension which will handle about 80% of the variance. In that community, a laboratory that has already developed a successful extension of NeuroCore would, as freeware or for a fee, offer its database schema to other people working in that area; the extensions required for any other laboratory working with a similar research paradigm would be minor, making it easier for colleagues to share and compare their data. However, researchers do not have to agree on all the appropriate extensions. Our goal is federation without conformity. Note that we do *not* take responsibility for providing protocols for all types of experiments. That would exceed our knowledge and resources. Rather, our

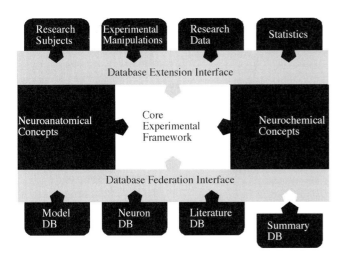

Figure 6 The USCBP NeuroCore database architecture.

task is to document NeuroCore and the tools for its extension, and clearly explain enough key examples to allow other researchers to program their own protocols and use the NeuroInformatics Workbench.

1.1.4 Visualization and Atlas-Based Databases

How are data from diverse experiments on the brains of a given species to be integrated? Our answer is to register the data—for example, the locations of cells recorded neurophysiologically, the tract tracings of an anatomical experiment, or the receptor densities revealed on a slice of brain in a neurochemical study—against a standard brain atlas for the given species, such as that for the rat brain developed at USC by Larry Swanson. Just as people have different faces, so do rats and other animals have different brains; therefore, there is a registration problem: given a location in an individual brain, what is the "best bet" as to the corresponding location in the "standard" brain?

The Swanson atlas (Swanson, 1998) contains 73 plates representing cross-sections of one half of the rat brain. These are not uniformly spaced, but were rather chosen to exhibit many crucial features of the rat's neuroanatomy. Each plate contains a photomicrograph of a stained brain section on the left and Swanson's representation of that section on the right, in which he draws boundaries separating different brain regions and labels the regions. We use the term "level" to refer to a two-dimensional representation of a slice of the rat brain obtained by pairing one of Swanson's drawings with its mirror image. Many of the curves dividing one nucleus from another correspond obviously to boundaries in the cell densities visible on the micrograph. Others cannot be seen from that particular micrograph and can only be revealed by a variety of staining techniques or by the incorporation of physiological and other data. It thus requires great skill on the part of the anatomist to draw those "non-obvious" divisions, and in fact even expert neuroanatomists may disagree. Thus, while there is much agreement between the Swanson atlas and the other leading atlas of the rat brain, the Paxinos-Watson atlas (Paxinos and Watson, 1998), there are also disagreements. Thus we have the future challenge of not only registering data against a particular choice of atlas but also facing the issue of how to update such datasets as future anatomical research resolves certain disagreements and leads to more reliable demarcation of boundaries.

Swanson has used his atlas as the basis for a personal database of PHAL (Phaseolus Vulgaris Leucoagglutinin) tract-tracing sections related to the projections of different regions of the hypothalamus. A tracer is injected into some region of the brain of interest, and this tracer is picked up by axons leading either into the given region or out of the given region. Successive sections through the brain may then reveal the stain which allows one to follow these fibers. In Swanson's laboratory, these observations of successive slices of different brains are meticulously drawn onto the different levels of the Swanson atlas, forming layers which can be shown in registration with the template for that level of the brain. Initially, all this work was done using Adobe Illustrator on a Macintosh, and the results were thus only available to someone who had access to all these files as a download onto their Macintosh. For us, the challenge was to replace this personal utility by a net-accessible database, in which the templates for different brain regions and the overlays from different experiments become elements in the Web-accessible database.

The solution to this problem is called NeuARt, a viewer for atlas-based neural data (the NeuroAnatomical Registration Viewer) which, though initially developed to register data against the Swanson atlas, is in fact a technology applicable to any atlas of the brain. For example, James Stone, formerly a member of USCBP and now at University of California, Davis, is adapting NeuARt to display data on the monkey brain gathered by Edward Jones. But, here, let us concentrate on the use of NeuARt with the Swanson atlas. The system allows one to view through a Web browser any level of the Swanson atlas together with any overlays retrieved from the database (Fig. 7). A Display Manager allows one to see these different results, and a Viewer Manager allows one to customize the Display Manager to one's needs. The Query Manager provides forms which make it easy to request anatomical information from our Informix database, the results of these queries are described textually by a Results Manager, and the user can maintain a set of results of interest. The Level Manager allows one to choose which level of the brain to examine, and the Active Set Manager then shows which results of the query have data relevant for that set. These can then be displayed by clicking on the appropriate elements.

NeuARt alone, however, does not solve the problem of transforming the results of an experiment into data

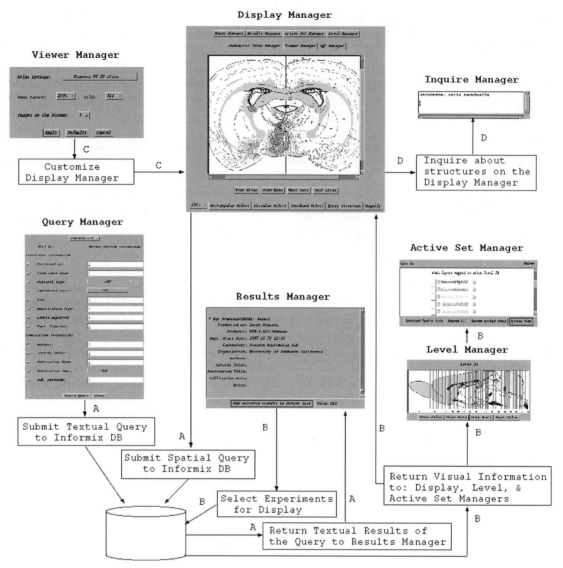

Figure 7 An overview of the NeuARt system. The Level Manager allows one to choose which level (i.e., drawing of a cross-section of the rat brain) of the Swanson atlas to examine. The Display Manager allows one to view through a Web browser any level of the Swanson atlas together with any overlays retrieved from the database. The Viewer Manager allows one to customize the Display Manager to one's needs. The Query Manager provides forms for requesting anatomical information from the database. The results of these queries are described textually by a Results Manager, and the user can maintain a set of results of interest. The Active Set Manager then shows which results of the query have data relevant for that set. These can then be displayed by clicking on the appropriate elements.

that can be overlaid against the atlas. Not only do different brains within a species differ, but also (even if we were using clones with identical brains) each brain will undergo different patterns of shrinkage as it is prepared for sectioning, and any actual slice made by the neuroanatomist will vary from those already used in the atlas. Thus, registering data against a level already in the atlas is not an optimal approach. We have thus produced a three-dimensional reconstruction of the rat brain by outlining the boundaries of each region in all 73 levels of the Swanson atlas and then using the Microstation CAD system to join up the outlines of a given region to form a three-dimensional representation as a surface bounding the region (Chapter 4.4). This surface can be rendered for viewing at different angles but, even more importantly for our present concern, the various surfaces can be sliced at arbitrary angles. Thus, given a particular slice of a particular brain containing data of interest—whether stains marking our fibers of passage, or stains representing density of chemical receptors, or marks indicating the position of cells encountered in a neurophysiological experiment—we match the slice not to the closest profile in the atlas, but rather to a whole variety of slices obtained from the three-dimensional atlas. We have used the warping algorithm developed by Fred Bookstein (1989), which provides a number called the "Procrustes distance" that indicates how far the landmarks on the original slice had to be moved to bring

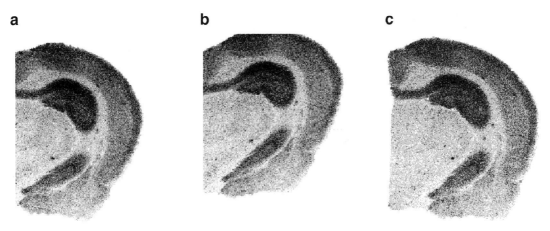

Figure 8 Improved registration results for matching against a three- rather than a two-dimensional atlas. (a) Experimental image. (b) Warped to closest profile in the atlas; Procrustes distance is 0.2135. (c) Warped to closest profile from resectioning; Procrustes distance is 0.0958, a numerical and visible improvement.

them into registration with landmarks on the slice from the atlas. We thus register the data slice to the atlas level which has the minimum Procrustes distance to yield our estimate of how best to embed this specific experimental data into our three-dimensional atlas of the brain. Fig. 8 demonstrates the improvement obtained by registration against the three-dimensional atlas.

We close this section by noting some of the other challenges for atlas-based databases. One is the issue of cytoarchitectonics, showing for each region of, for example, cerebral cortex, the distribution of cell bodies as seen in different layers through the cortex, a pattern that varies from position to position in the cortex. We have already spoken of registering the position of specific cells as identified during neurophysiological experiments. We can also link fine-grain neuroanatomical data to specific points in the brain to show what the characteristics of cells are as seen in that particular sub-area. For example, studies in neuromorphology may demonstrate the typical branching pattern of the dendrites and axons of a cell of given type in a given region and also show the distribution of synaptic spines along the various branches of the dendrites. An additional challenge to work on registering of brain sections is posed by the study of brains that have been damaged. As is well known by anyone who has watched television commercials, it is "easy" to morph any object into any other object; in particular, it is easy to register a brain section which has been lesioned against an intact brain section from the atlas. Thus, we must extend our registration technology to not only map to whole sections but to map to partial sections indicating not only what is the best slice for registration but also which sub-portion of that slice best matches the tissue of the experimental data that survived the lesion.

Given experiments on, for example, binding two different receptors, we may use registration to aid comparison of localization. The improved registration obtained with the three-dimensional atlas allows one to see subtle changes that were missed with less careful registration for analysis. Such results are important for the development of our atlas-based database of neurochemical data (Chapter 4.5). With good registration, it is easy to subtract one image from the other to isolate differences in, for example, two different ligand bindings on the same type of receptor (e.g., a map of AMPA-CNQX).

1.1.5 Data Management and Summary Databases

The USC Brain Project is developing a set of exemplary databases as a core for a larger federation of databases and will also develop tools such as NeuroCore for formatting neuroscience databases so that other people can build databases that are easy to federate with those at USC. However, no matter how much information we have in our own set of databases, there is going to be much relevant material "out there on the Web." Other groups will have interesting databases that are in a different format from NeuroCore. For example, people in neuroscience increasingly study genetic correlates of structure and function, with knock-out mice providing one exciting example. Thus, neuroscientists will want access to genome databases which are going to use different data structures, and this will pose a challenge for database federation. We also study techniques to manage and mine data from elsewhere and import them when we need them to augment our own databases.

Federation

No monolithic database will serve the needs of all neuroscientists; rather, neuroscience will rely upon a federation of databases (i.e., a set of databases linked in such a way as to allow queries to be answered by data gathered from any relevant database). Although private data could be stored in public databases with tags that limit access, many users will be concerned about security or will prefer the added control of keeping private data in a private database on a workstation whose contents are not accessible to general users of the Internet. Such lightweight personal databases might use relatively inexpensive, relatively widely available database managers, whereas the integrated public databases would use more powerful engines, such as the Informix DBMS used at present by the USC Brain Project. As papers are published, the data related to these publications should then be made available in Repositories of Empirical Data housed in integrated public databases which are accessible to a broad community of users via the Internet. Individual laboratories with the lightweight personal databases in which they develop new data or simulations will be linked to a public database in which the relevant results would eventually be published, with further potential access via the Internet to all the integrated public databases that serve the neuroscience community. We emphasize two different forms of access: one is to publish models or data, probably in a specific few integrated public databases; the other is to look for relevant data—whether for atlas data, other empirical data, articles in Article Repositories, or models—by broad searches across the Internet.

Federation is the interconnection of databases, with some variation in structure, so that they may be loosely coupled to support information sharing. This may involve more or less centralized control of information as there can be a spectrum of architectures for federated databases. The aim is to support information sharing between heterogeneous databases. The old pre-federation solution was full centralization, having an integrated database that subsumes the individual databases, replacing their individual schemas by one unified schema; however, the natural inclination of different user communities is towards heterogeneity. The problem, necessitating semantic models, is that terminology may differ from one database to another, and so the issue of matching the fields of one database to the fields of another database will be non-trivial and may depend more on negotiated agreements than on any automatic process. Our hope is that NeuroCore will develop into one interlingua so that many workers in neuroscience can communicate via this database structure whatever they want to export or import with other databases.

Federation provides the middle ground between the two extremes of integration with full centralization on the one hand and full autonomy on the other. One key aim is to support *discovery* in the sense of finding relevant data. This becomes very difficult in a too loosely coupled system, as there is no centralized knowledge. Because there will be databases that do not conform with our basic NeuroCore structure, research on database federation will be required to provide tools whereby some intermediate structure can be created to make "foreign"

data more readily accessible, maintaining information about the import and export schemas of the various databases and providing some "dynamic knowledge" of the available types of data as the pattern of sharing evolves over time.

This may be done "manually" by directly pointing (e.g., from a feature of a model to relevant laboratory data) or from a cell recording to atlas coordinates. This is useful in many cases, but often we would like to replace specific pointers by a generic description that can yield updated retrievals as the available data set changes. We want to avoid manual updating and "truth maintenance" to the extent possible.

Summary Databases: The Essential Notion Is the Clump

Journals are now available on-line, and a number of these journals provide the facility to link to "backup data sets." What we add is that Summary Databases may provide access to many different Article Repositories and Repositories of Empirical Data, and that "backup data" will not be isolated as appendices to specific articles but will be structured within Repositories of Empirical Data where they may more easily be collated with related data. A Summary Database might serve a large community, cover a general theme or a specialized theme, or be a personal database. In each case, the user needs tools to build the database and mechanisms to determine which users have access to a given class of data. We also need tools for merging (portions of) compatibly structured databases. For example, the author of a review article may simultaneously have (1) the article accepted for insertion in an article repository and (2) the personal Summary Database developed in compiling an article (with its assertions anchored by links to Article Repository clumps, Repository of Empirical Data data, and BMW models) merged into a public Summary Database serving the same community as the Article Repository. Taking an electronic file and adding it to an Article Repository is a well-understood process; much work remains to determine how to merge Summary Databases efficiently.

Whether an experimentalist is summarizing the fruit of multiple experiments, a reviewer is summarizing material in a variety of articles, or a modeler is providing the general implications of a set of modeling studies, the basic item in a Summary Database will be an *assertion* that can be supported, or contraverted, by the citation of specific data sets from a Repository of Empirical Data, specific extracts from an Article Repository, or specific simulation runs from a Model Repository. We use the term "clump" for the basic pieces of information which the summary thus refers to.

In other words, links to articles and databases will most usefully point to specific *clumps* of related material, rather than to the article or database as a whole. At present, a reference is usually to an entire article, or in some cases a specific page, table, figure, or equation. The notion of a *clump* generalizes this; a clump can be any set of sentences, parts of figures, entries in a table, etc. that provides the minimal description of a particular idea in an article or database.

In general, when we follow a link to an article repository, we would prefer to be sent to a highlighted clump in the article of interest, rather than to the first page of the article and then have to scroll through the article to find material of apparent relevance to the pointer. We thus need to provide a unique coordinate system that can identify portions of figures, videos, computer demos, etc., as well as portions of text. A clump can then be specified by giving its extended URL, the URL of the overall article together with the set of coordinate tuples that specify the constituents of the given clump. Currently, we have completed the task of extended URL definition for portions of a hypertext document and have provided the means to click on an index entry and be transferred to the relevant portion of the document with the desired clump of text shown highlighted. This work is part of USCBP's annotation technology. Selection of a clump involves generalized highlighting similar to normal click-and-drag highlighting but generalized to allow highlighting of several non-contiguous items within a given clump. Future work will provide appropriate extensions for figures, videos, computer demos, etc.

A clump may reside in the Article Repository, included in the set of indexed clumps which provide part of the extended hypertext of the article. More generally, it will reside in a Summary Database. The clump may be copied into the Summary Database if the owner of the original material grants permission. Alternatively, following the link from the Summary Database to the Article Repository may require a password and fee for access. As in a review article, the Summary Database may then contain a paraphrase or brief description to indicate the key point of the clump rather than its full content.

Model components may have explicit links to assertions, clumps, and laboratory data, as well as comparisons with elements of other models. When a model is consulted after its initial development, the assertions on which it is based can be used to anchor processes designed to discover new data which support these assertions or call them into question, thus allowing the user to judge the continuing validity of a model or to design paths whereby it may be updated.

Dynamic Classificational Ontologies

In philosophy, "ontology" studies "being as such," including the general properties of things. Quine (1953), however, saw ontology as concerning the question, "To the existence of what kind of thing does belief in a given theory commit us?" This question takes us halfway towards the definition of ontology used by database practitioners, a collection of concepts and their relationships used to describe a given application area and/or a database providing data about the given area. The key problem is that even if two databases record data describing similar aspects of the external world, the actual base concepts in each database might be quite different. "Personnel" might be an explicit concept in one database, but an implicit subset of "People" in another database. This raises a key issue for database federation, finding ways to translate between the different ontologies of different databases which contain related data necessary to fully answer a query. To address this, we developed the technique of dynamic classificational ontology (Chapter 5.2).

Basically, a *dynamic classificational ontology* is just a collection of interrelated terms, but we are really after concepts to which those terms refer and interrelationships to describe whatever information units we are trying to discuss. We start with a base ontology that describes these information units in a selective way. Perhaps it is given by one database that represents a certain set of research articles we are summarizing, experiments, protocols, and so on. However, as we add new articles or link to new databases, we need to generate a *concept thesaurus* which contains derived associations between concepts and the base ontology. To aid the discovery process of extracting relevant data in response to a query we extend the notion of thesaurus from "synonyms" (two ways of saying essentially the same thing) to "associated terms" which occur together with sufficient frequency that a search for one may fruitfully be enriched by a search for the other. For example, the terms "basal ganglia" and "dopamine" are commonly used together in research articles, so a search for articles on dopamine within a certain context can be improved by automatically searching for articles that use the term basal ganglia in that same context, even if the term dopamine does not appear. Our dynamic classificational ontology is one tool for updating the concept thesaurus and thus the derived ontology as use of the database federation proceeds. It is dynamic because the data are changing, so the ontology and the concept thesaurus will evolve in time, as well. Essentially, we employ a data-mining algorithm which counts concept co-occurrences and then takes advantage of common associations revealed in this way to aid further discovery of material relevant to our queries.

The Future of Publishing

We expect articles to continue to be basic units of scientific communication. More and more journals are now being placed on-line, and in many cases the publishers are allowing authors to augment the relatively short document that would correspond to a conventional hardcopy published article with various electronic appendices representing, for example, datasets that augment the figures and tables of the article or simulations that generated predictions within the article. This availability of more information on data and models is very welcome, but it is our view that the full utility of such augmented datasets cannot be reached unless the data are systematically embedded in databases structured to aid the integration of data from diverse laboratories and to provide support for data mining. It is thus our hypothesis that future publishing will link articles in electronic Article Repositories to relevant data in Repositories of Empirical Data and relevant models in Model Repositories such as BMW. To this end, we have set up an initial prototype for MOP (Model for On-line Publishing) which indicates some of the ways in which articles may be linked to datasets (Chapter 5.3). The display through the browser of a "journal page" is supplemented by an information area which can display any requested information that augments the basic article. A menu provides general facilities for locating supplementary information, and, when one chooses to explore a particular object, a submenu becomes available which lists the operations that are possible. For example, one may replace a figure, which is just a passive bit pattern, by a pop-up applet which allows one to treat the figure as a computational object. In one example shown in Chapter 5.3, each point on a scatter graph corresponds to a particular cell that has been examined in the given study. Clearly, presenting the data for all these cells within an article would overwhelm the balance of the article; however, the pop-up applet allows one to click on any single point in the display and have the choice of viewing general information (including anatomical data), raw data, or statistical data on the selected cell.

Our Model for On-line Publishing thus illustrates ways in which articles can be enriched by manifold links to the federated databases of neuroscience. In part, the success of our effort depends on the development of a data-sharing ethos. At present, many neuroscientists will not even share preprints, let alone data. They want to share papers only after they have been accepted for journal publication and view data as their personal property. This is frustrating because somebody might publish details about only one or two cells plus an average based on 100 cells when the details on those cells might provide crucial information for systematic understanding or contain outliers whose properties suggest new hypotheses. Improved data sharing requires us to surmount both a technological hurdle to make it easier for people to format, store, and retrieve data and a sociological hurdle in making people want to share data and getting pleasure because others analyzed their data to obtain new insights. A related issue is providing levels of security. Certainly there are some data one will never share, because they involve unpolished preliminary studies or early attempts with a new technique. Then there are the data to be kept confidential while writing up the primary papers. However, we believe that if one publishes a paper with tables and conclusions based on 100 cells, say, then one should let other scientists see the data on those 100 cells and have the opportunity to combine these data with other sets as the basis for new analyses which may yield far-reaching insights. There is nothing that the USC Brain Project can do directly to change the ethos, but we can at least provide tools which will make it increasingly easy to share data once an article is published.

Annotation Technology

Conventionally, we annotate hard-copy documents by marking pages and scribbling comments in the margins. Later, in writing up an article or proposal, we may search through a huge stack of papers to find relevant material and re-read the annotations to get ideas for the new document. Today, various word processors offer the capability to make in-line annotations of particular documents, but this is really no different from a footnote and, of course, requires having one's own copy of the electronic file for the document on which to make annotations. Thus, this approach is too limiting. You may often be reading a document and want to make a comment similar to one you have made on an earlier page. While you may have the tenacity to seek the relevant page if it is within the same document, it may exceed your patience to seek for the appropriate page in a pile of different documents. USCBP has developed Annotator (Chapter 5.4), a new Web-based annotation technology to solve this problem, and many more besides. We start by placing annotations in a database that is separate from the document. This allows a particular annotation to refer to multiple places in one document or even to multiple documents in different databases. Thus, one may solve the above problem by searching the annotation database for similar annotations to check whether or not they should be lumped with the proposed new annotation.

Annotator has the advantage that a single annotation can refer to different portions (as noted earlier, we call them "clumps") of a given document or even to relevant clumps in widely scattered documents, even at different Websites. It also solves the problem of annotating documents without having to own an electronic copy, so long as one has access to it over the Web. We use an extended URL to describe the location of each clump, thus making it possible for the user to retrieve the clump and view it together with the annotation even though no actual changes have been made to the annotated document, residing as it does on a different Website. Our annotation technology modifies the browser so that you can view both the annotations and the document referred to by the annotation, with the clump highlighted.

Once annotations are gathered within a database, rather than being embedded in the text of widely scattered documents, it becomes easy to search the annotations efficiently to bring related information together from these many documents. The extended URL methodology then makes it easy to view the clumps, whose relevance can then be determined.

The database of annotations so formed may be a personal database for private use or may be a database of annotations to be shared, whether between the members of a collaboratory or with a larger public. The important thing is that this technology is not just of interest to neuroscientists but is useful to anyone who uses the Internet. Everybody wants to use annotations; however, within the context of USCBP in particular and scientific databases in general, of particular importance is that a Summary Database is a particular example of an Annotation Database, where now each summary may be viewed formally as an annotation on all the clumps that it summarizes.

The NeuroScholar Project

Chapter 6.3 offers both a general framework for Summary Databases, known as Knowledge Mechanics, and an extensible implementation of this design within the domain of neuroscience (called NeuroScholar) that specifically deals with neuroanatomical connections from tract-tracing experiments. The key notion is that it is not sufficient to read, classify, and summarize a paper and then only record the summary. It is essential to record the reasoning that provides the basis for the chosen summary. The aim is to ensure that interpretations of data should be easy to follow. In particular, the way that data are selected as "reliable" or rejected as "unreliable" should be made explicit.

The process of building a representation of users' knowledge in NeuroScholar is accomplished by devising a suitable ontology or data model for it. Just as an individual's interpretations will change over time, individual users' knowledge models will need to be updated, adjusted, and maybe even completely reinvented. In order to keep track of all changes made to the database, detailed access logs will be maintained. Users will be encouraged to timestamp "versions" of their knowledge representation based on their particular outlook at a particular time.

Object primitives contain the classification of a concept and are classified according to "domain type" and "knowledge type." Domain types are based on a classification of the subject under consideration; in NeuroScholar, this is defined by the experimental method being used to obtain the results that support whatever classification is represented by the object in question. For example, this means that we differentiate between objects defined from "tract-tracing studies" and "electrophysiological studies." Fig. 9 illustrates the

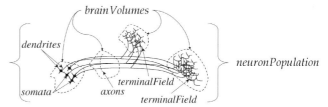

Figure 9 The general *neuronPopulation(interpretation, core)* object used in NeuroScholar.

most important composite object in NeuroScholar: a so-called *neuronPopulation(interpretation, core)* object. This defines a generalized population of neurons. A type of object of importance in this scheme is the *brainVolume(interpretation, core)* object, which is made up of various subobjects which contain relevant data describing the characteristics of those components. These *objects* are simply regions of brain tissue that are defined in terms of a specific publication's parcellation scheme, providing the geometrical substrate for the data in the system. According to this treatment, brain atlases are considered to be "just another publication" with a parcellation scheme expressed as *brainVolume(interpretation, core)* objects to which we link other publications' *brainVolume* objects with various relations. Each subobject of any given *neuronPopulation(interpretation, core)* object may be linked to a given *brainVolume(interpretation, core)* object in order to express its location.

Users' knowledge representations could be compiled into a global summary that represents their understanding of a specific phenomenon or system. This summary could provide the input to secondary analyses either for visualization (as was the case for the neuroanatomical connection databases that were the forerunners of NeuroScholar; Young, 1992) or to investigate organizational properties of the data. Techniques such as nonmetric multidimensional analysis (NMDS), nonparametric cluster analysis (MCLUS), and optimal set analysis (OSA) have been successfully applied to similar problems in the past. Analyses such as these may help expert users interpret their data and make experimental predictions.

Statistical approaches may also be used to generate accurate summaries by searching for optimal solutions that satisfy the many well-defined constraints that exist in the data. For example, consider a situation where the act of comparing data that originate from two different experiments is unreliable, but comparisons between data points from within the same experiment are reliable. In cases such as these, it would be possible to compile a set of metadata comprising all the relevant constraints which could then be analyzed globally to produce global constraints for the whole system. Methods like these were used to calculate finely graded connection weights for the rat visual system (Burns *et al.*, 1996). This was accomplished on the assumption that the density of labeling in tract-tracing studies is correlated with the anatomical strength of the connection, but different tracer chemicals have different sensitivities. Thus, comparisons of labeling density within the same experiment (or tentatively between experiments that used the same technique) reflected differences in connection strength that could not be inferred from comparisons between experiments. This general approach was also used to convert between parcellation schemes in macaque monkey (Stephan *et al.*, 2000).

The Neurohomology Database

The term "homology" is a central one in comparative biology, referring to characteristics of different species that are inherited from a common ancestor. Defining homologies between brain structures requires a process of inference from distinct clusters of attributes. Thus, we anchor our approach to neural homologies (Chapter 6.4) in the concept of *degree of homology*. To define a neural structure, neuroscientists use numerous attributes including gross morphology, relative location, cytoarchitecture, types of cell responses to different means of stimulation, and function. In similar fashion, we employ eight criteria for determining the degree of homology of two brain structures: the morphology of cells within each brain structure and the relative position, cytoarchitecture, chemoarchitecture (neurotransmitters that are found within a brain structure), myeloarchitecture, afferent and efferent connections, and function of each of a pair of brain structures from two species.

If two brain structures have common cell types, chemo- and cytoarchitectonics, and connectivity patterns, then one should expect that those two brain structures have the same function or related functions. This is the case for the primary visual area (area 17). In each major mammalian species, area 17 can be delimited on the basis of myeloarchitecture (heavy myelination) and cytoarchitecture (the presence of a granular layer IV), the presence of a single and systematic visuotopic map, a well-defined pattern of subcortical afferents, small receptive fields, and the presence of many orientation-selective neurons with simple receptive fields.

Chapter 6.4 presents not only a discussion of the homology criteria that can be established between pairs of brain structures across species but also introduces the Neurohomology Summary Database. This database contains three interconnected entities: *Brain Structures, Connections*, and *Homologies*. Fig. 10 shows the modules and relationships that are contained in the database. The various parts of the Neurohomology Database—Brain Structures, Connectivity Issues, and Homologies—can be accessed independently. We have designed the Web interface in independent parts to answer to queries from

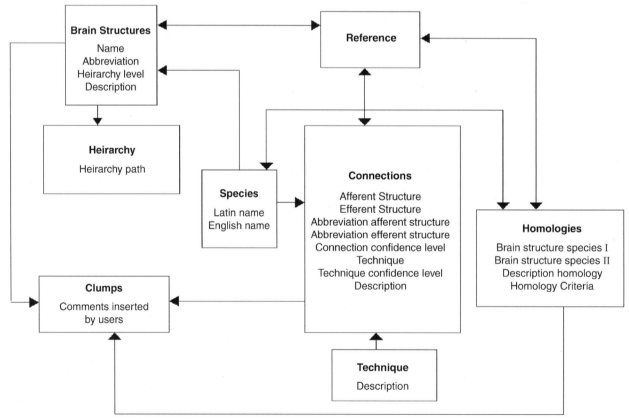

Figure 10 Schematic of the Neurohomology Database structure. A unidirectional connection denotes a 1 to n relationship, while a bidirectional connection denotes an n to n relationship.

a larger category of users. In this way, if users want to find if there is any homology between structures X and Y, from two different species, they can also find the definitions of structures X and Y, according to different sources, as well as the afferents and efferents of these two structures.

Brain Models on the Web

We need to be able to compare different models of related neural systems. At present, a paper describing a model will include graphs of a few simulation results, with an informal claim that the results are more or less consonant with available data. A goal of USCBP research in neuroinformatics is to provide tools that make it easier for researchers to import the code for each other's model and make explicit comparisons. As a result, the community might agree as to what is currently the best model of how the superior colliculus talks to the brainstem and then need the tools to readily modify other models so as to incorporate this model into larger schemes, just as the Fig. 1a model becomes a submodel of the Fig. 1b model. In other cases, comparison of two different models against a set of experimental data may show that one model provides insight into some data, while the other model seems more useful for other data.

The framework for this effort is provided by Brain Models on the Web (BMW; Chapter 6.2), a prototype model repository in which the hypotheses used in defining a model and the predictions made using the model can be linked to Summary Databases (Chapter 6.1) to aid the construction and testing of models. BMW is a step towards providing tools to aid modelers in making comparisons between models and data, learning from them, and constructing a new model that may be seen as an improved version of its precursors and also to provide tools for documenting these improvements—linking models to sets of empirical data—when the new model is placed in BMW. The aim is to have BMW provide modelers with the vehicle to polish their work collaboratively and to involve experimentalists in the collaboration through links to Summary Databases and Repositories of Empirical Data.

In summary: models should be modular so researchers can "mix and match" models of specific subsystems, BMW must provide tools so that comparisons can be made between models and empirical data, and the modeling environment must provide tools for model versioning and for keeping track not only of the new versions as compared to the old but also of the empirical data that justified that migration from one version to another. BMW is based on the idea that you publish models not as take it or leave it entities but with the modular structure made explicit with the links to empirical databases so that anyone with Web access can contribute to model development (subject to usual provisos about access and competence). At present, our task in BMW is simplified because models in the initial tranche "installment" (see *OED*, second edition) are all written in NSL. We also plan to address the issue of using modules that have been developed in different simulation environments. The idea is to develop a "wrapper technology" which enables modules to be "wrapped" in software that provides a uniform communication protocol to allow them to be used together as parts of a larger model. The only catch is that this may reintroduce some of the problems of platform dependence that we have been avoiding by developing a new version of NSL in Java.

Another aspect of BMW is to provide access not only for modelers who are doing the expert work on testing and developing the models but also for non-experts in modeling who are experimentalists who want to test if a model is good enough to justify investing time and resources in testing its predictions. This leads to the idea that we take experimental protocols and build user-interfaces that make it easy to conduct the simulation equivalent of the experiment described by the protocol. We have already developed a number of such interfaces, including that shown in Fig. 2. An interesting research challenge is to come up with a formalized language for protocol definition which makes it possibly to semi-automate the design of corresponding extensions of NeuroCore to store relevant empirical data, as well as the design of interfaces for generating appropriate input and display panels for simulation of experiments based on the protocol.

We provide three levels of membership for models documented in BMW:

1. *Level 1:* All modules of a model are stored in the database with protocol-related interfaces and links to Summary Databases, etc. to support hypotheses, test predictions, and compare models.
2. *Level 2:* Each model receives an exposition in HTML with links to code and to protocol-based interfaces for simulations related to standard experiments.
3. *Level 3:* Exposition and code of model available by ftp.

At present, BMW allows users to view, via the World Wide Web, tutorials, code, and sample simulations of models; in many cases, the model can be imported to a local site for detailed testing. Future work will not only expand the database but also develop tools allowing users around the world to comment on models, develop new versions, and contribute new models to BMW.

BMW will remedy the fact that journal publication of models often give insufficient information to recreate a model in its entirety and offer the results of a limited set of simulation runs. The further testing that BMW makes possible may serve to strengthen our confidence in a model or provide the basis for an analysis of strengths and weaknesses to be used in developing new models with greater validity.

The full development of the USC Brain Project will provide an environment which makes it easy for the user to pass from empirical data to related models and back again. Future BMW-based standards activity will provide modelers using a whole variety of simulation languages (not just NSL) with tools to develop interfaces that make it easy for non-programmers to run basic "experiments" with the models, to add to the database comments on the comparison of simulation results with available empirical data, and to install models, create versions of both models and parameter sets, and to freeze models in various "interesting" states for later analysis under varying conditions. A crucial aspect in all this is to catalyze a truly cumulative style of modeling in neuroscience by facilitating the *re-usability* of modules within current neural models, with the pattern of re-use fully documented and tightly constrained by the linkage with a federation of databases of empirical neuroscientific data.

1.1.6 The NeuroInformatics Workbench

The key contribution of USCBP is that not only has it made great strides in database design, visualization, and modeling for the neurosciences, but it has also designed and implemented an integrated architecture, the Neuro-Informatics Workbench, which will enable neuroscientists to use all these methodologies in an integrated neuroinformatics environment. The NeuroInformatics Workbench provides a suite of tools to aid the neuroscientist in constructing and using databases and then linking models and data. At present, the Workbench contains four main components: NeuroCore, for building neuroscience databases; NeuARt, for registering data against brain atlases; NSLJ and related tools, for neural simulation; and Annotator, for building databases of annotations on material distributed across the Web. This section provides a perspective on these varied contributions.

NeuroCore System for Neuroscience Database Construction

Our design of databases to capture time series data, such as records of cell firing or records of behavioral variables, emphasizes the inclusion of *protocol* information about what the experimentalist did to gather these data. NeuroCore is a system for constructing and using neuroscience databases, with the schema data (schema in the database sense of the organizing structure for database tables, etc.) for each "NeuroCore database" being an extension of the "NeuroCore Schema" which we have crafted as a core database schema readily adaptable to meet the needs of a wide variety of neuroscience databases (Chapter 3.2). NeuroCore also provides tools for data entry (JADE), retrieval of data (dbBrowser), inspection of different NeuroCore extensions (Schema Browser), and for data analysis (Datamunch) (Chapter 3.3). For example, if somebody has built a database with a schema that extends the NeuroCore database schema, the Schema Browser allows anyone who accesses that database to easily view the structure of these extensions. NeuroCore has been extended to serve the specific needs of three laboratories: one records data from the cerebellar system of behaving rabbits in a classical conditioning paradigm, another records data from hippocampal slices (Chapter 3.1), while a third stores neurochemical data (Chapter 4.5). The aim is to make these exemplary, so that other neuroscientists will be encouraged to adapt the schema to build federatable databases (which may be private to one laboratory or pool data from a whole community, subject to appropriate access restrictions).

NeuARt NeuroAnatomical Registration Viewer

"NeuARt" was originally short for "Neuroanatomy of the Rat", but it is much more generic than that, a general viewer for atlas-based neural data designed to anchor our work on managing spatial data. It is thus now the acronym for the NeuroAnatomical Registration Viewer. NeuARt is a viewer for atlas-based neural data, with a spatial index manager (Chapter 4.3). In designing NeuARt, the USCBP team emphasized a design closely linked to the Swanson atlas of the rat brain (Chapter 4.2), but the group of Ted Jones at UC Davis is, with the help of a programmer who was formerly a member of USCBP, now extending NeuARt for use with data on the monkey brain. We have also shown how to construct a three-dimensional brain atlas from an atlas of two-dimensional brain sections and to register data against the three-dimensional atlas (Chapter 4.4). This effort undergirds our development of an atlas-based NeuroCore database for neurochemistry (Chapter 4.5). We have also made progress on related Summary Databases, one for neural connections (Chapter 6.3) and one for inter-species neural homologies (Chapter 6.4).

NSL Neural Simulation Language

NSLJ is a modular, Java-based version of USC's NSL Neural Simulation Language, extended by an environment for neural simulation (Chapter 2.2; Weitzenfeld *et al.*, 2000). NSLJ is especially suited for large-scale simulation of neural systems, where we are looking at many brain regions with many, many cells involved. Java implementations enable clients to import a model easily and run it without worrying about what platform they have. Earlier versions of NSL have been used to develop models of basal ganglia, cerebellum, hippocampus, and other neural systems, as well as the interaction of these systems in learning and behavior, and a number of these have now been ported to NSLJ. Our work on EONS, a library of objects for modeling subsynaptic processes and other neural systems (Chapter 2.3), is at an earlier stage of development, but complements NSLJ by modeling the fine details of how cellular interactions change during LTP and LTD, etc. and has already been used in a number of other studies.

We have also developed a prototype Summary Database linked to the model repository Brain Models on the Web (BMW; see Chapters 6.1 and 6.2). The key idea is that BMW not only stores models but also provides the means to link models to data which support hypotheses and test predictions. All models must be firmly grounded in a careful analysis of a large body of relevant data, wherever it has been developed. Such data will come not only from our own Repositories of Empirical Data but also from the Repositories of Empirical Data of other laboratories where access has been granted and, most strongly, from the published literature at large, which is available in a variety of Article Repositories, an increasing amount being electronic, but much of it still available only in print libraries. The further development of NSLJ and EONS, and their integration with other simulators into a unified multi-modeling environment integrated with BMW and ancillary databases, is an ongoing focus of USCBP research. We will also focus on formal protocol analysis as a basis for providing search engines tuned to use protocols to match and compare sets of empirical and synthetic (model-generated) data. A future challenge is to build a modeling environment where we can go all the way from large-scale system analysis of many interacting regions down to the fine analysis of individual neurons.

Annotator

Annotation technology generalizes the notion of annotation from something placed within a document to a database entry which can refer to annotated data in many other databases. A summary, for example, can be seen as an annotation not on just one document but on all the documents it summarizes. Annotation technology is of general interest for the digital library area, and in fact for anybody who uses the Web, and thus will have much broader applications than simply in neuroscience. USCBP's Annotator allows the creation of paperless annotation databases in which the annotations are kept separate from the annotated document. This has the advantage that a single annotation can even refer to portions ("clumps") of documents at different Websites. It also solves the problem of annotating documents without having to own an electronic copy, so long as one has access to it over the Web. We use an extended URL to describe the location of different clumps, thus making it possible for the user to retrieve the clump and view it together with the annotation even though no actual changes have been made to the annotated document, residing as it does on a different Website.

All the components of the NeuroInformatics Workbench have been introduced earlier in this chapter and are fully described in later chapters of this book. All are available on our Website, some in the form of prototypes which provide proof of concept but are not yet ready for widespread use, while others (such as NSL) have already been released for public use. The reader can consult the "Available Resources" sections of each chapter for a status report as this book goes to press; the current status may always be found by turning to the USCBP Website, http://www-hbp.usc.edu.

Acknowledgments

The work of the USC Brain Project was supported in part by the Human Brain Project (with funding from NIMH, NASA, and NIDA) under the P20 Program Project Grant HBP: 5–P20–52194 for work on "Neural Plasticity: Data and Computational Structures" (M. A. Arbib, Director).

References

Arbib, M. A., Ed. (1995). *The Handbook of Brain Theory and Neural Networks*. A Bradford Book/The MIT Press, Cambridge, MA.

Bargas, J., and Galarraga, E. (1995). Ion channels: keys to neuronal specialization, in *The Handbook of Brain Theory and Neural Networks* (Arbib, M. A., Ed.). A Bradford Book/The MIT Press, pp. 496–501.

Bookstein, F. L. (1989). Principal warps: thin-plate splines and the decomposition of deformations. *IEEE Trans. Pattern Analysis and Machine Intelligence*. **11(6)**, 567–585.

Bower, J. M., and Beeman, D. (1998). *The Book of GENESIS: Exploring Realistic Neural Models with the GEneral NEural SImulation Systems*. 2nd ed., TELOS/Springer-Verlag, Berlin/New York.

Burns, G. A. P. C., and Young, M. P. (2000). Analysis of the connectional organisation of neural systems associated with the hippocampus in rats. *Philos. Trans. R. Soc. London B: Biol. Sci.* **255**, 55–70.

Burns, G. A. P. C., O'Neill, M. A. and Young, M. P. (1996). Calculating finely-graded ordinal weights for neural connections from neuroanatomical data from different anatomical studies, in *Computational Neuroscience Trends in Research* (J. Bower., Ed.). Boston, MA.

Dickinson, P. (1995). Neuromodulation in invertebrate nervous systems, in *The Handbook of Brain Theory and Neural Networks* (Arbib, M. A. Ed.). A Bradford Book/The MIT Press, Cambridge, MA, pp. 631–634.

Hill, A. V. (1936). Excitation and accommodation in nerve. *Proc. R. Soc. London B*. **119**, 305–355.

Hines, M. L., and Carnevale, N. T. (1997). The NEURON simulation environment, *Neural Computation*. **9**, 1179–1209.

Hodgkin, A. L., and Huxley, A. F. (1952). A quantitative description of membrane current and its application to conduction and excitation in nerve. *J. Physiol. London*. **117**, 500–544.

Koch, C., and Bernander, Ö. (1995). Axonal modeling, in *The Handbook of Brain Theory and Neural Networks* (Arbib, M. A. Ed.). A Bradford Book/The MIT Press, pp. 129–134.

Lapicque, L. (1907). Recherches quantitatifs sur l'excitation electrique des nerfs traitée comme une polarisation. *J. Physiol. Paris*. **9**, 620–635.

Payne, J. R., Quinn, S. J., olske, M., Gabriel, M. and Nelson, M. E. (1995). An information system for neuronal pattern analysis, *Soc. Neurosci. Abstr*. **21**, 376.4

Paxinos, G., and Watson, C. (1998). *The Rat Brain in Stereotaxic Coordinates*. 2nd ed., Academic Press, San Diego, CA.

Quine, W. V. O. (1953). *From a Logical Point of View*. Harvard University Press, Cambridge, MA.

Rall, W. (1995). Perspective on neuron model complexity, in *The Handbook of Brain Theory and Neural Networks* (Arbib, M. A. Ed.). A Bradford Book/The MIT Press, Cambridge, MA, pp. 728–732.

Stephan, K. E., K. Zilles, and Kötter, R. (2000). Coordinate-independent mapping of structural and functional data by Objective Relational Transformation (ORT). *Phil. Trans. R. Soc. London B.* **335**, 37–54.

Swanson, L. W. (1998). *Brain Maps: Structure of the Rat Brain*. Elsevier Science, Amsterdam.

Weitzenfeld, A., Alexander, A., and Arbib, M. A. (2000). *The NSL Neural Simulation Language*. The MIT Press, Cambridge, MA.

Young, M. P. (1992). Objective analysis of the topological organization of the primate cortical visual system. *Nature* **358**, 152–155.

CHAPTER 1.2

Introduction to Databases

Wen-Hsiang Kevin Liao and Dennis McLeod
*Computer Science Department, University of Southern California,
Los Angeles, California*

Abstract

Databases based on relational, object-oriented, and object-relational models represent significant advances in database technologies. In the context of general-purpose database management systems, the fundamentals of database models are examined. A historical perspective on the evolution of major database models is provided. The principal concepts underlying relational, object-oriented, and object-relational database models are presented with examples. Finally, a brief view of database federation issues is introduced that serves as the foundation for discussion later in the book.

Computerized databases are essential and inseparable components of most of today's information systems. Database systems are used at all levels of management, research, and production to provide uniform access to and control of consistent information. In the past few decades, we have witnessed enormous advances in database technologies. The goal of this chapter is to provide a brief and informal overview of state-of-the-art database systems and database federation issues. The emphasis is on important concepts of relational, object-based, and object-relational databases and the relationships among them.

In the first section, the components of databases and their functional capabilities are introduced, followed by a brief historical perspective on the developments of relational, object-based, and object-relational databases. The structures, constraints, and operations of the relational database model are then introduced, and an informal presentation of major SQL features is given. A discussion of the concepts of the object-based database model follows, and the principal concepts of the object-relational database model are presented and its relationships with relational and object-based databases are discussed. Finally, a brief view of database federation issues is introduced.

1.2.1 An Overview of Database Management

A *database* (DB) is a collection of structured (organized), interrelated information units (objects). An *information unit* is a package of information at various levels of granularity. An information unit could be as simple as a string of letters forming the name of an experimenter or a data value collected from an experiment. It could also be as complex as the protocol of an experiment, a published paper, the image of a rat brain, the audio clip of a speech, or the video clip of an experiment in a laboratory. Every database is a model of some real world system. At all times, the contents of a database are intended to represent a snapshot of the state of an application environment, and each change to the database should reflect an event (or sequence of events) occurring in the environment. A database can be of any size and of varying complexity. For example, a database can contain the information of only a few hundred people working on the same project, or it could contain the information of a bank, an airline company, or data collected from scientific experiments.

A general-purpose database management system (DBMS) can be viewed as a generalized collection of integrated mechanisms and tools to support the definition, manipulation, and control of databases for a variety of application environments. In particular, a general-purpose DBMS is intended to provide the following functional capabilities:

1. Support the independent existence of a database, apart from the application programs and systems that manipulate it.
2. Provide a conceptual/logical level of data abstraction.
3. Support the query and modification of databases.
4. Accommodate the evolvability of both the conceptual structure and internal (physical) organization of a database, in response to changing information, usage, and performance requirements.
5. Control a database, which involves the four aspects of semantic *integrity* (making sure the database is an accurate model of its application environment), *security* (authorization), *concurrency* (handling multiple simultaneous users), and *recovery* (restoring the database in the event of a failure of some type).

At the core of any database system is a *database model* (data model), which is a mechanism for specifying the structure of a database and operations that can be performed on the data in that database. As such, a database model should:

1. Allow databases to be viewed in a manner that is based on the meaning of data as seen by their users.
2. Accommodate various levels of abstraction and detail.
3. Support both anticipated and unanticipated database uses.
4. Accommodate multiple viewpoints.
5. Be free of implementation and physical optimization detail (physical data independence).

Abstractly speaking, a database model is a collection of generic structures, semantic integrity constraints, and primitive operations. The structures of a database model must support the specification of objects, object classifications, and inter-object relationships. The semantic integrity constraints of the database model specify restrictions on states of a database or transitions between such states so that the database accurately reflects its application environment. Some constraints are embedded within the structural component of a database model, while others may be expressed separately and enforced externally to the DBMS. The specification of a particular database constructed using these general-purpose structures and constraints can be referred to as a (conceptual) *schema*.

The operational component of a database model consists of a general-purpose collection of primitives that support the query and modification of a database; namely, given a database with an associated conceptual schema, the operations facilitate the manipulation of that database in terms of the schema. Such primitives may be embodied in a stand-alone end-user interface or a specialized language or embedded within a general-purpose programming language. Database-specific operations can be constructed using the primitives of the database model as building blocks.

1.2.2 Historical View of Key Database Developments

To provide a historical perspective on relational, object-based, and object-relational databases models and systems, Fig. 1 shows a 30-year timeline of their developments (McLeod, 1991). Note that there are many database models and systems being developed during this period of time. For the purpose of our discussion, only those relevant database models and systems are included in the figure.

The relational database model was first introduced in 1970 (Codd, 1970). It is based on a simple and uniform data structure, the relation, and has a solid mathematical foundation. Many commercial products of relational DBMSs were implemented in the 1980s. The descendants of these relational DBMSs are still dominating the database market around the world. The semantic/object-based database models adopt objects and classes as their basic structures and add many semantic primitives such as specialization/generalization, data abstraction, and encapsulation. Several semantic/object-based database systems were developed in the 1990s. The object-relational database model extends the relational database model with abstract data types and primitives from semantic/object-based data models. Its basic structure is still a relation, but the value contained in a cell of the relation could be of any abstract data type in addition to atomic data types like strings and numbers. The specialization/generalization of both relations and data

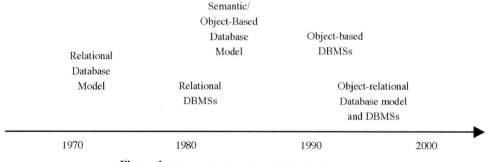

Figure 1 Historical view of key DBMS developments.

1.2.3 Relational Database Model

Simplicity and uniformity are the main characteristics of the relational database model. To introduce the relational database model, the basic structures and constraints of the relational model can be considered (Codd, 1970; Chen, 1976). The operational component of the relational model will be illustrated later in the section on the SQL query language. A relation can be crudely viewed as a table containing a fixed number of columns (attributes) and a variable number of rows (tuples). (Strictly speaking, a table depiction of a relation is an approximation; for example, rows and columns are unordered, and no duplicate rows are allowed.) We say a relation is n-ary if there are n columns in the table. Each column has a "domain" comprising the possible entries for that column. Thus, a relational database consists of a number of n-ary relations and a collection of underlying domains. Fig. 2 shows an example relation named FRIENDS which contains information on four friends. The relation is presented in tabular form. The relation has three columns: Name, Age, and Telephone. Each row represents information on a friend. For example, the first row of the relation indicates that Joe's age is 48 and his telephone number is 740-4312. The domain of a column specifies possible values a column entry may take. In this example, the domains of Name and Telephone columns are strings, and the domain of the Age column is numbers. In the relational model, the domains consist of atomic values, which may be built in (e.g., Numbers, Strings) or user-defined (e.g., Phone-numbers). Note that the user-defined domain is a subset of one of the built-in domains. For instance, the domain of the Phone-numbers column is the set of strings that represent meaningful telephone numbers. Each database may contain as many relations as necessary. The relational database model captures inter-record relationships in a uniform manner: by means of common data values. Each relation has a column or a collection of columns whose values taken together uniquely determine rows in the relation. Such a column or column collection is termed a *primary key* (logical unique identifier) and is underlined in the figure. In this example, the primary key is the Name column if no two friends have the same name.

Fig. 3 presents an example database which contains four relations. This database includes information on publications, authors, reviewers, and recommendations of publications by reviewers. By relating the common values of the Title columns in both the AUTHORS and RECOMMENDATIONS relations, it shows that the book "*The Metaphorical Brain 2*" is written by the author named Arbib and the book is recommended by two reviewers named Campbell and Oneil.

Fig. 3 also shows examples of three fundamental relational model constraints, which are termed *domain integrity, entity integrity,* and *referential integrity*. The domain integrity constraint specifies that the value of a column entry must come from the underlying domain of that column. The entity integrity constraint specifies the uniqueness of rows in a relation. The value(s) of column(s) of the primary key in a relation taken together uniquely determine the row in the relation. For example, the primary key of RECOMMENDATIONS is the combination of Title and Reviewer, which means that given the title of a publication and the name of a reviewer, there is at most a single value of Rating for that pair. Referential integrity constraints, indicated by dashed lines in Fig. 3, specify that values in some columns must occur as values in some other column. For example, the arrow from Title in RECOMMENDATIONS to Title in PUBLICATIONS means informally that for a recommendation to exist for a publication, that publication must exist in the PUBLICATIONS relation; more precisely, the title of the publication as indicated in RECOMMENDATIONS must also exist as the title of the publication in PUBLICATIONS.

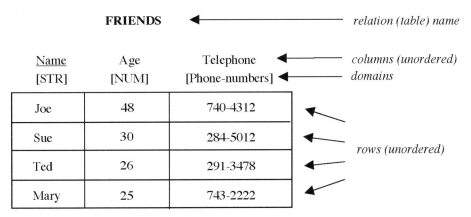

Figure 2 Tabular depiction of relations.

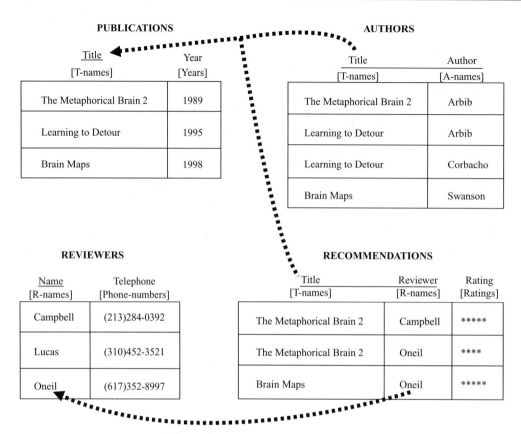

Figure 3 Example relational database.

1.2.4 SQL

The acronym SQL was originally an abbreviation for Structured Query Language. The SQL language consists of a set of primitives for defining, accessing, and managing data stored in a relational database (Date, 1990; Elmasri and Navathe, 1994). It is a high-level declarative language in that the user only specifies what the query result is and leaves the actual execution and optimization decisions to the DBMS. There are many commercial SQL implementations which comply to various SQL standards (SQL1, SQL2, or SQL3) and also provide their own extensions (Date and Hugh, 1997). The goal of this section is to present a brief and informal overview of some important primitives of standard SQL. The details of SQL can be found in Appendix B1 as well as in the references. The database in Fig. 3 will be used to illustrate using SQL primitives in defining tables, specifying constraints, updating the tables, and querying the tables. The SQL extensions for object-relational databases will be discussed later after the object-relational database model is introduced.

In Fig. 4, four CREATE TABLE statements are used to define the four tables in Fig. 3 with the specified names and specified named columns (with specified data types). Column names must be unique within their containing table. All four tables will initially be empty (i.e., contain no rows). There is a primary key specified for each of the four tables using the PRIMARY KEY keyword. The FOREIGN KEY keyword is used to indicate referential integrity constraints. Domain integrity, entity integrity, and referential integrity will be enforced by the underlying DBMS.

There are four basic SQL data manipulation operations: INSERT, DELETE, UPDATE, and SELECT. The first three operations are used to modify the contents of a table, and the SELECT statement is used to retrieve information from a database. The three INSERT statements in Fig. 5 illustrate how to add rows into the PUBLICATIONS table. In SQL syntax, entries for strings are quoted and entries for numbers are not quoted as shown in the examples. The UPDATE statement shows how to change the year of publication from 1993 to 1995 on the article entitled "Learning to Detour." The DELETE statement illustrates how to delete the row of publication "*Brain Maps*" from the PUBLICATIONS table. Note that the UPDATE and DELETE operations can be applied to several rows at once, although this is not shown in the examples. The same is also true in general for INSERT operations.

The SELECT operation has the general form "SELECT-FROM-WHERE" as illustrated in Fig 6. The SELECT clause indicates the columns to be retrieved and displayed. The FROM clause specifies the tables from which the results should be retrieved. The WHERE clause defines the criteria for the information

```
CREATE TABLE PUBLICATIONS          CREATE TABLE AUTHORS
    ( Title VARCHAR(80),                ( Title VARCHAR(80),
      Year INTEGER,                       Author CHAR(20),
      PRIMARY KEY (Title));               PRIMARY KEY (Title, Author),
                                          FOREIGN KEY (Title)
                                          REFERENCES PUBLICATIONS (Title));


    CREATE TABLE REVIEWERS         CREATE TABLE RECOMMENDATIONS
    ( Name CHAR(20),                    ( Title VARCHAR(80),
      Telephone CHAR(13),                 Reviewer CHAR(20),
      PRIMARY KEY (Name));                Rating CHAR(5),
                                          PRIMARY KEY (Title, Reviewer),
                                          FOREIGN KEY (Title)
                                          REFERENCES PUBLICATIONS(Title),
                                          FOREIGN KEY (Reviewer)
                                          REFERENCES REVIEWERS(Name));
```

Figure 4 Examples of SQL data definition.

```
INSERT INTO PUBLICATIONS VALUES               SELECT Title, Year, Author
    ( 'The Metaphorical Brain 2', 1989);      FROM PUBLICATIONS, AUTHORS
INSERT INTO PUBLICATIONS VALUES               WHERE PUBLICATIONS.Title = AUTHORS.Title
    ( 'Learning to Detour', 1993);
INSERT INTO PUBLICATIONS VALUES
    ( 'Brain Maps', 1998);

UPDATE PUBLICATIONS SET Year = 1995
    WHERE PUBLICATIONS.Title = 'Learning to Detour';

DELETE FROM PUBLICATIONS
    WHERE PUBLICATIONS.Title = 'Brain Maps';
```

Figure 5 Examples of SQL update.

Figure 7 An example of relational join operation.

that should be retrieved. Only the data that satisfy the predicate are returned as the result of the SELECT operation. If the WHERE clause is omitted, the result contains all rows of the tables in the FROM clause. The simple SELECT operation in Fig. 6 retrieves the title and rating of those publications that are recommended by Oneil. The operation returns two rows: *The Metaphorical Brain 2* with a four-star rating and *Brain Maps* with a five-star rating.

One of the important features of SQL is its support for the relational join operator. This operator makes it possible to retrieve data by "joining" two, three, or any number of tables, all in a single SELECT statement. Fig. 7 shows a query that retrieves the title, year of publication, and authors of each publication by joining PUBLICATIONS and AUTHORS on matching titles. The query produces a ternary (i.e., 3-ary) relation, with columns containing a title, the year of publication, and

an author of that publication. The join operation is illustrated in Fig. 8.

The SELECT operation can be nested. The result of a select statement can be used in another select statement. Fig. 9 shows a query that retrieves the title, year of publication, and authors of each publication that has been recommended by the reviewer named Oneil. This query is the same as the one in Fig. 7 except that only those publications recommended by Oneil are retrieved. The titles of publications recommended by Oneil are first selected in the inner select statement. A subset of the join on the PUBLICATIONS and AUTHORS relations on matching publication title is then selected corresponding to those publications recommended by Oneil in the inner select statement.

1.2.5 Object-Based Database Models

Object-based database models are placed in contradistinction to the relational database model. The term "object-based" refers to the class of database models that include those identified as "semantic" and "object-oriented" (Shipman, 1981; Hammer and McLeod, 1981; Fishman *et al.*, 1987; Hull and King, 1987). In what

Figure 6 General form of SELECT operations.

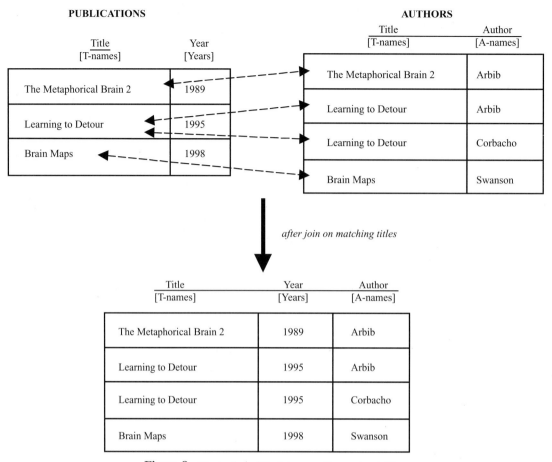

Figure 8 Illustration of the join operation on two tables.

```
SELECT Title, Year, Author
FROM PUBLICATIONS, AUTHORS
WHERE PUBLICATIONS.Title = AUTHORS.Title
    And PUBLICATIONS.Title in
       (SELECT Title
        FROM RECOMMENDATIONS
        WHERE Reviewer = 'Oneil' );
```

Figure 9 An example of relational database query.

follows, first the principal characteristics of object-based database models are examined. The main components of the object-based model are then discussed and illustrated using an example database which is an extension of the database previously used in illustrating the relational database model.

The term "object-based" refers to the following characteristics, as exhibited by a database model and a database system that embodies that model (McLeod, 1991):

1. *Individual object identity:* Objects in a database can include not only primitive (atomic) data values, such as strings and numbers, but also abstract objects representing entities in the real world and intangible concepts. Relationships among and classifications of such objects can themselves be considered as abstract objects in the database. Graphical, image, and voice objects can also be accommodated. Such "abstract" objects can be directly represented and manipulated.

2. *Explicit semantic primitives:* Primitives are provided to support object classification, structuring, semantic integrity constraints, and derived data. These primitive abstraction mechanisms support such features as aggregation, classification, instantiation, and inheritance.

3. *Active objects:* Database objects can be active as well as passive in that they exhibit behavior. Various specific approaches to the modeling of object behavior can be adopted, such as an inter-object message-passing paradigm or abstract data type encapsulation. The important point is that behavioral abstraction is supported, and procedures to manipulate data are represented in the database.

4. *Object uniformity:* All (or nearly all) information in a database is described using the same object model. Thus, descriptive information about objects, referred to here as *meta-data*, is conceptually represented in the same way as specific "fact" objects. Meta-data are considered dynamic, and can be modified in a manner analogous to that used to alter "fact" objects.

To provide a more substantive analysis of the concepts underlying object-based database models in general, and the notions of object identity and explicit semantic primitives in particular, the following main components of an object-based model are considered:

1. Objects are abstract, complex, or atomic entities that correspond to things in the application environment being represented in the database and may be at various levels of abstraction and of various modalities (media).
2. Inter-object relationships describe associations among objects. Such relationships are modeled as attributes of objects (logical mapping from one object to another) and their inverses, as well as by association objects (objects that represent relationships as entities).
3. Object classifications group together objects that have some commonalities. The term "object type" often refers to such classification, and the term "class" to the set of objects classified according to the type. (The term type and class are sometimes informally used somewhat synonymously.) Relationships among object classifications specify inter-classification associations (e.g., the subclass/superclass inter-class relationship supports specialization/generalization). Object classes or types can themselves be considered objects at a higher level of abstraction (e.g., with attributes of their own).

Classes and Attributes

To examine the fundamentals of object-based database models, consider again the example application environment involving publications, authors, reviewers, and recommendations. Fig. 10 illustrates the concepts of classes and attributes. Classes are classifications of objects, and attributes are mappings from one object class to another. In this figure, classes are indicated as ovals. Example classes include abstract entity classifications, such as PUBLICATIONS, AUTHORS, REVIEWERS, and RECOMMENDATIONS, as well as atomic value classes, such as Phone-numbers, T-names (title names), A-names (author names), Ratings, etc.

In Fig. 10, attributes are indicated by labeled directed arrows from the described class to the describing class (value class). Attributes are labeled with a name and are indicated as single valued (1) or multi-valued (m); attribute inverses are specified by a line connecting two small circles (or by two tracking small circles). An attribute A from class C1 to a class C2 means that an object classified as being of class C1 may be related to object(s) of class C2 via A; the inverse of A, which is always logically present, may be explicitly indicated and named. For example, class PUBLICATIONS has an attribute called has-authors that is multi-valued; for a given publication object, its authors are indicated by the values of this attribute, which relates the publication to zero or more objects of type AUTHORS. The inverse of this attribute

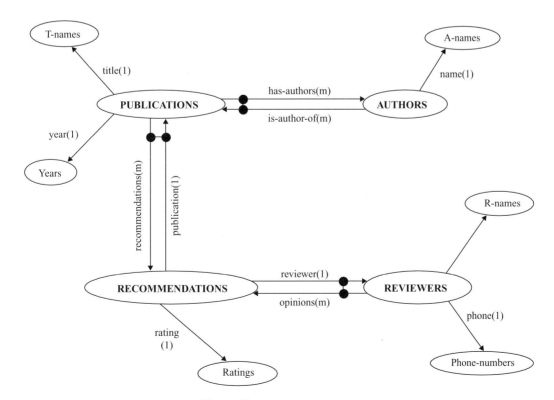

Figure 10 Classes and attributes.

is called is-author-of of AUTHORS, which is also multi-valued. Another example is the attribute year of PUBLICATIONS, which is single valued; its inverse is not explicitly present here but can be referenced as "the inverse of year of PUBLICATIONS" (the publications published in a given year), which is of course an attribute of class Years.

One-to-one, one-to-many, many-to-one, and many-to-many binary relationships between object classes can be expressed by attribute and attribute inverse pairs. The many-to-many relationships between classes PUBLICATIONS and AUTHORS are represented by the multi-valued attribute has-authors of PUBLICATIONS and its multi-valued inverse, is-author-of of AUTHORS. An example of a one-to-many relationship is indicated between the PUBLICATIONS and RECOMMENDATIONS classes by the single-valued attribute publication of RECOMMENDATIONS and its multi-valued inverse attribute recommendations of PUBLICATIONS; this means informally that an object of class RECOMMENDATIONS represents the evaluation of a single publication, and that a given publication may have many evaluations. An example of a one-to-one relationship, although not indicated as such in Fig. 8, might be between PUBLICATIONS and T-names; here the attribute called title of PUBLICATIONS is single valued, and its inverse (from T-names to PUBLICATIONS; e.g., called is-title-of) would also be single valued.

Fig. 11 illustrates the RECOMMENDATIONS class in more detail. Here RECOMMENDATIONS has three attributes (publication, reviewer, and rating), each of which is single valued. Objects of class RECOMMENDATIONS represent abstract entities that correspond to recommendations of publications by reviewers with a given rating; they in effect model a ternary relationship. It is also possible to consider a derived attribute called

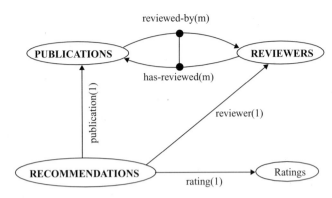

Figure 11 Association objects.

reviewed-by of PUBLICATIONS and its inverse (has-reviewed of REVIEWERS), both of which can be derived from information carried by the values of attributes of RECOMMENDATIONS. This is an example of derived data in general and derived attributes in particular; note that attribute and attribute inverse pairs are in a sense also derived or, more precisely, logically redundant information. A database system that supports an object-based database model must of course support this redundancy and maintain consistency in its presence.

Subtype and Inheritance

The concept of specialization (and its inverse, generalization) is an important kind of inter-class relationship, which is supported by subclass/superclass construct. Fig. 12 illustrates this relationship between a class and its subclass(es), using a large arrow from a class to a subclass. In the example, the class PUBLICATIONS has two subclasses, namely BOOKS and PAPERS. PAPERS in turn has two subclasses, namely JOURNAL PAPERS and CONFERENCE PAPERS. Attributes are inherited by a subclass from its superclass, and the subclass may

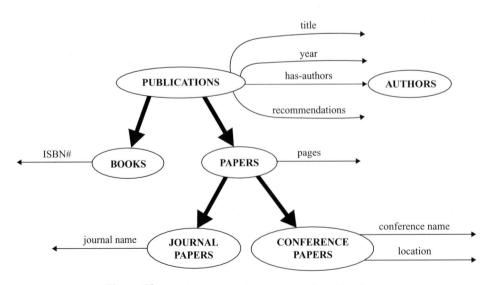

Figure 12 Subclasses, superclasses, and attribute inheritance.

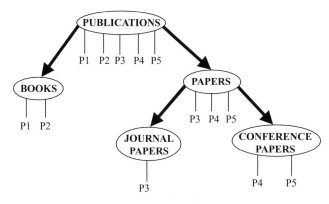

Figure 13 Subclasses and instances.

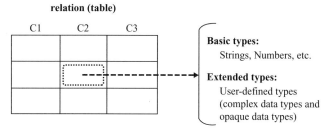

Figure 14 Extending domain types in a relation.

have additional attributes as well. Thus, BOOKS has its own attribute ISBN, as well as the attributes called title, year, has-authors, and recommendations, which are inherited from PUBLICATIONS. (For simplicity in the figure, the value classes of some attributes are omitted.)

The objects classified by a class may be termed the *instances* of that class. The set of instances of a subclass must be a subset of the set of instances of its superclass. Fig. 13 shows some instances of class PUBLICATIONS, indicated as P1, P2, P3, P4, and P5. Each instance of CONFERENCE PAPERS is an instance of three classes; for example, P5 can be viewed as a conference paper, a paper, and a publication. The subset constraint involving instances of a subclass must, of course, be enforced by the database system.

It is also possible for a class to have multiple superclasses. In this case, the specialization structure among classes is not necessarily a forest of trees but rather a directed acyclic graph. For example, suppose that FRIENDS and COLLEAGUES are defined as subclasses of PERSONS; the class FRIENDS AT WORK might then be defined as a subclass of both FRIENDS and COLLEAGUES. In this example, the instance subset constraint implies that the instances of FRIENDS AT WORK must be a subset of the instances of FRIENDS as well as a subset of the instances of COLLEAGUES. When multiple superclasses are permitted, some rules must be present in the database model to accommodate the problem of multiple inheritance (namely, the inheritance of attributes from more than one superclass).

1.2.6 Object-Relational Database Model

The object-relational database model is based on the relational database model with extensions of several key ideas from object-based database models (Rowe and Stonebraker, 1987; Stonebraker *et al.*, 1999). The basic structure of the object-relational model is still a relation, but the domains of a column in a relation are extended to include abstract data types in addition to atomic data types like strings and numbers (Fig. 14). Subclasses (subrelations) and inheritance are added to incorporate specialization/generalization inter-class (inter-relation) relationships. With the ability for users to define abstract data types with subclasses and inheritance, the object-relational database model has more capability for modeling complex data than the traditional relational database model. Unlike most object-based database models, the object-relational model maintains the top-level structure as relations and make it possible to use the standard SQL with extensions to define, access, and manipulate its databases. In what follows, extended data types and subclasses/inheritance of the object-relational database model will be further illustrated using Informix universal server DBMS (Informix Inc., 1997a,b).

The object-relational database model is an extension to the relational database model. By ignoring the additional types and constructs provided by the object-relational database model, one could just use an object-relational database system as a relational database system. In the relational database model, the domains of a column in a relation (table) are limited to basic atomic data types such as strings and numbers, etc. The object-relational database model eases the limitation to include user-defined extended types as the domains of columns in a relation. By doing so, it enables users to capture more semantics of complex application environments. There are two main categories of extended types. The first category is made up of the *complex data types* including row types and collections. A row type is a data type consisting of a record of values, which is analogous to a C structure or C++ class. Row types could be specifically named or created automatically as an anonymous (unnamed) row type during table creation using CREATE TABLE statements. In Fig. 15, the left side creates the PUBLICATIONS table with an unnamed row type. The right side creates a named row type PUBLICATIONS_T and uses it create the table PUBLICATIONS. In addition to using a named row type to create tables, you can also use it to define individual columns of a table. The collection types include set, list, and multiset types that can hold multiple values.

The second category of extended types is the *opaque type*. An opaque data type is a user-defined type to be used like an atomic data type. Opaque types are totally encapsulated. The database server knows nothing about the representation and implementation of an opaque type. All access to an opaque type is through the functions written by users. They are accessed through the functions defined on them. Once an opaque type is defined and implemented, it could be used just like built-in strings and numbers. For example, a complex number is not directly supported by most database system, but it can be implemented as an opaque type with input, output, and arithmetic functions. After complex number type is defined and implemented, operations on complex numbers can thus be supported in the database system. The object-relational database model also supports subclasses and inheritance. Subclasses can be applied to tables, row types, or opaque types. Both the data and functions defined on the superclasses are inherited by their subclasses; however, there is one restriction on creating sub-tables using named row types. If named row types are used in creating tables, the type hierarchy of the named rows must match the type hierarchy of the corresponding tables. Fig. 16 shows how to create PUBLICATIONS and BOOKS tables (see Fig. 12) using named row types. Note that the named row type BOOKS_T is a subtype of the type PUBLICATIONS_T. The corresponding BOOKS table is also a sub-table of the PUBLICATIONS table.

As a further extension to the basic relational database model, Informix as well as many other object-relational database management systems also add the capability of allowing multi-valued attributes. Thus, a many-to-many relationship can be expressed as set-valued attribute. Such set-valued attributes can be useful but introduce the problem that one must select which "side" to represent the relationship; using a separate table for many-to-many relationships avoids this problem by symmetrically representing the relationship as a pair of references to the tuples/objects involved in it. For example, we can represent the many-to-many relationship between authors and publications as a set-valued attribute for tuples in the AUTHORS relation or a set-valued attribute in the PUBLICATIONS relation. Alternatively, we can use a new table, say AUTHORING, which contains single-valued attributes that refer to the AUTHORS and PUBLICATIONS tuples; in the latter approach, there will be a tuple in AUTHORING for each binary pairing of an author and a publication; if there are attributes of this relationship, they can be placed in the AUTHORING relation.

1.2.7 An Overview of Federated Database Systems

With the rapid growth of computer communication networks over the last decade, a vast amount of diverse information has become available on the networks. Users often find it desirable to share and exchange information that resides either within the same organization or beyond organizational boundaries. As we shall explain in detail in Chapter 5.1, a federated database system is a collection of autonomous, heterogeneous, and cooperating component database systems (Heimbigner and McLeod, 1985; Sheth and Larson, 1990). Component database systems unite into a loosely coupled federation in order to achieve a common goal: to share and exchange information by cooperating with other components in the federation without compromising the autonomy of each component database system. Such environments are increasingly common in various application domains, including office information systems, computer-integrated manufacturing, scientific databases, etc.

The issues related to sharing and exchanging of information among federated database systems can be considered in three different phases (Kahng and McLeod 1996; Aslan 1998; Fang et al., 1996). First, the user of a component database in the federation needs to find out what information can be shared and identify the location

```
       CREATE TABLE PUBLICATIONS              CREATE ROW TYPE PUBLICATIONS_T
           ( Title VARCHAR(80),                   ( Title VARCHAR(80),
             Year INTEGER,                          Year INTEGER,
           PRIMARY KEY ( Title ) );              PRIMARY KEY ( Title ) );

                                                CREATE TABLE PUBLICATIONS
                                                    OF TYPE PUBLICATIONS_T;
```

Figure 15 Creating tables using named and unnamed row types.

```
       CREATE ROW TYPE PUBLICATIONS_T          CREATE TABLE PUBLICATIONS
           ( Title VARCHAR(80),                    OF TYPE PUBLICATIONS_T;
             Year INTEGER,
           PRIMARY KEY ( Title ) );

       CREATE ROW TYPE BOOKS_T                 CREATE TABLE BOOKS
           ( ISBN CHAR(13) )                       UNDER PUBLICATIONS;
           UNDER PUBLICATIONS_T;
```

Figure 16 Named row type hierarchy must match corresponding table hierarchy.

and content of relevant information units which reside in other component databases in the federation. This is categorized as the *information discovery phase*. Once the information units of interest have been identified and located, the next phase is to resolve the differences between the information units to be imported and the information units contained in the user's local component database. This is the *semantic heterogeneity resolution phase*. After the various kinds of heterogeneity have been resolved, the next phase is to actually import the information units into the user's local environment so they can be used in an efficient and integrated manner. This is called the *system-level interconnection phase*. In Chapter 5.1, we will discuss federated database systems in detail and introduce the approaches to the three layers of issues related to federated database systems.

References

Aslan, G. (1998). Semantic Heterogeneity Resolution in Federated Databases by Meta-Data Implantation and Stepwise Evolution, Ph.D. dissertation, University of Southern California, Los Angeles.

Chen, P. (1976). The entity-relationship model: toward a unified view of data. *ACM Trans Database Syst.* **1(1)**, 9–36.

Codd, E. F. (1970). A relational model of data for large shared data banks. *CACM.* **13(6)**, 37–387.

Date, C. J. (1990). *An Introduction to Database Systems.* Vol. 1, 5th ed. Addison-Wesley, Reading, MA.

Date, C. J., and Hugh, D. (1997). *A Guide to the SQL Standard.* 4th ed. Addison-Wesley, Reading, MA.

Elmasri, R., and Navathe, S. (1994). *Fundamentals of Database Systems.* 2nd ed. Benjamin/Cummings, Menlo Park, CA.

Fang, D., Ghandeharizadel, S. and McLeod, D. (1996). An experimental object-based sharing system for networked databases. *VLDB J.* **5**, 151–165.

Fishman, D., Beech, D., Cate, H. *et al.* (1987). Iris: an object-oriented database management system. *ACM Trans. Office Inf. Syst.* **5(1)**, 48–69.

Hammer, M., and McLeod, D. (1981). Database description with SDM: a semantic database model. *ACM Trans. Database System* **6(3)**, 351–386.

Heimbigner, D., and McLeod, D. (1985). A federated architecture for information management. *ACM Trans. Office Inf. Syst.* **3(3)**, 253–278.

Hull, R., and King, R. (1987). Semantic database modeling: survey, applications, and research issues. *ACM Computing Surveys* **19(3)**, 201–260.

Informix, Inc. (1997a). *Informix Guide to SQL: Tutorial Version 9.1.* Informix Press, Upper Saddle River, NJ.

Informix, Inc. (1997b). An introduction to Informix universal server extended features. Informix Press, Upper Saddle River, NJ.

Kahng, J., and McLeod, D. (1996). Dynamic classificational ontologies for discovery in cooperative federated databases. In *Proceedings of the First International Conference on Cooperative Information Systems.* Brussels, Belgium, pp. 26–36.

McLeod, D. (1991). Perspective on object databases. *Inf. Software Technol.* **33(1)**, 13–21.

Rowe, L., and Stonebraker, R. (1987). The POSTGRES data model. In Proceedings of the International Conference on Very Large Databases, September 1–4, 1987, Brighton, England, pp. 83–96.

Sheth, A., and Larson, J. (1990). Federated database systems for managing distributed, heterogeneous, and autonomous databases. *ACM Computing Surveys* **22(3)**, 183–236.

Shipman, D. (1981). The functional data model and the data language daplex. *ACM Trans. Database Syst.* **2(3)**, 140–173.

Stonebraker, M., Brown, P. and Moore, D. (1999). *Object-Relational DBMSs: Tracking the Next Great Wave.* Morgan Kaufmann, San Mateo, CA.

PART 2

Modeling and Simulation

CHAPTER 2.1

Modeling the Brain

Michael A. Arbib

*University of Southern California Brain Project and Computer Science Department,
University of Southern California, Los Angeles, California*

Abstract

Our work on developing Neuroinformatics tools and exemplary databases has been accompanied by a vigorous program of computational modeling of the brain. Amongst the foci for such work have been:

1. *Parietal-premotor interactions in the control of grasping:* We have worked with empirical data on the monkey from other laboratories and designed and analyzed PET experiments on the human to explore interactions between parietal cortex and premotor cortex in the control of reaching and grasping in the monkey, linking the analysis of the monkey visuomotor system to observations on human behavior.
2. *Basal ganglia:* We have examined the role of the basal ganglia in saccade control and arm control as well as sequential behavior and the effects of Parkinson's disease on these behaviors.
3. *Cerebellum:* We have modeled the role of cerebellum in both classical conditioning and the tuning and coordination of motor skills.
4. *Hippocampus:* Both neurochemical and neurophysiological investigations of long-term potentiation (LTP) have been related to fine-scale modeling of the synapse; we have also conducted systems-level modeling of the role of rat hippocampus in navigation, exploring its interaction with the parietal cortex. Our work on modeling mechanisms of navigation also includes a motivational component linking the role of hippocampus to a number of other brain regions.

This chapter provides a brief introduction to our systems-level modeling efforts. Chapter 2.2 presents the NSL Neural Simulation Language which has been the vehicle for much of our modeling at the level of systems neuroscience. Chapter 2.3 presents the EONS methodology for multi-level modeling, with particular attention to modeling synapses and the fine structures which constitute them, while Chapter 2.4 presents the Synthetic PET method which provides the bridge from neural modeling of large-scale neural systems to the data of brain imaging. Our Brain Models on the Web (BMW) model repository provides documentation (BMW is described in Chapter 6.2; the documentation of the models is on our Website), links to empirical data, and downloadable code for almost all the models described in this chapter.

2.1.1 Modeling Issues

Building on Arrays of Neurons

Brain Theory (Arbib, 1995) includes the use of computational techniques to model biological neural networks, but it also includes attempts to understand the brain and its function in terms of structural and functional" "networks" whose units are at scales both coarser and finer than that of the neuron. Modeling work for USCBP includes development of diverse models, many of which are incorporated in a database, Brain Models on the Web (BMW), which will advance Brain Theory.

While much work on artificial neural networks focuses on networks of simple discrete-time neurons whose connections obey various learning rules, most work in brain theory now uses continuous-time models that represent either the variation in average firing rate of each neuron or the time course of membrane potentials. The models also address detailed anatomy and physiology as well as behavioral data to feed back to biological experiments. As we saw in the "Levels of Detail in Neural Modeling" section in Chapter 1.1, the study of a variety of connected "compartments" of membrane in dendrite, soma, and axon can help us understand the detailed properties of individual neurons, but much can still be learned about the large-scale properties of interconnected regions of the brain by modeling each region as a set of subpopulations ("arrays") using the *leaky integrator* model for each neuron. As shown before, in this case the internal state of the neuron is described by a single variable, the *membrane potential, m(t)*, at the spike initiation zone. The time evolution of $m(t)$ is given by the differential equation:

$$\tau \frac{dm(t)}{dt} = -m(t) + \sum_i w_i X_i(t) + h$$

with resting level h; time constant τ, $X_i(t)$ the firing rate at the ith input; and w_i the corresponding synaptic weight. This model does not compute spikes on an individual basis – firing when the membrane potential reaches threshold – but rather defines the *firing rate* as a continuously varying measure of the cell's activity, approximated by a simple, sigmoid function of the membrane potential, $M(t) = \sigma(m(t))$.

Connections between these two-dimensional arrays of neurons are defined in terms of interconnection masks which describe the synaptic weights. Consider the following equation where A, B, and C are arrays of neurons and W is a 3×3 connection mask (Fig. 1b). The equations:

$$\tau_A = 10\text{ms}$$
$$S_A = C + W^*B$$

state that the membrane time constant for A, τ_A, is 10 msec, and that for each cell i,j in array A, the cell's input, $S_A(i,j)$, is the sum of the output of the i,j^{th} cell in C plus the sum of the outputs of the 9 cells in B centered at i,j times their corresponding weights in W. That is,

$$sv_j = \sum_{k=-d}^{d} w_{jk}\, uf_{j+k}$$

Thus, the *operator in "W^*B" indicates that mask W is spatially convolved with B.

Chapter 1.1 provides references to a number of alternative strategies for modeling neurons, but this is the level for much of the modeling in the USC Brain Project. A notable exception is the detailed modeling of synapses and subsynaptic components using the EONS methodology of Chapter 2.3, but it is the" "system → regions → arrays of leaky integrator neurons" that has dominated our modeling and motivated the approach to modular, object-oriented neural systems embodied in our NSL Neural Simulation Language (Chapter 2.2).

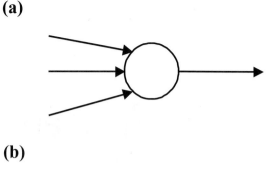

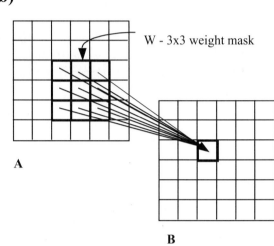

Figure 1 (a) An illustration of a basic neuron. A neuron receives input from many different neurons (*Sm*) which can affect the neuron's membrane potential (*m*), but it has only a single output (*M*), its firing rate. (b) Diagram showing the relationship between two arrays of neurons. In a uniformly connected network, each neuron in an array receives its inputs through the same synaptic weight mask. Here, the weight mask *W* applies a spatial convolution to input to *B* from a single array *A*, producing the result, $B = W^*A$. In general, the input to an array can come from many other arrays, each via its own mask.

Integration of Modeling and Experimentation

Analysis in neuroscience often goes to the extreme of focusing on one circuit or brain region and trumpeting it as "the" circuit implementing some specific function X, just as if we were to claim that Broca's area is "the" region for language. The other extreme is "holism," stressing that X may involve dynamic activity integrating all areas of the brain. Holism may be correct but seems to be useless as a guide to understanding. Our preferred approach is a compromise. We stress that any "schema" (see below) or function X involves the cooperative computation of many neural circuits (see Chapter 3 of Arbib *et al.*, 1998), but we then proceed by using a survey of the literature to single out a few regions for which good data are available correlating neural activity with the performance of function X. We then seek the best model available (from the literature, by research in our own group, or by a combination of both) which can yield a causally complete neural model that approximates both the patterns of behavior seen in animals performing X and the neural activity observed during these performances. As time goes by, the model may yield more and more insight into the data on X (some of which may be new data whose collection was prompted by the modeling), whether by conducting new analyses of the model, increasing the granularity of description of certain brain regions, or extending the model to include representations of more brain regions. Other challenges come from integrating separate models developed to explain distinct functions X, Y, and Z to derive a single model of interacting brain regions able to serve all these functions. As we come to understand more fully the roles of particular brain regions in serving particular functions, we can then turn to more extensive models. However, I fear that if we try to model" "everything all at once" we will understand nothing, as the map would then be co-extensive with the whole territory (Borges, 1975).

In the "Hierarchies, Models, and Modules" section of Chapter 1.1, we presented the following methodology:

For each brain region, a survey of the ... data calls attention to a few basic cell types. ... The modeler ... then creates one array of cells for each such cell type. The data tell the modeler what the activity of the cells should be in a variety of situations, but in many cases experimenters do not know in any quantitative detail the way in which the cell responds to its synaptic inputs, nor do they know the action of the synapses in great detail. In short, the available empirical data are not rich enough to get a model that would actually compute and so the modeler has to make a number of hypotheses about some of the unknown connections, weights, time constants, and so on to get the model to run. The modeler may even have to postulate cell types that experimenters have not yet looked for and show by computer simulation that the resulting network will indeed perform in the observed way when known experiments are simulated, in which case (1) it must match the external behavior; and (2) internally, for those populations that were based on cell populations with measured physiological responses, it must match those responses at some level of detail. What raises the ante is that (1) the modeler's hypotheses suggest new experiments on neural dynamics and connectivity, and (2) the model can be used to simulate experiments that have never been conducted with real nervous systems.

These ideas have motivated the efforts on Model Repositories and Summary Databases described in Part 6 of this volume, where we describe work to date on developing an environment that supports the linkage of data and models in the development, testing, and versioning of models. However, to date the development of models and the new modeling environment have proceeded in parallel. The result is that the models described below were developed "by hand" and then inserted into BMW. In the future, models will more and more be developed using the tools and environments of the NeuroInformatics Workbench to support the increased integration of modeling and experimentation.

Models, Modules, Schemas, and Interfaces

In general, we view a model as comprising a single "top-level module" composed of a number of different modules, which themselves may or may not be further decomposable. If a module is decomposable, we say this "parent module" is decomposed into submodules known as its "child modules;" otherwise, the module is a "leaf module."

MODULES AS BRAIN STRUCTURES

Fig. 1 of Chapter 1.1 (reproduced as Fig. 7 below) illustrated the modular design of a model by showing in (a) a basic model of reflex control of saccades involving two main modules, each of which is decomposed into submodules, with each submodule defining an array of physiologically defined neurons; (b) embedded this basic model into a far larger model which embraces various regions of cerebral cortex, thalamus, and basal ganglia. While the model may indeed be analyzed at this top level of modular decomposition, (c) showed how to further decompose basal ganglia to tease apart the role of dopamine in differentially modulating the direct and indirect pathways. This example introduces the case where the modules correspond to some physical structure of the brain – whether a brain region, an array of neurons, a neuron, or some subcellular structure. Here I want to emphasize two other kinds of module.

MODULES AS SCHEMAS

In modeling some complex aspect of the brain, we may want to use a detailed structural description for some parts of the model and functional descriptions for others. For example, in modeling the role of a parietal area in visually directed grasping, we may choose not to burden our model with a detailed representation of the actual brain regions (retina, thalamus, visual cortex, etc.) that process the visual input it receives but instead combine their functionality in a single abstract *schema*. In other cases, we may decompose a schema into finer schemas and simulate their interaction; some but not all of these subschemas will be mapped onto detailed neural structures. The starting point for schema theory as used here (Arbib, 1981) was to describe perceptual structures and distributed motor control in terms of coordinated control programs linking simpler perceptual and motor schemas, but these schemas provide the basis for more abstract schemas which underlie action, perception, thought, and language more generally.

In summary (see Arbib *et al.*, 1998, chap. 3), the functions of perceptual-motor behavior and intelligent action of animals situated in the world can be expressed as a network of interacting schemas. Each schema itself involves the integrated activity of multiple brain regions. A multiplicity of different representations – whether they be partial representations on a retinotopic basis, abstract representations of knowledge about types of object in the world, or more abstract "planning spaces" – must be linked into an integrated whole. Such linkage, however, may be mediated by distributed processes of "cooperative computation" (competition and cooperation). There is no one place in the brain where an integrated representation of space plays the sole executive role in linking perception of the current environment to action. We thus see that the modules that constitute any one of our models of (some aspect of) the brain may be designed to represent an actual structure of the brain, or a schema which captures a function which may possibly require interaction of multiple brain structures for its realization.

MODULES AS INTERFACES

The above types of modules represent the component structures and/or functions to be included in a model of some aspect of the brain. Other modules are designed to help the user interact with the model. For example, in the "Schematic Capture" section of Chapter 1.1, we related the key notion of an *experimental protocol* – which defines a class of experiments by specifying a set of experimental manipulations and observations – to the use of a *simulation interface* which represents an experimental protocol in a very accessible way, thus making it easy for the non-modeler to carry out experiments on a given model. Fig. 2 of Chapter 1.1 illustrated a particular user interface comprising two modules: One allows the user to set conditions for a particular simulation run of the model and to control the execution of that run; the other is a display module, showing the results of the simulation. The simulation interface allows one to explore hypotheses as to how specific cells work together to produce overall behavior and may also be understood as samplings of the interaction of multiple neural arrays, thus deepening our understanding of cellular activity from the systems point of view.

2.1.2 Parietal-Premotor Interactions in the Control of Grasping

This section presents the FARS (Fagg-Arbib-Rizzolatti-Sakata) model for grasping. It is named for the modelers Andy Fagg and myself and for the experimentalists Giacomo Rizzolatti and Hideo Sakata, whose work anchors the model. The model shows show how a view of an object may be processed to yield an appropriate action for grasping it and explains the shifting patterns of neural activity in a variety of brain regions involved in this visuomotor transformation.

The neurophysiological findings of the Sakata group on parietal cortex (Taira et al., 1990) and the Rizzolatti group on premotor cortex (Rizzolatti et al., 1988) indicate that parietal area AIP (the anterior intra-parietal sulcus) and ventral premotor area F5 in monkey (Fig. 2) form key elements in a cortical circuit which transforms visual information on intrinsic properties of objects into hand movements that allow the animal to grasp the objects appropriately (see Jeannerod et al., 1995, for a review). Motor information is transferred from F5 to the primary motor cortex (denoted F1 or M1), to which F5 is directly connected, as well as to various subcortical centers for movement execution. Discharge in most F5 neurons correlates with an action; the most common are grasping with the hand, grasping with the hand and the mouth, holding, manipulating, and tearing. Rizzolatti et al. (1988) thus argued that F5 contains a "vocabulary" of motor schemas (Arbib, 1981). The situation is in fact more complex, and" "grasp execution" involves a variety of loops and a variety of other brain regions in addition to AIP and F5.

Affordances (Gibson, 1979) are features of an object or environment relevant to actionvisual processing may exploit them to extract cues on how to interact with an object or move in the environment, complementing processes for categorizing objects or determining their identity. The FARS model (Fagg and Arbib, 1998) provides a computational account of what we call the *canonical system*, centered on the AIP → F5 pathway, showing how it can account for basic phenomena of grasping. The highlights of the model are shown in Figs. 3 and 4. Our basic view is that AIP cells encode *affordances for grasping* from the visual stream and send their neural codes on to area F5. As Fig. 3 shows, some cells in AIP are driven by feedback from F5 rather than by visual inputs so that AIP can monitor ongoing activity as well as visual affordances. Here we indicate the case in which the visual input has activated an affordance for a precision pinch, with AIP activity driving an F5 cell pool that controls the execution of a precision pinch.

What we show, however, is complicated by the fact that the circuitry is not for a single action, but for a behavior designed by Sakata to probe the time dependence of activity in the monkey brain. In the Sakata paradigm, the monkey is trained to watch a manipulandum until a go signal instructs it to reach out and grasp the object. It must then hold the object until another signal instructs it to release the object. In Fig. 3, cells in AIP instruct the set cells in F5 to prepare for execution of the Sakata protocol using a precision pinch. Activation of each pool of F5 cells not only instructs the motor apparatus to carry out the appropriate activity (these connections are not shown here), but also primes the next pool of F5 neurons (i.e., brings the neurons to just below threshold so they may respond quickly when they receive their own go signal) as well as inhibiting the F5 neurons for the previous stage of activity. Thus, the neurons that control the extension phase of the hand shaping to grasp the object are primed by the set neurons, and they reach threshold when they receive the first go signal, at which time they inhibit the set neurons and prime the flexion neurons. These pass threshold when receiving a signal that the hand has reached its maximum

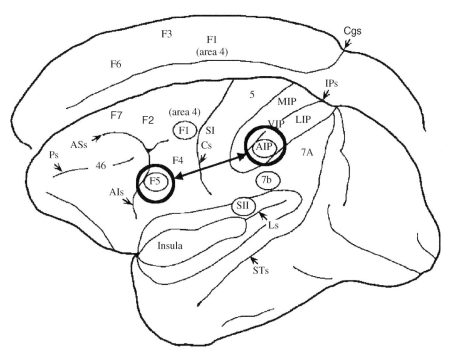

Figure 2 A side view of the left hemisphere of the Macaque monkey brain dominates the figure, with a glimpse of the medial view of the right hemisphere above it to show certain regions that lie on the inner surface of the hemisphere. The central fissure is the groove separating area SI (primary somatosensory cortex) from F1 (primary motor cortex, more commonly called MI). Frontal cortex is the region in front of (in the figure, to the left of) the central sulcus. Area F5 of premotor cortex (i.e., the area of frontal cortex just in front of primary motor cortex) is implicated in the elaboration of "abstract motor commands" for grasping movements. Parietal cortex is the region behind (in the figure, to the right of) the central sulcus. The groove in the middle of the parietal cortex, the intra-parietal sulcus, is shown opened here to reveal various areas. AIP (the anterior region of the intra-parietal sulcus) processes visual information relevant to the control of hand movements and is reciprocally connected with F5.

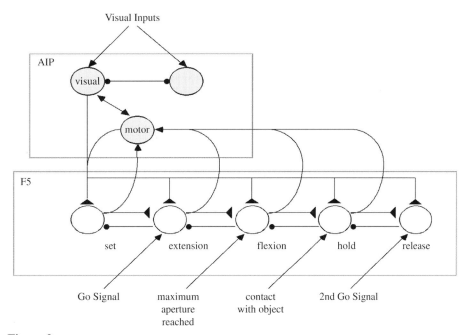

Figure 3 Hypothesized information flow in AIP and F5 in the FARS model during execution of the Sakata paradigm. This neural circuit appears as a rather rigid structure; however, we do not hypothesize that connections implementing the phasic behavior are hardwired in F5. Instead, we posit that sequences are stored in pre-SMA (a part of the supplementary motor area) and administered by the basal ganglia.

aperture. The hold neurons, once primed, will become active when receiving a signal that contact has been made with the object, and the primed release neurons will command the hand to let go of the object once they receive the code for the second go signal.

Karl Lashley (1951) wrote of "The Problem of Serial Order in Behavior," a critique of stimulus-response approaches to psychology. He noted that it would be impossible to learn a sequence such as A, B, A, C as a stimulus-response chain because the association "completing A triggers B" would then be interfered with by the association "completing A triggers C" or would dominate it to yield an infinite repetition of the sequence A, B, A, B, ... The generally adopted solution is to segregate the learning of a sequence from the circuitry that encodes the unit actions, the latter being F5 in the current study. Instead, another area has neurons whose connections encode an "abstract sequence" Q1, Q2, Q3, Q4, with sequence learning then involving learning that activation of Q1 triggers the F5 neurons for A, Q2 triggers B, Q3 triggers A again, and Q4 triggers C. In this way, Lashley's problem is solved. Other studies lead us to postulate that the storage of the sequence may be in the part of the supplementary motor area called pre-SMA (Luppino et al., 1993) with administration of the sequence (inhibiting extraneous actions while priming imminent actions) carried out by the basal ganglia (Bischoff, 1998; Bischoff-Grethe and Arbib, 2000).

We now turn to the crucial role of the inferotemporal cortex (IT) and prefrontal cortex (PFC) in modulating F5's selection of an affordance (Fig. 4). Here, the dorsal stream (from primary visual cortex to parietal cortex) carries, among other things, the information needed for AIP to recognize that different parts of the object can be grasped in different ways, thus extracting affordances for the grasp system which (according to the FARS model) are then passed on to F5, where a selection must be made for the actual grasp. However, the dorsal stream does not know "what" the object is; it can only see the object as a set of possible affordances. The ventral stream (from primary visual cortex to inferotemporal cortex), by contrast, is able to recognize what the object is. This information is passed to the prefrontal cortex, which can then, on the basis of the current goals of the organism and recognition of the nature of the object, bias F5 to choose the affordance appropriate to the task at hand. In particular, the FARS model represents the way in which F5 may accept signals from areas F6 (pre-SMA), 46 (dorsolateral prefrontal cortex), and F2 (dorsal premotor cortex) to respond to task constraints, working memory, and instruction stimuli, respectively (see Fagg and Arbib, 1988, for more details).

Synthetic PET

In Chapter 2.4, we will introduce the Synthetic PET method for analyzing simulation results obtained with a large-scale neural network model to predict the results of human brain-imaging experiments. In particular, we will describe Synthetic PET predictions based on the FARS model, their testing by studies using positron emission tomography (PET) imaging of human subjects, and the new insights that can be gained when modeling can use Synthetic PET to integrate findings from neurophysiology and neuroanatomy of monkey with brain imaging of human. Such studies both build on and contribute to our understanding of the homologies between brain regions of different species (Chapter 6.4).

2.1.3 Basal Ganglia

Control of Saccades

An early success in neural modeling came with the work of David Robinson and others who explained how the brain stem could serve as a saccade burst generator, acting as a control system to convert the current "error" in gaze (the angle between the direction of gaze and the position of a visual target) into, first, the motor neuron firing required for the burst of muscle activity that would turn the eye to the desired position and, second, the tonic activity to hold the eye in that position (see Robinson, 1981, for a review). The resultant control model was significant in that many of the variables could be related to the firing of specific classes of neurons in the brainstem; however, the early models were defective in that the position of the target was represented by a single error variable, rather than by the position of the target as imaged onto the retina. Thus, later models (e.g., Scudder, 1988) interposed a model of the superior colliculus as the device which could serve as the" "retinotopic mapper" converting position on the retina (retinotopy) into the sort of signal that could be used by the brainstem saccade burst generator.

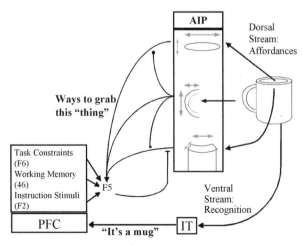

Figure 4 The role of inferotemporal cortex (IT) and prefrontal cortex (PFC) in modulating the F5' selection of an affordance.

Further studies had to address new data which showed an essential role for cerebral cortex in a variety of more complex behaviors. For example, a monkey can be trained not to respond to a target while a fixation light is on but then to saccade to the position of the target at the end of the fixation even if the target is no longer visible. This requires an additional" "schema" for target memory. Other data showed that if the animal is exposed to two targets of during the fixation period, then at the offset of the fixation he will saccade to the two targets in turn. Moreover, the neural representation of the second target after the first saccade is completed corresponds to the retinotopic coding that the second target would have had after the first saccade (but did not have, because it was no longer visible after that saccade had occurred). This requires yet another schema, this one for "remapping." The extension of the basic model thus required is shown in Fig. 5.

Dominey and Arbib (1992) used an extensive analysis of the literature to replace the schema model of Fig. 4 by the NSL model outlined in Fig. 6. Heavy outlines separate modules representing different brain regions, and each brain region is divided into arrays motivated by physiological data. For example, there are cells in different regions whose activity is best correlated with the onset of a saccade; other cells have activity best correlated with maintained activity or working memory between the presentation of a stimulus at the beginning of a delay period until the actual response. However, as noted earlier when discussing our general methodology, while a great deal of knowledge of the available data went into the construction of the model, the data are not rich enough to get a model that would actually compute so we, as modelers, had to make a number of hypotheses about missing connections, weights, and time constants to get the model to run. The model was then tuned so that the passage from the presentation of visual input to generation of eye movement matches the external behavior, and internally (for those populations that were based on cell populations with measured physiological responses), we match, at some level of detail, those responses. In particular, the model detailed how thalamocortical loops could implement target memory (which is thus a schema whose implementation is distributed across interacting brain regions), while remapping was conducted by circuitry intrinsic to the region LIP of posterior parietal cortex (shown as PP in Fig. 6), whose projections then accounted for related activity patterns in other brain regions. In particular, the model then showed how, when a new target representation was created by remapping, the result would replace the old representation in target memory.

A new model for the role of the basal ganglia (BG) in the control of saccades (Crowley, 1997) is shown in Fig. 7 (already shown as Fig. 1 of Chapter 1.1). The new model was motivated by two different issues, and our response to each issue illustrates the modular methodology of our approach to large-scale neural modeling. One was that new data about many different parts of the brain had been found during the early 1990s and we wanted to reflect those data in the model. There were certain hypotheses we made that proved to be robust, while there are others that are no longer consistent with the available data. For example, new data on superior

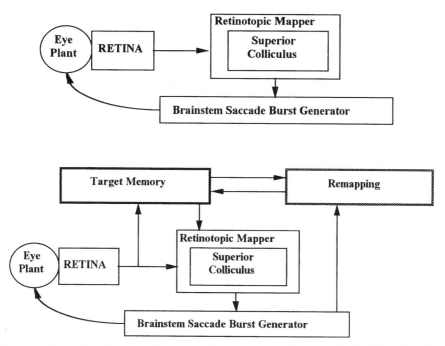

Figure 5 (Top) A modular description of *structures* involved in the reflex control of saccades. (Bottom) Extending the modular description which adds *schemas* for *Target Memory* and *Remapping* necessary to account for a variety of studies of saccades in monkeys.

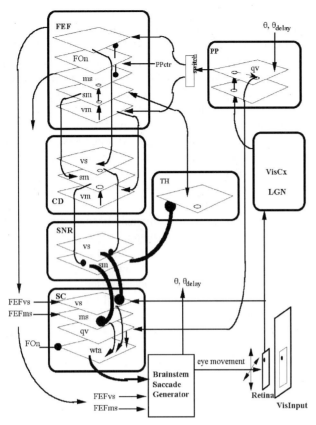

Figure 6 The modular design of the NSL model developed by Dominey and Arbib (1992) to explain the neural mechanisms for memory saccades and double saccades. The model includes the superior colliculus and brainstem, as well as cortical regions, thalamus, and basal ganglia. It represents each brain region by one or more layers of physiologically identified cell types. vs. = visual pre-saccade, ms = memory pre-saccade, sm = sustained memory, vm = visual response to memory target, qv = quasi-visual, wta = winner take all, PPctr = central element of PPqv, FEF = frontal eye fields, PP = posterior parietal cortex, CD = caudate nucleus, SNR = substantia nigra pars reticulata, SC = super colliculus, TH = thalamus (mediodorsal), Fon = fovea on (foveation).

colliculus led us to replace the superior colliculus of the Dominey-Arbib model by the one shown in Fig. 7a. The other was that we had earlier focused on data on normal function, but now we wanted to look at dysfunction. In particular, we wished to understand how dopamine depletion in the basal ganglia leads to the symptoms of Parkinson's disease. To this end, we extended the model of the direct pathway through the basal ganglia of the Dominey-Arbib model to include the indirect pathway and to model the differential effects of dopamine, providing a detailed representation of circuitry for all the regions shown in Fig. 7c. Fig. 8 shows one result obtained with this model. The graphical output shows saccade velocity in the double saccade task. The normal behavior of the model is on the left; on the right, we see the effect of cutting the dopamine in half in the model. By modeling the differential effect of the direct and indirect pathways within the basal ganglia, we find that with reduced dopamine there is only one saccade.

This begins to give us some insight into the bradykinesia and akinesia that are seen in Parkinson patients, as well as tying in with some of the available monkey data.

This work not only extended our analysis of a variety of brain regions, it also led (in USCBP work that used the Crowley model as a testbed for work on both NSL and BMW) to an important example of using the explicit description of an experimental protocol to design a simulation interface which allows non-expert users to easily conduct related simulation experiments. Fig. 2 of Chapter 1.1 shows the simulation interface for the double saccade protocol in which a monkey fixates a fixation point, during which time two brief targets are briefly flashed. After the fixation point is removed, the monkey saccades to the remembered position of the two targets in turn. The simulation interface contains two panels: one to set up and run the experiment, the other to observe a variety of displays of neural activity.

Control of Sequential Arm Movements

The modeling result of Fig. 8 is consistent with the finding that patients with diseases of the BG, particularly Huntington's disease and Parkinson's disease, do not have significant motor control difficulties when visual input is available (Bronstein and Kennard, 1985) but do have problems with specific forms of internally driven *sequences* of movements (Curra et al., 1997). This implies that the basal ganglia may be involved in assisting cortical planning centers in some fashion as well as providing sequencing information for cross-modal movements (e.g., simultaneous arm reach, hand grasp, head movement, and eye movement). We have thus extended the model of Fig. 7b to include the control of arm movements as well as saccades (Fig. 9). In each case, the *direct path*, involving projections directly from the striatum onto the BG output nuclei, is primarily responsible for providing an estimate of the next sensory state, based upon the current sensory state and the planned motor command, to cortical planning centers. The primary role of the *indirect path*, where the striatum projects onto the external globus pallidus (GPe), in turn projecting to the subthalamic nucleus (STN) and finally to the BG output structures, is to inhibit motor activity while cortical systems are either determining which motor command to execute or are waiting for a signal indicating the end of a delay period.

Because microstimulation of thalamic pallidal receiving areas tends not to elicit arm movement (Buford et al., 1996), the basal ganglia may not so much control the precise dynamics of movement as permit a movement to be made. That is, the disinhibition of the thalamus may increase the thalamic firing rate which in turn may provide the cortical regions enough input to reach threshold and fire. Without this disinhibition, which appears to be lacking in Parkinson's disease, cortical cells have a more

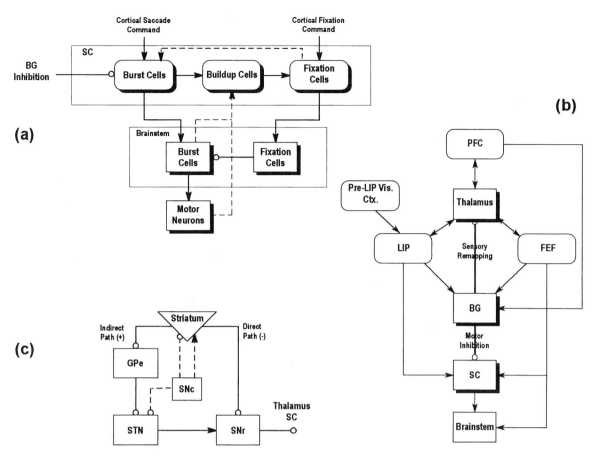

Figure 7 (a) A basic model of reflex control of saccades involves two main modules, one for superior colliculus (SC) and one for brainstem. Each of these is decomposed into submodules, with each submodule defining an array of physiologically defined neurons. (b) The model of (a) is embedded into a far larger model which embraces various regions of cerebral cortex (represented by the modules Pre-LIP Vis. Ctx., LIP, PFC, and FEF), thalamus, and basal ganglia. While the model may indeed be analyzed at this top level of modular decomposition, we need to further decompose BG, as shown in (c) if we are to tease apart the role of dopamine in differentially modulating (the two arrows shown arising from SNc) the direct and indirect pathways within the basal ganglia (Crowley, 1997).

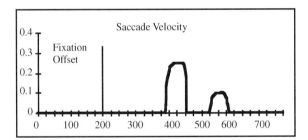 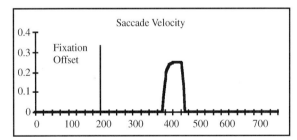

Figure 8 Double saccades with dopamine manipulation. Cutting dopamine in half caused a second saccade not to be generated due to increased inhibition in the SNr (substantia nigra pars reticulata) medial circuit of the model.

difficult time achieving threshold, leading to difficulties in movement performance. In Huntington's disease, therefore, one might suggest that too much thalamic disinhibition allows for indiscriminate motor activity, as cortical cells more easily reach threshold.

Although Fig. 9 shows a generic cortical projection to basal ganglia, this projection is composed of different areas of cortex. The simple box marked "Cortex" becomes a complex structure in our actual computer modeling, combining models of the prefrontal cortex (PFC), frontal eye fields (FEF), and the lateral intraparietal cortex (LIP) in our saccade model and two subregions of the supplementary motor area (SMA)pre-SMA and SMA-proper in our model of limb movements. Because we have already discussed the oculomotor system in some detail, we now focus on control of sequences of skeletomotor actions (Bischoff, 1998; Bischoff-Grethe and Arbib, 2000). Within the skeletomotor pathway, PFC is responsible for working memory and for sequence learning (Barone and Joseph, 1989; Jenkins

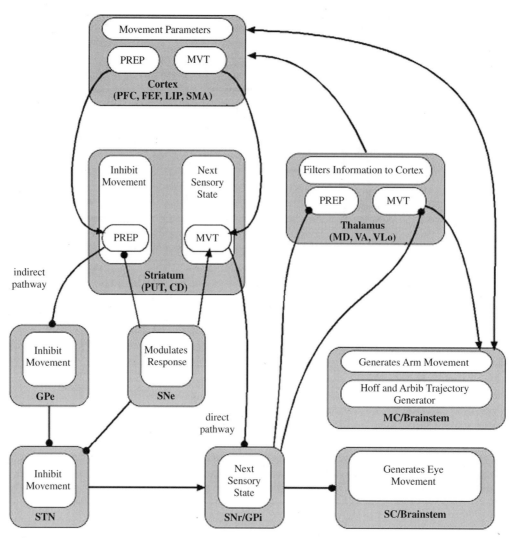

Figure 9 Model showing proposed functions for each component of the oculomotor and skeletomotor circuit. Although we have only shown the brain stem as divided into the movement-specific regions, we also follow the "loop hypothesis" that all other regions are subdivided into relatively distinct regions forming the separate oculomotor and skeletomotor circuits. Our key hypothesis is the shared functionality of the corresponding regions in the two circuits. GPe = external globus pallidus; Gpi = internal globus pallidus; MC = motor cortex; MVT = movement neurons; PREP = premovement neurons; SC = superior colliculus; SNc = substantia nigra pars compacta; SNr = substantia nigra pars reticulata; STN = subthalamic nucleus.

et al., 1994). PFC projects to the pre-supplementary motor area (pre-SMA). Because pre-SMA is responsible for preparing a movement sequence, particularly if visually guided, PFC may be a source for the sequences that pre-SMA "knows." Pre-SMA projects sequential information to both SMA-proper and to the basal ganglia's indirect pathway. SMA-proper is involved in the internal generation of sequences and repetitive movements.

SMA-proper neurons contain information on the overall sequence to be performed, keep track of which movement is next and which movement is currently being performed, project the current movement to be performed to MC and to the direct pathway of the basal ganglia, and project the preparations of the next movement of the sequence to another population of premovement neurons within MC and to the indirect (inhibitory) pathway of basal ganglia. The motor cortex, then, carries out the motor command and handles the fine tuning of the movement (e.g., target position, amplitude, force) partly based upon information provided by the basal ganglia. The motor cortex projects the motor parameters to both the brain stem and the basal ganglia's direct pathway. We postulate that the basal ganglia's two pathways perform two different roles: the indirect pathway inhibits upcoming motor commands from being performed while the current movement is in progress, while the direct pathway projects the next sensory state back to cortex. This informs SMA-proper and MC of the expected next state and allows these regions to switch to the next movement of the sequence.

In using this model to perform sequential movements (Bischoff, 1998; Bischoff-Grethe and Arbib, 2000), we provided the model with the location of three targets and the order in which contact should be made. SMA-proper contained neurons similar to those seen by Tanji and Shima (1994) representing an overall task sequence and the subsequences within the task. A trajectory generator (Hoff and Arbib, 1992) performed the movements as dictated by MC. Under normal conditions, the model successfully moved from target to target and exhibited a velocity similar to normal subjects. When we depleted dopamine, we encountered several behavioral results. As SNc reduced its ability to affect the direct and indirect pathways, there was an increase in the firing rates of STN and both sections of globus pallidus. The increase in inhibitory output led to a decrease in the MC firing rate, as MC was unable to completely overcome GPi's inhibition via ventrolateral pars oralis thalamus (VLo) projections. This reduced firing rate translated to a slower movement time. Thus, the natural balance between the two pathways is disturbed when dopamine is lost. In Parkinson's disease, this leads to an increase in activation of the indirect pathway (movement inhibition) and a decrease in effectiveness of the direct pathway (next sensory state). We also saw a slowdown in overall neural activity the basal ganglia and the motor cortex took longer to reach peak firing rates. The slowdown in neural activity, coupled with the increased inhibition of movement, was responsible for pauses between movements within the sequence. We therefore began to see pauses between movements from one target to the next; however, we also found that for each subsequent movement, the velocity was less. This is similar to the bradykinesia seen in Parkinson's disease patients when asked to trace the edges of a polygon (Agostino *et al.*, 1992). The model also exhibited a reduction in cortical firing rates. When dopamine was further depleted, the model was capable of performing only the movement towards the first target, reminiscent of the situation shown in Fig. 8. This model has also been used to reproduce results seen in normal and Parkinson's disease subjects in a reciprocal aiming task and normal behavior during conditional elbow movements (Bischoff, 1998).

2.1.4 Cerebellum

The architecture of the cerebellar circuitry is by and large well understood, as are the relationships between the cerebellum and other brain structures. The functions of the cerebellum include the learning of fine control of motor coordination and classical conditioning of motor responses, and it is widely believed that the same type of synaptic plasticity (i.e., a long-term depression, LTD, of synaptic transmission between the parallel fibers and the Purkinje cells in cerebellar cortex) is involved in both adaptation of skeletomotor and oculomotor skills and in learning of classically conditioned motor responses. The hypothesis postulates that repeated pairing of appropriate stimulus patterns reaching Purkinje neurons through the mossy fiber/parallel fiber pathways, with activation of the single climbing fiber which activates the Purkinje cell, results in LTD of the parallel fiber to Purkinje neurons synapses. However, the relative contribution of the cerebellar cortex and the deep cerebellar nuclei to the learning of skill and classical conditioning tasks has not been clearly understood. In particular, it has been proposed that a long-term potentiation (LTP) of synaptic transmission between the mossy fibers and the neurons of the deep cerebellar nuclei also contributes to such learning. In the models reported here, however, we concentrate on the parallel fiber to Purkinje neurons synapses as the loci of synaptic change.

Cerebellar Models of Motor Control and Coordination

Our models of cerebellar involvement in motor control hypothesize that the cerebellum does not act directly on the muscles but rather acts through motor pattern generators (MPGs) – circuits which combine, for example, trajectory or rhythmic control with local feedback circuitry. We view the cerebellum (combining cerebellar cortex and cerebellar nuclei) as being divided into *microcomplexes*. Each microcomplex is a general, modular computational module comprising a patch of cerebellar cortex and nuclear cells whose basic form is shown in Fig. 9. We do *not* hypothesize that the cerebellar microcomplex learns to *replace* the MPG but rather that the complex learns how to adjust its output to appropriately tune and augment the output of the MPG to which it belongs. Each microcomplex combines a patch of cerebellar cortex (Purkinje cells and Golgi cell) with the underlying set of cells in the cerebellar nucleus to which the Purkinje cells project and the inferior olive cells, which provide the climbing fibers that control learning by providing error signals in our models of motor learning and the CS (conditioned stimulus) in our models of classical conditioning Fig. 10. The "contextual input" is provided by the parallel fibers (each of which crosses a number of microcomplexes), the granule cell axons which provide a nonlinear combination of mossy fiber inputs. The job of the Purkinje cells is to learn to pair the parallel fiber output with a pattern of inhibition of the nuclear cells so as to ensure that these cells better tune the MPGs.

As Fig. 11 shows, the parallel fibers are long enough that their shared "contextual input" Purkinje cells may learn not only to tune individual MPGs to changing circumstances but also to coordinate multiple MPGs (such as those for reaching and grasping) so that a complex movement may be achieved in a smooth and integrated way. In short, we view the cerebellum as applying its compensations by modulating MPGs, whether cortical or subcortical, and this compensation occurs on multiple time scales. Further, the compensation patterns can

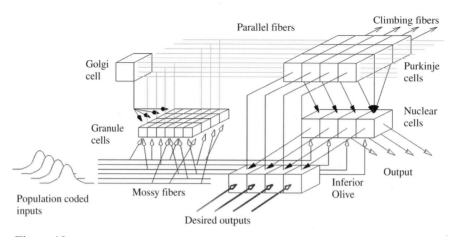

Figure 10 The microcomplex combines a patch of cerebellar cortex (Purkinje cells and Golgi cell) with the underlying set of cells in the cerebellar nucleus, to which the Purkinje cells project, and the inferior olive cells, which provide the climbing fibers which control learning. The parallel fibers (each of which crosses a number of microcomplexes) are the granule cell axons that provide a nonlinear combination of mossy fiber inputs. The job of the Purkinje cells is to learn to pair the parallel fiber output with a pattern of inhibition of the nuclear cells so as to ensure that these cells better tune the MPGs.

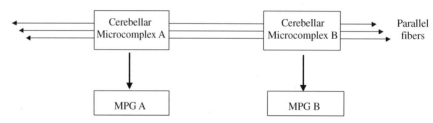

Figure 11 Hypothesis: The cerebellum adjusts the parameters of MPGs both to tune individual MPGs and to coordinate related MPGs.

be stored and recalled based on higher level task information. The critical test of the cerebellum is its ability to effectively compensate for complex nonlinear dynamics and to adapt that compensation as the dynamics change. Thus, a cerebellar model must be tested and developed in conjunction not only with an MPG model but also with a "plant model" which contains sufficient complexity to "challenge" the cerebellum model. However, we shall not get into the biomechanics in this chapter.

Our model posits a structure in which a microcomplex of cerebellar cortex and underlying cells forms the key module, but in this case we build a model system that integrates sensory-motor information for intersegmental coordination and nonlinear dynamic compensation. These microcomplexes in turn drive, tune, and coordinate motor pattern generators to create descending signals that impinge on motor neurons.

Schweighofer et al. (1996a,b), in modeling the role of cerebellum in prism adaptation and compensating for dysmetria in saccades, developed a computational model of the" "microcomplex" and showed how the microzone could act by adjusting the parameters of a motor pattern generator outside the cerebellum (in this case, in the brain stem). Schweighofer et al. (1998a) and Schweighofer et al. (1998b) showed how the cerebellum may compensate for Coriolis forces and other joint interactions in allowing coordinated control of multiple joints in reaching. Our saccade and throw models included no inhibitory interneurons in cerebellar cortex, but our reach model has brought in Golgi cells, basket cells, and stellate cells with realistic firing rates and interconnections. In this study, the model of the Purkinje cells has gone beyond the relatively simple leaky integrator neuron to include a set of spiking dendrites, so we are well on the way to combining a systems level view of cerebellar function with detailed cellular analysis (Fig. 12).

Arbib et al. (1994, 1995) were inspired by discussion of the data that have now been written up by Martin et al. (1996). This experimental work showed that prism adaptation for throwing a ball or dart involves (1) the cerebellum, and (2) (and this is the exciting part) task-specific sensorimotor transformations rather than a global adaptation of the visual field. Our work explained how this could take place, using the microcomplex developed in the saccade studies but now modulating a motor controller in premotor cortex. To ground our discussion of the role of cerebellum in motor control, we will devote the next section to an outline of the follow-up modeling conducted by Spoelstra and Arbib (1997). We then turn to a brief discussion of classical conditioning, but first we

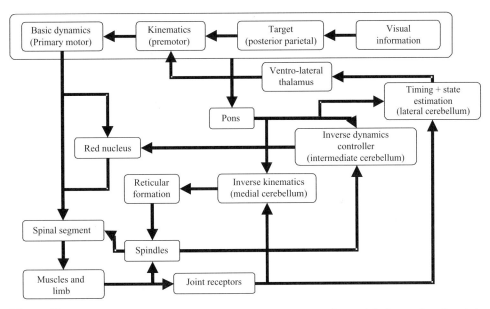

Figure 12 Diagram indicating the many components necessary to understand the interaction of cerebellum (with different parts involved in postural control and dynamic compensation), cerebral cortex, subcortical regions, spinal reflex circuitry, and the biomechanics of muscles and limb in the adaptive control of limb movements. (See Spoelstra *et al.*, 2000, for details.)

need to discuss the synaptic plasticity rule used in the models.

What is distinctive about the Purkinje cell is its two principal kinds of input. There are on the order of a hundred thousand parallel fibers, each making a single synapse with the Purkinje cell, and a single climbing fiber, which makes so many synapses with the Purkinje cell that a single spike arriving on a climbing fiber will cause the Purkinje cell to fire several times. Both Marr (1969) and Albus (1971) suggested that the climbing fiber provides the error signal for training the synapses from parallel fibers to Purkinje cells, so the Purkinje cell could learn to classify patterns of parallel fiber firing and thus of the states that these firings encoded. Where Marr suggested that the coincidence of parallel fiber and climbing fiber activity would yield an increase in the synaptic weight, Albus postulated that there would be a decrease. These conjectures inspired a great deal of work, first neurophysiological then neurochemical, by Masao Ito and his colleagues (see Ito, 1984, for a review) which finally established that the Albus hypothesis was correct: Active climbing fiber input to a Purkinje cell signals that any active parallel fiber synapse on that Purkinje cell should be weakened. What Ito and others in fact established was a phenomenon of long-term depression (LTD), a lowering of synaptic strength lasting several hours. In fact, modeling considerations have shown that this must be balanced by some form of long-term potentiation (LTP) to stop all synaptic strengths from drifting down towards 0 over time.

We later refined the Marr-Albus-Ito perspective not only by introducing equations that balance LTP and LTD, but also by attending to the fact that the delay between motion generation and error detection implies the need for a *short-term memory local to each synapse* to retain the locus of cerebellar activity involved in the movement. We posited that parallel fiber activity provides a synapse with a "window of eligibility" for modification of all synaptic strength on receipt of a climbing fiber input which peaks at about 200 msec, corresponding to the typical delay between the time the firing of a Purkinje cell modulates the generation of a movement and the time that the climbing fiber reports back to the cell on the visually sensed error in the movement. More detailed modeling shows that this eligibility can be represented by the concentration of a molecular species called a *second messenger* within the given synapse – with the concentration of the messenger largest for synapses involved in the movement related to the current climbing fiber "error signal."

What is of particular interest from a modeling perspective about this model is that a purely system-level model – linking overall human behavior to arrays of relatively simple neurons – led us to consider timing issues which then required us to modify our learning rules. This modification of the synapse adjustment rule then led us to consider new ideas about the possible neurochemistry of synaptic plasticity in the cerebellum. The model shows how work at one level of a hierarchy of levels for competition neuroscience can either depend on or will be stimulated by work at another. We thus see the linkage of data and modeling from the systems level to the neurochemical level, a key feature of the USC Brain Project. Moreover, because an IO cell typically spikes from only 2 to 6 times a second, it is inappropriate to represent its behavior by an average firing rate, so we instead use an integrate-and-fire model which explicitly generates spikes to be used in the synaptic learning rule.

Prism Adaptation for Throwing

A basic component of adaptive behavior is the adaptation of sensori-motor transformations. Prism adaptation is a paradigm for such adaptation, and the cerebellum is required for this. BMW includes a biologically based neural net model of this role of cerebellum (Spoelstra and Arbib, 1997), which builds on our earlier work (Arbib et al., 1994). The database for this modeling includes much data on the anatomy and physiology of cerebellum and other regions involved in motor control, as well as a body of behavioral data. The work of Martin et al. (1995, 1996) on prism adaptation of human eye-hand coordination in a number of throwing tasks shows the task specificity of the adaptation, as well as its dependence on the cerebellum. The behavior of interest here is the perturbation that we see in throwing a ball or dart at the target when a person puts on a pair of prism glasses (Fig. 13a). Most people are able to adjust over a number of trials, but this adaptation seems to be missing in people with certain types of cerebellar damage. A subject throws at a target, then dons 30° wedge prism glasses, causing him to throw 30° off target. With repeated throws, the subject improves until he once again throws on target. When the prisms are removed, the subject misses by almost 30° on the opposite side and has to readjust his aim. Fig. 13b shows that the model is able to reproduce this behavior. The systems-level circuit of the model is shown in Fig. 14.

It is beyond the scope of this overview to elaborate upon the details of the model, save to note that the cerebral cortex provides information about both where the arm is to be aimed during the throw and whether or not the subject is wearing prisms and uses this information to control throwing. The cerebellum samples the same information to provide an adaptive side loop that can modify this cortical control. As in other NSL models, brain regions are represented as two-dimensional arrays of neurons, and neurons (other than the inferior olive neurons) are modeled as leaky integrators with a positive real-valued output representing the instantaneous firing rate of the neuron.

What goes on in prism adaptation? One hypothesis is that it provides a global visual transformation, compensating for the global shift of the visual input provided by the prism; however, the data of Martin et al. suggest a set of "private" visuomotor transformations. Prism adaptation for throwing with the left hand does not transfer to throwing with the right hand. Even for throwing with the right hand, adaptation to prism throwing for throwing overarm will not transfer at all to underarm throwing in some subjects, will transfer slightly in others, and will transfer almost completely in others. The model of Spoelstra and Arbib reproduces this by varying the degree of overlap between the microcomplexes involved in overarm and underarm throwing, thus giving us new insight into the relation between sensory processing and motor control.

Classical Conditioning

The Thompson Laboratory at USC focuses on basic associative learning and memory using classical (Pavlovian) conditioning of discrete behavioral responses (e.g.,

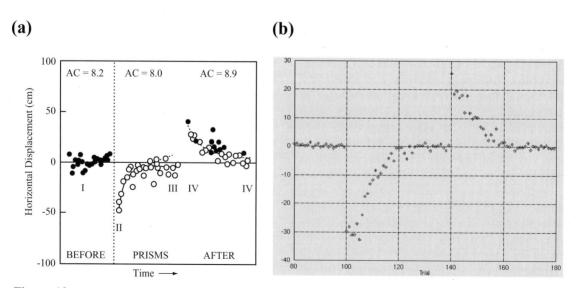

Figure 13 (a) Empirical data showing the accuracy of throwing in successive trials before donning prisms, then while wearing the prisms, and finally after doffing the prisms. Note the process of re-learning after removal of the prisms. (b) Reproduction of this behavior by the model in which adaptation is mediated by the cerebellum.

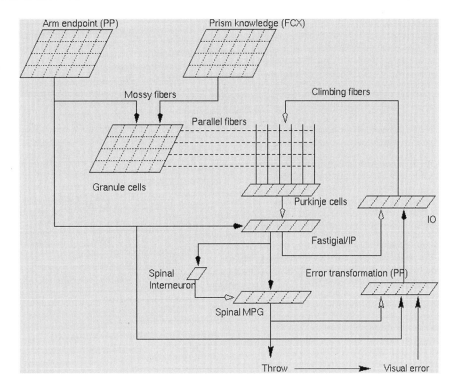

Figure 14 The structure of the prism adaptation model, based on the microcomplex of Fig. 10.

eyeblink) as a model system. They find that the cerebellum is essential for this form of learning and memory under all conditions. The hippocampus is not necessary for basic conditioning, but it is essential for *trace conditioning*, where the conditioned stimulus (CS) – in Thompson's study, a tone – ends before the unconditioned stimulus US – in Thompson's study, an airpuff to the rabbit eyeball – begins. If the unconditioned or training stimulus is sufficiently aversive, learned fear will develop associated with the CS and involving the amygdalar and hippocampal systems. The modeling of the eyeblink system described here focuses on how the cerebellar system learns to make specific behavioral responses that are most adaptive in dealing with the aversive event. In comparing our models of motor control to those for classical conditioning, we still view the cerebellar microcomplex as the building block for our modeling but now view the climbing fiber signal as the unconditioned stimulus (US) rather than the error signal. In a way, of course, an aversive US *is* an error signal—it is a mistake not to blink before the airpuff hits the eyeball.

This observation highlights the crucial role of *timing* in motor control, even when the controlled system is as apparently simple as an eyeblink. In his USCBP Ph.D. thesis, Jeffrey Grethe (1999) addressed the data in the NeuroCore database on classical conditioning by modeling the cerebellar microcomplex as a Hebbian cell assembly and studying its role in the production of the well-timed conditioned response (CR). We close our discussion of cerebellum with a review of this work and related models and data.

In Grethe's model (Fig. 15), the cerebellar cortex receives mossy fiber input from both pontine nuclei and recurrent projections from the interpositus nucleus, with the recurrent projection much smaller than the pontine sensory projection. The cerebellar cortex in the microcomplex, therefore, consists of two distinct populations of granule cells: those influenced by the recurrent mossy fiber connections from the deep nuclei and those that are not. Golgi cells in the cerebellar cortex inhibit granule cells surrounding that Golgi cell (Ito, 1984). In the microcomplex, both of the granule cell populations are inhibited by a Golgi cell that receives its input from the mossy fibers as well as the parallel fibers from the granule cells themselves. The Purkinje cells in the basic unit of the microcomplex receive input from both granule cell populations.

The interpositus consists of two cell populations: excitatory and inhibitory cells. Both the excitatory and inhibitory interpositus neurons generate recurrent collaterals within the nucleus; however, the excitatory cells generate the recurrent collateral projection to the cortex. In the microcomplex, the excitatory cells receive mossy fiber input and recurrent excitatory input as well as an inhibitory input from the Purkinje cell and the inhibitory interpositus neurons. The inhibitory interpositus cells of the microcomplex receive input from the excitatory cells and in turn inhibit these excitatory cells. This basic architecture allows the recurrent excitation in the excitatory interpositus cell to be modulated by the Purkinje cell in the cerebellar cortex. Each group of cells modeled in the cerebellar microcomplex is a "lumped" collection of

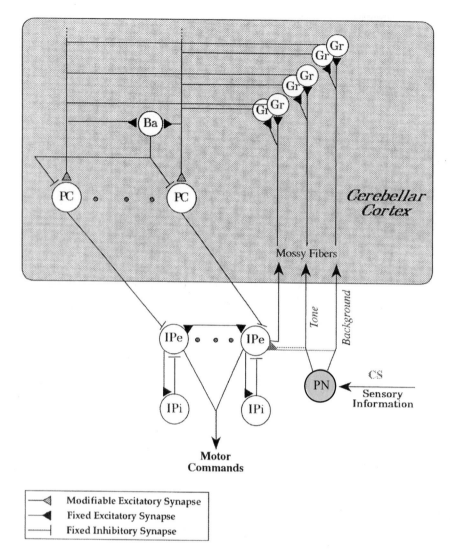

Figure 15 Cerebellar cell assembly. Recurrent activity in the cerebellum, between the cerebellar cortex and interpositus nucleus, is initiated and controlled by the cerebellar cortex. Sensory information is relayed to the cerebellar cortex and the interpositus nucleus through the pontine nuclei.

neurons of that type. For example, a modeled recurrent granule cell neuron is actually a representation of a pool of those neurons. The architecture of the model tries to preserve the topographic projections between the cerebellar cortex and the deep nuclei as well as the beam-like organization of Purkinje cells receiving input from the parallel fibers. Each Purkinje cell population modeled represents such a collection of Purkinje cells located on a common beam of parallel fibers. This allows the cerebellar microcomplex to be broken down into columns of topographically organized cells.

In the model of the microcomplex, the mossy fiber input consists of both sensory information from the pontine nucleus and a recurrent projection from the interpositus nucleus. The sensory information from the pontine nucleus is separated into two projections, a tone-responsive mossy fiber projection and a tone-unresponsive mossy fiber projection. The response of the tone-responsive mossy fibers was modeled after the activity of neurons in the pontine nuclei elicited by acoustic stimuli (Aitkin and Boyd, 1978). The majority of the units responded only at the onset of the tone stimulus and responded with a high-frequency burst of activation. An increase in the firing rate is applied 10 msec after the onset of the tone stimulus, which is consistent with onset latencies reported by Steinmetz (1990). The mossy fiber inputs to the model are the only elements whose firing rate are predetermined. All other elements in the model are determined by their inputs and neuron properties. In the model, only the first column of cells in the microcomplex receives the tone stimulus.

Purkinje cells tend to have high tonic firing rates (Eccles *et al.*, 1967, 1971). Berthier and Moore (1986) found that in the awake rabbit the range of Purkinje cell

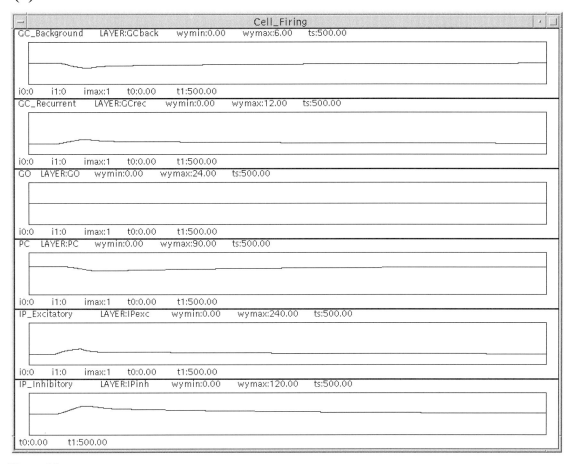

Figure 16 Model output of tone-responsive column. (a) Stability of cell assembly. A small stimulus was given that results in an initial activation of the cell assembly (as represented by the activity in the excitatory IP cell) that subsequently decays to background levels. (b) Bi-stability of cell assembly. In this example, a small stimulus was given that activates the cell assembly. After the stimulus onset, the activation of the recurrent feedback to the cerebellar cortex forces the Golgi cell to inhibit the granule cells and select the more active recurrent pathway. This example shows how the process of synaptic fatigue allows recovery from the activated state.

activity falls in the range of 40 to 100 Hz. In the model, the simulated Purkinje cell rate of 60 Hz falls well within this range. The high tonic activity of Purkinje cells is the result of integrating the low-frequency inputs from the granule cells. The tonic firing rate of neurons in the interpositus was 60 Hz, which corresponds to the mean rate of 60 Hz observed in awake rabbits.

Grethe did not study the actual learning process but sought to study what state of the microcomplex would support an appropriately timed conditioned response. The synaptic weights from the recurrent granule cell population to the Purkinje cell are assumed to have undergone LTD, whereas the synapses from the non-recurrent granule cell population are slightly potentiated. This normalization of the weights is necessary to keep the Purkinje cell firing normally during baseline activity. This synaptic normalization has been used in our other models of the cerebellum. The synaptic weights from the tone-responsive mossy fiber to tone-responsive interpositus cell has undergone LTP.

A microcomplex must be stable despite the inherent noisiness of the input as well as able to initiate itself given a salient enough input. To examine the issue of stability, a tone-responsive cerebellar microcomplex will be considered (i.e., the first column of cells in the microcomplex is tone responsive, all other columns are not). Given a weak input at 10 msec into the trial, the trained microcomplex shows no prolonged response (Fig. 16a). Once the stimulus terminates, the activity in the microcomplex (represented by the collective activation of the excitatory interpositus cells) returns to its baseline level; however, we need to be able to activate this microcomplex. In order to accomplish this, a nonlinear threshold must be present that allows the activity of the microcomplex to grow after this threshold has been reached. A candidate nonlinear characteristic was found in the Purkinje cell by Llinás and Mühltethaler (1988). If the input to the Purkinje cell drops below a certain level, the Purkinje cell ceases to fire. This nonlinearity is modeled in the cerebellar microcomplex by defining the proper squashing

(b)

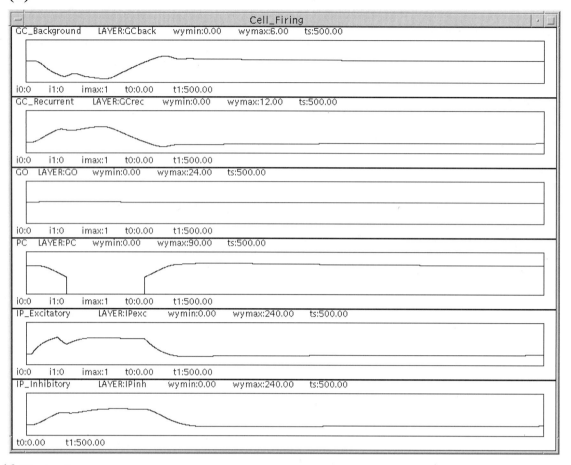

Figure 16 (*Continued*)

function for the Purkinje cell activity so that when the activity of the Purkinje cell falls below a certain threshold it ceases to fire.

Now, if a salient enough stimulus is given to the microcomplex, the microcomplex activates (Fig. 16b). This activation is controlled by the granule cell-Golgi cell network. Given a stimulus the excitatory interpositus cells activate the recurrent connection to the cerebellar cortex. If this activation is strong enough, the recurrent mossy fiber input to the recurrent granule cell population causes these granule cells to activate. This mossy fiber input and subsequent granule cell activation begins to activate the Golgi cell. It is important to note, however, that the Golgi cell activity varies only slightly from its baseline firing rate. This is due to the feedforward and feedback nature of its influence on the granule cells. The small tone-responsive input to the Golgi cell through the mossy fibers will only slightly activate the Golgi cell; however, this slight activation will already cause the Golgi cell to begin inhibiting the granule cell populations. The Golgi cell then, through its continued inhibitory influence on the granule cells, selects for the most active granule cell subpopulation the recurrent granule cell population. Because the parallel fiber synapses from the recurrent granule cell population to the Purkinje cell have been depressed relative to the synapses from the background granule cell population, the Purkinje cell begins to slow its firing rate. If this process is able to cause the Purkinje cell to reach the threshold where the neuron ceases firing, the microcomplex has then reached an activated state, which will allow the activity in the microcomplex to reverberate.

Once the microcomplex has been activated, however, a process must exist that is responsible for deactivating the microcomplex. Various mechanisms have been proposed to stop regenerative activity in the nervous system. Synchronized thalamocortical oscillations spontaneously appear and disappear. The appearance of these oscillations results from the activation of a small number of cells that in turn recruit a progressively larger number of neighboring neurons. These synchronized oscillations then cease spontaneously. This cessation of activity is partially due to a persistent activation of a hyperpolarization-activated cation conductance (Bal and McCormick, 1996); however, a more widely accepted phenomenon that could be responsible for the deactivation of a microcomplex is synaptic fatigue. Kaplan *et al.*

(1991) extended the notion of the microcomplex by including use-dependent synaptic fatigue. Experimental evidence exists that suggests that many synapses exhibit decreased efficacy with repeated use. The synaptic fatigue in the model is due to an activity-dependent depletion of synaptic resources in the excitatory interpositus neuron synapses. Once an excitatory interpositus cell has crossed a firing threshold where the resources for synaptic transmission can no longer be replenished, the synaptic efficacy of the synapses begins to decrease. The synaptic resources are then replenished once the neurons activity falls below a certain level and the resources are able to be replenished.

The microcomplex itself is stable in that it is able to return to its baseline firing characteristics after activation; however, in order to test the overall stability and robustness of the system, one has to look further. In order to evaluate the model, noise was injected at the level of the inputs to the model (i.e., because all activation in the model is derived from the inputs, noise added at this stage will affect all levels of processing). The model is actually very robust when noise is added to the system.

The microcomplex is now able to produce an activation in response to mossy fiber input, but how is the timing of this response controlled? Once the microcomplex has been initiated, the rest of the cells in the microcomplex must also activate to produce the response. Once a tone-responsive interpositus cell (in the first column) forces the first Purkinje cell to cease firing, the microcomplex is given the necessary recurrent excitatory drive to begin influencing other Purkinje cells in the microcomplex. This causes a cascade of Purkinje cells that cease to fire over a period of time. Each time a Purkinje cell shuts off, it endows the microcomplex with more excitatory drive that in turn causes more Purkinje cells to cease firing. This cascade in turn determines the timing and topography of the response of the interpositus neurons; therefore, the overall timing of the response is dependent upon the onset time of the Purkinje cell bi-stability. This bi-stability is controlled by the synaptic strength between the recurrent granule cell population and the Purkinje cell. When a column in the microcomplex activates, activation in the granule cell population shifts from a balanced state (all populations firing at their baseline levels) to one in which the recurrent granule cell population is firing at a higher rate, whereas the background granule cell population is firing at a lower rate. This allows the recurrent population to have a greater influence on the Purkinje cell. If the synapses in the projection are depressed (through a process such as LTD), then the Purkinje cell will not receive as much input and will begin to slow its firing rate. If this shift of balance continues long enough until the Purkinje cell reaches the bi-stability threshold, then that Purkinje cell will cease firing and release the system further to allow other Purkinje cells to follow. The overall timing of the response is, therefore, dependent on how quickly the Purkinje cells in the microcomplex reach this bi-stability threshold, which is dependent upon the synaptic strength (i.e., the level of synaptic depression at the recurrent granule cell to Purkinje cell synapse). The latencies are completely controlled by the level of depression in the recurrent granule cell to Purkinje cell projection (with more depression at 225 msec than at 400 msec). It is interesting to note that this implies that with learning (LTD) the microcomplex will produce activations at earlier and earlier latencies – the response will continue to move forward in time. This is what is actually observed in a typical classical conditioning experiment.

2.1.5 Hippocampus, Parietal Cortex, and Navigation

Our models of the roles of parietal cortex in saccades and grasping are firmly grounded in data from *monkey* neurophysiology and comprise networks of biologically plausible neurons implemented on computers, yielding many simulation results. By contrast, our models of the role of *rat* parietal cortex and hippocampus in navigation address data from rat neurophysiology but are strongly motivated by data from primate neurophysiology whose implications for analogous properties of the rat brain have yet to be tested.

Cognitive Maps and the Hippocampus

To use a road map, we must locate (the representations of) where we are and where we want to go, and then find a path that we can use as we navigate towards our goal. We use the term *cognitive map* for a mental map together with these processes for using it. Thus, the "place cells" found in rat hippocampal CA3 and CA1 (O'Keefe & and Dostrovsky, 1971) provide only a "you are here" signal, not a full cognitive map. Moreover, a given place cell will have a "place field" in a highly familiar environment, with up to 70% probability. This suggests that the hippocampus is dynamically tuned to a" "chart" of the current locale, rather than providing a complete "atlas" with a different place cell for every place in the rat's "entire world." This suggest two alternatives (at least):

1. The different "charts" are stored elsewhere and must be "reinstalled" in hippocampus as dictated by the current task and environment.
2. The cells of hippocampus receive inputs encoding task and environment which determine how sensory cues are used to activate a neural representation of the animal's locale.

In either case, we see that these "charts" are highly labile. Noting the importance of parietal systems in representing the personal space of humans, we seek to understand cognitive maps in a framework that embraces parietal cortex as well as hippocampus.

Although some hippocampal cells fire when rats drink water or approach water sources (Ranck, 1973), O'Keefe & and Conway (1978) did not find the role of food or water to be markedly different from other cues that identify the location of a place field. Eichenbaum et al. (1987) recorded from rats repetitively performing a sequence of behaviors in a single odor-discrimination paradigm and found *goal-approach cells* which fired selectively during specific movements, such as approach to the odor port or to the reward cup. Despite these and related findings, there is still no evidence that hippocampus proper can simultaneously encode the rat's current location and the goal of current navigation.

Taxon and Locale Systems: An Affordance Is Not a Map

O'Keefe & and Nadel (1978) distinguished the *taxon (behavioral orientation) system* for route navigation (a *taxis* is an organism's response to a stimulus by movement in a particular direction), and the *locale system* for map-based navigation and proposed that the locale system resides in the hippocampus. We have already qualified the latter assertion, showing how the hippocampus may function as *part of* a cognitive map. Here, we want to relate taxis to the notion of an affordance. Just as a rat may have basic taxes for approaching food or avoiding a bright light, say, so does it have a wider repertoire of affordances for possible actions associated with the immediate sensing of its environment. Such affordances include "go straight ahead" for visual sighting of a corridor, "hide" for a dark hole, "eat" for food as sensed generically, "drink", and the various turns afforded by, for example, the sight of the end of the corridor. Because the rat's behavior depends more on smell than on vision, we should add "olfactory affordances," but relevant data are sparse. Both normal and hippocampal-lesioned rats can learn to solve a simple T-maze in the absence of any consistent environmental cues other than the T-shape of the maze. If anything, the lesioned animals learn this problem faster than normals. After criterion was reached, probe trials with an eight-arm radial maze were interspersed with the usual T-trials. Animals from both groups consistently chose the side to which they were trained on the T-maze. However, many did not choose the 90° arm but preferred either the 45° or 135° arm, suggesting that the rats had solved the T-maze by learning to rotate within an egocentric orientation system at the choice point through *approximately* 90°. This leads to the hypothesis of an *orientation vector* being stored in the animal's brain but does not tell us where or how the orientation vector is stored. One possible model would employ coarse coding in a linear array of cells, coded for turns from −180° to +180°. From the behavior, one might expect that only the cells close to the preferred *behavioral* direction are excited and that learning "marches" this peak from the old to the new preferred direction. However, it requires a simpler learning scheme to "unlearn"−90°, say, by reducing the peak there, while at the same time "building" a new peak at the new direction of +90°. If the old peak has "mass" $p(t)$ and the new peak has "mass" $q(t)$, then as $p(t)$ declines toward 0 while $q(t)$ increases steadily from 0, the center of mass $\frac{(-90)p(t) + 90q(t)}{p(t) + q(t)}$ will progress from −90 to +90, fitting the behavioral data.

The determination of movement direction is easily modeled by "rattification" of the Arbib and House (1987) model of frog detour behavior. There, prey were represented by excitation coarsely coded across a population, while barriers were encoded by inhibition whose extent closely matched the retinotopic extent of each barrier. The sum of excitation was passed through a winner-take-all circuit to yield the choice of movement direction. As a result, the direction of the gap closest to the prey, rather than the direction of the prey itself, was often chosen for the frog's initial movement. The same model serves for behavioral orientation once we replace the direction of the prey (frog) by the direction of the orientation vector (rat), while the barriers correspond to the absence of affordances for movement.

Hippocampal-Parietal Interactions in Navigation

McNaughton *et al.* (1989) found cells in rat posterior parietal cortex with location specificity that were dependent on visual input for their activation; 40% of the cells had responses discriminating whether the animal was turning left, turning right, or moving forward (we call these MLC cells). Some cells required a conjunction of movement and location (e.g., one parietal cell fired more for a right turn at the western arm of a cross-maze than for a right turn at the eastern arm), and these firings were far greater than for all left turns. Another parietal cell fired for left turns at the center of the maze but not for left turns at the ends of the arms or for any right turns. Turn-direction information was varied, with a given cell responding to a particular subset of vestibular input, neck and trunk proprioception, visual field motion, and (possibly) efference copy from motor commands.

McNaughton and Nadel (1990) offered a model of rat navigation with four components (Fig. 17a): B is a spatio-sensory input; A, in hippocampus, provides a place representation; L, posited to be parietal, outputs a movement representation; and AL provides a place/movement representation by means of the parietal MLC cells. In the model, A both transforms visual input from B into place cell activity and responds to input from AL by transforming the neural code of (prior place, movement) into that for the place where the rat will be after the movement. But this model becomes untenable, as MLC cells are *not* place/movement cells in the sense of

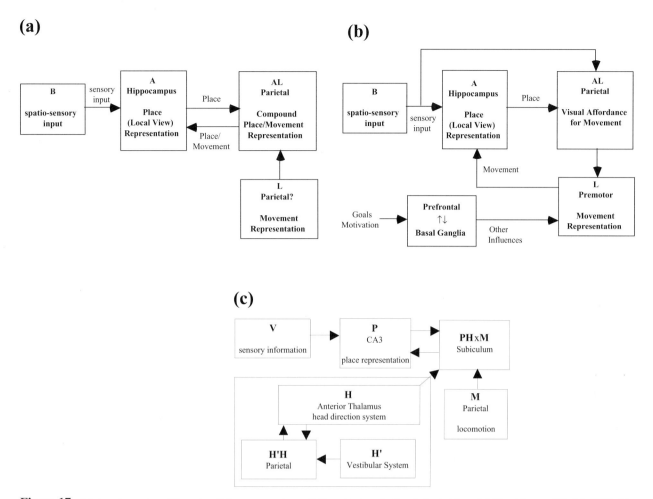

Figure 17 (a) A systems view of the role of hippocampus in spatially guided navigation (adapted from McNaughton and Nadel, 1990). (b) A recasting of the systems view of (a) which makes explicit that parietal cortex provides" "affordances" rather than explicit place information (ArbibEá et al., 1997). (c) The McNaughton et al. 1996) systems-level view of the role of the hippocampus in navigation, redrawn from the perspective of (a).

"place" in "place cells." Examples of such a cell's "place/movement field" are "left turn at end of arm" or "move ahead along arm."

To proceed, we pursue analogies with monkey parietal cortex. We first *hypothesize* (for future experimental testing) that the parietal cortex of the rat also contains *affordance cells* for locomotion and navigation, akin to those documented in the monkey for reaching and grasping. Fig. 17b thus has the cells in AL (parietal cortex) code for affordances and includes a direct link from B to AL that is absent in Fig. 17a. Because "grip affordance cells" in AIP project to "grip premotor cells" in premotor region F5, we postulate analogously (Fig. 17b) that L is premotor and driven by AL. Once again, the current place is encoded in A, but we now posit that movement information from L enables A to compute the correct transformation from current place to next place. In either model, then, the loop via AL can, in absence of sensory input, update the place representation in A, with the cycle continuing until errors in this *expectation* accumulate excessively.

Fig. 17b also adds links from brain areas involved in goals and motivation – the navigation of a hungry rat is different than that of a sated rat. In analogy to the monkey, it posits (without further analysis) that interactions between prefrontal cortex and basal ganglia *choose* the rat's next movement on the basis of its current position relative to its goal. We shall say more of this when we discuss the World Graph model.

We now return to the MLC cells. Some fire prior to the rat's execution of a turn and correspond to the affordance cells of Fig. 17b, but others do not fire until the turn commences. These, then, are not signaling affordances, and we now describe a supplementary circuit (Paul Rothemund, USC term paper, Spring 1996) that includes our model of these cells. To ground this circuit, we turn to reports from the *monkey* literature of neurons that detect optic flow and have traces similar to those for MLC cells whose responses start at the initiation of a turn. These include MST neurons selective for expansion, contraction, and rotation (Graziao *et al.*, 1994). Sakata *et al.* (1994) identified rotation-sensitive

(RS) neurons of the posterior parietal association cortex which, unlike MST neurons which seem specific for rotation in a fronto-parallel plane, were mostly sensitive to rotation in depth.

A neuron responding to a focus of expansion could seem to code for straight ahead, and a neuron encoding translational flow or rotation in depth could code for a turn. Some of the MLC cells which fire at a turn require head orientation in addition to visual input. We suggest that "parietal left turn" neurons combine input from an "MST left turn neuron" with vestibular input, efference copy, or somatosensory input. Such a "parietal left turn" neuron can thus function in the dark but does not code place as well as movement. A neuron that does code for a "place-specific left turn" might be constructed from a "parietal left turn" neuron and hippocampal input. We postulate that, though based on monkey data, this circuitry will (with different anatomical loci) be instantiated in rat brain, as well.

McNaughton et al. (1996) have proposed a new systems-level model that incorporates head direction information in the implementation of an angular path-integrator module. Note that head direction here is *not* absolute but is established with respect to sensory (e.g., visual) cues. It can be maintained moderately well if the animal moves in the dark but will be changed if the sensory cues are rearranged. Fig. 17c presents this model redrawn to emphasize its similarity with Fig. 17a. Box PHxM, previously PxM and assumed to be parietal, is now assumed to be implemented by the subiculum, based on the report by Sharp & and Green (1993) that some cells in the subiculum and dorsal presubiculum have broad, but significant, directional tuning in situations where directionality is absent from hippocampal cells.

McNaughton et al. (1996) view head direction as a point on a circle centered on the rat and assign each head-direction cell a location on this circle. This head-direction ring (H) has local Gaussian excitatory connections from a cell to its neighbors. Another layer of neurons (H'H) receives information about the current location from H and information about rotational motion from the vestibular system and other sources of such information (H'). These cells encode the interaction between current location and the sign of rotation and feed this information to cells on the appropriate side of the current focus of activity in the direction circle. Cells with these response properties have been observed in the rat posterior cortex (Chen et al., 1994a,b). The posterior cortex is now assumed to implement the H'H system – which is almost identical to the dynamic remapping module postulated for LIP in the Dominey-Arbib model described before.

Taxon-Affordances (TAM) and World Graph (WG) Models

To move beyond these rather abstract diagrams, we developed two models of the rat navigational system (e.g., see Guazzelli et al., 1998). The first, our TAM model of the rat's Taxon-Affordances system, explains data on the behavior of fornix-lesioned animals. The second, our World Graph (WG) model, explains data on the behavior of control rats and is also used for the study of cognitive maps. Together, these models shed light on the interactions between the taxon and locale systems and allow us to explain motivated spatial behavior. In order to describe both models, however, we first consider the notion of motivation and its implementation as a motivational schema with a dual role: setting current goals and providing.

So far, most models of spatial learning and navigation do not incorporate the role of motivation into their computation. By contrast, the WG theory (Arbib and Lieblich, 1977; Lieblich and Arbib, 1982) posits a set $d_1, d_2, \ldots, d_k$ of discrete drives to control the animal's behavior. At time t, each drive d_i in $(d_1, \ldots, d_k)$ has a value $d_i(t)$. The idea is that an appetitive drive spontaneously increases with time towards d_{max}, while aversive drives are reduced towards 0, both according to a factor ad intrinsic to the animal. An additional increase occurs if an incentive $I(d,x,t)$ is present, such as the aroma of food in the case of hunger. Drive reduction $a(d,x,t)$ takes place in the presence of some substrate – ingestion of water reduces the thirst drive. If the animal is at the place it recognizes as being node x at time t, and the value of drive d at that time is $d(t)$, then the value of d for the animal at time $t + 1$ will be

$$d(t+1) = d(t) + a_d \mid d_{max} - d(t) \mid -a(d,x,t) \mid d(t) \mid \\ + I(d,x,t) \mid d_{max} - d(t) \mid$$

which is the drive dynamics incorporated into the motivational schema.

Moreover, our motivational schema is not only involved in setting current goals for navigation; it is also involved in providing reinforcement for learning. Midbrain dopamine systems are crucially involved in motivational processes underlying the learning and execution of goal-directed behavior. Schultz et al. (1995) found that dopamine neurons in monkeys are uniformly activated by unpredicted appetitive stimuli such as food and liquid rewards and conditioned reward-predicting stimuli; they hypothesized that dopamine neurons may mediate the role of unexpected rewards to bring about learning. In our work, reward information is used to learn *expectations* of future reinforcements. Furthermore, the motivational schema implements the idea that the amount of reward is dependent on the current motivational state of the animal. If the rat is extremely hungry, the presence of food might be very rewarding, but, if not, it will be less and less rewarding. In this way, the influence of a reward $r(t)$ in TAM and the WG model will depend on the value of the animal's current dominant drive $d(t)$ (the maximum drive amongst all drives at a particular time t), its corresponding maximum value, and the actual drive reduction, according to:

$$r(t) = \frac{d(t)}{d_{max}} a(d, x, t)$$

In this way, reward will bring about learning in the TAM and the WG models depending on the animal's current motivational state. On the other hand, if the rat reaches the end of a corridor and does not find food, the drive-reduction factor $a(d,x,t)$ is set equal to -0.1 times the normal value $a*(d,x)$ it would have if food were present, and the motivational schema uses the above formula to generate a negative value of $r(t)$, a frustration factor, that will cause unlearning in TAM and in the WG model.

We have already outlined the basic ideas of the Taxon-Affordances model, based on the Arbib and House (1987) model of frog detour behavior. The full TAM model employs a reinforcement learning system, which tries to influence the behavior of the rat to maximize the sum of the reinforcement that will be received over time. However, rather than detailing the learning rule here, we turn to the World Graph model, which is rooted in the original model of Arbib and Lieblich (1977). In order to describe the WG model, we first describe the kinds of inputs presented to it, how these are processed, and how these inputs are finally incorporated into the model dynamics.

The WG model receives three kinds of inputs. These are given by three different systems: the place layer, the motivational schema, and the action selection schema. The WG model is composed of three distinct layers. The first layer contains coarse coded representations of the current sensory stimuli. This layer comprises four different perceptual schemas which are responsible for coding head direction, walls, and landmarks (bearings and distances), respectively. Perceptual schemas incorporate "what" and "where" information, the kind of information assumed to be extracted by the parietal and inferotemporal cortices (Kolb et al., 1994; Mishkin et al., 1983) to represent the perception of environmental landmarks and maze walls. In this model, perceptual schemas, like affordances, employ coarse coding in a linear one-dimensional array of cells to represent pertinent sensory input.

In the WG model, two distinct perceptual schemas are responsible for presenting landmark information to the network: one to code landmark bearing and the other to code landmark distance. However, because the experiments here are landmark free, the importance of landmarks for navigation and their further WG modeling will be described in another paper. Moreover, walls are also responsible for the formation of the rat's local view. O'Keefe and Burgess (1996) show place cells responsive to the walls of a box environment that modify their place fields in relation to different configurations of the environment. Muller and Kubie (1987) observed that, in some cases, place cells had a crescent-shaped field and were assumed to be related to the arena wall. In the WG model, if the wall is close to the rat (less than a body length in the current implementation), it will be represented in the "wall perceptual schema."

Cells that show a unimodal tuning to head direction independent of location and relative to a "compass direction" have been reported in several areas of the rat brain, including the post-subiculum (Taube et al., 1990a,b) and the posterior parietal cortex, PPC (Chen et al., 1994a,b). Head direction cells provide the animal with an internal representation of the direction of its motion. This information, like the other perceptual inputs in the WG model, will be coarse coded as a bump of activity over a linear array of cells, the head direction perceptual schema.

Each perceptual schema projects in turn to its respective feature detector layer. Each of these layers is organized in a two-dimensional matrix. Each node in the perceptual schema connects to all nodes in its corresponding feature detector layer (i.e., the two are fully connected). The feature detector units then interact through a local competition mechanism to contrast-enhance the incoming activity pattern; that is, a neuron in a feature detector layer produces a non-zero output (G_j) if and only if it is the most active neuron within a neighborhood. The adjustments to the connection weights between any pair of perceptual schema-feature detector layers are done using a normalized Hebbian learning algorithm. This ensures that the same set of feature detectors is activated by increasing the connection strength from active units (V_i) in the perceptual schema to active feature detectors, thereby increasing their response level the next time. This rule is captured in the following connection strength update equation:

$$\Delta_{w_{ij}} = \alpha V_i G_j w_{ij}$$

where Δw_{ij} is the change to the connection strength w_{ij} and α is the learning rate parameter.

Placed atop the feature detector layers is the Place Layer. This layer will keep track of the feature detector winners and will activate a different place/node if the set of winners that becomes active in the feature detector layers does not activate any previously activated node in the place layer. (The matching of winners is based on a pre-determined threshold.) In the WG theory, there is a node for each distinct place/situation the animal experiences. This is implemented in the WG model by a functional layer, called the World Graph layer, placed above the Place Layer. In this way, for each "new" set of feature detector winners, a "new" node is activated in the place layer and, consequently, a node x is created in the world graph layer of the WG model.

In the WG theory, there is an edge from node x to node x' in the graph for each distinct and direct path the animal has traversed between a place it recognizes as x

and a place it recognizes as x'. This is implemented in the WG model as a link between two nodes of the World Graph layer. Suppose, for example, a node x is currently active at time t. By deciding to move north, the animal activates a "different" set of feature detectors and so a distinct node x' in the world graph at time $t + 1$, and a link will be created from node x to node x'. Appended to each link/edge will be sensorimotor features associated with the corresponding movement/path. In the current model, these features consist of a motor efference copy (e.g., movement towards north) supplied to the world graph via a feedback connection from the action selection schema.

Edge information allows the animal to perform goal-oriented behaviors. The WG model learns expectations of future reward by the use of temporal differences learning in an actor-critic architecture (Barto *et al.*, 1983; Sutton, 1988). Expectations of future reinforcement are associated with pairs of nodes/edges, not with state/actions as is the case in TAM. For this reason, when a particular node of the world graph is active and, for example, the simulated animal is hungry, all it needs to do is select the edge containing the biggest hunger-reduction expectation and follow its direction toward the node it points to. In the WG model, each drive will have an associated drive-reduction expectation value. Thus, if the rat is thirsty, it will not base next node selection on hunger-reduction expectations, but on thirst-reduction expectations.

The integrated model, TAM-WG, joins together TAM and the WG model. The result is depicted in Fig. 18. In TAM-WG, the decision to turn to a certain angle (the orientation vector) is given by a winner-take-all process performed over the integration of fields produced by the available affordances (*AF*), drive-relevant stimuli (*ST*), rewardness expectation (*RE*), map information from the world graph (*MI*), and a curiosity level (*CL*) – to be described below. *MI*, as well as *RE*, contains an associated noise factor – *noise_mi*. If a node x is active in the world graph, there will be expectations associated with each edge coming from node x. Because associated with each edge is the direction it points to, a peak of activity with the magnitude of the expectation associated with that node/edge pair will be coarse coded over a linear array of neurons in the location respective to the edge's direction. In TAM-WG, then, the total input I_{in} to the action selection schema becomes:

$$I_{in}(i,t) = AF(i,t) + ST(i,t) + (RE(i,t) + noise\_re) \\ + (MI(i,t) + noise\_mi) + CL(i,t)$$

Eventually, the selected motor outputs will enable the system to reach the goal object. When a selected action is executed, it will produce consequences in the animal's internal state and in the way it perceives the world (*cf.* the action-perception cycle; Arbib, 1972). The internal state will alter the drive levels, which will in turn influence the selection of the next node in the world graph, and so on.

A crucial aspect of the WG theory and the TAM-WG model is that an animal may go to places that are not yet represented in its world graph. To see how this occurs, we must clarify what is meant by a node x' immediately adjacent to a node x. Certainly, if there is an edge from x to x' in the world graph, then x' is immediately adjacent to x. However, the WG theory also considers as adjacent to x those world situations x' not yet represented in the graph, but that can be reached by taking some path from x. In the WG model, these situations will have a value attached to them, which can be seen as a "curiosity level" attached to the unknown. The curiosity level is based on the value of the animal's biggest drive at a particular given time.

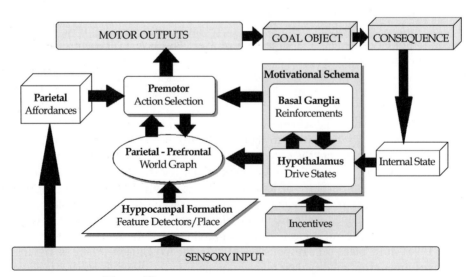

Figure 18 Integrated TAM-WG model of rat navigation.

2.1.6 Discussion

Chapter 2.2 presents the NSL Neural Simulation Language which (whether in earlier or current versions) has been the vehicle for much of the modeling at the level of systems neuroscience discussed in this chapter. It should be noted that the NSL of Chapter 2.2 is NSLJ, an implementation of NSL 3.0 in Java which extends the object-oriented approach implemented in C++ for NSL 2.1. Because the development of the models reported in this chapter was contemporaneous with the development of NSLJ, the current implementations of these models reflect this heterogeneity – they are implemented in NSL2.1 or NSLJ, C++ or Java. However, all are being re-implemented into NSLJ or, at least, into a form that will allow the modular description and links to data that test and support the model which will be offered by the Brain Models on the Web (BMW) model repository (Chapter 6.2). BMW provides documentation, links to empirical data, and downloadable code for almost all the models described in this chapter. Chapter 2.4 presents the Synthetic PET method, which provides the bridge from such neural-level modeling, grounded in the data of animal neurophysiology, to the data of human brain imaging.

Much attention has been paid by other groups to multi-compartment modeling (e.g., the GENESIS and NEURON neural simulation environments) using the Hodgkin-Huxley equation and detailed models of receptor kinetics and plasticity to model membrane properties in dendrite, soma, and axon to help us understand small neural circuits. One of the goals for future work by the USC Brain Project is to develop a set of BMW Standards so that many different neural simulation environments (NSL, GENESIS, NEURON, etc.) can have their models structured for inclusion in BMW.

Another effort of the USC Brain Project has focused on modeling at an even finer level than that provided by either GENESIS and NEURON. Chapter 2.3 will present the work on EONS (Elementary Objects of the Nervous System), with particular attention to EONS models of synapses and the fine structures which constitute them. Not only can EONS simulations take into account the position of transmitter release and the position of receptor macromolecules, but they can also accommodate the irregular geometry of the pre- and postsynaptic membranes based on electron micrograph data.

Acknowledgments

This work was supported in part by a Program Project (P20) grant from the Human Brain Project (P01MH52194) and in part by the National Science Foundation (IRI-9221582). My thanks to Peter Dominey, Michael Crowley, and Amanda Bischoff-Grethe for their work on modeling the basal ganglia and to Tom McNeill (a founding faculty member of USCBP) for his unfailing counsel during the development of these models; to Nicolas Schweighofer and Jacob Spoelstra for their work on modeling the cerebellum and to Tom Thach for discussion of relevant empirical data; to Andrew Fagg for his work on the FARS model and to Giacomo Rizzolatti and Hideo Sakata for their discussion of relevant data; to Paul Rothemund for suggesting the role in navigation of parietal systems for detection of visual motion; and to Alex Guazzelli, Mihail Bota, and Fernando Corbacho for valuable discussion on the linkage of hippocampus to other systems in the control of navigation.

References

Agostino, A., Berardelli, A., Formica, A., Accornero, N., and Manfredi, M. (1992). Sequential arm movements in patients with Parkinson's disease, Huntington's disease and dystonia. *Brain* **115**, 1481–1495.

Aitkin, L. M., and Boyd, J. (1978). Acoustic input to the lateral pontine nuclei. *Hearing Res.* **1**, 67–77.

Albus, J. (1971). A theory of cerebellar function. *Mathematical Biosci.* **10**, 25–61.

Arbib, M. A. (1997). From visual affordances in monkey parietal cortex to hippocampo-parietal interactions underlying rat navigation. *Phil. Trans. R. Soc. London B.* **352**, 1429–1436.

Arbib, M. A., Ed. (1995). *The Handbook of Brain Theory and Neural Networks*. Bradford Books/The MIT Press, Cambridge, MA.

Arbib, M. A. (1981). Perceptual structures and distributed motor control. In *Handbook of Physiology, Section 2. The Nervous System.* Vol. II. *Motor Control, Part 1* (V. B. Brooks, Ed.). American Physiological Society, pp. 1449–1480.

Arbib, M. A. (1972). *The Metaphorical Brain*. Cybernetics as Artificial Intelligence and Brain Theory. Wiley-Interscience, New York.

Arbib, M. A., and House, D. H. (1987). Depth and detours, an essay on visually-guided behavior. In *Vision, Brain, and Cooperative Computation* (Arbib, M. A., and Hanson, A. R. Eds.). A Bradford Book/The MIT Press, Cambridge, MA, pp. 129–163.

Arbib, M. A., and Lieblich, I. (1977). Motivational learning of spatial behavior. In *Systems Neuroscience*. (Metzler, J. Ed.). Academic Press, New York, pp. 221–239.

Arbib, M. A., Érdi, P., and Szentágothai, J. (1998). *Neural Organization*: Structure, Function and Dynamics. A Bradford Book/The MIT Press, Cambridge, MA.

Arbib, M. A., Schweighofer, N., and Thach, W. T. (1995). Modeling the Cerebellum: From Adaptation to Coordination, in *Motor Control and Sensory-Motor Integration*: Issues and Directions (Glencross, D. J., and Piek, J. P., Eds.). North-Holland Elsevier Science, Amsterdam, pp. 11–36.

Arbib, M. A., Schweighofer, N., and Thach, W. (1994). Modeling the role of cerebellum in prism adaptation. In *From Animals to Animats 3* (Cliff, D., Husbands, P., Meyers, J., and Wilson, S., Eds.). The MIT Press, Cambridge, MA, pp. 36–44.

Arbib, M. A., Boylls, C. C., and Dev, P. (1974). Neural models of spatial perception and the control of movement. In *Kybernetik und Bionik/Cybernetics* (Oldenbourg, R., Ed.). pp. 216–231.

Bal, T., and McCormick, D. A. (1996). What stops synchronized thalamocortical oscillations? *Neuron.* **17**, 297–308.

Barone, P., and Joseph, J.-P. (1989). Prefrontal cortex and spatial sequencing in macaque monkey. *Exp. Brain Res.* **78**, 447–464.

Bartha, G., and Thompson, R. (1995). Cerebellum and conditioning. In *The Handbook of Brain Theory and Neural Networks* (Arbib, M. A., Ed.). The MIT Press, Cambridge, MA, pp. 169–172.

Barto, A. G., Richard, S., and Anderson, C. W. (1983). Neuronlike adaptive elements that can solve difficult learning control problems. *IEEE Trans. Systems, Man, Cybernetics*, SMC-**13**, 834–846.

Berthier, N. W., and Moore, J. W. (1986). Cerebellar Purkinje cell activity related to the classically conditioned nictitating membrane response. *Exp. Brain Res.* **63**, 341–350.

Bischoff, A. (1998). Modeling the Basal Ganglia in the Control of Arm Movements, Ph. D. thesis, Department of Computer Science, University of Southern California, Los Angeles.

Bischoff-Grethe, A., and Arbib, M. A. (2000). Sequential movements: a computational model of the roles of the basal ganglia and the supplementary motor area.

Borges, J. L. (1975). Of exactitude in science. In *A Universal History of Infamy* Penguin Books, New York, p. 131,.

Bronstein A,M., and Kennard, C. (1985). Predictive ocular motor control in Parkinson's disease. *Brain* **108**, 925–940.

Buford, J. A., Inase, M., and Anderson, M. E. (1996). Contrasting locations of pallidal-receiving neurons and microexcitable zones in primate thalamus. *J. Neurophysiol.* **75**, 1105–1116.

Chen, L. L., Lin, L. H., Green, E. J., Barnes, C. A., and McNaughton, B. L. (1994a). Head-direction cells in the rat posterior cortex. 1. Anatomical distribution and behavioral modulation. *Exp. Brain Res.* **101(1)**, 8–23.

Chen, L. L., Lin, L. H., Barnes, C. A., and McNaughton, B. L. (1994b). Head-direction cells in the rat posterior cortex. 2. Contributions of visual and ideothetic information to the directional firing. *Exp. Brain Res.* **101(1)**, 24–34.

Crowley, M. (1997). Modeling Saccadic Motor Control: Normal Function, Sensory Remapping and Basal Ganglia Dysfunction, Ph. D. thesis, Department of Computer Science, University of Southern California, Los Angeles.

Curra, A., A. Berardelli, R. Agostino, N. Modugno, C. C. Puorger, N. Accornero, and M. Manfredi (1997). Performance of sequential arm movements with and without advance knowledge of motor pathways in Parkinson's disease. *Movement Disord.* **12**, 646–654.

Dominey, P. F., and Arbib, M. A. (1992). A cortico-subcortical model for generation of spatially accurate sequential saccades. *Cerebral Cortex.* **2**, 135–175.

Eccles, J. C., Faber, D. S., Murphy, J. T., Sabah, N. H., and Táboríková, H. (1971). Afferent volleys in limb nerves influencing impulse discharge in cerebellar cortex. I. In mossy fibers and granule cells. *Exp. Brain Res.* **1**, 1–16.

Eccles, J. C., Ito, M., and Szentágothai, J. (1967). *The Cerebellum as a Neuronal Machine.* Springer-Verlag, Berlin.

Eichenbaum, H., Kuperstein, M., Fagan, A., and Nagode, J. (1987). Cue-sampling and goal-approach correlates of hippocampal unit activity in rats performing an odor-discrimination task. *J. Neurosci.* **7**, 716–732.

Fagg, A. H., and Arbib, M. A. (1998). Modeling parietal-premotor interactions in primate control of grasping. *Neural Networks* **11**, 1277–1303.

Georgopoulos, A., Kettner, R., and Schwartz, A. (1988). Primate motor cortex and free arm movements to visual targets in three-dimensional space. II. Coding of the direction of movement by a neuronal populaion. *J. Neurosci.* **8**, 2928–2937.

Gibson, J. J. (1979). *The Ecological Approach to Visual Perception.* Houghton Mifflin, Boston, MA.

Graziao, M. S. A., Andersen, R. A., and Snowden, R. J. (1994). Selectivity of area MST neurons for expansion, contraction, and rotation motions. *Invest. Ophthalmol. Visual Sci.* **32**, 823–882.

Grethe, J. S. (1999). Neuroinformatics and the Cerebellum: Towards an Understanding of the Cerebellar Microzone and Its Contribution to the Well-Timed Classically Conditioned Eyeblink Response, Doctoral dissertation, University of Southern California, Los Angeles.

Guazzelli, A., Corbacho, F. J., Bota, M., and Arbib, M. A. (1998). Affordances, motivation, and the World Graph theory. *Adaptive Behav.* **6**, 435–471.

Hoff, B., and Arbib, M. A. (1992). A model of the effects of speed, accuracy, and perturbation on visually guided reaching. In *Control of Arm Movement in Space, Neurophysiological and Computational Approaches.* (Caminiti, R., Johnson, P. B., and Burnod, Y., Eds.), Springer-Verlag, Berlin.

Humphrey, D. (1979). The cortical control of reaching. In *Posture and Movement.* (Humphrey, D. R. and Talbot, D., Eds.). Raven Press, New York, pp. 51–112.

Ito, M. (1984). *The Cerebellum and Neural Control.* Raven Press, New York.

Jeannerod, M., Arbib, M. A., Rizzolatti, G., and Sakata, H. (1995). Grasping objects: the cortical mechanisms of visuomotor transformation. *Trends Neurosci.* **18**, 314–320.

Jenkins, I. H., Brooks, D. J., Nixon, P. D., Frackowiak, R. S. J., and Passingham, R. E. (1994). Motor sequence learning: a study with positron emission tomography. *J. Neurosci.* **14**, 3775–3790.

Kaplan, S., Sonntag, M., and Chown, E. (1991). Tracing recurrent activity in cognitive elements (TRACE): a model of temporal dynamics in a cell assembly. *Connection Sci.* **3(2)**, 179–206.

Kolb, B., Buhrmann, K., McDonald, R. and Sutherland, R. J. (1994). Dissociation of the medial prefrontal, posterior parietal, and posterior temporal cortex for spatial navigation and recognition memory in the rat. *Cerebral Cortex.* **4(6)**, 664–680.

Lashley K. S. (1951). The problem of serial order in behavior. In *Cerebral Mechanisms in Behavior.* (Jeffress, L. A., Ed.). John Wiley & Sons, New York, pp. 112–146.

Lieblich, I., and Arbib, M. A. (1982). Multiple representations of space underlying behavior. *Behavioral Brain Sci.* **5**, 627–659.

Llinás, R. R., and Mühlethaler, M. (1988). Electrophysiology of guinea-pig cerebellar nuclear cells in the *in vitro* brainstem cerebellar preparation. *J. Physiol.* **404**, 241–258.

Luppino, G., Matelli, M., Camarda, R., and Rizzolatti, G. (1993). Corticocortical connections of area F3 (SMA-proper). and area F6 (pre-SMA). in the macaque monkey. *J. Comp. Neurol.* **338**, 114–140.

Marr, D. (1969). A theory of cerebellar cortex. *J. Physiol.* **202**, 437–470.

Martin, T., Keating, J., Goodkin, H., Bastian, A., and Thach, W. (1996). Throwing while looking through prisms. 1. Focal olivocerebellar lesions impair adaptation. *Brain* **119(suppl. 4)**, 1183–1198.

Martin, T., Keating, J., Goodkin, H., Bastian, A., and Thach, W. (1995). Throwing while looking through prisms. 1. Focal olivocerebellar lesions impair adaptation. *Brain* **119(suppl. 4)**, *1183–1198.*

McNaughton, B. L., and Nadel, L. (1990). Hebb-Marr networks and the neurobiological representation of action in space. In *Neuroscience and Connectionist Theory.* (Gluck, M. A., and Rumelhart, D. E., Eds.). Lawrence Erlbaum Assoc., Norwood, NJ, chap. 1, pp. 1–63.

McNaughton, B. L., Barnes, C. A., Gerrard, J. L., Gothard, K., Jung, M. W., Knierim, J. J., Kudrimoti, H., Qin, Y., Skaggs, W. E., Suster, M., and Weaver, K. L. (1996). Deciphering the hippocampal polyglot: the hippocampus as a path integration system. *J. Exp. Biol.* **199(pt. 1)**, 173–185.

McNaughton, B. L., Leonard, B., and Chen, L. (1989). Cortico-hippocampal interactions and cognitive mapping: a hypothesis based on reintegration of the parietal and inferotemporal pathways for visual processing. *Psychobiology* **17**, 230–235.

Mishkin, M., Ungerleider, L. G., and Mack, K. A. (1983). Object vision and spatial vision: two cortical pathways. *Trends Neurosci.* **6**, 414–417.

Muller, R. U., and Kubie, J. L. (1987). The effects of changes in the environment on the spatial firing of hippocampal complex-spike cells. *J. Neurosci.* **7(7)**, 1951–1968.

O'Keefe, J., and Burgess, N. (1996). Geometric determinants of the place fields of hippocampal neurons. *Nature* **381(6581)**, 425–8.

O'Keefe, J., and Conway, D. H. (1978). Hippocampal place units in the freely moving rat: why they fire when they fire. *Exp. Brain Res.* **31**, 573–590.

O'Keefe, J., and Dostrovsky, J. (1971). The hippocampus as a spatial map: preliminary evidence from unit activity in the freely moving rat. *Exp. Brain Res.* **34**, 171–175.

O'Keefe, J., and Nadel, L. (1978). *The Hippocampus as a Cognitive Map.* Clarendon Press, Oxford.

Ranck, J. B. (1973). Studies on single neurons in dorsal hippocampal formation and septum in unrestrained rats. I. Behavioral correlates and firing repertoires. *Exp. Neurol.* **41**, 461–535.

Rizzolatti, G., Camarda, R., Fogassi, L., Gentilucci, M., Luppino, G., and Matelli, M. (1988). Functional organization of inferior area 6 in the Macaque monkey. II. Area F5 and the control of distal movements. *Exp. Brain Res.* **71**, 491–507.

Robinson, D. A. (1981). The use of control systems analysis in the neurophysiology of eye movements. *Ann. Rev. Neurosci.* **4**, 463–503.

Sakata, H., Shibutani, H., Ito, Y., Tsurugai, K., Mine, S., and Kusunoki, M. (1994). Functional properties of rotation-sensitive neurons in the posterior parietal association cortex of the monkey. *Exp. Brain Res.* **101(2)**, 183–202.

Schultz, W., Romo, R., Ljungberg, T., Mirenowicz, J., Hollerman, J. R., and Dickinson, A. (1995). Reward-related signals carried by dopamine neurons. In *Models of Information Processing in the Basal Ganglia.* (Houk, J. R., *et al.*, (Eds.). The MIT Press, Cambridge, MA, pp. 233–248.

Schweighofer, N., Arbib, M. A., and Kawato, M. (1998a). Role of the cerebellum in reaching quickly and accurately. I. A functional anatomical model of dynamics control. *Eur. J. Neurosci.* **10**, 86–94.

Schweighofer, N., Spoelstra, J., Arbib, M. A., and Kawato, M. (1998b). Role of the cerebellum in reaching quickly and accurately. II. A detailed model of the intermediate cerebellum, *Eur. J. Neurosci.* **10**, 95–105.

Schweighofer, S., Arbib, M. A., and Dominey, P. F. (1996a). A Model of Adaptive Control of Saccades: I. The model and its biological substrate. *Biol. Cybernetics* **75**, 19–28.

Schweighofer, S., Arbib, M. A., and Dominey, P. F. (1996b). A model of adaptive control of saccades. II. Simulation results. *Biol. Cybernetics* **75**, 29–36.

Scudder, C. A. (1988). A new local feedback model of the saccadic burst generator. *J. Neurophysiol.* **59**, 1455–1475.

Sharp, P. E., and Green, C. (1994). Spatial correlates of firing patterns of single cells in the subiculum of the freely moving rat. *J. Neurosci.* **14**, 2339–2356.

Spoelstra, J., and Arbib, M. A. (1997). A computational model of the role of the cerebellum in adapting to throwing while wearing wedge prism glasses. In *Proceedings of the 4th Joint Symposium on Neural Computation.* Vol. 7. Los Angeles, CA, pp. 201–208.

Spoelstra, J., Arbib, M. A., and N. Schweighofer (2000). Cerebellar control of a simulated biomimetic manipulator for fast movements. *Biol. Cybernetics*, in press.

Steinmetz, J. E. (1990). Classical nictitating membrane conditioning in rabbits with varying interstimulus intervals and direct activation of cerebellar mossy fibers as the CS. *Behav. Brain Res.* **38**, 97–108.

Sutton, R. (1988). Learning to predict by the methods of temporal differences. *Machine Learning* **3**, 9–44.

Taira, M., S. Mine, Georgopoulos, A. P., Murata, A., and Sakata, H. (1990). Parietal cortex neurons of the monkey related to the visual guidance of hand movement. *Exp. Brain Res.* **83**, 29–36.

Tanji, J., and Shima, K. (1994). Role for supplementary motor area cells in planning several movements ahead. *Nature* **371**, 413–416.

Taube, J. S., Muller, R. U., and Ranck, J. B. Jr. (1990a). Head-direction cells recorded from the postsubiculum in freely moving rats. I. Description and quantitative analysis. *J. Neurosci.* **10**, 420–435.

Taube, J. S., Muller, R. U., and Ranck, J. B. Jr. (1990b). Head-direction cells recorded from the postsubiculum in freely moving rats. II. Effects of environmental manipulations. *J. Neurosci.* **10(2)**, 436–47.

CHAPTER 2.2

NSL Neural Simulation Language

Michael A. Arbib, Amanda Alexander, and Alfredo Weitzenfeld
*University of Southern California Brain Project and Computer Science Department,
University of Southern California, Los Angeles, California*

The NSL Neural Simulation Language provides a platform for building neural architectures (modeling) and for executing them (simulation). NSL is based on object-oriented technology and provides modularity at all model development levels. In this chapter, we discuss these basic concepts and how NSL takes advantage of them. NSL, now in its third major release, is a neural network simulator which is both general and powerful, designed for users with diverse interests and programming abilities.

For novice users interested only in an introduction to neural networks, we provide user-friendly interfaces and a set of predefined artificial and biological neural models. For more advanced users well acquainted with the area who require more sophistication, we provide evolved visualization tools together with extensibility and scalability. We provide support for varying levels in neuron model detail, which is particularly important for biological neural modeling. In artificial neural modeling, the neuron model is very simple, with network models varying primarily in their network architectures and learning paradigms. While NSL is not particularly intended to support detailed single neuron modeling, as opposed to systems such as GENESIS and NEURON primarily designed for this task, NSL does provide sufficient expressiveness to support this level of modeling.

The Neural Simulation Language (NSL) has evolved for over a decade. The original system was written in C (NSL 1) in 1989, with a second version written in C++ (NSL 2) in 1991 and based on object-oriented technology. Both versions were developed at the University of Southern California by Alfredo Weitzenfeld, with Michael Arbib involved in the overall design. The present version, NSL 3, is a major release completely restructured over former versions both as a system as well as the supported modeling and simulation, including modularity and concurrency. It provides a powerful neural development environment supporting the efficient creation and execution of scalable neural networks, incorporating a compiled language, NSLM, for model development, and a scripting language, NSLS, for model interaction and simulation control. It offers rich graphics and a full mouse- driven window interface supporting creation of new models as well as their control and visualization.

NSL 3 includes two different environments, one in Java (NSLJ, developed at USC by Amanda Alexander's team) and the other in C++ (NSLC, developed at ITAM in Mexico by Weitzenfeld's team), again with Arbib involved in the overall design. Both environments support similar modeling and simulation and are described fully in Weitzenfeld *et al.* (2001). In this chapter, we focus on NSLJ, the version developed by the USC Brain Project team and available from our Website. We offer a free download of the complete NSLJ system, including full source code as well as forthcoming new versions. We also provide free and extensive support for downloading new models from our Websites, where users may contribute with their own models and may criticize existing ones (see Chapter 6.2; http://www-hbp.usc.edu/Projects/BMW.htm). The advantages of Java include:

1. *Portability*: Code written in Java runs without changes "everywhere."
2. *Maintainability*: Java code requires maintaining one single software version for different operating systems, compilers, and platforms.
3. *Web-oriented*: Java code runs on the client side of the Web, simplifying exchange of models without the owner of the model having to provide a large server on which other people can run simulations.

In summary, NSL is especially suitable for use in an academic environment where NSL simulation and model development can complement theoretical courses in both biological and artificial neural networks and for use in a research environment where scientists require rapid model prototyping and efficient execution. NSL may easily be linked to other software tools, such as additional numerical libraries, or hardware, such as robotics, by doing direct programming in either Java or C++. NSL supports the design of modular neural networks which simplify modeling and simulation while providing better model extensibility.

2.2.1 Modeling and Simulation of Neural Networks

Neural network simulation is an important research and development area extending from biological studies to artificial applications. In this book, we stress biological neural networks designed to model the brain in a faithful way at a level of detail appropriate to the data under analysis.

Modeling and Simulation

Modeling develops a neural architecture which can explain and reproduce anatomical and physiological experimental data. It involves choosing appropriate data representations for neural components, neurons, and their interconnections, as well as network input, control parameters, and network dynamics specified in terms of a set of mathematical equations. The neuron model varies depending on the details being described. Neuron models can be very sophisticated biophysical models, such as compartmental models based on the Hodgkin-Huxley model. When behavioral analysis is desired, simpler neuron models such as the analog *leaky integrator* model may be appropriate. The particular neuron model chosen defines the dynamics for each neuron, yet a complete network architecture also involves specifying interconnections among neurons as well as specifying input to the network and choosing appropriate parameters for different tasks using the neural model specified. Moreover, many models involve learning, whose dynamics must be specified in the model architecture.

Simulation, then, consists of using the computer to see how the model behaves for a variety of input patterns and parameter settings. A simulation may use values pre-specified in the original formulation of the model but will in general involve specifying one or more aspects of the neural architecture that may be modified by the user. Simulation also involves analyzing the results, both visual and numerical, generated by the simulator; on the basis of these results, one can decide if any modifications are necessary in the network input or parameters. If changes are required, these may be interactively specified and the model re-simulated, or the changes may require more structural modifications at the neural architecture level and the model will have to be re-compiled. Simulation also involves selecting one of the many approximation methods used to solve neural dynamics specified through differential equations.

In addition, the simulation environment may need to change when moving a model from the development phase to the test phase. When models are initially simulated, good interactivity is necessary to let the user modify inputs and parameters as necessary. As the model becomes more stable, simulation efficiency is a primary concern where model processing may take considerable time—possibly hours or even days for the largest networks to process. In our system, modularity, object-oriented programming, and concurrency play an important part in building neural networks as well as in their execution. We briefly review these key ideas.

Modularity

Modularity is today widely accepted as a requirement for structuring software systems. As software becomes larger and more complex, being able to break a system into separate modules enables the software developer to better manage the inherent complexity of the overall system. As neural networks become larger and more complex, they too may become difficult to read, modify, test, and extend. Moreover, when building biological neural networks, modularization is further motivated by taking into consideration the way we analyze the brain as a set of different brain regions. The general methodology for making a complex neural model of brain function is to combine different modules corresponding to different brain regions. To model a particular brain region, we divide it anatomically or physiologically into different neural arrays. Each brain region is then modeled as a set of neuron arrays, where each neuron is described, for example, by the leaky integrator, a single-compartment model of membrane potential and firing rate. (However, one can implement other, possibly far more detailed, neural models.) For example, Fig. 1 shows the basic components in a model describing the interaction of the *Superior Colliculus (SC)* and the saccade generator of the *Brainstem* involved in the control of eye movements. In this model, each component or module represents a single brain region.

Structured models provide two benefits. The first is that it makes them easier to understand, and the second is that modules can be reused in other models. For example, Fig. 2 shows the two previous *SC* and *Brainstem* modules embedded into a far more complex model, the *Crowley-Arbib* model of basal ganglia. Each of these modules can be further broken down into submodules, eventually reaching modules that take the form of neural arrays. Fig. 3 shows how the single *Prefrontal Cortex* (*PFC*) module can be further broken down into four submodules, each a crucial brain region involved in the control of movement.

There are, basically, two ways to understand a complex system. One is to focus in on some particular subsystem, some module, and carry out studies of that in detail. The other is to step back and look at higher levels of organization in which the details of particular modules are hidden. Full understanding comes as we cycle back and forth between different levels of detail in analyzing different subsystems, sometimes simulating modules in isolation, at other times designing computer experiments that help us follow the dynamics of the interactions between the various modules. Thus, it is important for a neural network simulator to support modularization of models. This concept of modularity is best supported today by object-oriented languages and the underlying modeling concepts described next.

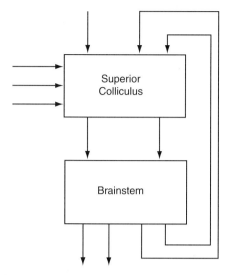

Figure 1 The diagram shows two interconnected modules, the *Superior Colliculus (SC)* and the *Brainstem*. Each module is decomposed into several submodules (not shown here), each implemented as an array of neurons identified by their different physiological response when a monkey makes rapid eye movements.

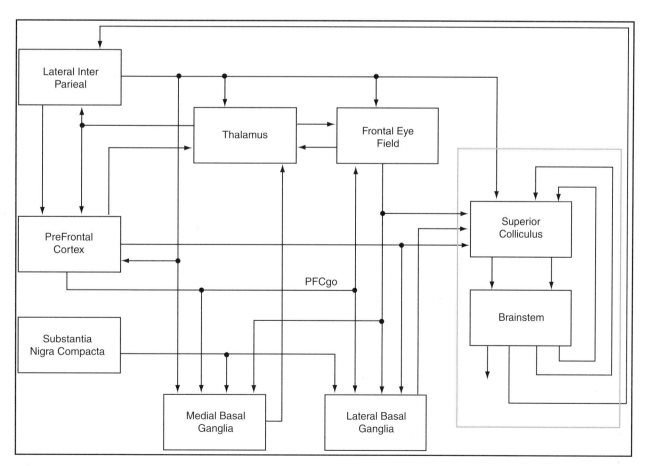

Figure 2 The diagram shows the *SC* and *Brainstem* modules from Fig. 1 embedded in a much larger model of interacting brain regions.

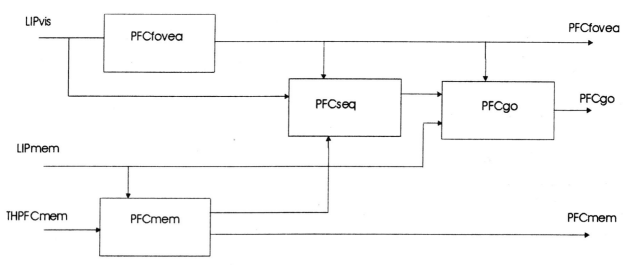

Figure 3 The *Prefrontal Cortex (PFC)* model is further decomposed into four submodules.

Object-Oriented Programming

Object-oriented technology has existed for more than 30 years; however, only in this past decade have we seen it applied in so many industries. What makes this technology special is the concept of the *object* as the basic modularization abstraction in a program. Prior to object-orientation, a complete application would be written at the data and function level of abstraction. Data and functions would be global to a program, and any changes to them could potentially affect the complete system, an undesired effect when large and complex systems are being modified. To avoid this problem, an additional level of abstraction is added—the object. At the highest level, programs are made exclusively out of objects interacting with each other through pre-defined object interfaces. At the lowest level, objects are individually defined in terms of local data and functions, avoiding global conflicts that make systems so difficult to manage and understand. Changes inside objects do not affect other objects in the system so long as the external behavior of the object remains the same. Because there is usually a smaller number of objects in a program than the total number of data or functions, software development becomes more manageable. Objects also provide abstraction and extensibility and contribute to modularity and code reuse. These seemingly simple concepts have a great repercussion in the quality of systems being built, and its introduction as part of neural modeling reflects this. Obviously, the use of object-orientation technology is only part of writing better software. How the user designs the software or neural architectures with this technology has an important effect on the system, an aspect which becomes more accessible by providing a simple to follow yet powerful modeling architecture such as that provided by NSL.

Concurrency in Neural Networks

Concurrency can play an important role in neural network simulation, both in order to model neurons more faithfully and to increase processing throughput (Weitzenfeld and Arbib, 1991). We have incorporated concurrent processing capabilities in the general design of NSL for this purpose. The computational model on which NSL is based has been inspired by the work on the Abstract Schema Language (ASL; Weitzenfeld, 1992), where *schemas* (Arbib, 1992) are *active* or concurrent objects (Yonezawa and Tokoro, 1987) resulting in the ability to concurrently process modules. NSL software is currently implemented on serial computers, emulating concurrency. Extensions to NSL and its underlying software architecture will implement genuine concurrency to permit parallel and distributed processing of modules in the near future.

2.2.2 NSL Modules and Simulation

As an object-oriented system, NSL is built with modularization in mind. As a neural network development platform, NSL provides a modeling and simulation environment for large-scale, general-purpose neural networks by the use of modules that can be hierarchically interconnected to enable the construction of very complex models. NSL provides a modeling language, NSLM, to build/code the model and a scripting language, NSLS, to specify how the simulation is to be executed and controlled. Modeling in NSL is carried out at two levels of abstraction—*modules* and *neural networks* somewhat analogous to object-orientation in its different abstraction levels when building applications. Modules define the top-level view of a model, hiding its internal complexity. A complete model in

NSL requires the following components: (1) a set of modules defining the entire model, (2) neurons comprised in each neural module, (3) neural interconnections, (4) neural dynamics, and (5) numerical methods to solve the differential equations.

Modules and Interconnections

At the highest level model architectures are described in terms of *modules* and *interconnections*. We describe in this section these concepts as well as the *model*, representing the main module in the architecture, together with a short overview of scheduling and buffering involved with modules.

Modules in NSL correspond to objects in object orientation in that they specify the underlying computational model. These entities are hierarchically organized, as shown in Fig. 4. Thus, a given module may either be decomposed into a set of smaller modules or may be a "leaf module" that may be implemented in different ways, where neural networks are of particular interest here (Fig. 5). The hierarchical module decomposition

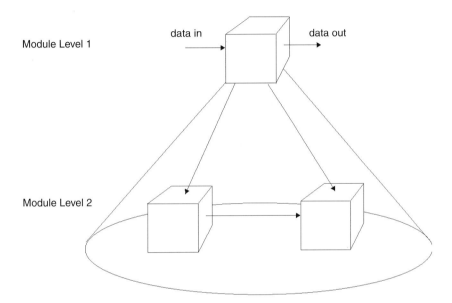

Figure 4 The NSL computational model is based on hierarchical modules. A module at a higher level (Level 1), is decomposed (dashed lines) into submodules (Level 2). These submodules are themselves modules that may be further decomposed. Solid arrows show data communication among modules. (There are also port inputs and outputs, not shown, at Level 2 which correspond to the "data in" and "data out" ports at Level 1.)

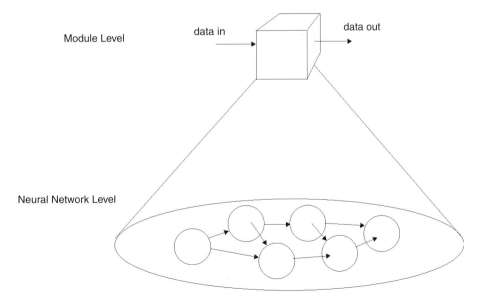

Figure 5 A module in NSL implemented by neural networks made of multiple neurons. (There are port inputs and outputs at the Neural Network Level as well, not shown, which correspond to the "data in" and "data out" ports at the Module Level.)

results in what is known as *module assemblages*—a network of *submodules* that can be seen in their entirety in terms of a single higher level module. These hierarchies enable the development of modular systems where modules may be designed and implemented independently of each other following both top-down and bottom-up development.

The *module*, the basic component in a model architecture, is somewhat analogous to the *object* in object-oriented applications. The external portion of the module is the part of the module seen by other modules. The internal portion is not seen by other modules—this makes it easier to create and modify modules independently from each other—and defines the actual module behavior. This behavior need not be reducible to a neural network; it could also be a module doing something else (e.g., providing inputs or monitoring behavior).

The most important task of a module's external interface is to permit communication between different modules. As such, a module in NSL includes a set of input and output *data ports* (we shall call them simply *ports*). The port represents an entry or exit point where data may be sent or received to or from other modules, as shown in Fig. 6. For example, the *Maximum Selector* model architecture incorporates a module having two input ports, *s_in* and *v_in*, together with a single output port, *uf*, as shown in Fig. 7.

Data sent and received through ports is usually in the form of numerical values. These values may be of different numerical types while varying in dimension. In the simplest form, a numerical type may correspond to a float, double, or integer. Dimensions correspond to a single scalar, a one-dimensional array (*vector*), a two-dimensional array (*matrix*), or higher dimensional arrays. For example, in the *Ulayer* module shown in Fig. 7, *v_in* is made to correspond to a scalar type, while *s_in* and *uf* both correspond to vector arrays (the reason for this selection will become clear very soon). In terms of implementation, the NSL *module* specification has been made as similar as possible to a *class* specification in object-oriented languages such as Java and C++ in order to make the learning curve as short as possible for those who already have a programming background. The actual code for specifying these constructs may be found in Chapter 3 of Weitzenfeld *et al.* (2001) or on our Website.

The general module definition includes the *structure* representing module attributes (data) and the *behavior* representing module methods (operations). The former specifies the ports of the module, as well as the internal variables of the module. The behavior spells out how values read in through the input ports and/or internal variables may be operated upon both to update the internal variables and to provide values as appropriate to the output ports. The *behavior* of a module will include a number of specific methods called by the simulator during model execution. These methods are used for different purposes (e.g., initialization, execution, termination).

Interconnections between modules are achieved by interconnecting output ports in one module to input ports in another module. Communication is unidirectional, flowing from an output port to an input port. Fig. 8 shows the interconnections between the two modules of a simple model, the *Maximum Selector* model, which we will describe below to introduce basic features of the NSL simulation system. In the example, a connection is made from output port *uf* in *Ulayer* to input port *v_in* in *Vlayer*; additionally, output port *vf* in *Vlayer* is connected to input port *v_in* in *Ulayer*. In general, a single output port may be connected to any number of input ports, whereas the opposite is not allowed (i.e., connecting multiple output ports to a single input port). The reason for this restriction is that the input port could receive more than one communication at any given time, resulting in inconsistencies. Again, the actual code for specifying these connections may be found in Chapter 3 of Weitzenfeld *et al.* (2001) or on our Website.

The output to input port connection is considered "same level module connectivity." By contrast, in "different level module connectivity," an output port from a module at one level is *relabeled* (we use this term instead of *connected* for semantic reasons) to an output

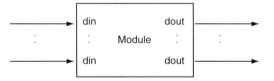

Figure 6 The NSL computational model is based on the module concept. Each module consists of multiple input ($din_1, \ldots, din_n$) and output ($dout_1, \ldots, dout_m$) *data ports* for unidirectional communication. The number of input ports does not have to be the same as the number of output ports.

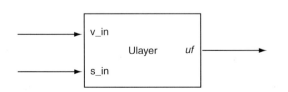

Figure 7 The ULayer module of the *Maximum Selector* model has two input ports, *s_in* and *v_in*, and a single output port, *uf*.

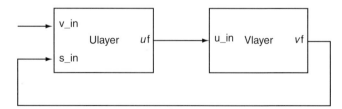

Figure 8 Interconnections between modules *Ulayer* and *Vlayer* of the *Maximum Selector* model.

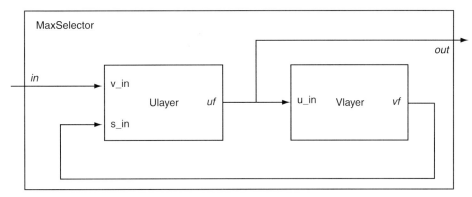

Figure 9 *Maximum Selector* model architecture contains a *MaxSelector* module with two interconnected modules, *Ulayer* and *Vlayer*.

port of a module at a different level. Alternatively, an input port at one level module may be *relabeled* to an input port at a different level. For example, in Fig. 9 we introduce the *MaxSelector* module, containing an input port *in* and an output port *out*, encapsulating modules *Ulayer* and *Vlayer*. *MaxSelector* is considered a higher level module than the other two as it contains—and instantiates—them. In general, relabeling lets input and output ports forward their data between module levels. (This supports module encapsulation in the sense that a module connected to *MaxSelector* should not connect to ports in either *Ulayer* or *Vlayer* nor be able to get direct access to any of the modules private variables.) Relabelings, similar to connections, are unidirectional, where an input port from one module may be relabeled when viewed as an input port at a different level.

We close this section with a brief mention of scheduling and buffering. *Scheduling* specifies the order in which modules and their corresponding methods are executed, while buffering specifies how often ports read and write data in and out of the module. NSL uses a multi-clock-scheduling algorithm where each module's clock may have a different time step although it is synchronized between modules during similar time steps. During each cycle, NSL executes the corresponding simulation methods implemented by the user. *Buffering* relates to the way output ports handle communication. Because models may simulate concurrency, such as with neural network processing, we have provided *immediate* (no buffer) and *buffered* port modes. In the immediate mode (sequential simulation), output ports immediately send their data to the connecting modules. In the buffered mode (pseudo-concurrent simulation), output ports do not send their data to the connecting modules until the following simulated clock cycle. In buffered mode, output ports are double buffered. One buffer contains the data that can be seen by the connecting modules during the current clock cycle, while the other buffer contains the data being currently generated that will only be seen by the connected modules during the next clock cycle.

Modules and Neural Networks

Some modules will be implemented as neural networks where every neuron becomes an element or attribute of a module, as shown in Fig. 10. Note that although neurons also may be treated as modules, they are often treated as elements inside a single module (e.g., one representing an array of neurons) in NSL. We thus draw neurons as spheres instead of cubes to highlight the latter possibility.

There are many ways to characterize a neuron. The complexity of the neuron depends on the accuracy needed by the larger model network and on the computational power of the computer being used. The GENESIS (Bower and Beeman, 1998) and NEURON (Hines and Carnevale, 1997) systems were designed specifically to support the modeling of a single neuron which takes account of the detailed morphology of the neuron in relation to different types of input. The NSL system was designed to let the user represent neurons at any level of desired detail; however, we will focus here on the simulation of large-scale properties of neural networks modeled with relatively simple neurons.

We consider the neuron shown in Fig. 10 to be "simple" as its internal state is described by a single scalar quantity, membrane potential *mp*; its input is *S*, and its output is *mf*, specified by some nonlinear function of *mf*.

The neuron may receive input from many different neurons, while it has only a single output (which may "branch" to affect many other neurons or drive the net

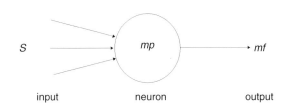

Figure 10 Single compartment neural model represented by a value, *mp*, corresponding to its membrane potential and a value, *mf*, corresponding to its firing, the only output from the neuron. *S* represents the set of inputs to the neuron.

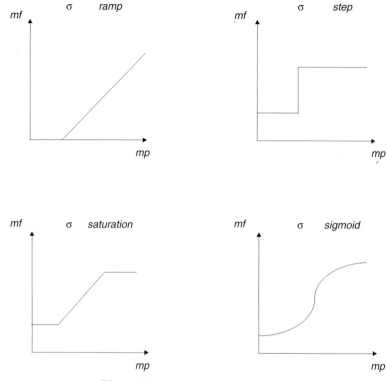

Figure 11 Common threshold functions.

work's outputs). The choice of transformation from S to mp defines the particular neural model utilized, including the dependence of mp on the neuron's previous history. The membrane potential mp is described by a simple first-order differential equation:

$$\tau \frac{dmp(t)}{dt} = f(S, mp, t)$$

depending on its input s. The choice of f defines the particular neural model utilized, including the dependence of mp on the neuron's previous history. For example, the *Leaky Integrator* model defines the membrane potential mp with the first-order differential equation

$$\tau \frac{dmp}{dt} = -mp + s$$

While neural networks are continuous in their nature, their simulated state is approximated by discrete time computations. For this reason, the NSL code for updating the membrane potentials of neurons will include the choice of an integration or approximation method to approximate the solution of differential equations.

The *average firing rate* or output of the neuron mf is obtained by applying some "threshold function" to the neuron's membrane potential:

$$mf = \sigma(mp)$$

where σ usually is described by a non-linear function. Basic threshold functions defined in NSL include *step*, *ramp*, *saturation* and *sigmoid*, whose behaviors are plotted in Fig. 11.

Synapses, the links among neurons, are—in biological systems—complex electrochemical systems and may be modeled in exquisite detail (see Chapter 2.3 on EONS). However, many models have succeeded with a very simple synaptic model, with each synapse carrying a connection weight that describes how neurons affect each other. The most common formula for the input to a neuron is given by:

$$sv_j = \sum_{i=0}^{n-1} w_{ji} \, uf_i$$

where uf_i is the firing of neuron u_i whose output is connected to the jth input line of neuron v_j, and w_{ji} is the weight for that link (up and vp are analogous to mp, while uf and vf are analogous to mf).

While module interconnections are specified in NSLM via an *nslConnect* method call, doing this with neurons would in general be prohibitively expensive considering that there may be thousands or millions of neurons in a single neural network. Instead, we use mathematical expressions similar to those used for their representation. Instead of describing neurons and links on a one-by-one basis, we extend the basic neuron abstraction into *neuron arrays* and *connection masks* describing spatial arrangements among homogeneous neurons and their connections, respectively. We consider uf_i the output from a single neuron in an array of neurons and sv_j the input to a single neuron in another array of neurons. If mask w_{jk} (for $-d \leq k \leq d$) represents the synaptic weights from the uf_{j+k} (for $-d \leq k \leq d$) elements to sv_j, for each j, we then have:

$$sv_j = \sum_{k=-d}^{d} w_{jk}\, uf_{j+k}$$

where the same mask w is applied to the output of each neuron uf_{i+k} to obtain input sv_j. In NSLM, the equation is described by a single array *convolution* operation (operator "@"):

$$sv = w@uf$$

This kind of representation results in great conciseness, an important concern when working with a large number of interconnected neurons. Note that this is possible as long as connections are regular; otherwise, single neurons would still need to be connected separately on a one-by-one basis. This also suggests that the operation is best defined when the number of v and u neurons is the same, although a non-matching number of units can be processed using a more complex notation. There are special considerations with convolutions regarding edge effects—a mask centered on an element at the edge of the arrays extends beyond the edge of the array—depending on how out-of-bound array elements are treated. The most important alternatives are to treat edges as zero, wrap around array elements such as if the array was continuous at the edges, or replicate boundary array elements.

Simulation

Simulation involves interactively specifying aspects of the model that tend to change often, in particular parameter values and input patterns. Also, this process involves specifying simulation control and visualization aspects. For example, Fig. 12 shows five snapshots of the Buildup Cell activity after the simulation of one of the submodules in the *Superior Colliculus* of the *Crowley-Arbib* model shown in Fig. 1. We observe the activity of single neurons, classes of neurons, or outputs in response to different input patterns as the cortical command triggers a movement up and to the right. We see that the cortical command builds up a peak of activity on the Buildup Cell array. This peak moves towards the center of the array where it then disappears (this corresponds to the command for the eye moving towards the target, after which the command is no longer required).

Not only is it important to design a good model, but it is also important to design different kinds of graphical output to make clear how the model behaves. Additionally, an experiment may examine the effects of changing parameters in a model, just as much as changing the inputs. One of the reasons for studying the basal ganglia is to understand Parkinson's disease, in which the basal

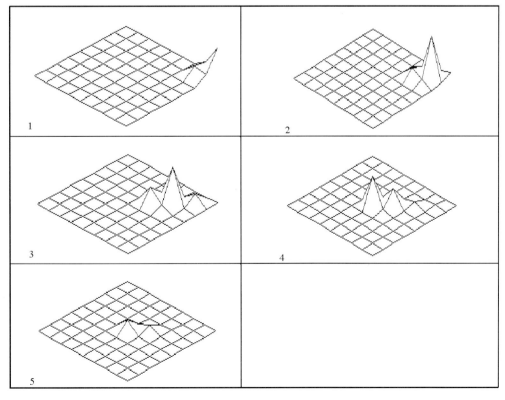

Figure 12 An example of Buildup Cell activity in the *Superior Colliculus* model of Fig. 1.

ganglia is depleted of a substance called dopamine, the depletion of which is a prime correlate of Parkinson's disease. The model of Fig. 2 (at a level of detail not shown) includes a parameter that represents the level of dopamine. Fig. 13 shows the response—saccade velocity, the behavior of a single cell, and the behavior of another single cell—to the brief presentation of a pair of targets. The "normal" model yields two saccades, one each in turn to the positions at which the two targets appeared; the "low-dopamine" model only shows a response to the first target, a result which gives insight into some of the motor disorders of Parkinson's disease patients.

Visualizing Model Architectures with the Schematic Capture System

There are two ways to develop a model architecture: by direct programming in NSLM or by using the *Schematic Capture System* (SCS), a visual programming interface to NSLM. SCS provides graphical tools to build hierarchical models. In SCS, the user graphically connects icons representing modules into what we call a *schematic*. Each icon can then be decomposed further into a schematic of its own. The benefit of having a schematic capture system is that modules can be stored in libraries and easily accessed by the schematic capture system. As more modules are added to the NSL model library, users will benefit by being able to create a rich variety of new models without having to write new code. The success of this will obviously depend on having good modules and documentation. When coming to view an existing model, the schematics make the relationship between modules much easier to visualize, besides simplifying the model creation process. As modules are summoned to the screen and interconnected, the system automatically generates the corresponding NSL module code.

Fig. 14 shows an example of a schematic. The complete schematic describes a single higher-level module, where rectangular boxes represent lower level modules or submodules. These modules can be newly defined modules or already existing ones. The lines describe connections among submodules, while small arrows describe input ports and straight lines describe output ports of modules. Pentagon-shaped boxes represent input (when lines come out from a vertex) and output (when lines

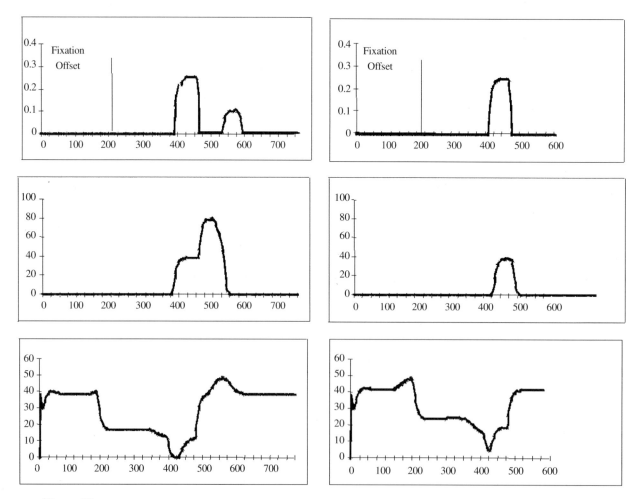

Figure 13 Cutting dopamine in half caused the second saccade not to be generated in response to a double saccade stimulus.

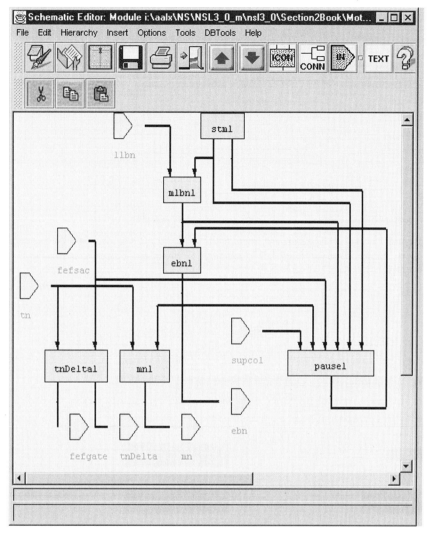

Figure 14 Schematic Editor showing the Motor Module for the *Dominey-Arbib* model. Pentagons in the shape of arrows show the input and output ports of the module.

point to a side) for the higher level module whose schematics are being described.

To select a model from those already existing in the NSL *Model Library*, we must first open the library by calling the Schematic Capture System. The system initially presents the *Schematic Editor* (SE) window. To execute an existing model, we select "Simulate Using Java" (or "Simulate Using C++") from the "Tools" menu. SCS then presents a list of available models. Once we choose the particular model, the system brings up the NSL Executive window presented in Fig. 15 together with an additional visualization window particular to this model, shown in Fig. 16. At this point we are ready to simulate the selected model. Yet, before we do that, we will quickly introduce the NSL Simulation Interface.

The NSL Executive window, shown in Fig. 15, is used to control the complete simulation process such as visualization of model behavior. Control is handled either via mouse-driven menu selections or by explicitly typing textual commands in the "nsls" shell. Because not all commands are available from the menus, the "nsls" shell is offered for more elaborate scripts.

The top part of the window (or header) contains the window name, NSL Executive, and the *Window Control* (right upper corner) used for iconizing, enlarging, and closing the window. Underneath the header immediately follows the *Executive Menu Bar*, containing the menus for controlling the different aspects involved in a simulation. The lower portion of the window contains the *Script window*, a scrollable window used for script command entry, recording, and editing. The Script Window is a superset of the pull-down menus in that any command that can be executed from one of the pull-down menus can also be typed in the Script window, while the opposite is not necessarily so. Furthermore, commands can also be stored in files and then loaded into the Script window at a later time. The Script window supports two levels of commands. The basic level allows *Tool Command Language* (TCL) commands (Ousterhout, 1994), while the second level allows NSLS

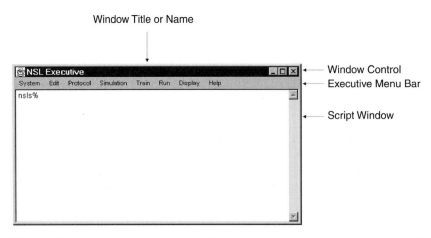

Figure 15 The NSL Executive window. The top part of the window contains the title and underneath the title is the Executive Menu Bar. The larger section of the window contains the Script Window.

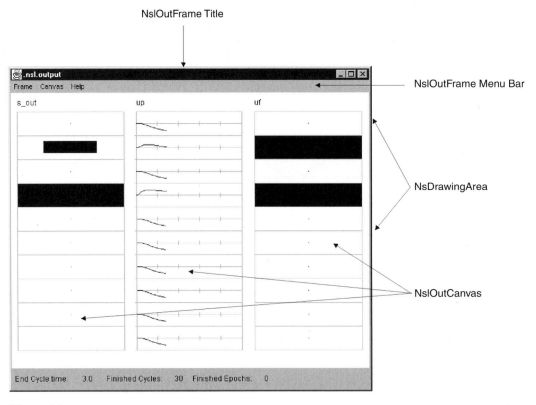

Figure 16 *MaxSelector* model NslOutFrame. The first canvas displays an area-level graph for the input variable s_out. The second canvas displays a temporal graph for the internal variable *up*, and the third canvas displays an area-level graph for the output variable *uf*.

commands. The NSLS commands have a special "nsl" prefix keyword to distinguish them from TCL commands. While there is a single NSL Executive window per simulation, there may be any number of additional output windows containing different displays. For example, the *Maximum Selector* model brings up the additional frame shown in Fig. 16.

The top part of the window contains the title or frame name and the very bottom of the frame contains the *Status line* or *Error message line*. When there are no errors or warnings, the status line displays the current simulation time. In the middle, the frame contains the *NslDrawingArea*. In this example, the drawing area contains three *NslOutCanvases*: the first and third

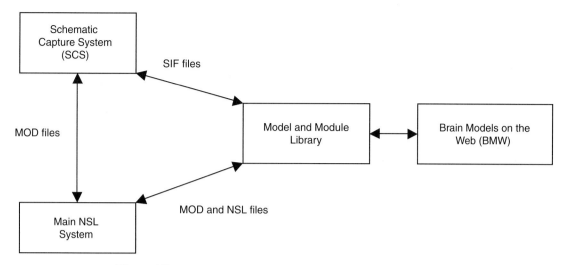

Figure 17 Schematic Capture System and its relation to the NSL system.

correspond to *Area* graphs while the second corresponds to a *Temporal* graph.

2.2.3 The NSL System

The complete NSL system is made of three components: the *Main System*, the *Schematic Capture System*, and the *Model/Module Library*, as shown in Fig. 17. Three file types are used as communication between the three components:

1. *MOD* files describing NSL models, executed by the Main System, stored in the Model Library, and optionally generated from SCS
2. *NSL* files describing NSL model simulation, executed by the Main System and stored in the Model Library
3. *SIF* files storing schematics information about the model stored in the Model Library as well.

The figure also shows BMW (Brain Models on the Web). This is not part of NSL but is a model repository under development by the USC Brain Project in which model assumptions and simulation results can be explicitly compared with the empirical data gathered by neuroscientists.

Main System

The Main NSL System contains several subsystems: the *Simulation subsystem*, where model interaction and processing takes place, and the *Window Interface subsystem*, where all graphics interaction takes place, as shown in Fig. 18. Note that we are now discussing the subsystems or modules that comprise the overall simulation system, not the modules of a specific neural model programmed with NSL. But, in either case, we take advantage of the methodology of object-oriented programming.

The Simulation subsystem is composed of:

1. *Control*, where external aspects of the simulation are controlled by the *Script Interpreter* and the Window Interface
2. *Scheduler*, where the model specified in *MOD file* gets executed
3. *Model Compiler*, where NSLM code is compiled and linked with NSL libraries to generate an executable file
4. *Script Interpreter*, where the simulation gets specified and controlled using NSLS, the script language

The Window Interface subsystem is composed of:

1. *Graphics Output*, consisting of all NSL graphic libraries for instantiating display frames, canvases, and graphs
2. *Graphics Input*, consisting of NSL window controllers to interact with the simulation and control its execution and parameters

Model/Module Library

NSL-developed models and modules are hierarchically organized in libraries. NSL supports two library structures. The first is called the *legacy* or *simple structure* as shown in Table 1. The second structure is a new one that the Schematic Capture System (SCS) builds and maintains. It is pictured in Table 2. The difference between the two is in how they manage modules. The legacy system gives a version number only to models, not modules. The SCS system gives version numbers to both models and modules and contains an extra directory, called the *exe* directory, for executables specific to different operating systems for the C++ version of the software.

There are several reasons for maintaining both systems. For the new system, we decided that it is better to

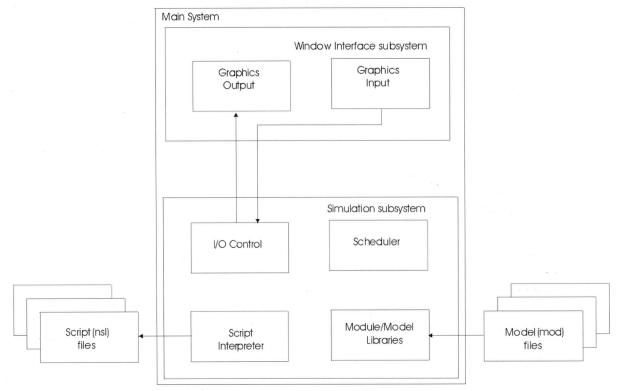

Figure 18 NSL's Main System composed of the Simulation and Window Interface subsystems

Table 1 NSL Model Library Hierarchy Organization for the Legacy System

	nsl3_0	
	Library Name	
	Model Name	
	Version Number	
io	src	doc

Table 2 NSL Model and Module Library Hierarchy Organization for the New System

	nsl3_0		
	Library Name		
	Model or Module Name		
	Version Number		
io	src	exe	doc

The general organization of legacy architecture is shown in Table 1, where the levels correspond to directories. The root of the hierarchy is "nsl3_0" the current system version. A library name is defined to distinguish between libraries. Obviously, there may be multiple model libraries. Each library may contain multiple *models* identified by their corresponding name. Each model is then associated with different implementations, each identified by its corresponding *version number* (version numbers start at 1_1_1). The bottom level of the directory hierarchy contains the directories where the actual model or module files are stored. The *io* directory stores input and output files, usually in the form of NSLS script files. The *src* directory contains source code that must be compiled and is written in the NSLM modeling language; this directory also includes files produced from the compilation including executables. The *doc* directory contains any documentation relevant to the model, including theoretical background, why certain values were chosen for certain parameters, what is special about each of the protocols, how to perform more sophisticated experiments, relevant papers, etc. All models given in the NSL book (Weitzenfeld *et al.*, 2000) were originally developed using the legacy system; thus, as of early 2000, this is what will be found at our Website: http://www-hbp.usc.edu/Projects/nsl.htm.

SCS allows the user to create new libraries as well as add new revisions to existing models and modules. The user can browse and search the libraries for particular

version (that is, create new versions of) the modules so that the user can try different versions of a module in his/her model and so that the modules can be shared more easily between models. Larger models will typically share modules and thus require the SCS management system. For the legacy/simple system, if you are not using SCS, it is easier to manage all of the module files in one directory, *src*. Also, if the modules are not intended to be shared or contributed to Brain Models on the Web (BMW), then they do not necessarily need to be versioned.

models or modules. When building a schematic, the user has the choice of choosing the most recent modification of a model or module or sticking with a fixed version of that model or module. If the user chooses a specific version, this is called "using the *fixed* version." If the user specifies "0_0_0" then the most current version of the module would be used instead. This is called "using the *floating* version." Each individual library file stores *meta-data* describing the software used to create the corresponding model/module. Because SCS generates and manages the NSLM source code files, it also provides a utility for generating the compilation scripts necessary to build the model

2.2.4 Simulating a Model – The Maximum Selector Model

If a model is a discrete-event or discrete-time model, the model equations explicitly describe how to go from the state and input of the network at time t to the state and output after the event following t is completed or at time $t + 1$ on the discrete time scale, respectively. However, if the model is continuous-time, described by differential equations, then simulation of the model requires that we must replace the differential equation by some discrete-time numerical method (e.g., Euler or Runge-Kutta) and choose a simulation time step (t so the computer can go from state and input at time t to an *approximation* of the state and output at time $t + \Delta t$. In each case, the simulation of the system proceeds in steps, where each *simulation cycle* provides the updating for one time step of the chosen type. Simulation involves the following aspects of model interaction: (1) simulation control, (2) visualization, (3) input assignment, and (4) parameter assignment.

Simulation Control

Simulation control involves the execution of a model. The Executive window's "Simulation," "Train," and "Run" menus contain options for starting, stopping, continuing, and ending a simulation during its training and running phase, respectively.

Visualization

Model behavior is visualized via a number of graphics displays. These displays are drawn on canvases, *NslOutCanvas*, each belonging to a *NslOutFrame* output frame. Each NslOutFrame represents an independent window on the screen containing any number of NslOutCanvases for interactively plotting neural behavior or variables in general. NSL canvases can display many different graph types that display NSL numeric objects—objects containing numeric arrays of varying dimensions. For example the *Area* graph shown in Fig. 16 displays the activity of a one-dimensional object at every simulation cycle. The size of the dark rectangle represents a corresponding activity level. On the other hand, the *Temporal* graph shown displays the activity of a one-dimensional objects as a function of time (in other words, it keeps a history).

Input Assignment

NSL supports input via custom-designed input windows as well as by the default user interface using script commands in the NSLS language using the *Script window*. The Script window also allows one to save and load script files.

Parameter Assignment

Simulation and model parameters can be interactively assigned by the user. Simulation parameters can be modified via the "Options" menu while model parameters are modified via the Script window. Additionally, some models may have their own custom-designed window interfaces for parameter modification.

2.2.5 Maximum Selector Model

The remaining sections of this chapter illustrate model simulation using the *Maximum Selector* model. It is a very simple model but allows us to briefly present a number of key features of the NSL environment. The models described in BMW offer far fuller use of the NSL system, or of related approaches to modular architecture of large-scale neural systems. To get one of the models, just visit our Website at http://www-hbp.usc.edu/Projects/nsl.htm. From that page, a list of models executable from NSL will be displayed, and you can pick and choose which one you wish to download.

The *Maximum Selector* neural model (Amari and Arbib, 1977) is an example of a biologically inspired neural network and is shown in Fig. 19. The network is based on the *Didday* model for prey selection (Didday, 1976) and is more generally known as a *Winner Take All* (WTA) neural network. The model uses competition mechanisms to obtain, in many cases, a single winner in the network where the input signal with the greatest strength is propagated along to the output of the network. External input to the network is represented by s_i (for $0 \leq i \leq n-1$). The input is fed into neuron u, with up_i representing the membrane potential of neuron u, while uf_i represents its output. uf_i is fed into neuron v as well as back into its own neuron. vp represents the membrane potential of neuron v, which plays the role of inhibitor in the network. w_m, w_{ui}, and w_i represent

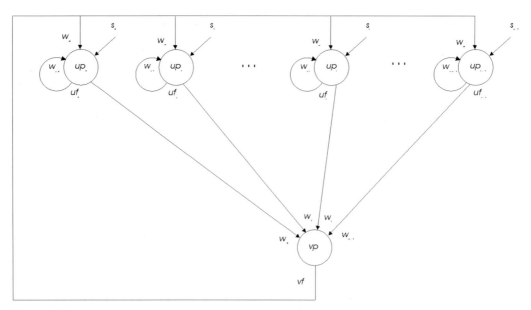

Figure 19 The neural network architecture for the Maximum Selector (Didday, 1976; Amari and Arbib, 1977)[AU8] where s_i represents input to the network and up_i and vp represent membrane potentials, while uf_i and vf represent firing rates; w_m, w_{ui}, and w_i correspond to connection weights.

connection weights, whose values are not necessarily equal.

The neural network is described by the following set of equations:

$$\tau_u \frac{du_i(t)}{dt} = -u_i + w_u f(u_i) - w_m g(v) - h_1 + s_i \quad (2.1)$$

$$\tau_v \frac{dv}{dt} = -v + w_n \sum_{i=1}^{n} f(u_i) - h_2$$

where w_u is the self-connection weight for each u_i, w_m is the weight for each u_i for feedback from v, and each input s_i acts with unit weight; w_n is the weight for input from each u_i to v. The threshold functions involve a *step* for (fu_i):

$$f(u_i) = \begin{cases} 1 & u_i > 0 \\ 0 & u_j \le 0 \end{cases} \quad (2.2)$$

and a *ramp* for $g(v)$:

$$g(v) = \begin{cases} v & v > 0 \\ 0 & v \le 0 \end{cases}$$

Again, the range of i is $0 \le i \le n-1$, where n corresponds to the number of neurons in the neural array u. Note that the actual simulation will use some numerical method to transform each differential equation of the form $\tau\, dm/dt = f(m,s,t)$ into some approximating difference equation of the form $m(t + t) = F(m(t), s(t), t)$ which transforms state $m(t)$ and input $s(t)$ at time t into the state $m(t + \Delta t)$ of the neuron one "simulation time step" later.

As the model equations get repeatedly executed, with the right parameter values, u_i values receive positive input from both their corresponding external input and local feedback. At the same time negative feedback is received from v. Because the strength of the negative feedback corresponds to the summation of all neuron output, as execution proceeds only the strongest activity will be preserved, resulting in many cases in a "single winner" in the network.

To execute the simulation, having chosen a differential equation solver and a simulation time step (or having accepted the default values), the user would simply select "Run" from the NSL Executive's Run menu, as shown in Fig. 20. We abbreviate this as Run → Run.

The output of the simulation would be that as shown in Fig. 21.

The resulting statistics on how long the simulation took to run are displayed in the Executive window's shell, as shown in Fig. 22.

Recall that NSLS is the NSL scripting language in which one may write a script file specifying, for example, how to run the model and graph the results. The user may thus choose to create a new script, or retrieve an existing one. In the present example, the user gets the system to load the NSLS script file containing preset graphics, parameters, input, and simulation time steps by selecting "System → Nsls file …", as shown in Fig. 23.

From the file selection pop-up window we first choose the "nsl" directory and then *MaxSelectorModel*, as shown in Fig. 24. Alternatively, the commands found in the file could have been written directly into the Script window, but it is more convenient the previous way.

Simulation control involves setting the duration of the model execution cycle (also known as the delta-t or simulation time step). In all of the models we will present, we will preset the simulation control parameters within the model; however, to override these settings the user

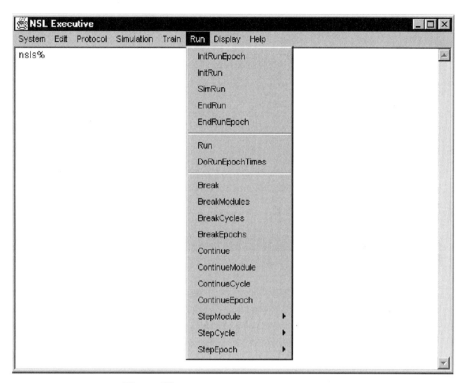

Figure 20 The "Run → Run" menu command.

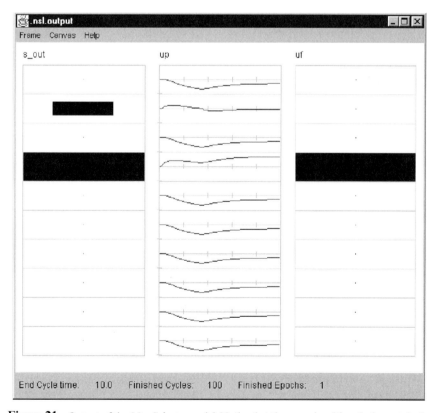

Figure 21 Output of the *MaxSelector* model. Notice that the second and fourth elements in the "maxselector.u1.up" membrane potential layer are affected by the input stimuli; however, the "winner take all" circuit causes the fourth element to dominate the output, as seen in the firing rate, "maxselector.u1.uf."

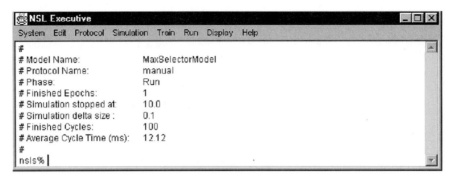

Figure 22 Executive window output from *Maximum Selector* model run.

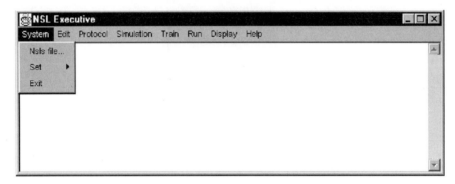

Figure 23 Loading a "nsls" script file into the Executive.

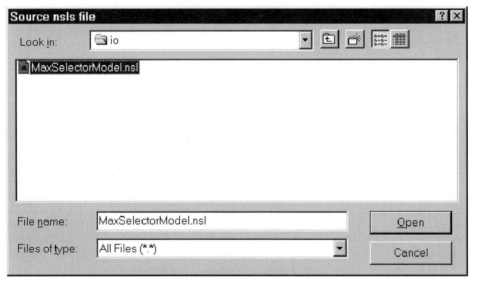

Figure 24 The MaxSelectorModel script loaded into the Executive.

can select System → Set → RunEndTime and System → Set → RunDelta, as shown in Fig. 25.

A pop-up window appears showing the current parameter value that may be modified by the user. In this model we have set the *RunEndTime* to 10.0, as shown in Fig. 26, and *RunDelta* to 0.1 giving a total of 100 execution iterations. These values are long enough for the model to stabilize on a solution.

To execute the actual simulation, we select "Run" from the "Run" menu, as we did in Fig. 20. The user may stop the simulation at any time by selecting the "Run" menu and then selecting "break." We abbreviate this as Run → Break. To resume the simulation from the interrupt point, select Run → Continue.

The model output at the end of the simulation is shown in Fig. 21. The display shows input array *S_out*

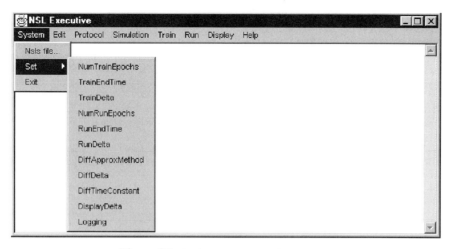

Figure 25 Setting system control parameters.

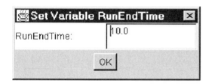

Figure 26 RunEndTime parameter setting.

with an Area type graph (i.e., the area of the black rectangle codes the size of the corresponding input), while array *up*, with a Temporal type graph, shows the time course for *up*. The last canvas shows another Area type graph for *uf* at the end of the simulation. The largest input in *S_out* determines the only element of *S_out* whose activity is still positive at the end of the simulation as seen in *uf*—the network indeed acts as a maximum selector.

The *Maximum Selector* model example is quite simple in that the input *S_out* is constant. In the example chosen, it consists of only two different positive values (set to 0.5 and 1.0) while the rest are set to zero (total of 10 elements in the vector). In general, input varies with time. Because input is constant in the present case, it may be set similarly to any model parameter. To assign values to parameters, we use the "nsl set" command followed by the variable and value involved. If we are dealing with dynamic input, we have different alternatives for setting input. One is to specify a "nsl set" command with appropriate input values every time input changes. Another alternative is to specify the input directly inside the model description or through a custom interface.

Parameters whose values were not originally assigned in the model description, or that we may want to modify, are specified interactively. Two parameters of special interest in the model are the two thresholds, *hu* and *hv*. These two parameters are restricted as follows: $0 \leq hu$, and $0 \leq hv < 1$. (For the theory behind these restrictions, see Arbib, 1989, Sec.4.4.). Their initial values are set to 0.1 and 0.5, respectively. To exercise this model, the user can change both the input and parameter values to see different responses from the model. We suggest trying different combinations of input values, specifying different number of array elements receiving these values. In terms of parameters, we suggest changing values for *hu* and *hv*, including setting them beyond the mentioned restrictions. Every time parameters or input changes, the model should be reinitialized and executed by selecting the "run" menu option.

2.2.6 Available Resources

As mentioned in this chapter, the USC Brain Project Website contains the links to the NSL, SCS, and BMW home pages. The NSL software, documentation, and an exemplary set of models written in NSL can be found from the page: http://www-hbp.usc.edu/Projects/nsl.htm. The SCS software, documentation, and models can be found from the main NSL home page. The models that have been formatted and annotated for BMW can be found from http://www-hbp.usc.edu/Projects/bmw.htm. In addition, the USCBP Website contains a general modeling page that contains some models that are neither in the NSL flat file repository nor the BMW repository. These models can be found from the page: http://www-hbp.usc.edu/Thrusts/modeling.htm.

Acknowledgment

This research was supported in part by P20 Program Project grant 5–P20–52194 from the Human Brain Project (NIMH, NIDA, and NASA) and in part by a grant from NSF-CONACyT. This chapter is based on material from Weitzenfeld et al. (2001).

References

Arbib, M. A. (1992). Schema theory, in *The Encyclopedia of Artificial Intelligence.* 2nd ed. (S. Shapiro, Ed.), pp. 1427–1443, Wiley-Interscience, New York.

Bower, J. M., and Beeman, D. (1998). *The Book of GENESIS: Exploring Realistic Neural Models with the GEneral NEural SImulation Systems* 2nd ed.. TELOS/Springer-Verlag, Berlin/New York.

Hines, M. L., and Carnevale, N. T. (1997). The NEURON simulation environment. *Neural Computation* **9**, 1179–1209.

Hodgkin, A. L., and Huxley, A. F. (1952). A quantitative description of membrane current and its application to conduction and excitation in nerve. *J. Physiol. London,* **117**, 500–544.

Rall, W. (1959). Branching dendritic trees and motoneuron membrane resistivity. *Exp. Neurol.* **2**, 503–532.

Weitzenfeld, A. (1992). A Unified Computational Model for Schemas and Neural Networks in Concurrent Object-Oriented Programming, Ph.D. thesis, Computer Science, University of Southern California, Los Angeles.

Weitzenfeld, A., and Arbib, M. A. (1991). A concurrent object-oriented framework for the simulation of neural networks, in Proceedings of ECOOP/OOPSLA 1990 Workshop on Object-Based Concurrent Programming. *SIGPLAN, OOPS Messenger* **2(2)**, 120–124.

Weitzenfeld, A., Alexander, A. and Arbib. M. A. (2001). *The NSL Neural Simulation Language*. The MIT Press, Cambridge, MA.

Yonezawa, A., and M. Tokoro, Eds. (1987). *Object-Oriented Concurrent Programming*. The MIT Press, Cambridge, MA.

CHAPTER 2.3

EONS: A Multi-Level Modeling System and Its Applications

Jim-Shih Liaw,[1] Ying Shu,[2] Taraneh Ghaffari,[1] Xiaping Xie,[1] and Theodore W. Berger[1]

[1]Department of Biomedical Engineering, University of Southern California, Los Angeles, California
[2]Department of Computer Science, University of Southern California, Los Angeles, California

2.3.1 Introduction

A major unresolved issue in neuroscience is how functional capacities of the brain depend on (emerge from) underlying cellular and molecular mechanisms. This issue remains unresolved because of the number and complexity of mechanisms that are known to exist at the cellular and molecular level, their nonlinear nature, and the complex interactions of these processes. Thus, the capability of simulating neural systems at different levels of organization is essential for an understanding of brain function. To this end, we have undertaken the development of a multi-level modeling system, Elementary Objects of Nervous Systems (EONS), to provide a framework for representing structural relationships and functional interactions among different levels of neural organization.

In EONS, object-oriented design methodology is adopted to form a hierarchy of neural models, from molecules, synapses, and neurons to networks. Each model (e.g., for a synapse or a neuron) is self-contained and any reasonable degree of detail can be included. This provides the flexibility to compose complex system models that reflect the hierarchical organization of the brain from objects of basic elements. Our aim is to make EONS objects available for use by models created with other modeling tools. The EONS library is thus separated from the user interface in order to increase its portability.

For computational models to be useful both as interactive research and teaching tools and as means for technology transfer, it is critically important that a closed-loop interaction between computational and experimental studies be provided. This requires a well-structured modeling framework that parallels biological nervous systems and a mechanism for mapping between neurobiological system parameters and model parameters for a direct comparison and interpretation of experimental manipulations and model simulations. To provide a link with experimental studies, we have developed a prototype for protocol-based simulation in which information about a model is organized in the same way as the specifications of an experiment. This prototype has been developed in conjunction with the NeuroCore component of the USC Brain Project focusing on the design of a protocol-based database for experimental data. Together with NeuroCore, NSL (Chapter 2.2), and BMW (Chapter 2.4), EONS will provide an interactive environment for integrating computational and experimental studies.

This chapter is organized into three sections. The EONS object library is presented in the first section, as well as two example models for illustrating the principles and functionalities of EONS. The first example is a detailed model of a synapse that includes the biophysics of the membrane, various channels, and a spatial representation of the synapse for calculating the distribution of particles through the diffusional process. The second example is a neural network that demonstrates how a concise representation of various synaptic processes can be derived from a detailed study for the incorporation of synaptic dynamics into neural networks. The second section describes the protocol-based simulation scheme, and the conclusion and future development of EONS are given in the last section.

2.3.2 EONS Object Library

The complexity of the nervous system is enormous due to its vast number of components ranging from molecules, synapses, and neurons to brain regions. Dealing with such complexity calls for a modeling environment that provides a high degree of flexibility for choosing a set of biologically consistent constituents for studying a specific aspect of the brain. Object-oriented methodology provides a flexible scheme for representing the hierarchical organization of the nervous system.

Using object-oriented design methodology, we have developed objects for different neural components, from networks, neurons, dendrites, and synapses down to molecules. Various degrees of detail can be included in any of these objects. An example of a hierarchical representation of a neuron is shown in Fig.1. A neuron can be represented as a simple integrate-and-fire object containing a set of synaptic weights, membrane potential, threshold, and action potential as its attributes. Or, it can be a complex object composed of three objects representing dendrite, soma, and axon; each of these in turn can be a composite of multiple objects. A critical requirement for allowing flexible representation of objects and their composition is a predefined input/output specification, both in terms of its type and semantics. EONS reinforces this by closely following the property of the corresponding biological elements.

For example, one can define a class of neurons that includes a set of dendritic spines and a set of axon terminals. Each neuron receives as input a real number representing the membrane potential of the dendrite and generates a real number representing the membrane potential along the axon. An object representing spine or axon terminals can incorporate any features, as long as it conforms to this I/O specification. For instance, an axon terminal can be represented as a simple relay (identity function) at one end of the spectrum, or it can include various voltage- and ligand-gated channels, calcium diffusion, and the cascade of molecular interactions leading to neurotransmitter release. Likewise, a spine object can be represented simply as a number (synaptic weight), or it can include the kinetics of various receptor channels, etc.

One issue when constructing neural networks models with non-homogeneous elements is finding a way to keep track of the connectivity among these elements. For example, a synaptic element may be comprised of multiple molecular objects (e.g., vesicles, receptor, release site, etc.) which may not be uniformly distributed for each synapse. This precludes the use of convolution, a highly efficient method for updating neural networks used by NSL (Chapter 2.2).

We have developed a systematic indexing scheme to handle such complexity: Given neural networks composed of layers of neurons and the connectivity matrix for each layer, EONS will "grow" the synapses between neurons by automatic generation of bi-directional synap-

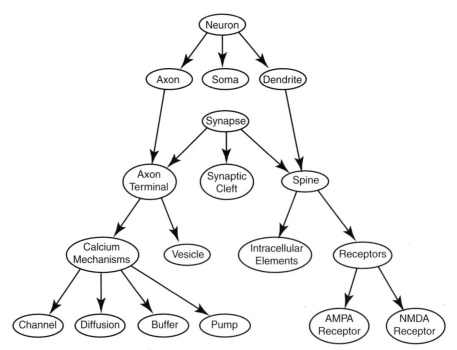

Figure 1 A hierarchy of neuron components.

tic pointers according to the connectivity matrix. This is implemented by the following procedure: From the connectivity matrix, the number of axon terminals and dendritic spines needed for each neuron is counted and instantiated. Next, a bi-directional pointer between each axon terminal and dendritic spine pair, as specified in the connectivity matrix, is set up by going through each neuron in the network.

Moreover, EONS also contains objects representing the network, layers, neurons, axon terminals, and spines. Initially, these objects only contain the specification of the input and output types. These objects, together with the synaptic indexes, form a basic skeleton for building neural models such that various neural processes can be incorporated at each level in a flexible manner. We will illustrate the flexibility of EONS for multi-level model development by the following two examples: a detailed model of a synapse which incorporates morphology revealed by electron microscopy and a dynamic synapse neural network.

A Detailed Model of a Synapse

Synapse, the contact between neurons, is a complex system composed of a multitude of molecular machinery. Typically, it consists of an axon terminal, the end point of the input neuron, and a postsynaptic spine on the dendrite of the second neuron. The axon terminal and the spine is separated by a space called synaptic cleft. Each of the synaptic processes is tightly controlled by a large number of molecular interaction, which themselves are subject to change under a variety of conditions.

To study issues at the synaptic level, we have constructed a model of a glutamatergic synapse which includes three main objects representing the axon terminal, the synaptic cleft, and the postsynaptic spine (Fig. 2A). The basic axon terminal module contains objects of ion channels, calcium mechanisms, vesicle, molecular processes, and intrinsic membrane properties (Fig. 2B). The parameter values that the model used for these objects are based on experimental data and are consistent with those used in many existing models. We have conducted an intensive series of computer simulations of the synaptic model to address a number of unresolved issues regarding synaptic transmission and plasticity (Gaffari et al., 1999; Liaw and Berger, 1999; Xie et al., 1995, 1997). Here, we illustrate a particular extension of this model: the incorporation of morphological details.

Changes in synaptic morphology consistently have been observed as a consequence of a variety of experimental manipulations, including associative learning (Black et al., 1990; Federmeier et al., 1994; Kleim et al., 1994, 1997), environmental rearing conditions (Bhide and Bedi, 1984; Diamond et al., 1964; Globus et al., 1973; Turner and Greenough, 1983, 1985; Volkmar and Greenough, 1972), and increased synaptic "use" induced by direct electrical (Buchs and Muller, 1996; Desmond and Levy, 1983, 1988; Geinisman et al., 1993; Wojtowicz et al., 1989), as well as the processes of normal development and aging (De Groot and Bierman, 1983; Dyson and Jones, 1984; Harris et al., 1992). Some structural changes involve morphological profiles corresponding to "perforated" or "partitioned" synapses. A defining property of perforated and partitioned synapses is the finger-like formations of postsynaptic membrane and cytoplasm enveloped by presynaptic membrane and cytoplasm at single or multiple locations along the synaptic cleft. The pre- and postsynaptic specializations are correspondingly discontinuous, resulting in multiple, compartmentalized synaptic zones within the same end-terminal region.

A model of partitioned synapses was developed to study the impact of structural modifications on neurotransmitter release and synaptic dynamics. Specifically, a

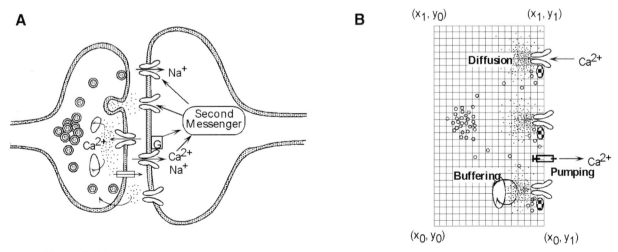

Figure 2 (A) A schematic representation of a synapse. **(B)** A schematic representation of a model of the axon terminal.

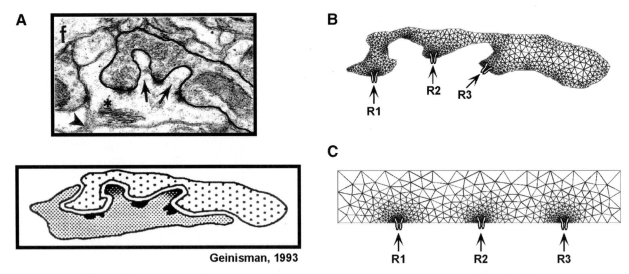

Figure 3 Partitioned synapse: EM data and the model. (From Geinisman *et al.*, 1993. With permission.)

serial section electron micrograph of a hippocampal partitioned synapse (Fig. 3A) after high-frequency stimulation was used to specify the geometry of the model (Fig. 3B).

The general structure of our model of the axon terminal consists of a finite element representation of a two-dimensional geometry, such as that of a partitioned presynaptic terminal. The geometric representation of the model is developed using MATLAB with the "constructive solid geometry" modeling paradigm. The two-dimensional geometry is then discretized by generating a triangular mesh (Fig. 3B). The mesh is refined until the smallest boundary element is just smaller than 0.5 nm wide in order to minimize the approximation error. After the terminal boundary is accurately specified, calcium channels and neurotransmitter release sites are assigned as time-varying boundary conditions. We have extended the mathematical functionality such that boundary conditions can be specified at any location on the boundary. In the case of the partitioned synapse, the calcium channels and release sites are placed (10 nm apart) at positions corresponding to the location of PSDs in the electron micrograph picture. An additional step is taken to refine the mesh such that the triangles near the calcium channels and release sites are higher in number and smaller in size to increase the accuracy of the solution (Fig. 3B). The kinetics of calcium conductance and neurotransmitter release were optimized with experimental data. The diffusion coefficient is specified to include the appropriate rate of calcium buffering. Given the discretized mesh representation of the terminal, the boundary and initial conditions, and the coefficients of the diffusion equation, a numerical solution to the diffusion equation is obtained using the MATLAB partial differential equation toolbox (The MathWorks, 1995).

A series of computer simulations is conducted to test the hypothesis that local structure in the partitioned synapse affects calcium diffusion and distribution and hence the probability of neurotransmitter release. In the first set of simulations, a rectangular model of a symmetrical presynaptic terminal (Fig. 3C) is used as the control for comparison with the partitioned synapse model. There are three equally spaced calcium channel release sites in the control case. Simulations of the symmetrical model show an equal release probability for all of the release sites for both a single depolarization step and a train of random inputs (i.e., a sequence of depolarization steps over 200 msec, shown in Fig. 4B).

Simulations of the partitioned synapse model (see Fig. 3B) in response to a single depolarization step show that the probability of release is different for the three release sites. The peak amplitude of release probability for site 1 is 32% higher than that of site 2 and 19% higher than that of site 3 (Fig. 4A). The results indicate that the differences in the release probability within the time course of a single response are a consequence of unequal morphological partitioning. Next, we simulate the partitioned synapse model with the random train of inputs. The peak response of site 1 to a sequence of depolarization steps (Fig. 4B) varied with respect to site 3. For example, the peak response of site 3 was higher than that of site 1 in response to the group of six consecutive pulses beginning with the second pulse (at 15 msec). However, the relative peak amplitudes of the two sites reversed from the eighth pulse to the end of the input train, revealing different temporal dynamics within a single synapse.

A series of computer simulations was conducted to test the hypothesis that morphological alterations lead to variations in synaptic dynamics by systematically examining a variety of spatial features. For instance, the effect of a nearby membrane wall was examined by moving release site 3 in the partitioned synapse approximately 300 nm to the right (Fig. 5). The reversal in the relative

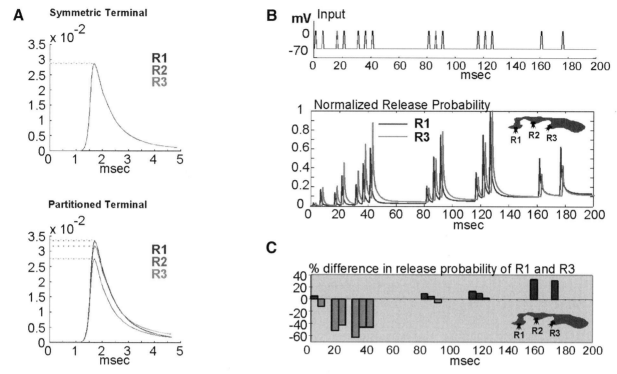

Figure 4 Release probability in the symmetric terminal and the partitioned synapse.

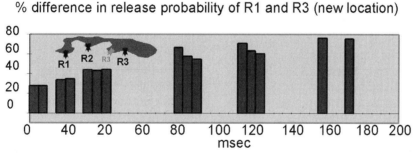

Figure 5 Effect of proximity to adjacent membrane wall on release probability.

peak amplitudes between release sites 1 and 3 seen in Fig. 4 is eliminated. Instead, the probability of release is always higher in release site 1 than release site 3 at the new location.

Dynamic Synapse Model

In the previous section, we presented a detailed model of a glutamatergic synapse. Though such a comprehensive model of a synapse is necessary for studying synaptic transmission, its complexity precludes the details being carried over to a neural network of any significant size. Instead, one must extract essential principles of the synaptic processes, which can be gained only through a careful and in-depth study of that particular level, to provide proper abstraction and simplification necessary not only for the feasibility of modeling, but also, more importantly, for insight regarding the emergence of higher level functionalities from cellular or molecular mechanisms.

Here we show one example, the development of the concept of the *dynamic synapse*, which captures the essential principles of neurotransmitter release with only a few equations. This concise mathematical representation is extracted from simulations of the detailed synaptic models described in the previous section. With the systematic indexing scheme of EONS, we have constructed a neural network model incorporating dynamic synapses which is used to perform the task of speech recognition.

A straightforward implementation of the dynamic synapse is by expressing the potential of neurotransmitter release as a weighted sum of the activation of presynaptic mechanisms:

$$R_i(t) = \sum_m K_{i,m}(t) F_{i,m}(Ap(t)) \quad (1)$$

where $R_i(t)$ is the potential for release for presynaptic terminal i at time t. $F_{i,m}$ is the m'th mechanism in terminal i, $K_{i,m}$ is the coefficient (weight) for mechanism $F_{i,m}$, and Ap indicates the occurrence ($Ap = 1$) or non-occurrence ($Ap = 0$) of an action potential. Any number of synaptic mechanisms (e.g., facilitation, augmentation, PTP, modulation, etc.) can be included in this equation. Note that in this implementation, R is deterministic, though it can be made stochastic by, say, expressing it as a probability variable and adjusting all the other parameters to be consistent with the probability constraints (e.g., probability of release ≤ 1).

One simple expression, known as the leaky-integrator equation, can be used to describe the dynamics of various synaptic mechanisms:

$$\tau_{i,m} \frac{dF_{i,m}(t)}{dt} = -F_{i,m}(t) + Ap(t) \quad (2)$$

where $\tau_{i,m}$ is the time constant of mechanism m in terminal i. Other equations can also be used to express synaptic mechanisms, such as the alpha function, without deviation from the spirit of a dynamic synapse. Typically, $F_{i,m}$ is a function of the presynaptic action potential, $Ap(t)$. That is, the magnitude of $F_{i,m}$ varies continuously based on the temporal pattern of the spike train. Usually, $Ap(t)$ originates from the same neuron that gives rise to the particular axon terminal; however, $Ap(t)$ may also be the action potential generated by some other neuron, such as the case of feedback modulation. Furthermore, $F_{i,m}$ can also be a function of other type of signals. For example, it can be a function of the synaptic signal generated by the same axon terminal i (e.g., via presynaptic receptor channels) or a function of the synaptic signals produced by other synapses (e.g., via axon-axonal synapses).

A quanta of neurotransmitter is released if R is greater than a threshold (θR) and there is at least one quanta of neurotransmitter available (N_{total}):

If($P_R > \theta_R$ & $N_{total} Q$) then $N_R = Q$ and N_{total}
$$= N_{total} - Q. \quad (3)$$

where N_R is the concentration of neurotransmitter in the synaptic cleft. Note that the reduction of available neurotransmitter represents an underlying cause of synaptic depression. The neurotransmitter is cleared from the synaptic cleft at an exponential rate with time constant τ_{N_t}:

$$N_R = N_R e^{-t/\tau_{N_t}} \quad (4)$$

Furthermore, there is a continuous process for replenishing neurotransmitter:

$$\frac{dN_{total}}{dt} = \tau_{rp}(N_{max} - N_{total}) \quad (5)$$

where N_{max} is the maximum amount of available neurotransmitter and τ_{rp} is the rate of replenishing neurotransmitter.

The postsynaptic potential (*PSP*) is a function (*G*) of the concentration of the neurotransmitter in the synaptic cleft:

$$PSP_j(t) = \sum_n W_{j,n}(t)^* G_{j,n}(N_i(t)) \quad (6)$$

where $W_{j,n}$ denotes the efficacy of mechanism n at the postsynaptic site j to respond to signals from presynaptic terminal i. Typical postsynaptic mechanisms are conductances generated by various types of receptor channels. They can be expressed in a way similar to presynaptic mechanisms (e.g., Eq. (2) or the alpha function). Different kinetics of receptor channels can be approximated by adjusting the time constants in Eq. (2) (or the alpha function).

The postsynaptic potentials can be integrated at the soma of an "integrate-and-fire" model neuron:

$$\tau_v \frac{dV(t)}{dt} = -V(t) + \sum_s PSP_s(t) \quad (7)$$

where V is the somatic membrane potential and τ_v is the time constant. An action potential is generated (i.e., $Ap = 1$) if V crosses some threshold.

Dynamic Learning Algorithm

The process of neurotransmitter release is dynamically mediated by complex molecular and cellular mechanisms. Furthermore, these mechanisms themselves are dynamic and under constant regulation and modulation. How can such second-order regulation be expressed mathematically? An even more intriguing question is how can such regulation be utilized for learning and memory? One plausible answer is to modify the presynaptic mechanisms by the correlation of presynaptic and postsynaptic activation. The biological plausibility is based on the existence of molecules that are attached to both pre- and post-synaptic membranes (e.g., those that are thought to bind the two membranes together). These molecules can provide a means for conveying the activation signal of the postsynaptic neuron to the presynaptic terminal. The theoretical plausibility is rooted in the notion of the Hebbian rule. More specifically, the contribution of each synaptic mechanism to the release of neurotransmitter or the activation of postsynaptic potentials is changed as shown below:

$$\Delta K_{M,i} = \alpha_M * L_{M,i} * Ap_j \quad (8)$$

where $K_{M,i}$ is the co-efficient of synaptic mechanism M in synapse i, αM is the learning rate for M, $L_{M,i}$ is the activation level of M in synapse i, and Ap_j ($= 0$ or 1) indicates the occurrence or non-occurrence of an action potential of postsynaptic neuron *j*. Note that the mechanisms in Eq. (8) also include those in the postsynaptic site, which is equivalent to the conventional Hebbian rule for changing "synaptic efficacy" or "synaptic weight."

By changing the contribution of each synaptic mechanism to the process of neurotransmitter release, the transformation function of the axon terminal is modified. And, by basing such changes on the correlation of pre- and postsynaptic activation, an axon terminal can learn to extract temporal patterns of action potential sequences that occur consistently. That is, each axon terminal will extract different statistical regularities (invariant features) embedded in the spike train, whereas correlated changes in the postsynaptic mechanisms (i.e., the synaptic efficacy or weight) determine how such presynaptic signals should be combined.

Speech Recognition Case Study

We have demonstrated the computational capability of dynamic synapses by performing speech recognition from unprocessed, noisy raw waveforms of words spoken by multiple speakers with a simple neural network consisting of a small number of neurons connected by dynamic synapses (Liaw and Berger, 1996, 1999). The neural network is organized into two layers, an input layer and an output layer, with five neurons in each layer, plus one inhibitory interneuron, each modeled as an "integrate-and-fire" neuron (Fig. 6). The input neurons receive unprocessed, raw speech waveforms. Each input neuron is connected by dynamic synapses to all of the output neurons and to the interneuron, with a total of $5 \times 6 = 30$ dynamic synapses in the model. There is a feedback connection from the interneuron to each presynaptic terminal providing the inhibition. We experimented with two learning procedures for this small neural network.

The first experiment involved unsupervised learning. Given a set of N words spoken by M speakers, the goal is to train the neural network to converge to N distinct output patterns at the output neurons. The network is trained following Hebbian and anti-Hebbian rules. During learning, the presentation of the raw speech waveforms is grouped into blocks in which the waveforms of the same word spoken by different speakers are presented to the network a total of 20 times. Within a presentation block, the Hebbian rule is applied. In this way, the responses to the temporal features that are common in the waveforms will be enhanced, while the idiosyncratic features will be discouraged. When the presentation first switches to the next block of waveforms of a new word, the anti-Hebbian rule is applied. This enhances the differences between the response to the current word and the response to the previous, different word. As the learning process progresses, the output patterns of neural network gradually converges to N distinct clusters, each corresponding to a given word. Our dynamic synapse neural network has successfully generated invariant output patterns for $N = 12$ different words (readers are referred to Liaw and Berger, 1996, 1999, for more details).

The second experiment applied supervised learning to making use of the invariant features obtained in experiment 1 to achieve word recognition by forcing each output neuron to respond preferentially to a specific word. The results demonstrated that, by making use of the set of invariant features that were extracted dynamically during unsupervised training, the network can readily perform word recognition. We then tested the robustness of our model when the speech signals are corrupted with progressively increasing amount of noise. The results showed that our model is extremely robust against noise, retaining an 85% performance rate with SNR = −30 when the recognition threshold is dynamically adjusted. This performance is better than human listeners tested with the same speech data set.

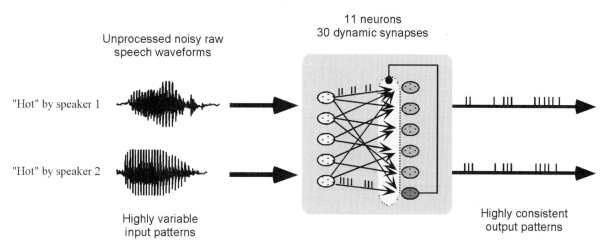

Figure 6 A dynamic synapse neural network model for speech recognition.

2.3.3 Protocol-Based Simulation

A tight coupling with experimental studies is critically important for the success of a computational model. This is particularly important given that experimental and computational approaches can complement each other. For example, it is difficult, if not impossible, to physically translocate certain molecules in a biological neuron with current experimental methods. This can be easily achieved with a computational model with an appropriate spatial representation. On the other hand, it is very difficult to solve the diffusion equation over a complex three-dimensional space, a process that can readily be manipulated experimentally. To increase the synergy between empirical and computational neuroscience, a communication language for these two disciplines is needed. One of the fundamental characteristics of biological experiments is the testing of certain hypotheses. The experimental procedures for testing a hypothesis can be fully characterized by a set of protocols that specify the conditions of the experiment and the manipulations of system parameters. Correspondingly, we have developed a protocol-based simulation scheme to facilitate the communication between these two disciplines.

Furthermore, one of the major difficulties in integrating the experimental and modeling approaches is finding relevant data to constrain model parameters such that comparisons of the simulation results and experimental data can be grounded. Currently, the search for relevant data has been done manually in an *ad hoc* manner. The protocol-based simulation scheme can establish a link to experimental databases to alleviate the data searching problem. We will illustrate this simulation scheme using a case study of the formulation and testing of a specific hypothesis that receptor channel aggregation on the postsynaptic membrane is the cellular mechanism underlying the expression of long-term potentiation (Xie et al., 1997).

In the protocol-based scheme, information of a simulation is organized into two levels, namely, a meta level which spells out the hypothesis being tested and a procedural level which contains initial conditions and manipulations of model parameters. The formal specification of a simulation manipulation is given below:

Definition: Let simulation manipulation = $[T, P]$, where

$T = [(T_{1,1}, T_{1,2}, \ldots T_{1,n}); (T_{2,1}, T_{2,2}, \ldots T_{2,m}); \ldots (T_{k,1}, T_{k,2}, \ldots T_{k,t})]$. $P = [(P_x, V_{1,1}, V_{1,2}, \ldots, V_{1,n}); (P_y, V_{2,1}, V_{2,2}, \ldots V_{2,m}) \ldots (P_z, V_{k,1}, V_{k,2}, \ldots V_{k,t}]$.

T is a list of points in time where corresponding particular parameter manipulations take place. P is a list of manipulations applied to model parameters during the period T. For example, $(T_{1,1}, T_{1,2}, \ldots)$ together with $(P_x, V_{1,1}, V_{1,2}, \ldots)$ specify that parameter P_x in a simulation is changed from initial value to $V_{1,1}$ at time $T_{1,1}$ and to $V_{1,2}$ at time $T_{1,2}$, and so forth.

Model parameters are grouped into two classes, namely, *conditions*, which include those parameters that remain constant throughout the simulation, and *manipulations*, which contain the time-stamped changes of parameters as defined above. Many default values can be used in the condition part to simplify the specification process. A series of simulations to test a specific hypothesis is organized into blocks, with each block consisting of multiple trials. Manipulations that remain constant for all trials in a block are specified once at the block level, whereas those that vary from trial to trial are specified on a trial-basis. This simulation organization closely matches the structure of an experimental protocol in NeuroCore. We illustrate the protocol-based simulation scheme by the following example of using the synaptic model mentioned in the previous section to study molecular/cellular mechanisms underlying long-term potentiation (LTP).

Simulation Protocols for Testing Alternative Hypotheses of LTP

Long-term potentiation (LTP) is a widely studied form of use-dependent synaptic plasticity expressed robustly by glutamatergic synapses of the hippocampus. Although there is a convergence of evidence concerning the cellular/molecular mechanisms mediating the induction of NMDA receptor-dependent LTP, there remains substantial debate as to the underlying molecular/cellular mechanisms. Here, we illustrate three hypotheses regarding possible mechanisms and the corresponding simulation protocols for testing these hypotheses.

The key data set for testing alternative hypotheses is provided by the experimental component of USCBP (electrophysiological experiment conducted by Xie in Berger's lab; see Xie *et al.*, 1997). The main finding of the experiment is that during the initial phase of LTP, AMPA receptor-mediated responses evoked by presynaptic stimulation are increased; however, AMPA responses to focal application of agonist are not affected.

The following simulation protocols for testing alternative hypotheses use a synaptic cleft object and a kinetic model object from the EONS library (Fig. 1). The synaptic cleft object contains a spatial representation of the space between the pre-and postsynaptic membranes as a two-dimensional array (20 nm wide by 1000 nm long; Fig. 7A). The PDE method in the EONS library is applied to solve the diffusion of neurotransmitter in the synaptic cleft. The object contains a simple kinetic scheme:

$$A + B \underset{K_{-1}}{\overset{K_1}{\rightleftharpoons}} AB \underset{K_{-2}}{\overset{K_2}{\rightleftharpoons}} AB^*$$

where A represents the concentration of neurotransmitter at the position where the receptor B is located on the postsynaptic membrane, AB represents the probability that B is bound to the neurotransmitter, and AB^* represents the probability that B is bound and open. K_1, K_{-1}, K_2, and K_{-2} are the binding, unbinding, opening, and closing rates, respectively. The amplitude of excitatory postsynaptic potential (EPSP) is proportional to AB^*.

The following simulation protocols demonstrate how the above synaptic model can be used to test alternative hypotheses regarding mechanisms underlying LTP.

Hypothesis 1: An increase in the neurotransmitter release is the underlying mechanism.

The corresponding simulation protocol is constructed by varying the concentration of agonist, A, across different trials (see Table 1).

Table 1

Trial #1	$[T_{1,1}, T_{1,2}, T_{1,3}); (A, 0, 100, 0)]$
Trial #2	$[T_{1,1}, T_{1,2}, T_{1,3}; (A, 0, 200, 0)]$
Trial #3	$[(T_{1,1}, T_{1,2}, T_{1,3}); (A, 0, 400, 0)]$

Hypothesis 2: Changes in binding rate constant (K_1) are the basis for synaptic plasticity.

The corresponding simulation protocol is similar to that for testing Hypothesis 1 except that the binding rate constant, K_1, is varied across trials (see Table 2).

Table 2

Trial #1	$K_1 = 0.1$
Trial #2	$K_1 = 0.2$
Trial #3	$K_1 = 0.4$

Hypothesis 3: Changes in the distribution of receptor (B) are the mechanisms underlying synaptic plasticity.

The key component of the protocol is the variation of the location of the receptor channel, B, from trial to trial (see Table 3).

Table 3

Trial #1	Loc-of-B = 0 nm from release site
Trial #2	Loc-of-B = 20 nm from release site
Trial #3	Loc-of-B = 40 nm from release site

The Results of Three Simulation Protocols

For each simulation, the perfusion of neurotransmitter (A) into the synapse is simulated by allowing glutamate to diffuse into the synaptic cleft from the edges. The time-evolution of A is calculated by solving the diffusion equation. The results from the three simulation protocols are given below.

Protocol 1

The simulation result based on this protocol showed that the magnitude of EPSC is higher when the amount of agonist is increased. This is inconsistent with the experimental data described above showing that when agonist is directly applied AMPA response does not show an increase after LTP induction.

Protocol 2

The simulation result based on this protocol showed that the magnitude of EPSC is higher when the binding rate is increased. This is inconsistent with the experimental data described above.

Protocol 3

The simulation result based on this protocol showed that the magnitude of EPSC is similar when the receptor (B) is located in different places. This is due to the uniform distribution of agonist throughout the synaptic cleft after perfusion, making the difference in receptor location irrelevant. The simulation result is consistent with the experimental data.

In an additional block of trials, the above simulation condition is changed slightly such that the neurotransmitter is released from a specific presynaptic release site. Under this condition, the simulations show that the resulting EPSC is 42% larger in magnitude when the release site and receptors are in complete alignment with the release site (i.e., Loc-of-B = 0 nm) compared to the EPSCs simulated when B is located 40 nm away (Fig. 7B and C). Furthermore, the rising and decaying phases of the EPSC become faster, in agreement with experimental data (further examination revealed that the receptor is bound at the 40 nm location with a probability of 0.23; therefore, a higher probability for receptor binding can be achieved when the receptor moves closer to the release site). The results suggest that hypothesis 3 is the most likely candidate for explaining the molecular/cellular mechanisms underlying LTP. Thus, by constructing a model that includes appropriate representation of the synapse and a protocol for mapping experimental data and model simulation, one can formulate multiple biologically based hypotheses and identify the one that is most consistent with experimental observations.

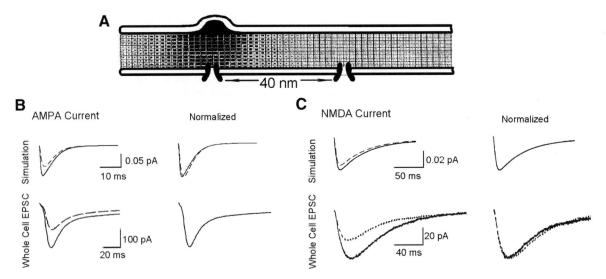

Figure 7 (A) A model that includes representations of synaptic cleft and postsynaptic receptors. Comparisons of simulation results with experimental data are provided in **(B)** and **(C)**.

2.3.4 Conclusion

This chapter has provided an introduction to the multi-level modeling system, EONS. In EONS, object-oriented design methodology is employed to facilitate the construction of a hierarchy of neural models ranging from cellular mechanisms, synapses, and neurons to networks. Each model (e.g., for a synapse or a neuron) is self-contained, and various degrees of detail can be included. This provides a way to compose complex system models from objects of basic elements. One objective is to make EONS objects available for use by models created with other modeling tools. For this purpose, the EONS library is separated from the user interface to increase its portability. Two examples were presented; one is a detailed model of a synapse in the hippocampus with complex geometry, and the other is a neural network incorporating dynamic synapses.

Currently, the EONS library contains objects for voltage-gated ion channels; receptor channels; kinetics of molecular interaction; synaptic models; dynamic synapses, neurons, and neural networks; and methods for building complex two-dimensional geometry and solving diffusion over such geometry. Though these objects can be combined to build more complex models, at the present this task has to be carried out manually. One future development plan involves using the Schema Capture System of NSL (Chapter 2.2) to provide a graphic user interface for object composition.

In order to enhance the communication between modelers and experimentalists, we are developing protocol-based simulation for people in the two disciplines to interact with each other more easily. The simulation protocols closely match the structure of an experimental protocol in NeuroCore. One advantage of using similar protocols for both experiments and model simulation is to help make explicit the assumptions adopted in developing the model. Furthermore, queries for searching experimental databases can be adapted for retrieving models stored in databases. Future developments will aim at providing a mechanism for composing models with proper objects based on the specifications derived from a user-defined simulation protocol.

Acknowledgments

This research was supported by the University of California Human Brain Project (with funding from NIMH, NASA, and NIDA), NIMH (MH51722 and MH00343), NCRR Biomedical Simulations Resource (P41 RR01861), ONR (N00014-98-1-0259), and the Human Frontiers Science Organization.

References

Bhide, P. G., and Bedi, K. S. (1984). The effects of a lengthy period of environmental diversity on well-fed and previously undernourished rats. II. Synapse-to-neuron ratios. *J. Comp. Neurol.* **227**, 305–310.

Black, J. E., Isaacs, K. R., Anderson, B. J., Alcantara, A. A., and Greenough, W. T. (1990). Learning causes synaptogenesis, whereas motor activity causes angiogenesis, in cerebellar cortex of adult rats. *Proc. Nat. Acad. Sci. USA.* **87**, 5568–72.

Buchs, P. A., and Muller, D. (1996). Induction of long-term potentiation is associated with major ultrastructural changes of activated synapses. *Proc. Nat. Acad. Sci. USA.* **93**, 8040–8045.

De Groot, D. M., and Bierman, E. P. (1983). The complex-shaped "perforated" synapse, a problem in quantitative stereology of the brain. *J. Microsc.* **131**, 355–360.

Desmond, N. L., and Levy, W. B. (1983). Synaptic correlates of associative potentiation/depression: an ultrastructural study in the hippocampus. *Brain Res.* **265**, 21–30.

Desmond, N. L., and Levy, W. B. (1988). Synaptic interface surface area increases with long-term potentiation in the hippocampal dentate gyrus. *Brain Res.* **453**, 308–314.

Diamond, M. C., Krech, D., and Rosenzweig, M. R. (1964). The effects of an enriched environment on the histology of the rat cerebral cortex. *J. Comp. Neurol.* **123**, 111–120.

Dyson, S. E., and Jones, D. G. (1984). Synaptic remodelling during development and maturation: junction differentiation and splitting as a mechanism for modifying connectivity. *Brain Res.* **315**, 125–137.

Federmeier, K., Kleim, J. A., Anderson, B. J., and Greenough, W. T. (1994). Formation of double synapses in the cerebellar cortex of the rat following motor learning. *Soc. Neurosci. Abstr.* **20**, 1435.

Gaffari, T., Liaw, J.-S., and Berger, T. W., (1999). Consequence of morphological alterations on synaptic function. *Neurocomputing.* **26/27**, 17–27.

Geinisman, Y., de Toledo-Morrell, L., Morrell, F., Heller, R. E., Rossi, M., and Parshall, R. F. (1993). Structural synaptic correlate of long-term potentiation: formation of axospinous synapses with multiple, completely partitioned transmission zones. *Hippocampus.* **3**, 435–446

Globus, A., Rosenzweig, M. R., Bennett, E. L., and Diamond, M. C. (1973). Effects of differential experience on dendritic spine counts in rat cerebral cortex. *J. Comp. Physiol. Psychol.* **82**, 175–181.

Harris, K. M., Jensen, F. E., and Tsao, B. (1992). 3-dimensional structure of dendritic spines and synapses in rat hippocampus at postnatal day-15 and adult ages: implications for the maturation of synaptic physiology and long-term potentiation. *J. Neurosci.* **12**, 2685–2705.

Kleim, J. A., Napper, R. M. A., Swain, R. A., Armstrong, K. E., Jones, T. A., and Greenough, W. T. (1994). Selective synaptic plasticity in the cerebellar cortex of the rat following complex motor learning. *Soc. Neurosci. Abstr.* **20**,1435.

Kleim, J. A., Vij, K., Ballard, D. H., and Greenough, W. T. (1997). Learning-dependent synaptic modifications in the cerebellar cortex of the adult rat persist for at least 4 weeks. *J. Neurosci.* **17**, 717–721.

Liaw, J.-S., and Berger, T. W. (1996). Dynamic synapse: a new concept of neural representation and computation. *Hippocampus.* **6**, 591–600.

Liaw, J.-S., and Berger, T. W. (1999). Dynamic synapse: harnessing the computing power of synaptic dynamics. *Neurocomputing* **26/27**, 199–206.

Shu, Y., Xie. X., Liaw, J.-S., and Berger, T. W. (1999). A protocol-based simulation for linking computational and experimental studies. *Neurocomputing* **26/27**, 1039–1047.

Turner, A. M., and Greenough, W. T. (1983). Synapses per neuron and synaptic dimensions in occipital cortex of rats reared in complex, social or isolation housing. *Acta Stereol. (Suppl.)* **2**, 239–244.

Turner, A. M., and Greenough, W. T. (1985). Differential rearing effects on rat visual cortex synapses. I. Synaptic and neuronal density and synapses per neuron. *Brain Res.* **329**, 195–203.

Volkmar, F. R., and Greenough, W. T. (1972). Rearing complexity affects branching of dendrites in the visual cortex of the rat. *Science* **176**, 1145–1147.

Wojtowicz, J. M., Marin, L., and Atwood, H. L. (1989). Synaptic restructuring during long-term facilitation at the crayfish neuromuscular junction. *Can. J. Physiol. Pharmacol.* **67**, 167–171.

Xie, X., Liaw, J.-S., Baudry, M., and Berger, T. W. (1995). Experimental and modeling studies demonstrate a novel expression mechanism for early phase of long-term potentiation. *Computational Neural Syst.* 33–38.

Xie, X., Liaw, J.-S., Baudry, M., and Berger, T. W. (1997). Novel expression mechanisms for synaptic potentiation: alignment of presynaptic release site and postsynaptic receptor. *Proc. Nat. Acad. Sci. USA.* **94**, 6983–6988.

CHAPTER 2.4

Brain Imaging and Synthetic PET

Amanda Bischoff-Grethe[1] and Michael A. Arbib[2]
[1]Center for Cognitive Neuroscience Dartmouth College, Hanover, New Hampshire
[2]University of Southern California Brain Project and Computer Science Department,
University of Southern California, Los Angeles, California

Abstract

The advent of brain imaging techniques (PET, positron emission tomography; fMRI, functional magnetic resonance imaging) has led to new understanding of brain function. For the first time researchers can examine activation in the human brain not only during motor behavior, but also during cognitive behavior. Brain imaging has provided a way to study mental disorders and how various medications influence both function and activation. Unlike neurophysiological recordings of individual neurons where activity is recorded over milliseconds, imaging represents activation over a time period of seconds or even minutes. Knowing that the motor cortex was active during a particular task still does not tell us at what point the motor cortex was active, nor does it tell us the effect of excitatory and inhibitory connections to it. Synthetic PET was created as a way to bridge this gap by predicting functional activation based upon activity collected from models of brain function. Thus, it serves neuroscientists in multiple ways: it provides a way to validate models; it allows scientists to relate neuron activity to regional activation; and, finally, it allows us to study tasks which may normally be difficult to examine with traditional imaging or neurophysiological techniques. Synthetic PET therefore becomes a useful tool within the Neuroinformatics Workbench. In this chapter, we will discuss the Synthetic PET paradigm and provide examples of its function from models stored within BMW (Chapter 6.2). While our work to date has focused on the PET methodology, we stress that future work will pay increasing attention to fMRI data and the development of a corresponding methodology of Synthetic fMRI.

2.4.1 PET Imaging and Neurophysiology

Before delving into the methodology behind Synthetic PET, let us first review the biological basis of data collection in imaging. In brain imaging using PET, small doses of a radioactive tracer such as $H_2^{15}O$ water are injected into the subject prior to imaging during the performance of specified tasks. After a delay period of 10 to 20 seconds, the radioactive tracer reaches the brain. Its spatial distribution is determined using detectors sensitive to gamma photons generated by positron annihilation events. Typically, the first 45 to 90 seconds after injection of $H_2^{15}O$ produce a number of counts in each location which correlate almost linearly with the absolute regional cerebral blood flow (rCBF) (Herscovitch et al., 1983; Raichle et al., 1983). PET scans less than 45 seconds are usually not acquired, as the insufficient counts obtained are unable to overcome Poisson noise inherent in the detection method. In MRI, the subject lies within a strong magnetic field which causes the hydrogen atoms within his body to align their spin axes along the direction of the applied field. When brief pulses of radio waves are applied, the spin axes are perturbed; their spin axes tip away from their orientation with the magnetic field. When the pulse is turned off, the atoms realign with the magnetic field, releasing energy in the form of radio waves. These radio waves are the basis of the MRI image. The time interval used to acquire activity images varies among techniques: for PET, rCBF imaging is between 45 and 90 seconds; for fMRI at 1.5 Tesla using echo planar imaging (EPI), acquisition can be on the order of 1 or 2 seconds.

Neurophysiology operates on a different level. It is here that we use electrodes to study the firing pattern of individual neurons during a variety of tasks. Given the time course of the paradigm and the firing behavior of individual neurons, researchers can hypothesize not only how a brain region might be involved with a task, but also what the underlying cell populations may be doing in order to accomplish the task. In behavioral paradigms, neurons have been known to alter firing rates during pre-movement conditions, in expectation of reward, during movement, and at the end of movements. One can also determine the response of a neuron under varying neurochemical conditions; the application of an antagonist, for example, may affect the firing rate of a particular population of neurons. These responses are studied on the order of milliseconds, as opposed to up to several seconds in imaging. The drawback of neural recording, however, is that the explanation of the neuronal response is not clear in relation to the task at hand. A pre-movement neuron in one study may either be considered a predictor of behavior or an inhibitor of movement. There is also no true indicator that a population of recorded neurons represents a larger population with the same functional purpose. Conversely, the lack of a particular firing behavior does not mean it does not exist; it merely did not exist within the sample set. Still, neural recordings have greatly enhanced our understanding of brain function under various conditions. With the increasing popularity of computational models of neuronal behavior during task performance, activation patterns from similar tasks can be used to make predictions about behavior, leading to further experimental research.

Given these two methods of neuroscientific study, how, then, can we relate them not only to each other, but to computer models of neuron function as well? Despite the increased interest in brain imaging, it is still not clear exactly what the activations represent. It is impossible to distinguish excitatory and inhibitory cellular activity based on blood flow changes. After all, inhibition is an energy requiring task (Ackermann et al., 1984). We also do not know the time course of the activation or how the individual neurons contribute to the activation, i.e., is activation due to pre-movement response or motor response in a brain area known to contain both? The purpose of Synthetic PET, then, is to attempt to bridge this gap and provide us with a mechanism for linking large-scale behavior with individual cellular behavior. Within the Neuroinformatics Workbench, Synthetic PET will allow us to visually compare behavioral responses of models stored within BMW to real imaging results and time series data.

2.4.2 Defining Synthetic PET

Synthetic PET is a technique by which neural models based upon primate neurophysiology are used to predict and analyze results from PET imaging taken during the performance of a variety of human behaviors. It is based upon the hypothesis that PET is correlated to the integrated synaptic activity within a region (Brownell et al., 1982) and reflects neural activity in regions *afferent* to the region studied, rather than the intrinsic neural activity of the region alone.

Given the neural network representation described in the section "Building on Arrays of Neurons" of Chapter 2.1, the question is how to map the activity these networks simulate into predictions of activity to be recorded in corresponding regions within the human brain. There are two issues:

1. *Localization:* Each array within the neural network model represents a neural population that has been identified both physiologically and anatomically in the monkey. A Synthetic PET comparison requires explicit homologies (e.g., that region A_m in the monkey is homologous to region A_h in the human brain). (See Chapter 6.4 for details on how we determine homologies between species and how these homologies are stored within the database.) These regions should carry out the tasks under consideration in the same fashion. Sometimes the homology is well defined; at other times, its existence is not known. Thus, one of the aspects of Synthetic PET is to enable one to compare the results of a human brain study with those of a synthetic brain study, allowing one to test hypotheses of homologies across species. This in turn may lead to further studies to support or refute the claim.

2. *Modeling activation:* PET is said to measure rCBF. We hypothesized (Arbib et al., 1995) that the counts acquired within PET scans may be correlated to local synaptic activity within a particular region, calling this the "raw PET activity." PET studies do not typically work directly with these values but rather with the comparative values of this activity within a given region during two different tasks or behaviors. We therefore define our computation in two stages:

a. Compute $rPET_A$, the simulated value of raw PET activity for each region A of the neural network while it is used to simulate the monkey's neural activity for a given task.
b. Compare the activities computed for two different tasks by "painting" this difference on the region A_h of a human brain (as represented by the Talairach Atlas; Talairach and Tournoux, 1988). The result is a Synthetic PET comparison presenting our prediction of human brain activity based upon a neural network model constrained by monkey neurophysiology and known functional neuroanatomy.

The *synthetic raw activity* $rPET_A$ associated with a cell group A is defined as:

$$rPET_A = \int_{t_0}^{t_1} \sum_B W_{B \to A}(t) dt, \quad (1)$$

where A is the region of interest, the sum is over all regions B that project to the region of interest; $w_{B \to A}(t)$ is the synaptic activity (firing rate $\times$ | synaptic strength |) summed over all the synapses from region B to region A at time t; and the time interval from t_0 to t_1 corresponds to the duration of the scan.

The comparative activity $PET_A(1/2)$ for task 1 over task 2 for a given region is then calculated as:

$$rPET_A(1/2) = \frac{rPET_A(1) - rPET_A(2)}{rPET_A(2)}, \quad (2)$$

where $rPET_A(i)$ is the value of $rPET_A$ in condition i. This allows us to compare the change in PET_A from task 2 to task 1. In the current study, we use a different measure, defining the change in *relative synaptic activity* for region A from task 1 to task 2, with $max(rPET_A(1), rPET_A(2))$ replacing $rPET_A(2)$ in the denominator of Eq. (2):

$$rPET_A(1/2) = \frac{rPET_A(1) - rPET_A(2)}{max(rPET_A(1), rPET_A(2))} \quad (3)$$

This gives us a more robust measure of relative activity.

Synthetic PET results can be displayed in several ways, from using a graph or table to displaying the results on the region A_h homologous to A on a Talairach Atlas by converting the A values to a color scale. The resulting images then predict the results of human PET studies. Note, however, that we are displaying the *synaptic* activity of region A and not the neural activity of region A. As a computational plus, we may also collect the contributions of the excitatory and inhibitory synapses separately, based on evaluating the integrals in Eq. (1) over one set of synapses or the other. Using simulated PET we can tease apart the different factors that comprise the measure of synaptic activity for independent study. This can provide a more informed view of the actual PET data that are collected, possibly shedding light on apparent contradictions that arise from interpreting rCBF simply as cell activity.

2.4.3 Example: A Model of Saccadic Eye Movements

In order to demonstrate Synthetic PET, our example uses a neural network simulation originally designed by Dominey and Arbib (1992) for exploring how the cortex and basal ganglia interact with the superior colliculus and brainstem to produce a variety of saccadic movements. For the generation of a Synthetic PET we will concentrate on two kinds of saccades: *simple saccades*, in which the monkey fixates on a spot of light (the fixation point) that disappears upon the appearance of another spot of light (the target point) in another location, and *memory saccades*, in which the target point is briefly illuminated during the fixation period and the monkey saccades to the target's location upon removal of the fixation point. (See, for example, Hikosaka and Wurtz, 1983a,b, for details of simple and memory saccade studies in nonhuman primate.) We shall first briefly review the model (further details are provided in Chapter 2.1) before presenting the Synthetic PET comparing simple vs. memory saccades.

Model Description

Saccades are produced by the model via the interaction of neural networks representing the superior colliculus (SC), posterior parietal cortex (PP), frontal eye fields (FEF), the caudate (CD) and substantia nigra pars reticulata (SNr) of the basal ganglia, mediodorsal thalamus (MD), lateral geniculate nucleus (LGN), and the saccade generator of the brainstem (Fig. 1). These regions are further broken down into neuronal populations. Their interactions, both within a region and between regions, are responsible for generating succadic

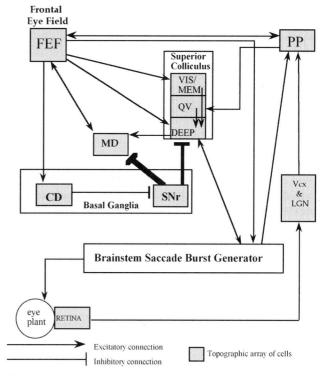

Figure 1 An overview of the Dominey and Arbib (1992) model of saccade generation. CD, caudate; DEEP, deep layer of superior colliculus; FEF, frontal eye field; LGN, lateral geniculate nucleus; MD, mediodorsal thalamus; MEM, superior colliculus memory cells; PP, posterior parietal cortex; QV, quasi-visual cells; SNr, substantia nigra pars reticulata; VIS, superior colliculus visual cells; Vcx, visual cortex.

behavior. The saccade generator performs a spatiotemporal transformation from the SC and FEF motor error maps, thus generating the eye movements needed to reposition the eye and hold it at a new vertical and horizontal gaze angle. For further details of the neurophysiology and the model constraints, please see Dominey and Arbib (1992).

The flow of activity through the model is as follows: The retina projects to PP cells via the LGN and the visual cortex and also provides input to the superficial superior colliculus (SCsup), a region that generates reflexive saccades to visual targets not recently fixated. PP codes future eye movements and has reciprocal connections with FEF and projections to FOn[AU4], cells which respond to visual stimulation of the fovea. Either FEF or SC activity is sufficient for commanding the execution of a saccade. FEF also projects to SC via the basal ganglia. The basal ganglia disinhibit SC and the thalamus for the purpose of saccades and spatial memory, respectively. CD receives projections from FEF which can trigger the release of SNr's inhibition of SC. The caudate and SNr provide an indirect link between FEF and SC, allowing FEF to selectively modulate the tonic inhibition of SNr on SC and thalamus via CD. The caudate contains a large number of neurons which are responsive to visual saccade targets and to remembered targets; the substantia nigra contributes to the dual inhibition of SC and the thalamus to allow cortex to selectively manage the inhibitory mask on SC via the FEF to CD pathway, thus preventing SC from evoking unintended saccades. Finally, SC applies a winner-take-all strategy to its various arrays, generating signals which convey the saccade metrics to the brainstem saccade generator.

Generating Synthetic PET

The Synthetic PET viewer is a Java JDK 1.2 applet designed to run within your current Internet browser. There are two main ways to generate Synthetic PET: (1) through the results generated by a NSL model within BMW, and (2) manually. In the first method, a NSL model, when executed, will automatically generate the synaptic data needed by Synthetic PET. Users planning to use the data at a later time must choose within BMW the option to save the data output within a database; this will store not only the time series data generated by the model but also the overall synaptic activity during the course of the trial (or trials). That is, during an experiment, the negative input, positive input, and overall input into a given region are saved. By saving these data separately, it allows us to view how negative and/or positive input determines the overall activation of a given region. Although we currently have manually coded the saccade model to generate the overall synaptic activity, there are plans for NSL itself to automatically calculate and save the synaptic activity for future use with Synthetic PET.

In order to use the synaptic data to paint atlas slices, the user need only select the "Generate Synthetic PET" option within BMW. The user will be queried as to which model to use. This will bring up an applet displaying the slices with the appropriate blobs of color, representing synaptic activation, within it. Users wishing to manually color the slices may do so using the controls along the top of the Synthetic PET viewer. To add a previously undefined ellipsoid to the display, the user presses the button titled "Add Blob." This button brings up a dialog box for entering in the required information. The user provides the region, the center coordinates of the ellipsoid, the ellipsoid's dimensions, an activation strength between 0.0 and 1.0 (correlated to the region with the maximum change in activation), and a color for display purposes. Pressing the submit button processes the entry, and the appropriate location of the atlas is colored.

We now consider the results generated by the saccade model. Using Eq. (3), the Synthetic PET viewer calculates the comparative data and determines the color and size of an ellipsoid. The size and center of the ellipsoid are based upon neurological data. The ellipsoid color is generated from Eq. (2); a negative change in activation correlates with a green ellipsoid, while a positive change in activation correlates with a red ellipsoid. This ellipsoid is then centered on coordinates based upon the homologous region in the human, A_h's location in the Talairach Atlas (Fig. 2; see color insert). The Synthetic PET viewer assumes the user wishes to display all regions represented within the model and for all connections (both positive and negative). In our example, areas showing activation increases are in red; areas decreasing their activation are green. The values computed from Eq. (3) are FEF: −23.6%; CD: 142.8%; MD: −10.5%; PP: 257.9%; SC: −22.0%; SNr: 208.2%; Vcx: 2.8%. The full color range of the blobs corresponds to ±257%. Because PP has the largest change in activation, we correlate this range with the highest intensity of color. That is, if the color intensity were to range from 0 to 1, PP would have an activation strength of 1.0, whereas FEF, which has a lower change in activation, would have an activation strength of 0.09. By pressing a mouse button over a blob, the user can view the Talairach coordinates, the name of the region, and its activation strength (Fig. 3; see color insert). Pressing the mouse button over any area within a slice will always return the Talairach coordinates of the mouse pointer. If the user wishes to view a slice more closely, double clicking the mouse over the slice will add a new display with a larger view of the chosen slice. This is particularly helpful when examining a slice that contains several blobs of activity. The user may have multiple frames open at any given time.

The user may choose to manipulate the data being displayed in various ways. Because the current method of highlighting a region is an ellipsoid which may overlap into other regions, the user may hide or show individual ellipsoids. To do so, the user must first ensure that the

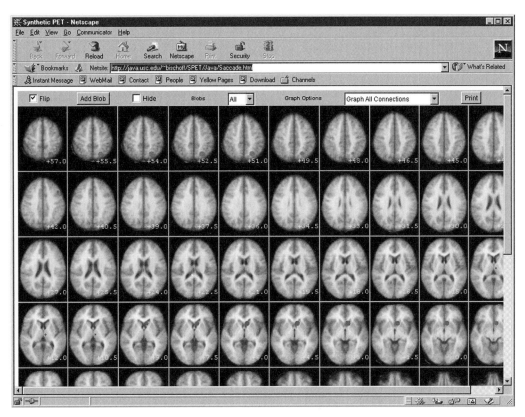

Figure 2 Synthetic PET comparison generated for the saccadic movement model. The top portion of the frame provides a checkbox for flipping the atlas orientation, adding new blobs, viewing or hiding specific blobs, and for choosing graph options. In the saccade example, we have chosen to view all the blobs and activation changes related to both positive and negative synaptic input. Red activation: overall increase in activity. Green activation: overall decrease in activity. (see color plates.)

checkbox on the toolbar marked "Hide" has been checked. To the right of the checkbox is a pull-down menu of all the blobs that exist in the dataset. These include both the blobs of activation input from the database and those the user has hard coded using the "Add Blob" dialog. The user chooses the blob to hide, and upon releasing the mouse button, the blob is hidden from view. The user can also examine the activation strengths with respect to connectivity. A menu allows the user to choose to view all connections, positive connections, and negative connections. By choosing to view only positive connections, for example, the user will see only the regions whose synaptic activation receives some (or only) excitation; regions with no positive connections (such as SNr) will not be highlighted. Similarly, choosing to view only regions receiving inhibition will result in highlighting regions receiving some (or only) inhibition. It is with this option that SNr's activation will be displayed.

2.4.4 Synthetic PET for Grasp Control

Chapter 2.1 introduced the FARS (Fagg-Arbib-Rizzolatti-Sakata) model of parietal-premotor interactions in grasping. The model focuses on the roles of several intra-parietal areas (AIP, anterior; PIP, posterior; and VIP, ventral intra-parietal), inferior premotor cortex (F4 and F5), presupplementary motor area (F6), frontal cortex (area 46), dorsal premotor cortex (F2), inferotemporal cortex (IT), the secondary somatosensory cortex (SII), and the basal ganglia (BG). However, in this chapter we focus on the contributions of AIP, F5, F6, F2, and the BG.

Model Description

Here we go beyond the description provided in Chapter 2.1 to note that the *current context* used by F5 to select among available grasps may include task requirements, position of the object in space, and even obstacles. When the precise task is known ahead of time, it is assumed that a higher level planning region predisposes the selection of the correct grasp. In the FARS model, it is area F6 that performs this function. However, we can also imagine a task in which the grasp is not known prior to presentation of the object and is only determined by an arbitrary instruction stimulus made available during the course of the trial (e.g., an LED whose color indicates one of two grasps). The dorsal premotor cortex (F2) is thought to be responsible for the association of arbitrary instruction stimulus (IS) with the preparation of motor

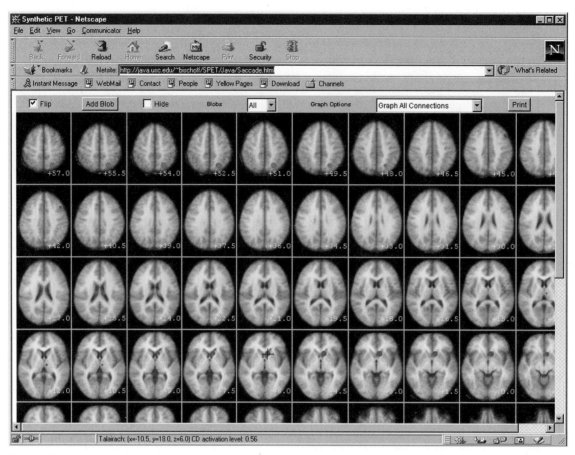

Figure 3 An example showing the crosshair (on slice Z = +6.0) aligned over an activation blob generated by the saccade model. The message bar on the browser provides information regarding the Talairach location of the crosshair. It also indicates that it is aligned over a blob, in this case the caudate (CD), with a relative activation level of 0.56 out of a possible 1.0 score. (See color plates).

programs (Evarts et al., 1984; Kurata and Wise, 1988; Wise and Mauritz, 1985). In a *conditional task* in which a monkey must respond to the display of a pattern with a particular movement of a joystick, some neurons in F2 respond to the sensory-specific qualities of the input, but others specifically encode which task is to be executed on the basis of the instruction—they thus form *set cells* which encode the motor specification until the go signal is received (Fagg and Arbib, 1992; Mitz et al., 1991). We therefore implicate F2 as a key player in this grasp association task. What is particularly interesting about this type of conditional task is that alone neither the view of the object (with its multiple affordances), nor the IS is enough to specify the grasp in its entirety. The visual input specifies the details of all the possible grasps; the IS specifies only the grasp mode—and not the specific parameters of the grasp (such as the aperture). F5 must combine these sources of information in order to determine the unique grasp that will be executed.

Fig. 4a presents an outline of the neural regions involved in the FARS model of grasp production. The precision pinch and power grasp pools of AIP receive inputs from both the dorsal and ventral visual pathways (pathways not shown; more details may be found in Fagg and Arbib, 1998). The pools in F5 and AIP are connected through recurrent excitatory connections. Affordances represented by populations of units in AIP excite corresponding grasp cells in F5; active F5 units, representing a selected grasp, in turn support the AIP units that extract the motorically relevant features of the objects. In monkey, the number of neurons in F5 involved in the execution of the precision pinch is greater than the number observed for any other grasp (Rizzolatti et al., 1988). The model reflects this distribution in the sizes of the precision and power pools in *both* F5 and AIP.

When the model is presented with a conditional task in which the choice of grasp depends on which IS was recently presented, how is a unique grasp selected for execution? According to the FARS model, AIP first extracts the set of affordances that are relevant for the presented object (say, a cylinder). These affordances, which also encode the diameter of the cylinder, activate the corresponding *motor set* cells in F5; however, because there are multiple affordances, several competing subpopulations of F5 set cells achieve a moderate level of activation. This competition is resolved only when the IS is presented. This instruction signal, mapped to a grasp mode preference by the basal ganglia (connections

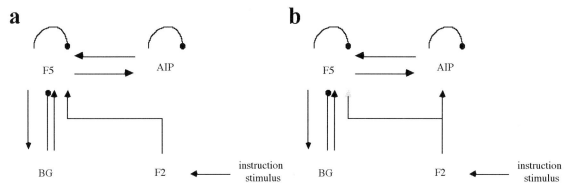

Figure 4 (a) A schematic view of the FARS model's architecture. Arrows indicate excitatory connections between regions; filled circles indicate inhibitory connections. The precision pinch and power grasp pools in F5 and AIP are connected through recurrent excitatory connections. The precision pinch pool contains more neurons than other grasps, which affects the Synthetic PET measure in these and downstream regions. F6 represents the high-level execution of the sequence; phase transitions dictated by the sequence are managed by BG. F2 biases the selection of grasp to execute as a function of the presented instruction stimulus. (b) An updated functional model, influenced by comparison of Synthetic PET data with actual PET data. In the revised model, the information from F2 flows (primarily) into the circuit through a projection into AIP. AIP: anterior intra-parietal area; BG: basal ganglia; F2: dorsal premotor cortex; F5: inferior premotor cortex; F6: presupplementary motor area.

not shown in the figure), is hypothesized to arrive at F5 via F2. The signal increases the activation level of those F5 cells that correspond to the selected grasp, allowing them to win the competition over the other subpopulations.

Learning from the Comparison of Synthetic PET and Human PET Results

We have conducted several Synthetic PET experiments with the FARS model to predict what we might expect when PET studies are performed in the human. We describe one of them in which we examine the effects of knowing which grasp to use prior to the onset of recording (non-conditional task), as compared with only being told which grasp to use after a delay period (conditional task). In the latter task, an instruction stimulus in the form of a bi-colored LED informs the subject which grasp should be used. The most significant change in PET activity predicted by the model is the level of activity exhibited by area F2 (dorsal premotor cortex). Its high level of activity in the conditional task is due to the fact that this region is only involved when the model must map an arbitrary stimulus to a motor program. In the non-conditional task, the region does not receive IS inputs, thus its synaptic activity is dominated by the general background activity in the region. The additional IS inputs in the conditional task have a second-order effect on the network, as reflected in small changes in synaptic activity in F5, BG, and AIP. There is increased synaptic activity in F5 due to the additional positive inputs from F2. These inputs also cause an increase in the region's *activity level*, which is passed on through excitatory connections to both AIP and BG (recall Fig. 4a).

The Synthetic PET experiments were tested in human PET (see Grafton *et al.*, 1998, for details of the protocol and conditional task results). Subjects were asked to repeatedly perform grasping movements over the 90-second scanning period. The targets of grasping were mounted on the experimental apparatus shown in Fig. 5. Each of three stations mounted on the apparatus consisted of both a rectangular block that could be grasped using a power grasp and a pair of plates (mounted in a groove on the side of the block; see inset of Fig. 5), which could be grasped using a precision pinch (thumb and index finger). A force-sensitive resistive (FSR) material, mounted on the front and back of the block, detected when a solid power grasp had been established. The two plates were attached to a pair of mechanical microswitches, which detected when a successful precision pinch had been executed. For each station, the block and plates were mounted such that the subject could grasp either one without requiring a change in wrist orientation. A bi-colored LED at each station was used to instruct the subject as to the next target of movement. A successful grasp of this next target was indicated to the subject by a change in the color of the LED. The subject then held the grasp position until the next target was given. Targets were presented every $3 +/- 0.1$ seconds.

Four different scanning conditions were repeated three times each. In the first, subjects repeatedly performed a power grasp to the indicated block. The target block was identified by the turning on of the associated LED (green in color). When the subject grasped the block, the color of the LED changed from green to red. For the second condition, a precision pinch was used. The target was identified in the same manner as the first condition. In the third grasping condition (conditional task), the initial color of the LED instructed the subject

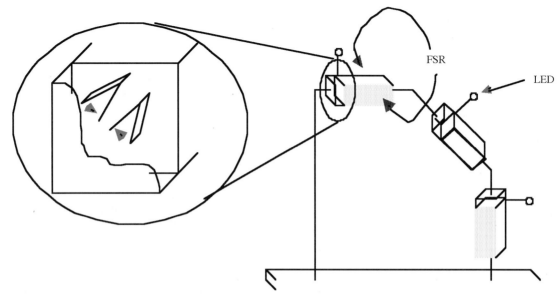

Figure 5 Apparatus used in PET experiment. Each of three stations can be grasped in two ways: precision pinch of the two plates in the grove (inset) or power grasp of the block. The side of the blocks are covered with a force-sensitive resistive (FSR) material; light-emitting diodes (LEDs), depending upon the task, indicate both the goal and type of grasp.

to use either a precision pinch (green) or a power grasp (red). When contact was established, the LED changed to the opposite color. In the fourth (control) condition, the subjects were instructed to simply fixate on the currently lit LED and not make movements of the arm or hand (prior to the scan, the arm was placed in a relaxed position). The lit LED changed from one position to another at the same rate and variability as in the grasping tasks. Prior to scanning, subjects were allowed to practice the tasks for several minutes.

Conditional vs. Non-Conditional Task

The model predicts that the conditional task should yield much higher activation in F2, some activation of F5, and a slight activation of AIP. The human experiment confirmed the F2 result but failed to confirm the predictions for F5. Furthermore, in human we see an activation of the inferior parietal cortex, along the intra-parietal sulcus, which is perhaps an AIP homologue.

Can we make use of the negative F5 result to further refine the model? Consider the functional connectivity of these regions in the model (Fig. 4a). In the model, the strength of the projection from F2 to F5 is essentially a free parameter. In other words, there is a wide range of values over which the model will correctly perform the conditional and non-conditional tasks. The implication is that, by tuning this parameter, we can control this projection's contribution to the synaptic activity measure in F5. However, the difference in AIP synaptic activity from the non-conditional to the conditional task will always be less than the difference observed in F5. Why is this the case? By increasing the projection strength from F2 to F5, we observe an increase in both F5 synaptic *and* cell activity. The increase in F5 cell activity, however, is attenuated by local, recurrent inhibitory connections. Thus, the excitation that is then passed on to AIP via F5 does not reflect the full magnitude of the signal received from F2.

The conclusion is that, although we can adjust the free parameter to match one or the other observations in the human experiment (of either F5 or AIP changes), the model cannot reflect both at the same time. One possibility for repairing this problem in the model is to reroute the F2 information so that it enters the grasp decision circuitry through AIP (or both AIP and F5), rather than exclusively through F5 (Fig. 4b). This would yield an increase in activity in AIP due to F2 activation with only an attenuated signal being passed on to F5, resulting in only a small increase in F5 synaptic activity. Note that we do not necessarily assume that there exists a direct cortico-cortical connection from F2 to AIP or F5, but only that there is a functional connection (which potentially involves multiple synapses).

The low-level details of the FARS grasping model (Fagg and Arbib, 1998) were derived primarily from neurophysiological results obtained in monkey. The synthetic PET approach extracts measures of regional synaptic activity as the model performs a variety of tasks. These measures are then compared to rCBF (regional cerebral blood flow) observed during human PET experiments as the subjects perform tasks similar to those simulated in the model. In some cases, the human results provide confirmation of the model behavior. In other cases, where there is a mismatch between model prediction and human results, it is possible (as we have shown) to use these negative results to further refine and constrain the model and, on this basis, design new experiments for both primate neurophysiology and human brain imaging.

2.4.5 Discussion

The Synthetic PET methodology allows users to make specific predictions in PET experiments based upon neural network models of behavior. Because models typically rely upon neurophysiology, neuroanatomy, and behavioral data, Synthetic PET provides a way of bridging the gap between different temporal resolutions; models can be compared with actual PET studies of the same kind. This allows the modeler to further validate his research and can provide suggestions for experimentalists for future research regarding a particular phenomenon. It also provides a way to separate the contributions of positive and negative connections, a powerful method towards the understanding of the contribution of individual regions to another area's overall activation level.

Our current implementation of Synthetic PET focuses upon coloring the Talairach Atlas with elliptical blobs. In the future, we plan to add other atlases, such as rat or monkey, to its functionality. PET is increasingly used in research with nonhuman primates; the Synthetic PET method would then provide another method of analysis in comparing models based upon monkey data to the data gathered with conventional PET. Furthermore, we plan to add the capability of generating a Synthetic PET image based directly upon single-unit recordings. While the activation generated would be limited to the regions which were recorded, it will still provide a way of examining how neurons with different temporal characteristics can contribute to the activation of a region under a particular paradigm. As described in Chapters 3.1 and 3.2, the NeuroCore database stores neurophysiological data and includes features for manipulation of these data. Synthetic PET would be able to connect with the data and would contain the routines necessary for calculating blobs of activation.

We plan to improve our algorithm in the future for the modeling of fMRI data as well. This would provide yet another link to the neurophysiological data by presenting us with a visual time course of activation changes. The current Synthetic PET technique is based upon the hypothesis that PET is correlated to the integrated synaptic activity within a region. Both to improve our Synthetic PET algorithms and to develop Synthetic fMRI, we will have to investigate possible correlates with neural firing and synaptic plasticity, as well as synaptic activity. We must also note that we will need to extend our basic neural models to better understand the link between blood flow (hemodynamics), neural metabolism, and the information processing role of these neurons. Further, we plan to improve our representation of activation from that of an ellipsoid to one which more closely corresponds to the shape of the region under study. This will improve the conditions of comparing Synthetic PET/fMRI to conventional PET/fMRI. By representing the region's shape itself within the database, it becomes possible to be more specific as to the location of the activation (e.g., ventrolateral pallidum as opposed to the entire pallidum will be highlighted). However, it is worth stressing that as a region becomes more active the increase of the active area we see in its image may not reflect an increase of the population of neurons engaged in the activity so much as the increased rCBF in regions through which blood vessels pass en route to the activated region. We also need to add a stochastic analysis to account for the variation in PET activity seen in the same subject on different trials.

The current implementation of Synthetic PET is stand-alone in that it does not interface with the NSL environment (Chapter 2.2). Future development plans include providing crosstalk between the two programs. For example, a user running an experiment in NSL should be able to link the model's results to a Synthetic PET image upon completion of the experiment. The information required by Synthetic PET should be part of a NSL program and its results; our current implementation involves manually calculating and entering the required information into the database.

Finally, Synthetic PET only displays brain slices and a blob of activity related to a given region's activation. In Arbib *et al.* (1995) we used a line graph to illustrate activation differences between two tasks. Within Synthetic PET, a click of a button would display this graph in a separate window so that the current task may be compared to another task's results (such as simple vs. memory saccades, or internally vs. externally guided motor sequencing). We could thus view the material in three forms (returning to the saccade model for these particular figures): integration over inhibitory connections (Fig. 6a; see color insert), over excitatory connections (Fig. 6b; see color insert), and over all connections (Fig. 6c; see color insert). This kind of presentation allows us to refer back to the simulation to infer reasons for the results which may seem contradictory when the PET values are related to total cell firing in a region (reflecting the *difference* between excitatory and inhibitory input) as opposed to the *summed* contributions of the absolute values of the excitatory and inhibitory inputs. For example, the negative activity in the thalamus decreases when comparing simple saccades to memory saccades, while the positive activity strongly increases; overall, however, there is a decrease in synaptic activity in MD thalamus when both positive and negative connections are considered. In order to generate a graph illustrating the connections, the user merely presses a button labeled "Graph Activity." This will

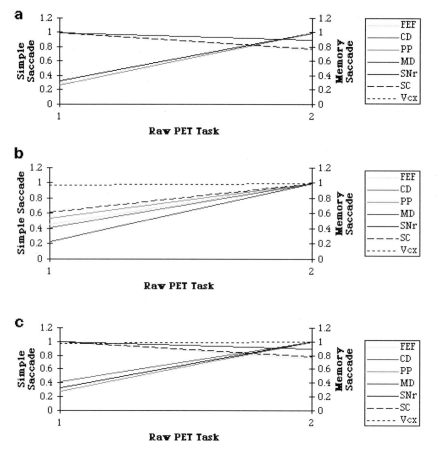

Figure 6 The raw activity rPET$_a$ associated with each cell group A in the neural network model for saccade generation obtained by integrating: **(a)** over inhibitory connections, **(b)** over excitatory connections, and **(c)** over all connections. Left column: rPET$_a$ values for the simple saccade task. Right column: values for the memory saccade task. Note that in B, CD and Vcx coincide, as do SC and FEF. Similarly, FEF and SC coincide in C. CD: caudate; FEF: frontal eye fields; PP: posterior parietal cortex; SC: superior colliculus; SNr: substantia nigra pars reticulata; MD: mediodorsal thalamus; Vcx: visual cortex. (see color plates.)

display a Matlab graph similar to that seen in Fig. 6 (see color insert).

As a fully developed tool, Synthetic PET will allow experimenters to integrate data collected via various methods. With the Neuroinformatics Workbench supplying techniques for storing both modeled, behavioral, and neurophysiological data, Synthetic PET allows us to link the various aspects of our database together. As we have seen with the saccade example, results can be counterintuitive, as the predictions are based upon both excitatory and inhibitory synaptic activity. In the future, an imaging database will provide yet another method of comparing modeled and physiological data with that collected by PET or fMRI methodologies. This would make it possible to determine if PET activation seen is largely based upon excitatory or inhibitory synaptic activity. Finally, Synthetic PET is a method that allows interaction between researchers within the neurosciences who share interests in a particular behavior and strive for a more rigorous understanding of the hypothesized homologies across species.

References

Ackermann, R. F., Finch, D. M., Babb, T. L. and Engel, J. J. (1984). Increased glucose metabolism during long-duration recurrent inhibition of hippocampal pyramidal cells. *J. Neurosci.* **4**, 251–264.

Arbib, M. A., Bischoff, A., Fagg, A. H. and Grafton, S. T. (1995). Synthetic PET: analyzing large-scale properties of neural networks. *Human Brain Mapping* **2**, 225–233.

Brownell, G. L., Budinger, T. F., Lauterbur, P. C. and McGeer, P. L. (1982). Positron tomography and nuclear magnetic resonance imaging. *Science* **215**, 619–626.

Dominey, P. F., and Arbib, M. A. (1992). A cortico-subcortical model for generation of spatially accurate sequential saccades. *Cerebral Cortex* **2**, 153–175.

Evarts, E. V., Shinoda, Y., and Wise, S. P. (1984). *Neurophysiological Approaches to Higher Brain Function*. Wiley-Interscience Press, New York.

Fagg, A. H., and Arbib, M. A. (1998). Modeling parietal-premotor interactions in primate control of grasping. *Neural Networks* **11**, 1277–1303.

Fagg, A. H., and Arbib, M. A. (1992). A model of primate visual-motor conditional learning. *J. Adaptive Behav.* **1**, 3–37.

Grafton, S. T., Fagg, A. H. and Arbib, M. A. (1998). Dorsal premotor cortex and conditional movement selection: a PET functional mapping study. *J. Neurophysiol.* **79**, 1092–1097.

Herscovitch, P., Markham, J. and Raichle, M. (1983). Brain blood flow measured with intravenous $H_2^{15}O$. I. Theory and error analysis. *J. Nuclear Med.* **24**, 782–789.

Hikosaka, O., and Wurtz, R. H. (1983a). Visual and oculomotor functions of monkey substantia nigra pars reticulata. I. Relation of visual and auditory responses to saccades. *J. Neurophysiol.* **49**, 1230–1253.

Hikosaka, O., and Wurtz, R. H. (1983b). Visual and oculomotor functions of monkey substantia nigra pars reticulata. III. Memory continent visual and saccade responses. *J. Neurophysiol.* **49**, 1268–1284.

Kurata, K., and Wise, S. P. (1988). Premotor cortex of rhesus monkeys: set-related activity during two conditional motor tasks. *Exp. Brain Res.* **69**, 327–343.

Mitz, A. R., Godshalk, M. and Wise, S. P. (1991). Learning-dependent neuronal activity in the premotor cortex. *J. Neurosci.* **11**, 1855–1872.

Raichle, M. E., Martin, W. R. W. and Herscovitch, P. (1983). Brain blood flow measured with intravenous $H_2^{15}O$. II. Implementation and validation. *J. Nucl. Med.* **2**, 790–798.

Rizzolatti, G., Camarda, R., Fogassi, L., Gentilucci, M., Luppino, G. and Matelli, M. (1988). Functional organization of inferior area 6 in the macaque monkey. II. Area F5 and the control of distal movements. *Exp. Brain Res.* **71**, 491–507.

Talairach, J., and Tournoux, P. (1988). *Co-planar Stereotaxic Atlas of the Brain*. Thieme Medical Publishers, New York.

Wise, S. P., and Mauritz, K. H. (1985). Set-related neuronal activity in the premotor cortex of rhesus monkeys: effects of changes in motor set. *Proc. R. Soc. London.* **223**, 331–354.

PART 3

Databases for Neuroscience Time Series

CHAPTER 3.1

Repositories for the Storage of Experimental Neuroscience Data

Richard F. Thompson,[1] Jeffrey S. Grethe,[2] Theodore W. Berger,[3] Xiaping Xie[3]

[1]Biological Sciences/LAS, University of Southern California, Los Angeles, California
[2]Center for Cognitive Neuroscience, Dartmouth College, Hanover, New Hampshire
[3]Biomedical Engineering Department, University of Southern California, Los Angeles, California

3.1.1 Introduction

The advancement of science depends heavily upon the ability of researchers to collect, store, analyze, and share their data efficiently. In order to help achieve these goals, informatics research has been playing an increasingly important role in various scientific communities. For example, molecular biologists have benefited greatly from the development of large-scale genome databases and new advanced genetic sequence search tools (Chen and Markowitz, 1994a,b; Frenkel, 1991). These new tools have allowed researchers, in this scientific community as well as other communities, to efficiently store, analyze, and share their sequence data. In the last decade attention has also focused on the development of informatics tools for the neuroscience community.

One of the problems encountered by many neuroscientists is the coherent storage, organization, and retrieval of extremely diverse and large datasets. Current technology allows neuroscientists to collect massive amounts of data. In order to help the neuroscientist manage these complex datasets, we have developed a novel database schema and implementation that allow datasets from diverse labs to be stored coherently in a single framework. The formal structure of that framework, called NeuroCore, will be detailed in Chapter 3.2, while various applications devised to help users interact with NeuroCore databases will be described in Chapter 3.3. In this chapter, we present the key ideas behind the development of NeuroCore and illustrate its application in the design of two databases, one for "Cerebellum and Classical Conditioning" (*in vivo* studies, Fig.1), and the other for "Long-Term Potentiation (LTP) in the Hippocampal Slice" (*in vitro* studies, Fig.2).

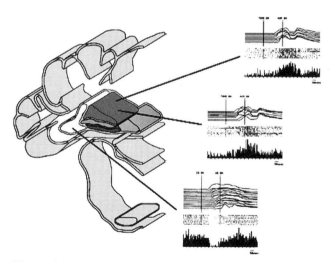

Figure 1 Unit recordings from a behavioral experiment: classical conditioning of the eye-blink response in the rabbit. In this experiment recordings are made from cerebellar cortex and nuclei and other structures while the animal performs certain behaviors as various external stimuli are presented. The data consist of unit recordings from various brain regions as well as recordings of the subject's behavioral response.

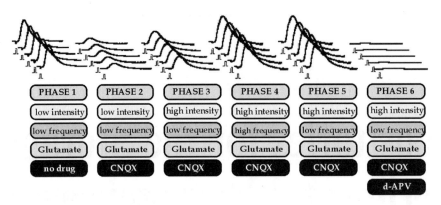

Figure 2 Recordings from a hippocampal slice. In this experiment various chemical manipulations and various types of electrical stimulation are applied to a brain slice at different instances. The data consist of field potential recordings from various regions of the hippocampus.

During experiments on learning (note that the current considerations are not restricted to learning studies), neural activity can be monitored, recorded, and then later analyzed to examine the contribution of varying brain structures to performance or learning in the intact animal or some more abstract markers of neural activity in a slice preparation. For each research subject used in an experiment (whether intact animal or neural slice), data can be collected from dozens of single neurons or even synapses. The data collected from each of these neurons or synapses typically consists of data from multiple individual training trials.

Over the years, hundreds of thousands of cells have been recorded from across a variety of paradigms, subjects, and conditions. Some interesting issues that point to the need for an experimental database are as follows:

1. How does one efficiently store data from neurons which have been recorded from multiple research subjects under a variety of conditions?
2. How can one analyze datasets of neurons that have been collected across subjects and even from different experiments?
3. How can researchers re-analyze large sets of data quickly and efficiently?
4. How can future researchers effectively use all the data from a variety of experiments even if they use different paradigms from current studies?
5. How can a group of researchers effectively collaborate on a research project using databases and Internet technology?

The goal in building a neuroscience database is to aid the researcher in the storage, retrieval, and analysis of electrophysiological and behavioral data, as well as anatomical and neurochemical data, from the critical brain regions involved in some behavior, memory phenomenon, or other function of interest. This database should also be a tool that is used by the researcher at all stages of the scientific process, from the development of a new experiment through the collection and analysis of the data generated by this experiment.

3.1.2 Protocols: A Data Model To Address Schema Complexity in Neuroscience

Probably the most important consideration in the design of any neuroscientific database is the complexity and richness of the data that need to be stored. Neuroscientists collect data comprised of many types, from the storage of simple tabular data in both numeric and textual form to more complex data types such as images, videos, and time-series data. One of the problems with many of these types of data is that there is no standard internal structure for representing them with most common commercial database systems. For example, the standard neurophysiological recording (time-series data) that consists of hundreds to thousands of individual measurements taken at regular or irregular time intervals cannot be easily or efficiently stored in standard relational database systems. However, a more daunting problem is how to represent and store the immense variety of experimental preparations that generate these types of data.

For example, we might want to store data in our database that are collected from unit recordings during a behavioral experiment as well as field potential recordings done in the hippocampal slice (Figs. 1 and 2). These data are actually quite similar in that they consist mainly of time-series recordings, even though the recordings will quite likely have been made by different recording electrodes which have their own unique properties. However, the methods used to generate the data in these two paradigms are quite different.

Our solution to coherently storing these datasets is to "associate" the experimental data with the information regarding the experimental protocol that was used in collecting the data. This protocol combines information on the preparation used, the types of experimental manipulations, and stimuli applied, as well as the kinds of observations and measurements made during the experiments. More specifically, the protocol provides a generic, parametric description of a class of experiments, while a specific experiment must be described by "filling in the parameters" to show, for example, how the preparation was stimulated and what observations were recorded as a result.

In essence, data without the definition of the experimental protocol used to generate them are useless. One of the problems in specifying and storing protocols and data is their heterogeneity. Each researcher and/or laboratory works with protocols that might be specific to their research focus. The data they collect and analyze will also contain attributes that are important to their specific research goals.

This is quite clear by examining the very simple case outlined above (Figs. 1 and 2), where one wishes to store both *in vivo* and *in vitro* time-series recordings. Both of these situations require that time-series data be stored; however, each of these data records must contain additional information. For single-unit recordings that are made in a behaving animal, one must record the precise three3-dimensional coordinate of the location of the recording electrode so that one can later reconstruct (through the use of marking lesions and general histology) the exact anatomical region in which the electrode was located when the data were being recorded. When recording from a hippocampal slice, one does not have a similar coordinate system, but one is able to directly note exactly in what region and even in what cell layer the recordings are being made. Therefore, in designing the database architecture, a very important principle has to be taken into account: One cannot hope to describe *a priori* all the experimental protocols and research data that neuroscientists will want to incorporate in their own databases. It would be foolish to believe that one could. Due to this constraint, any database designed for the neuroscience community needs to be easily modified to meet a specific researcher's needs without requiring major modifications in the database's overall structure.

An Extendible Database

As we shall show in detail in Chapter 3.2, the key idea of NeuroCore is to provide a core database schema (i.e., a set of tables to be used in the database) comprising tables whose entries will be needed in almost all neuroscience experiments, together with machinery that makes it easy to add extensions to the core that specify data entries specific to the needs of a particular laboratory or protocol (as the latter may be shared by several laboratories). The key notion in being able to relate extensions to a core database structure is that of inheritance: Object-relational databases (see Chapter 1.2) allow types and tables to be placed in inheritance hierarchies. For example, in Fig. 3, a hierarchy of people is pictured, where PERSON is the top-level "parent" table containing the two columns *Name* and *SSN* (Social Security Number). In inheritance hierarchies, the children of a parent table inherit all the columns, primary key

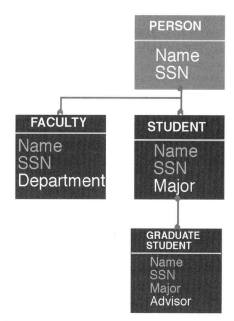

Figure 3 Example of an inheritance hierarchy. In this example, the PERSON table is a parent table. Both FACULTY and STUDENT are children of the PERSON table and inherit all the columns and constraints from the PERSON table. The STUDENT and FACULTY tables also define extra columns specific to themselves.

relations, and rules of the parent. In this case, both FACULTY and STUDENT inherit the *Name* and *SSN* columns from PERSON. However, both tables define new columns that are to contain data specific to their table (e.g., STUDENT defines the column *Major*). When selecting data from a parent table, the select function will also return data from the children. All information concerning the relations that exist in the inheritance hierarchy is stored in the system tables of the database. This allows easy reconstruction of the inheritance hierarchies through the system metadata, thereby allowing users to easily view the extensions added to the core database.

3.1.3 Considerations of the User Community

One of the major considerations that needs to be taken into account when designing any database system is the user community that will be interacting with the system. This is especially true of the neuroscience community where researchers from diverse fields and backgrounds (e.g., biology, psychology, engineering, and computer science) need to store, analyze, and share their data. By examining the neuroscience user community, the scope of a database for neuroscience experiments falls into one of three categories:

1. A database for a single experimenter or laboratory
2. A database for a collaboratory or research community
3. A publicly accessible database

Each of these database's users has different needs. The single experimenter needs to be able to securely store his work in progress as well as have sophisticated analysis and statistical tools to examine his data. A group of researchers collaborating on a project needs these tools as well as methods for the group to securely share the data being collected and analyzed at remote sites. Once the data collection and analysis is complete the data is then published and made publicly accessible to other researchers. The researchers who access this public data can be categorized as follows:

1. A researcher working on the same or a similar problem
2. A researcher interested in the work
3. A user who might just be browsing

For the researcher working in the same field, tools need to be available to allow researchers to examine and re-analyze the data using their own statistical methods and techniques. For both researchers, complete reference to the protocols and methods used to collect and analyze the data must be available. For all users, interfaces must be available that seamlessly allow users to gain access to the data stored in these databases from the familiar environment of an on-line journal article (for more discussion of this topic please see Chapter 5.3). In order to develop a database for the neuroscientific community all aspects of this varied community need to be addressed. The development of NeuroCore began with the task to build a database for the storage of *in vivo* neurophysiological recordings (Grethe *et al.*, 1996; Arbib *et al.*, 1996).

3.1.4 Building a Time-Series Database for *In Vivo* Neurophysiology: Cerebellum and Classical Conditioning

For many years, psychologists and neurobiologists have been searching for the substrates underlying learning and memory. One paradigm that has been extremely useful in examining these substrates is that of classical conditioning. From his research on this form of learning, Thompson (1986, 1990; Thompson and Krupa, 1994; Thompson *et al.*, 1997) has proposed an essential brain circuit, based on the cerebellum, that is responsible for this form of associative learning. This circuit now allows one to examine in detail the processes involved during classical conditioning of the nictitating membrane response. We here describe the issues involved in constructing a NeuroCore database to meet the needs of the Thompson laboratory.

Current evidence argues very strongly that the essential memory trace for the classically conditioned response is formed and stored in the cerebellum (Clark and Lavond, 1993; Clark *et al.*, 1992; Krupa *et al.*, 1993). There are actually two locations in the cerebellum where these memory traces appear to be formed: the anterior interpositus nucleus and lobule HVI of the cerebellar cortex itself. In examining the role that these two structures play in the acquisition and performance of the conditioned response, researchers rely heavily on neurophysiological data from neurons within these two regions. This section will use an experiment performed by Joanne Tracy, as part of her Ph.D. dissertation (Tracy, 1995), as a case study to examine how the NeuroCore database can be extended for a specific experiment. The dataset from this experiment concerns single-unit recordings from neurons in the deep cerebellar nuclei in rabbits trained using the classical conditioning paradigm.

Experimental Preparation

When storing data from any experiment, the data must be embedded in a framework that is based on information from the protocol which generated the data. During a typical classical conditioning experiment, a puff of air (the unconditioned stimulus, US) is paired with some neutral stimulus such as a light or tone (the conditioned stimulus, CS) (Fig. 4). The US by itself is able to elicit an eyeblink (the unconditioned response, UR) from the beginning. Over training the animal is able

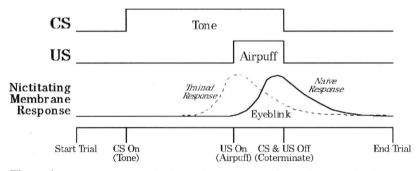

Figure 4 Depiction of a typical training trial where the subject receives two stimuli: a tone (CS) and an airpuff (US). Notice that the trained response (CR) predicts the onset of the airpuff (US), whereas the naïve response (UR) is just a reflexive response to the airpuff on the cornea.

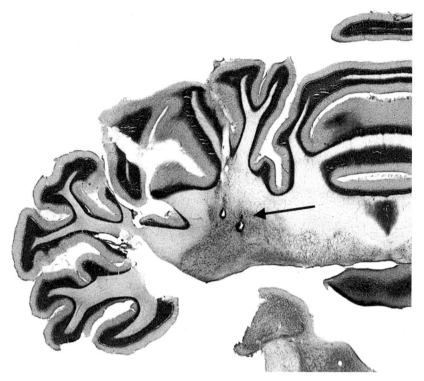

Figure 5 Single-unit electrode marking lesions in the interpositus from a stained rabbit cerebellar brain section. The two marking lesions can be seen to the left of the arrow.

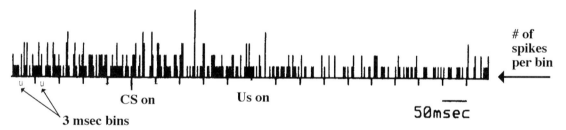

Figure 6 Sample of a single trial of single-unit data in binned format. Each bin records the number of times a neuron fired within that time period.

to associate the CS with the blinking of the eye and is able to blink in anticipation of the airpuff. This response is called the conditioned response (CR). Each training day consists of a number of these conditioning trials split into several training blocks.

When neuronal activity will be recorded, recording electrodes are lowered into the cerebellar cortex or into one of three deep cerebellar nuclei. These recordings are made under a number of conditions. Prior to removal of an electrode from a recording site, a marking lesion is made which is easily seen in a stained tissue section (Fig. 5). The actual recording sites are then determined post mortem via brain section and mapped onto a standard brain atlas diagram.

Data

A typical classical conditioning experiment generates a variety of data that must be stored. These data can consist of raw unit and behavioral data, data from post mortem histological sections, and statistical data generated from the raw data.

ELECTROPHYSIOLOGICAL AND BEHAVIORAL DATA

The electrophysiological and behavioral data collected during these experiments are typically digitized ("binned") data, including both single-unit data (record of activity from one neuron), multiple-unit activity (record of activity from many neurons near the recording electrode), and the movement of the nictitating membrane (Figs. 6 and 7).

Once the raw data have been gathered, researchers will want to analyze the data with various statistical methods. A summary over many cells and/or animals is then usually entered in a spreadsheet (Fig. 8). One of the goals, then, in developing this database is to give the researcher one location where all these data can be stored, indexed, and cross referenced; however, all these

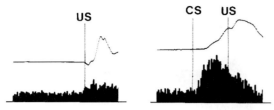

Figure 7 Sample of multiple-unit data from the interpositus. The top trace in each figure represents the behavior, whereas the bottom trace displays the unit activity from a group of neurons in the interpositus.

data must be stored in relation to from where these recordings where taken.

HISTOLOGICAL DATA

The most important histological data for an electrophysiological experiment is an accurate record of recording sites as determined from examination of electrode marking lesions (Fig. 5). Researchers also normally note the exact coordinate where a recorded cell is located. For example, Tracy (1995) stored all relevant electrode placement and cell information in a spreadsheet that was used for further analysis (Fig. 8).

Extensions to the NeuroCore Database

Each set of extensions to the base NeuroCore database described in the following sections is accompanied by a schema diagram which shows all the tables and relations involved in that particular aspect of the database. A more comprehensive discussion of NeuroCore itself can be found in Chapter 3.2. The legend for all these diagrams can be found in Fig. 9. In the figures that follow, the shaded boxes with a solid border will indicate super tables from the core schema that can be extended. The shaded boxes with the dashed border will show a variety of extension tables for the current paradigm, while other boxes will indicate various core and support tables within the NeuroCore database structure. It should be noted that a group of experimentalists may choose a set of extensions as defining an "extended core" for their investigations so that the extensions required for a specific protocol become very simple indeed. In order to develop the extensions necessary for NeuroCore to be

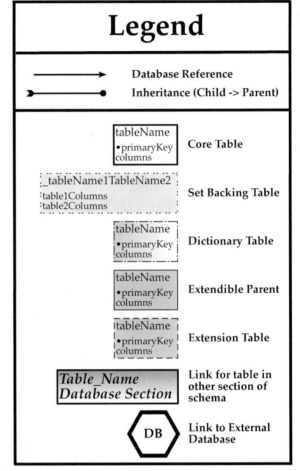

Figure 9 Legend for all schema diagrams. These diagrams represent the database schema by displaying the actual tables and their contents with the relations (primary key, foreign key, or inheritance) that exist between the tables. For a more detailed description of the table classes contained within NeuroCore see Chapter 3.2.

able to store the above described experiment, four distinct extensions need to be developed:

1. Definition of the research subject
2. Definitions of the protocol used in the experiment
3. Definitions of the research data being collected
4. Definitions of the statistical data being generated

The first step in creating the extensions to the NeuroCore database is to define the research subject that will

rabbit #	cell #	anterior	lateral	ventral	latency	duration	baseline freq.	response
91-289	1	1.0A	6.0L	19				
91-289	2	1.0A	6.0L	19.5	177	543	33.2	CR ++
91-323	3	1.0A	6.0L	19.7	69	636	30.4	CR ++
91-323	4	0.5	5	20.8	111	93	32.4	CR + UR +
91-323	5	0.5	5	20.6				UR+

Figure 8 Sample Summary Spreadsheet – Data from a few interpositus neurons. Each neuron is indentified by its ID and anatomical location. Various statistical measures are associated with each neuron.

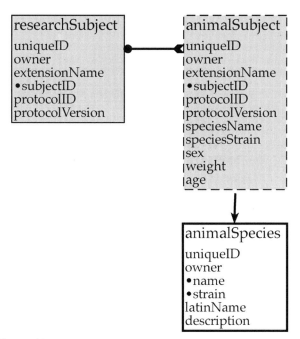

Figure 10 Research subject extensions. In order to store data concerning animal research subjects, an extension table (*animalSubject*) was added to the *researchSubject* table that allows the researcher to store the necessary information concerning the research subjects.

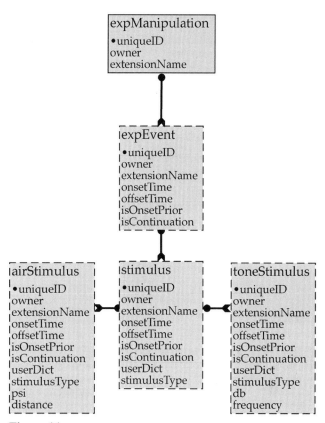

Figure 11 Experimental manipulation extensions. For the classical conditioning experiment, the tone and airpuff stimuli must be represented in the database as timed experimental events; however, each stimulus has its own parameters.

be used during the experiments. This extension (*animalSubject*) is created under the *researchSubject* table (Fig. 10). In the conditioning experiment we wish to store, the typical research subject is a New Zealand white albino rabbit. For our purposes, the only additional data necessary to be stored for each research subject are its species, sex, age, and weight.

To be able to store the experiment described above in a coherent fashion that will be easy to retrieve and analyze at a later date, the data collected must be "associated" with an experimental protocol. The protocol for the experiment contains a set of manipulations, the actual stimuli and conditions that comprise the specific experiment. These manipulations are created as children of the *expManipulation* table (Fig. 11). In defining the protocol for the standard classical conditioning paradigm, information regarding two distinct stimuli, the tone and airpuff, needs to be stored. In developing the extension for these two stimuli, it must first be noted that each of these stimuli contain explicit timing information (i.e., the onset and offset times) that is defined in the *expEvent* table. The *stimulus* table (a child of the *expEvent* table) is then able to define the general qualities of what constitutes a stimulus (e.g., what type of stimulus is it—a CS or US). The specific data for each stimulus are then stored in either the *airStimulus* or *toneStimulus* tables. This set of extensions illustrates a very nice property of inheritance: the ability to group tables into hierarchical collections where each child table further refines the information stored. This allows one to then query the database at intermediate levels of complexity. For example, one could search for all experiments that used two conditioned stimuli and one unconditioned stimulus, even though one does not know the exact nature of each stimulus.

Once we have defined the manipulations that were used to generate the data we need to describe the tables that will hold the actual research data (Fig. 12). These tables are created as children of the *researchData* table. For this experiment we need to store time-series data (defined by the *timeSeriesData* table) related to neuronal activity as well as the behavior being performed. For the unit data (defined by the *unitData* table), we need to record the actual anatomical location and coordinates where the electrode was located. However, when storing data from single cells (defined by the *singleUnitData* table), one also needs to store more information regarding the cell being recorded from (e.g., the ID of the cell and also what type of spikes were being recorded).

The last information that must be stored in the database is the statistical data generated from the analysis of the raw data discussed above. Various statistical methods are used to analyze single-unit data. The tables that define these statistics are created as children of the *expMetadata* table (Fig. 13). For the experiments to be stored in this database, two distinct classes of statistical data were generated. The first class of statistics analyzes

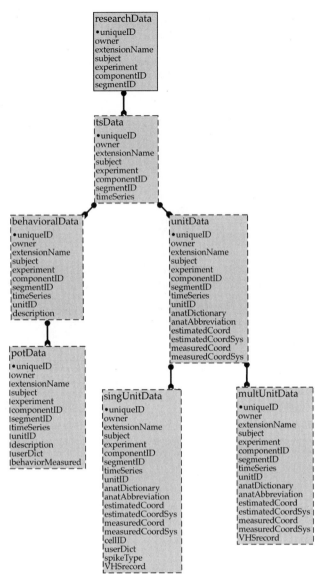

Figure 12 Research data extensions. For many neurophysiological experiments, researchers must store the neuronal unit data, single- as well as multiple-unit data, along with the behavioral data acquired from the research subject during the execution of the experimental task.

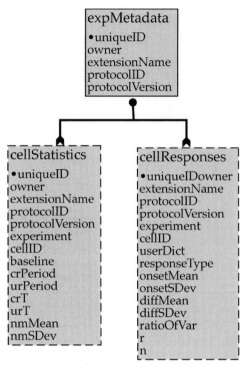

Figure 13 Metadata Extensions. For the experiment outlined above, two forms of statistical data were generated for each cell recorded. Cell statistics describe the general responsive properties of the cell to the task, whereas, cell responses characterizes the relation of the cell's responses to the behavioral response.

the general response properties of a neuron (defined in the *cellStatistics* table) and includes such measures as the baseline firing frequency and the significance level of the behavioral and unit response during various portions of the training trial. The second class of statistics (defined in the *cellResponses* table) analyzes the exact details of all the onset and offset responses produced by a given cell. It should be noted that all statistical data stored in the database are linked to the raw data through the Neuro-Core core tables that were used to generate these statistics.

Data Input

Once the definition of the database is complete, one must begin to populate the database. Most of the data to be entered in this database consist of legacy data; data already collected and stored as individual data files. In the case of the Thompson laboratory, much legacy data had been collected using routines written in the language "forth." In order to facilitate the entry of this legacy data into the database, a special conversion program was written that allows the conversion of forth datafiles (that contain the raw data) into SQL code (standard query language for databases; introduced in Chapter 1.2) that can be read by the database, so that the data contained in these files are stored in the database. This utility uses a datafile descriptor to read in the data from the forth datafile. All necessary variables and information regarding the experimental protocol are defined in the datafile descriptor. Using this information, the utility extracts the time-series data segments from the forth datafile which are then formatted into SQL statements that can be inserted into the database. The first step in the design of a user interface for data input was the implementation of a Web-based interface (Fig. 14) to aid the experimenter in constructing a descriptor file for a specific datafile.

After the legacy data have been stored, data for a new experiment can be entered using JADE (Java Applet for Data Entry; see Chapter 3.3). JADE is a Web-enabled interface that allows a researcher to enter experiment information (such as found in the datafile descriptor) directly into the database. After the experiment has

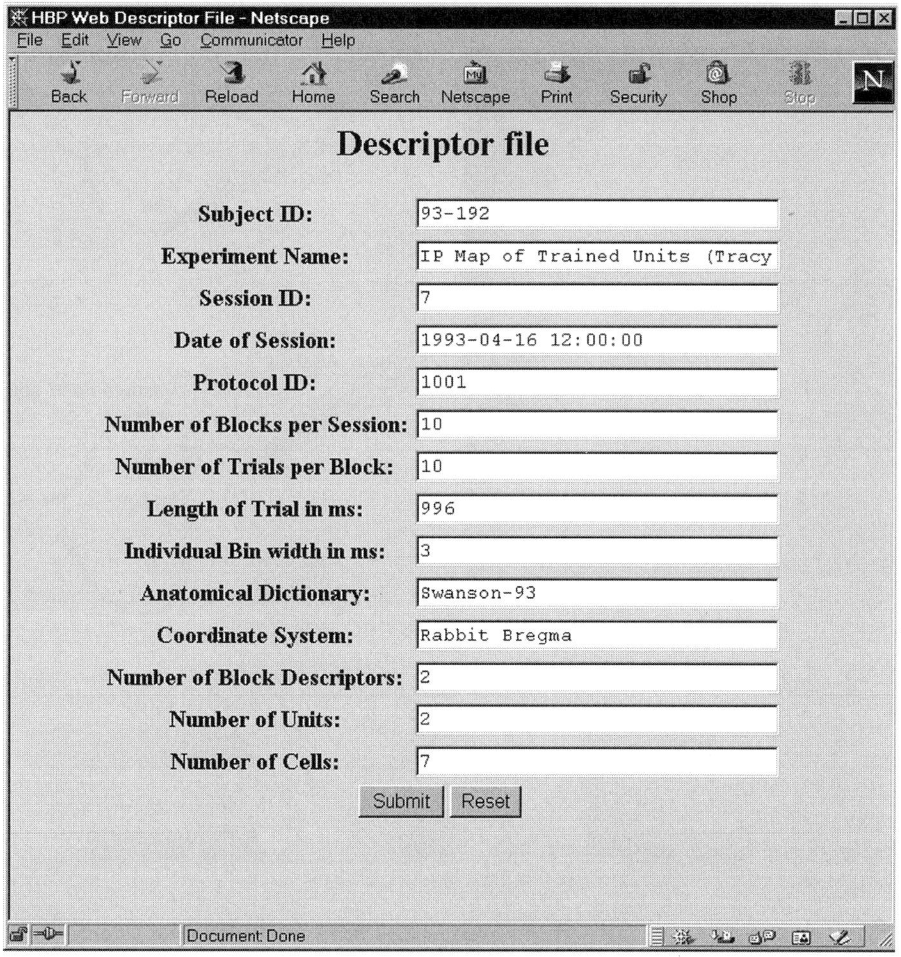

Figure 14 HTML-based interface used to generate descriptor files for data files that need to be entered into the database. The example above illustrates the top-level input required to format all subsequent data input forms. The descriptor file is used by a bulk loader to parse the data file and convert it into SQL statements that can be sent directly to the database.

been conducted, the datafile can be uploaded to the database server and the conversion program will enter the data into the database. JADE has the flexibility to adapt to any NeuroCore database associated with a laboratory. JADE prompts the user for experiment information based upon the tables in the database; therefore, new tables created to extend the needs of a laboratory can be recognized by JADE and used to enter experiment information without creating a new data entry tool. This exemplifies the benefits of having a standard core structure that is shared among various databases.

3.1.5 Building a Time-Series Database for *In Vitro* Neurophysiology: Long-Term Potentiation (LTP) in the Hippocampal Slice

In the current section we will use a series of experiments performed in Dr. T. W. Berger's lab as a case study to examine how the NeuroCore database can be extended for *in vitro* experiments. Brief trains of high-frequency stimulation to monosynaptic excitatory pathways in the hippocampus cause an abrupt and sustained increase in the efficiency of synaptic transmission. This effect, first described in detail in 1973 (Bliss and Lomo, 1973), is called long-term potentiation (LTP). LTP is a widely studied form of use-dependent synaptic plasticity expressed robustly by glutamatergic synapses of the hippocampus. The initial stage of LTP expression, typically identified as short-term potentiation (STP), is characterized by a rapid decay in the magnitude of potentiation to an asymptotic, steady-state level. Although there is a convergence of evidence concerning the cellular/molecular mechanisms mediating the induction of N-methyl-D-aspartate (NMDA) receptor-dependent STP and LTP, there remains substantial debate as to whether the expression of potentiation reflects change in presynaptic release mechanisms or postsynaptic receptor-channel function. Because of the synaptic co-existence of AMPA (α-amino-3-hydroxy-5-methyl-4-isoxazole proprionic acid) and NMDA glutamatergic receptor sub-

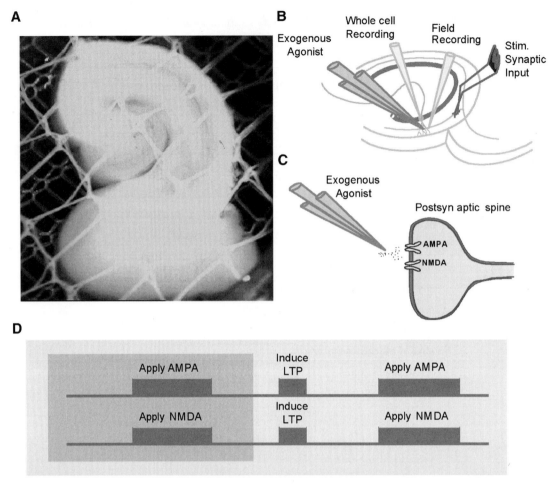

Figure 15 Demonstration of experimental paradigm. **(A)** Transverse slice of the hippocampus of a rabbit. **(B)** Placement of electrodes. **(C)** A multi-barrel pippette is used to focally apply agonists, such as AMPA and NMDA, to glutamate receptors on the postsynaptic spine. **(D)** One of the experimental protocols used in this study.

types, substantial differences in the magnitude of LTP expressed by AMPA and NMDA receptors would favor a mechanism that is postsynaptic in origin.

Experimental Preparation

Unlike the previous experiment, where neural activity was being analyzed with respect to the subject's behavioral performance, the current study aims to examine the detailed functioning of individual neurons. This is accomplished through the use of a slice preparation (Fig. 15), where a slice of tissue is extracted and all stimulation and recording is done directly on this tissue. This allows the experimenter much greater control of the conditions in the tissue slice (e.g., the balance of chemicals in the surrounding medium) as well as much finer control of the stimulation and recording electrodes used in the preparation. Fig. 15a demonstrates the placement of various electrodes in dentate gyrus of the hippocampal slice. In these experiments, we investigated the potential differential expression of STP by AMPA and NMDA receptors and found that the decay time course of STP is markedly different for AMPA and NMDA receptor-mediated excitatory postsynaptic potentials (EPSPs). Furthermore, during both STP and LTP, we found evidence of a differential responsivity of AMPA and NMDA receptors to focal application of their respective agonists. These results strongly support a postsynaptic expression mechanism of STP and LTP.

Data

Electrophysiological Data

A typical LTP experiment generates time-series data that consist of hundreds to thousands of individual EPSPs or excitatory postsynaptic currents (EPSCs) (Fig. 16a) taken at regular or irregular time intervals. The individual responses with duration of 200 msec, at a sampling rate of 10k Hz each, are then digitized and stored on the hard drive of a PC. Each of them also is parameterized and expressed in terms of its amplitude, onset slope, and area (Fig. 16b) in another data file. Data can also be gathered from cells after focal application of various agonists (Fig. 17). In this case, changes in

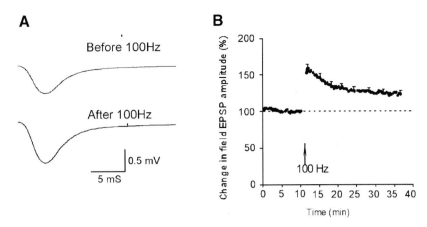

Figure 16 (A) Individual field EPSPs. (B) The amplitude of each EPSP was measured and represented as a point in this graph. A complete experiment lasts tens of minutes and consists of hundreds of individual EPSPs. Data from several experiments with the same protocol were grouped and statistically analyzed.

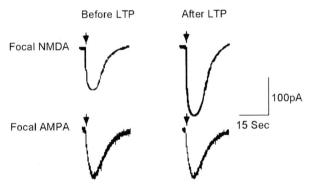

Figure 17 A representative graph showing the change in membrane conductance of a granule cell in response to focal application of glutamate receptor agonists, NMDA and AMPA. Membrane conductance in response to focal application of NMDA, but not AMPA, is increased during STP.

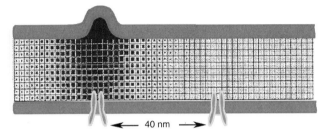

Figure 18 The effect of redistribution of receptors. Neurotransmitter release from a vesicle into the synaptic cleft. Because the cleft is extremely narrow, the concentration of glutamate (dark blue) at the postsynaptic membrane is highly localized.

membrane conductance of a granule cell in response to focal application of glutamate receptor agonists, NMDA and AMPA, were observed.

In this study, we also gathered continuous data containing information on the membrane conductance change during focal drug application. The duration of each of such record usually lasts for more than 20 seconds (Fig. 17). Raw data from this experiment are stored in a customized binary format. A conversion program has been developed to allow conversion of the experimental protocol and data to an SQL-ready format using a similar data input program described for the previous classical conditioning experiment.

IMAGING OR GRAPHIC DATA

More electrophysiological experiments are now seen combined with imaging techniques to reveal electrically triggered biochemical cascades, such as intracellular calcium redistribution, or morphological changes of the synapse following LTP induction. We routinely use graphic illustration (Fig. 15a,b) to record the accurate sites of recording and stimulating electrodes, drug applying pipettes, and surgical cuts in the brain slice. Working hypotheses underlying our specific research are often expressed graphically for clarity and simplicity (Fig. 18).

Extensions to the NeuroCore Database

In this section, we will discuss three distinct extensions that were developed for the storage of data from experiments performed on a hippocampal slice preparation:

1. The definition of the research subject
2. The definitions of the protocol used in the experiment
3. The definitions of the research data being collected

The first step in creating the extensions to the NeuroCore database is to define the research subject that will be used during the experiments. This extension (*animalSubject*) is created under the *researchSubject* table (Fig. 19). As was the case with the conditioning experiment, the typical research subject is a New Zealand white

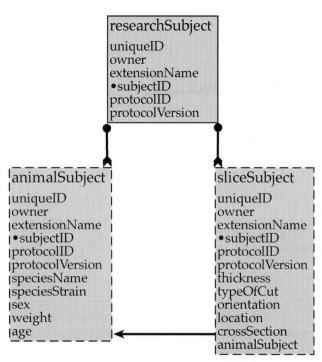

Figure 19 Database extensions for the research subject. In the case of a slice preparation, one must define a table to store information regarding the actual slice as well as the animal that was used to obtain the slice.

albino rabbit; however, in the LTP experiment, we are not dealing with a behaving animal, we are only dealing with a hippocampal slice. Hippocampal slices were prepared from male New Zealand white rabbits. Animals first were anesthetized with 5% halothane, and the skull overlying the parietal cortex was then removed bilaterally. The hippocampal formation and overlying neocortex from each hemisphere were extracted and gently separated. Both hippocampi were sectioned into blocks while being washed with cold, oxygenated medium, and slices of tissue (500 μm thick) then were cut perpendicular to the longitudinal axis using a vibratome. During the experiment, slices were superfused with medium consisting of (in mM): 126 NaCl; 5 KCl; 1.25 NaH$_2$PO$_4$; 26 NaHCO$_3$; 10 glucose; 2 CaCl$_2$; 0.1 or 1.0 MgSO$_4$, aerated with 95% O$_2$/5% CO$_2$ and maintained at 32°C. As one can see, the preparation and maintenance of a slice preparation is a complicated protocol by itself. The first step in being able to handle this data is to create the table that will store information regarding our various slices (*sliceSubject*). It is important to note that this table only contains the basic description of the slice as well as a reference to the animal subject that was used to obtain the slice. The critical information regarding the actual protocol used in the preparation of the slice is stored as a set of manipulations (defined as extensions of the *expManipulation* table) that belong to a specific protocol referenced by the *sliceSubject* table.

To be able to store the experiment described above in a coherent fashion that will be easy to retrieve and analyze at a later date, the data collected must be "associated" with an experimental protocol. The protocol for the experiment contains a set of manipulations (i.e., the actual stimuli) and conditions that comprise the specific experiment. These manipulations are created as children of the *expManipulation* table (Fig. 20). Unlike the previous classical conditioning experiment where two rather simple stimuli had to be represented, the current experiment requires a more complex representation of the stimuli, which in this case is either focal application of chemicals or electrical stimulation through an electrode. Focal application of NMDA (500 μM) was obtained by microinjection into the middle 1/3 dendritic region of the granule cell (single pulse, 5 psi, 100 msec in duration, pH 7.3) via a multi-barrel pipette (Fig. 15a–c). Chemical stimulation is stored as part of the *chemicalFocal* table. This table defines the general parameters of the chemical application; however, the specifics of which chemicals were actually involved is referenced by *chemicalFocal*'s parent table, *chemicalManip*. It is through this table that all chemical manipulations are related (through the *chemicalSet* table) to the specific chemicals (in the *chemicalDict*) and their concentration that comprise a particular chemical manipulation. The information regarding the chemical bath in which the slice remains during the entire experiment is also stored in a similar fashion through the *chemicalBath* table. Electrical stimulation to the perforant path input to dentate granule cells was activated using a bipolar nichrome stimulating electrode placed in the medial 1/3 of the *s. moleculare* to evoke field EPSPs. The information regarding the stimulation electrode and the pattern of electrical stimulation is stored as part of the *pulseElectrode* table.

Once we have defined the manipulations that were used to generate the data we need to describe the tables that will hold the actual research data (Fig. 21). These tables are created as children of the *researchData* table. As with the classical conditioning experiment, we need to store time-series data (defined by the *timeSeriesData* table) related to cellular activity in the hippocampal slice. In this experiment, data could be collected using either a whole cell electrode or an extracellular microelectrode. Whole cell electrodes (glass 7052 1.65-mm OD) were filled with (in mM): 120 cesium gluconate; 5 KCl; 2 MgSO$_4$; 10 N-2-hydroxy-ethyl-piperazine-N-2-ethanesulfonic acid (HEPES); 0.1 CaCl$_2$; 1.0 BAPTA; and 3.0 ATP-Mg (resistance: 6–9 MΩ;). Field EPSPs were recorded in the *s. moleculare* of the dentate gyrus using microelectrodes (glass thin-wall 1.5-mm OD) filled with 2.0 M NaCl (resistance: 1–2 MΩ). When field EPSPs and whole cell EPSCs were recorded simultaneously, the extracellular recording electrode was placed in the medial 1/3 of the *s. moleculare*, 200 μm from the whole cell pipette. Fig. 15a demonstrates the placement of various electrodes in dentate gyrus of the hippocampal slice. As with our experimental protocol, the exact specification of the internal chemical makeup of the electrodes is of

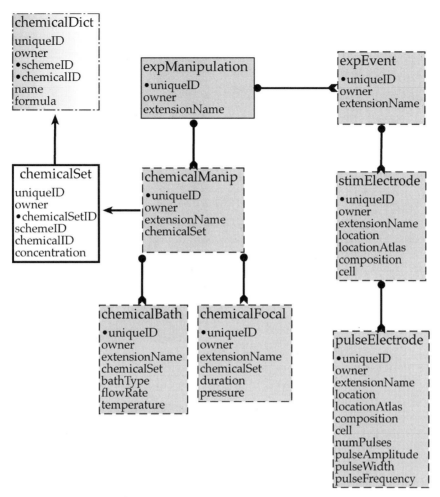

Figure 20 Database extensions for the experimental manipulations. Of great importance in the specification of any slice paradigm is the specification of the chemical manipulations that are an integral part of the protocol.

importance. Therefore, the *recordingElectrode* table used to store information regarding the various electrodes must also reference the *chemicalSet* table in order to define the chemical concentration within the electrode.

3.1.6 Building a Database for Human Neuroimaging Data

The previous two experiments have shown how NeuroCore can be extended to accommodate neurophysiological experiments. However, during the design of NeuroCore, it was extremely important to design a database framework in which a neuroscience researcher could store any type of neuroscience experiment. Recently, NeuroCore has been adopted as the foundation for an experimental database concerned with storing neuroimaging data (Grethe and Grafton, 1999). Neuroimaging has seen an explosive growth over the past decade with the proliferation of positron emission tomography (PET) and functional magnetic resonance imaging (fMRI) paradigms.

Data

Neuroimaging data consist of three-dimensional images of the human brain. During an experiment, researchers collect both anatomical reference scans (Fig. 22) and a number of image series associated with the experimental task. In PET experiments, each trial is usually represented by a single image volume. However, in fMRI experiments, each trial consists of a large number of image volumes (Fig. 23) that are similar to an electrophysiological time-series except that the data being stored at each time point are now an image volume instead of a numerical value. Most neuroscientific databases being developed to store neuroimaging data tend to store these data as either an external file or an internal binary object. In both cases, the database is unable to manipulate the data by itself. They must first be exported to an application that can then process the data. For users to be able to automatically analyze and efficiently mine through large volumes of imaging data located at various data sites, the data must be in a format that the database can access and manipulate itself. Just

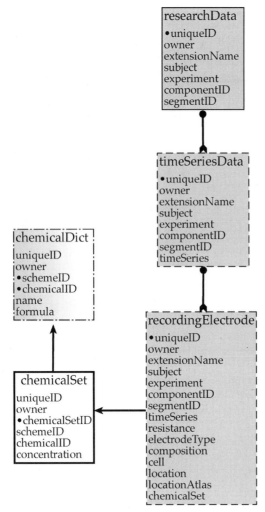

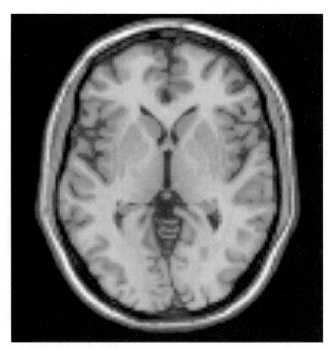

Figure 22 A single slice from an MRI anatomical volume.

Figure 21 Database extensions for the research data. The recording electrodes used in the experiment must also reference their internal chemical composition in a similar fashion as the chemical manipulations discussed in Fig. 20.

as a TimeSeries Datablade (Appendix A2) has been developed for electrophysiological data, a new datatype is being developed to store neuroimaging time-series in an efficient format. In addition to the data associated with the large number of experiments being performed by researchers in this field, each experiment consists of a massive quantity of image data. For example, a typical imaging experiment might be comprised of thousands of image volumes from tens of subjects.

The analysis of neuroimaging data is a lengthy process involving many steps and intermediate results. In order to examine how a researcher arrived at his conclusions, the neuroimaging database must be able to store all versions of the data with information regarding the protocol or procedure that was used to transform the data at each step in the analysis process (Fig. 24). This allows for users to examine how archival imaging data was processed and analyzed to obtain certain results. Furthermore, the database structure allows users to reprocess and reanalyze datasets using new methods without alter

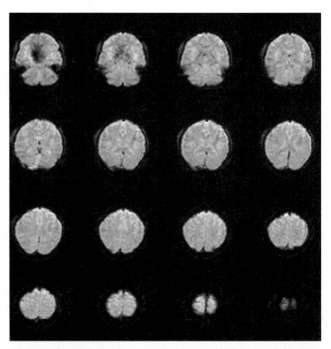

Figure 23 A single volume from a set of echo planar images (EPI) taken during an fMRI scan. This represents one time point in the functional scan.

ing past analyses. The neuroimaging database is currently under development and is an important new avenue in the development of the NeuroCore system as a whole. An interesting connection between the imaging data and physiological data can be found in the discussion of synthetic PET (see Chapter 2.4). Such technologies can hopefully provide a bridge between

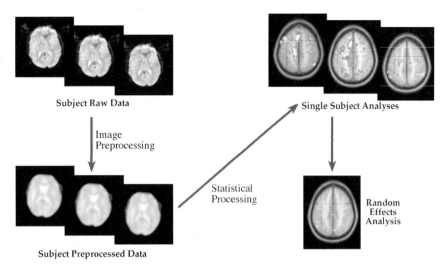

Figure 24 Statistical data flow in the neuroimaging database. All raw imaging data are stored in the database, so any analyses can be tracked back to the original data that generated them. One also has the option of storing any data generated during the processing and analyzing of the data. For example, when analyzing a set of fMRI data, one must first preprocess the data (alignment, normalization, smoothing). After the data have been preprocessed, one can apply various statistical analyses.

neurophysiological and imaging data through the use of computational models (see Chapter 2.1 for more discussion of computational models).

3.1.7 Discussion

The previous section introduced a database schema for the storage of neuroscientific data, which will be further described in Chapter 3.2, and specific extensions to this structure from various experiments. Currently only a few such database systems other than NeuroCore are under development (Payne *et al.*, 1995; Gardner *et al.*, 1997). The three major contributions from the development of NeuroCore are as follows:

1. *Tools for building extendible databases around a structured core.* This allows for the easy extension of the database by individual laboratories without the need to develop a complete database from scratch, as well as data sharing between NeuroCore databases as the core from each of the databases is identical. The common core also allows tools and extension sets (a collection of extensions developed for a specific experimental paradigm) to be easily shared between databases.

2. *Importance of the scientific protocol in storing experimental data.* Most neuroscientific databases only give a cursory description of the protocol for a given data record in the database. In order for data in a database to be useful to all users, they must contain all the specifics regarding the protocol used in the generation of the data.

3. *Complexity of neuroscientific protocols.* Most scientific databases developed currently view scientific protocols as being process-oriented (i.e., an input is processed by a manipulation that generates an output—in the molecular biology community, for example, an electrophoretic gel is labeled and generates a labeled separation; see Chen and Markowitz, 1995). Protocols in the neurosciences are much more complicated and consist of a hierarchy of manipulations that can be ordered, as well as specifically timed in relation to one another.

4. *Storage of complex data in a usable format.* Neuroscientific databases being developed to store neurophysiological time-series data tend to store these data as either an external file or an internal binary object. In both cases, the database is unable to manipulate the data; they must first be exported to an application that can then process the data. For users to be able to analyze and mine their data, the data must be in a format that the database can access.

Future Work

The major effort in developing NeuroCore to date has been in the construction of the core schema (Chapter 3.2) and the implementation of lab-specific extensions and interfaces that were discussed in this chapter. Work has also begun on general purpose user interfaces for the researcher to interact with the database (Chapter 3.3); however, for this database to become accepted by the neuroscience community as a whole, various developments must occur.

HANDLING OF HETEROGENEOUS DATA

With the database having been tested on a few select lab datasets, the database and extension capabilities need to be tested on a larger scale. The Thompson database currently consists of data from 312 cells from the paired training experiment in Tracy's thesis (Tracy, 1995). However, this database now has now been tested with data from other laboratories that use different experimental protocols and collect their data in varying ways and store them in differing formats. NeuroCore has also been used for the storage of neurochemical data (Chapter 4.5) and neuroimaging data (current chapter) and as a basis for the NeuARt system (Chapter 4.3).

Extend the Database To Handle New Datatypes

Neuroscientists collect various datasets that are highly structured and complex (e.g., time-series data, video, MRI and PET images). The current version of the database has been extended to handle time-series data through the development of a time-series datablade (a datablade is a new datatype and associated functions added to the database system; see Chapter 1.2 and Appendix A2). In order to be able to handle a larger variety of neuroscience data, new datablades and core support tables need to be developed to handle some of the other complex datatypes collected by neuroscience researchers (e.g., three-dimensional images obtained through MRI or PET). Such development is accompanying the work being done at Emory University and Dartmouth College on the neuroimaging database discussed earlier in this chapter. One could store these objects as simple binary large objects; however, this would not allow the objects to be manipulated within the database itself. Therefore, one could not query the data specifically but only the metadata associated with it. It is important to construct these various datablades so that users can interact directly with their datasets stored in the database.

Development of "Commercial Grade" User interfaces

Most of the user interfaces developed to date are lab specific and perform a specific task or query. For example, in order to load more of the raw data into the database, various tools were constructed: a bulk loader, which takes as input a definition file and a data file and produces the SQL code necessary to load that data into the database, and an HTML interface coupled to a cgi-bin application that produces the proper definition files for the bulk loader. This specific interface has been developed for our specific laboratory data and can be extended for similar types of data; however, it is not meant as a tool for the general neuroscience community. Work has begun on various general user interfaces (e.g., a protocol viewer/editor and a generic input applet), but these interfaces are in their very early stages of development. It is imperative that a consistent look and feel for all user interfaces are adopted so that users can work with all the tools available without having to learn a new interface for each tool.

Data Mining

There is no point in storing all these data in a database unless one begins to develop the statistical and analytical methods necessary to explore the data in a useful fashion. Just being able to retrieve the data is not extremely useful; however, being able to retrieve the data due to some statistical or analytical measure and then display the results is of great value. For example, many neuroscientists have developed their own techniques for analyzing single unit data (Berthier and Moore, 1986,1990; King and Tracy, 1998; Tracy, 1995). Being able to include these methods within a database would allow researchers to compare methods across the same dataset. One such tool in development at USC is DataMunch (see Chapter 3.3). The types of tools required for data mining depend on the scientific community that will be accessing the database. With the addition of significant amounts of raw unit data over the next few months, serious investigation of data-mining techniques related to electrophysiological unit recording can take place.

References

Arbib, M. A., Grethe, J. S., Wehrer, G. L., Mureika, J. R., Tracy, J., Xie, X., Thompson, R. F. and Berger, T. W. (1996). An on-line neurophysiological and behavioral database for the neuroscientist. *Soc. Neuroscie. Abstr.* **22**, 359:316.

Berthier, N. W., and Moore, J. W. (1990). Activity of deep cerebellar nuclear cells during classical conditioning of nictitating membrane extension in rabbit. *Exp. Brain Res.* **83**, 44–54.

Berthier, N. W., and Moore, J. W. (1986). Cerebellar Purkinje cell activity related to the classically conditioned nictitating membrane response. *Exp. Brain Res.* **63**, 341–350.

Bliss T. V., and Lomo, T. (1973). Long-lasting potentiation of synaptic transmission in the dentate area of the anaesthetized rabbit following stimulation of the perforant path. *J. Physiol. (London)* **232(2)**, 331–356.

Chen, I. A., and Markowitz, V. M. (1995). An overview of the object-protocol model (OPM) and the OPM data management tools. *Inf. Syst.* **20(5)**, 393–417.

Chen, I. A., and Markowitz, V. M. (1994a). Mapping Object-Protocol Schemas into Relational Database Schemas and Queries (OPM version 2.4). Lawrence Berkeley Laboratory Technical Report, LBL-33048.

Chen, I. A., and Markowitz, V. M. (1994b). The Object-Protocol Model (version 2.4). Lawrence Berkeley Laboratory Technical Report, LBL-32738.

Clark, R. E., and Lavond, D. G. (1993). Reversible lesions of the red nucleus during acquisition and retention of a classically conditioned behavior in rabbits. *Behav. Neurosci.* **107(2)**, 264–270.

Clark, R. E., Zhang, A. A. and Lavond, D. G. (1992). Reversible lesions of the cerebellar interpositus nucleus during acquisition and retention of a classically conditioned behavior. *Behav. Neurosci.* **106(6)**, 879–888.

Frenkel, K. A. (1991). The human genome project and informatics. *Communications of ACM.* **34**, 11.

Grethe, J. S., and Grafton, S. T. (1999). An on-line experimental database for the storage and retrieval of neuroimaging data. *Soc. Neurosci. Abstr.* **25**, 104.52.

Grethe, J. S., Wehrer, G. L., Thompson, R. F., Berger, T. W. and Arbib, M. A. (1996). An extendible object-relational database schema for neurophysiological and behavioral data. *Soc. Neurosci. Abstr.* **22**, 359.17.

King, D. A. T., and Tracy, J. (1998). DataMunch: a MATLAB-based open source code analysis tool for behavioral and spike-train data, http://www.novl.indiana.edu/~dmunch/.

Krupa, D. J., Thompson, J. K. and Thompson, R. F. (1993). Localization of a memory trace in the mammalian brain. *Science* **260(5110)**, 989–991.

Payne, J. R., Quinn, S. J., Wolske, M., Gabriel, M., Nelson, M. E. (1995). An information system for neuronal pattern analysis. *Soc. Neurosci. Abstr.* **21**, 376.4.

Thompson, R. F. (1990). Neural mechanisms of classical conditioning in mammals. *Phil. Trans. R. Soc. London B*. **329**, 161–170.

Thompson, R. F. (1986). The neurobiology of learning and memory. *Science* **233**, 941–947.

Thompson, R. F., and Krupa, D. J. (1994). Organization of memory traces in the mammalian brain. *Ann. Rev. Neurosci.* **17**, 519–549.

Thompson, R. F., Bao, S., Berg, M. S., Chen, L., Cipriano, B. D., Grethe, J. S., Kim, J. J., Thompson, J. K., Tracy, J. and Krupa, D. J. (1997). Associative learning. In *The Cerebellum and Cognition, International Review of Neurobiology*. Vol. 41 (Schmahmann, J., Ed.). Academic Press, New York, pp. 151–189.

Tracy, J. (1995). Brain and Behavior Correlates in Classical Conditioning of the Rabbit Eyeblink Response, Ph.D. thesis, University of Southern California, Los Angeles.

Design Concepts for NeuroCore and NeuroScience Databases

Jeffrey S. Grethe,[1] Jonas Mureika,[2] and Edriss N. Merchant[2]

[1]Center for Cognitive Neuroscience, Dartmouth College, Hanover, New Hampshire
[2]University of Southern California Brain Project,
University of Southern California, Los Angeles, California

3.2.1 Design Concepts for Neuroscience Databases

In Chapter 3.1, we gave a review of our design of NeuroCore to provide extendible databases for repositories of experimental neuroscience data. In this chapter, we offer a more formal presentation of NeuroCore. In designing the NeuroCore database, four specific design concepts were taken into account:

1. Due to the heterogeneity in the data and the methods used to collect the data in a neuroscience laboratory, the database should be easily modified to meet a specific laboratory's need while remaining compatible with other databases.
2. In order for a researcher to fully integrate these new data repositories into their own laboratory, the database should be an experimental tool as well as a data storage, retrieval, and analysis system.
3. Due to the diversity of the neuroscience community, there will not be a single monolithic database to store all neuroscience data. This is in contrast to the molecular biology community, where a few monolithic databases are being used to store data concerning genetic sequences.
4. Like the molecular biology community, however, any tools developed should foster intra- as well as inter-laboratory collaborations.

These four design principles lay out the specific requirements that any neuroscience database framework must fulfill. The following sections give an overview of the major components of the NeuroCore database and how they relate to these design criteria.

3.2.2 The Three Main Components of the NeuroCore Database

In order for the database to be easily modified to meet a specific laboratory's need, a core database schema, with a standard core structure, was created that can be extended as each researcher sees fit (see Fig. 1). By exploiting the object-oriented properties of new object-relational database technologies (i.e., the definition of new data types and the ability to construct inheritance hierarchies), we were able to create a database schema that is easily modified without requiring major changes to its core structure. The NeuroCore database consists of four primary components:

1. Core database
2. Database extension interface
3. Database federation interface
4. New datatypes for the database created as datablades (see Appendix A2 for more information on the Neuroscience Time-Series Datablade)

Each of these three components is discussed briefly in the following sections.

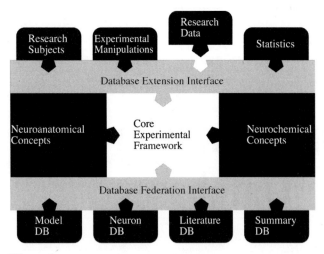

Figure 1 Overview of the NeuroCore database. The core database consists of a description of the experimental framework as well as constructs to define the neuroanatomical and neurochemical concepts needed in defining a neuroscience experiment. In order for individual laboratories to be able to store their own data, an extension interface is attached to the core schema to allow users to extend the database as they see fit.

Core Database Schema

The core schema of NeuroCore consists of five basic components:

- Core experimental framework
- Neuroanatomical concepts
- Neurochemical concepts
- Database federation interface
- Database extension interface

The following sections will give an overview of each of these areas of the NeuroCore database. For a more detailed description of each area, including a description of the database tables used in the implementation of the database, please visit the USCBP website at http://www-hbp.usc.edu.

Core Experimental Framework

The core experimental framework in NeuroCore contains the database tables that store the critical information needed for storing an experiment. The most important information is the hierarchical description of what constitutes an experiment. Attached to this hierarchical definition of an experiment are mechanisms to store the protocols involved in the experiment. Also included in this core framework are the description of the research subjects involved in the experiment as well as the description of the researchers involved in the experiment.

Neuroanatomical Concepts

The neuroanatomical section of the NeuroCore database contains the basic neuroanatomical information needed to provide a framework for storing specific neuroanatomical data as well as providing the proper anatomical framework to store any neuroscience experiment within NeuroCore. The storage of neuroanatomical concepts within the NeuroCore database can be accomplished in three areas:

- Information regarding single cells and their associated cell types that have been stored in the database
- The storage of multiple anatomical dictionaries that can be used to reference atlas locations within the database
- The storage of graphical anatomical atlases

Neurochemical Concepts

The neurochemical section of the NeuroCore database contains the basic neurochemical information needed to provide a framework for storing neurochemical data within NeuroCore. The NeuroCore database allows for the storage of chemical dictionaries that can be used to reference chemicals from within the database.

Database Federation Interface

The federation interface enables each NeuroCore database to be able to share information with other databases. To this end, three basic considerations were taken into account:

- The core database structure needs to contain information concerning the basic terminology used in structuring and storing the data (e.g., neuroanatomical and neurochemical dictionaries).
- All data stored in the database should contain information regarding the unit of measurement on which the data are based (e.g., cm vs. mm).
- Support for various levels of security is necessary so that researcher's can store and share their data securely.

Database Extension Interface

The database extension interface provides mechanisms whereby researchers can augment their database to store necessary information regarding their particular experiments. The core database contains the database tables necessary to describe the core information concerning any neuroscience experiment (e.g., the experiment table, which contains information on the name of the experiment, the type of the experiment, and the researchers involved in the experiment). However, the core database also contains "parent" super-tables that allow researchers to attach their own table definitions as "child" sub-tables to the core database in an inheritance hierarchy. Recall the definition of inheritance in Chapter 1.2 (Introduction to Databases) and the description of inheritance and its relation to NeuroCore in Chapter 3.1. This allows the database to be extended so that it can store relevant information concerning the research subjects (e.g., the *researchSubject* table allows researchers to attach their own tables as sub-tables to this table to describe the subjects used in their experiments). A standard behavioral neuroscience laboratory might have a table describing animal subjects and another describing human subjects used in their laboratory, the experimental data collected, the experimental protocols used, and any annotations or statistics (metadata) normally included in the experiments. This allows the researcher to add data specific to their laboratory in a fairly simple manner without the need to change the core database structure. Examples of specific extensions to the core database were given in Chapter 3.1, Repositories for the Storage of Experimental Neuroscience Data, whereas some interfaces designed to help users work with these databases are described in Chapter 3.3, NeuroCore On-Line Notebook.

Neuroscience Time-Series Datablade

The last component of the NeuroCore database is the Neuroscience Time-Series Datablade (see Appendix A2 for further details). Time-series data cannot normally be stored efficiently with standard commercial databases as they have no standard mechanisms for storing and manipulating physiological time series (most commercial databases would need to store all the individual data points from a time-series recording as individual entries in a table). In order to allow for the efficient storage of such data, a new datatype was created to store neurophysiological time-series data. This new datatype, created in Informix™ as a datablade, stores the actual time-series data as an opaque data type (internally stored as a large binary object in the database) that can be manipulated in the database.

3.2.3 Detailed Description of the NeuroCore Database

Overview of Table Descriptions

In developing the core database structure, various aspects of what constitutes a neuroscience experiment were considered. Not only does the core schema need to incorporate the generalities of an experiment, it must also allow the user to capture information regarding neuroanatomy, neurochemistry, and neurophysiology in a structured format. Each of the following sections describes an aspect of NeuroCore necessary to provide a structure that will allow any user to extend the core database structure so that the extended database can capture *any* neuroscience experiment.

The NeuroCore database consists of objects, concepts, or entities that need to be stored for a neuroscientific experiment and the relations that exist between these objects. The objects in the database are implemented as tables which consist of one or more columns that define what constitutes the object described by the table. Each instance of an object, a single data entry, is entered as a row (tuple) in one of these tables. The tables themselves can be categorized as follows.

CORE TABLE

The core tables in the schema are all the tables that allow for the storage of data related to specific concepts that are part of the core framework. One example of such a core table is the table that stores information regarding experiments that are stored in the database.

SET BACKING TABLE

The set backing tables in the schema allow a many-to-many relationship to be defined between data across tables. For instance, to associate the person table with the laboratory table, we first realize that many people can be associated within a laboratory and similarly many laboratories can be associated with a particular person. To implement this situation, a set backing table is used to hold the multiple entries (see Chapter 1.2 for more information on set backing tables).

DICTIONARY TABLE

Defines the terminology that is needed to build a structured framework for the storage of neuroscience data. For example, the names of neuroanatomical structures or neurochemical compounds that need to be defined are stored in dictionary tables so they can be referenced from other tables.

EXTENDIBLE PARENT TABLE AND EXTENSION TABLE

The extendible parent and extension tables comprise the framework that allows for individual users to tailor the database to suit their own needs. The extendible parent table is a super table that allows users to attach their own extension tables to the core database structure. For example, all tables that define what research data are to be stored in the database are extension tables and are children of (attached to) a single extendible parent table.

The relations between the objects implemented in the database schema can be classified into two general categories, as follows.

Inheritance Relations This relation describes the parent/child (super-table/sub-table) relations in the database that form the object-oriented structure (inheritance hierarchy) of the database. These relations are the necessary link between the extension tables and the extendible parent tables that form the basis for adding extensions to the database.

Foreign Key-Primary Key Relations The foreign key-primary key relation is a standard construct that allows one database entry in a specified table to relate to one or more entries in another table (e.g., a set of email

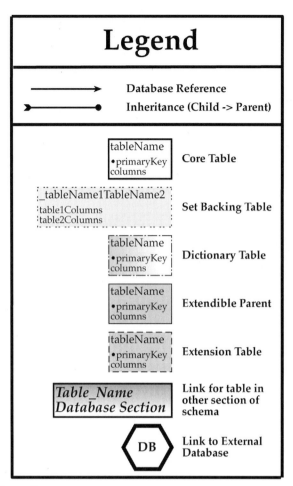

Figure 2 Legend for all schema diagrams. These diagrams represent the database schema by displaying the actual tables and their contents with the relations (primary key-foreign key or inheritance) that exist between the tables.

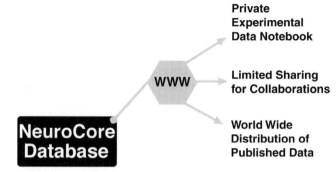

Figure 3 Various levels of access taken into consideration in the design of NeuroCore. Access to the database can be limited to individual researchers or can be granted to allow either a set of collaborators or the general public to view the data.

addresses is associated with one person; a research subject has only one species designation). Relations that map one entry to one or more entries (1-1 or 1-n) are captured by the definition of the objects themselves. However, relations that may map a set of entries to another set of entries (n-m) are defined through the addition of *set backing tables*. Each aspect of the core described in the following sections is accompanied by a schema diagram (see legend in Fig. 2) which shows all the tables and relations involved in that particular aspect of the core database.

Top-Level of Core Database

All tables in the core database schema are created as sub-tables (children) of a top-level database object (*redObject*; see Fig. 4). This allows all tables in the core schema to have certain necessary or global information defined for them at a higher level. Each section of the core schema falls under its own super-table (parent) which is a child of the *redObject* table. The *redObject* table itself just defines that each tuple will have a uniqueID. Three tables are then used to partition the database into the following categories:

- *databaseMetadata*: all the necessary information regarding the tables, columns, relations in the NeuroCore database
- *contactDB*: all the information regarding researchers and laboratories that needs to be stored in the database
- *databaseCore*: the parent for all the experiment related tables in the database. This section of the database can further be partitioned into the following categories:
  - *experimentCore*: contains a hierarchical description of the experiment and the framework for storing the protocols used within this hierarchy and the research subjects used in the experiments
  - *neuroChemCore*: defines the necessary neurochemical constructs needed by the database
  - *neuroAnatCore*: defines the necessary neuroanatomical constructs needed by the database
  - *coreExtension*: the interface layer where users can add extensions to tailor the database to meet their own needs

This hierarchical structure with a single root object is also useful in another regard; one can associate a global relation from any tuple in the database to a specified table, as the uniqueID defined by the *redObject* table is unique across all tables in the NeuroCore database. For example, we may want to be able to associate a literature reference found in an external article repository (AR) to specific tuples in the database through the use of the *_redReference* table. This is important because a literature reference will relate to a specific set of tuples in the database (e.g., authors, animal subjects, research data, etc.) which are distributed across many different tables. Each of the tuples can also be related to different sets of references. It is important to remember that the NeuroCore database is not designed to store all neuroinformatics data (e.g., published articles, neural models, summaries and annotations). NeuroCore, therefore, needs to be able to refer to external federated databases that contain this information. Article repositories are a prime example of this concept. Many scientific journals being published today are also being published in an electronic format (e.g., the *Journal of Neuroscience* published by the Society for Neuroscience is now available electronically through the society's Web site); therefore, they will need to be able to be referenced from within NeuroCore and will also need to be able to reference data stored within NeuroCore (see Chapter 3.3 for more information regarding the interaction between article and data repositories).

Another very important aspect for any experimental database is the issue of database security. For a database to be widely accepted in the neuroscience community, researchers must be able to decide who can retrieve, view, and analyze their data. Unlike standard database security constraints where one can set access privileges for entire tables or even, in some cases, specific columns found in certain tables, a scientific database needs to be able to define access privileges at a much finer scale, on the level of individual tuple entries. Sharing of data in a scientific community can occur at three distinct levels (Figure 3).

PRIVATE

Private indicates data that the researcher does not want made public to anyone. This would include data that have been recently collected or are currently being analyzed and have not yet been published. Information stored as private would form a researcher's personal experimental database. This is the default access privilege for all newly entered tuples, and the only access privileges given to these tuples would be to the owner.

COLLABORATORY

In certain instances, it will be useful for the researcher to share information with a select group of individuals and/or other groups. This method of sharing requires that a specific set of researchers be given access to a limited set of tuples throughout the database.

PUBLIC

Once data have been formally published, they can be made available to all interested individuals. In order to share data at these various levels, a tuple-level security system (i.e., security at the level of individual data entries) has been implemented where each tuple in the database can be given numerous security conditions (Fig. 5). There are three access privileges for a tuple that can be granted to any user:

- Select: allows the user to view the information stored in that tuple
- Update: allows the user to update information in the tuple
- Delete: allows the user to delete the entire tuple

Each grantor who receives any of these privileges can also be given the permission to grant these privileges to other users. When allowing users to grant access to certain individuals and groups (collections of individuals), one also needs to be able to explicitly revoke privileges to a particular user. The *redSecurity* table stores all information regarding tuple access permissions (tupleAuth of each user stored in the *databaseUser* table). In addition to being able to define individual users, one can define groups of users (a group is also a database user) through the *_userGroup* table. Each user is also assigned to a specific *userClass* and *userStatus*. Each user's personal information is stored in the contact portion of the database (Fig. 7) and is referenced through the *_userPerson* table. A complete security log is also maintained as part of the *updateLog* table.

For each table created as part of the NeuroCore database, certain information needs to be stored so that all tuples and their relations with other tables can be fully described (Fig. 6). Including the release information stored in the *schemaVersion* table, information is stored that defines the functionality of all tables within a NeuroCore database and all columns within these tables (*databaseTables* and *databaseColumns*). The general information contained in these tables allows users to query this information to discover the description and purpose of each table along with its contents. Along with the general description of each table column, each column's unit of measurement is referenced (with the *measurementTypes*, *measureSystems*, and *measurementUnits* tables that are referenced through the *_dbColumnUnits* table), which can aid in the conversion of data from one format to another format. This becomes very important when one begins dealing with multiple databases residing in a federation. In order to share data from these databases, we must know what units of measurement have been used to store the data and in what format the data have been stored (this is especially important for image data, which are stored in a specific format and also have specific dimensions).

The last piece of information that must be stored in order to fully describe any NeuroCore database are the relations that exist between tables in the database. All relations in the database are described in the *databaseRelation* table (each set of columns in a relation is referenced through the *relationColumns* table, whereas specific relation types are defined by the *relationTypes* table). By storing all the information concerning the tables and their relations, one is able to completely reconstruct the database from the information contained in the metadata tables. One is then able to create tools that can build browsing, query, and input interfaces dynamically based on the information contained in the metadata. This affords users of NeuroCore databases a simple mechanism through which they can federate their databases. This also makes examining and extending a NeuroCore database easier, as one knows what columns

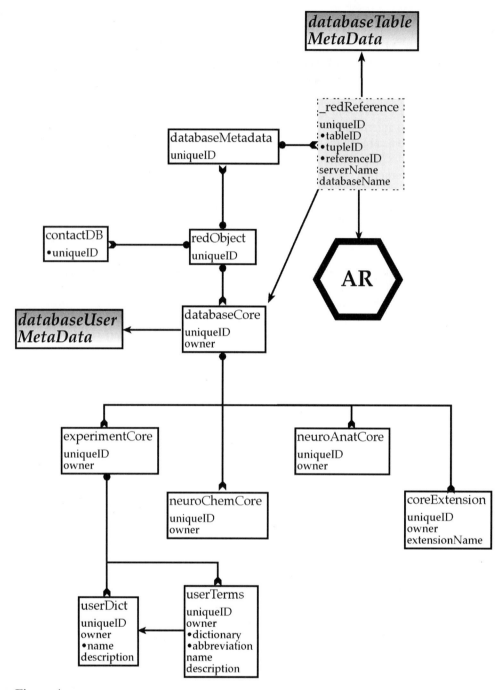

Figure 4 Top-level schema diagram for the NeuroCore database. The NeuroCore database can be subdivided into three main sections: database metadata, contact database (core) and the experiment core which contains information about the generalities of an experiment, neuroanatomical and neurochemical information as well as all the extensions added by the user.

a table contains and its data type and unit of measurement and this table's relationships to the rest of the database.

NeuroCore Schema for Contact Information

The contact portion of the core schema stores all the relevant information regarding the people and labs involved in the experiments being stored, as well as all of the users of the database (Fig. 7). Each researcher and the laboratory(s) with which they are involved are stored within the *person* and *lab* tables, respectively. The information concerning which laboratories are associated with each researcher is stored in the *_labPerson* table. The contact information (email address, mailing address, and phone number) for each researcher is stored in the *netAddress*, *address*, and *phone* tables, respectively. Each *contact* (being a *lab* or a *person*) can have multiple

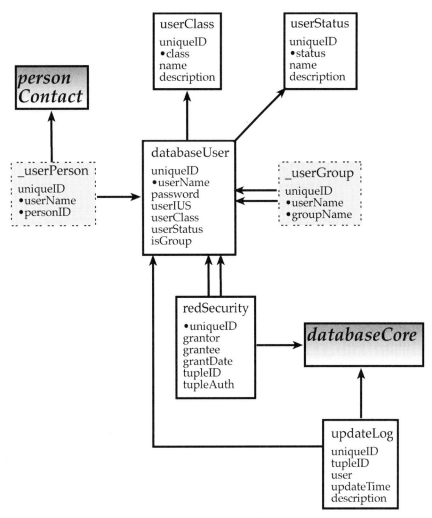

Figure 5 Schema diagram for security in the NeuroCore database. All tables depicted are sub-tables of the *databaseMetadata* table (Fig. 4). For an experimental database, each tuple in the database needs to be able to have security permissions assigned to it.

addresses, phone numbers, and electronic addresses associated with it. Relations between the contacts and tables in the *experimentCore* are handled by the *_orLogPerson*, *_hypPerson*, and *_expPerson* tables.

NeuroCore Schema for a Scientific Experiment

Each experiment that is described in a NeuroCore database consists of three distinct components:

- Hierarchical description of the experiment structure
- Description of the research subjects involved in the experiment
- Framework for describing the protocol used to generate the data

The following three sections describe each of these distinct components in more detail.

Hierarchical Description of Experiment

Fig. 8 depicts the experimental portion of the core database. To begin storing the definition of an experiment, one first enters general descriptive information in the *experiment* table. Any researchers involved in the experiment are also entered in the *_expPerson* table. If one also wishes to describe any hypotheses being tested by this experiment, one enters similar information in the *hypothesis* and *_hypPerson* tables, respectively. All hypotheses related to a specific experiment are entered in the *_expHypothesis* table. Researchers may then also associate various keywords located in the *userTerms* table (see also Figure 4) with their experiment through the *_expKeyword* table. The complete structure of the experiment is then stored in a hierarchical fashion.

Each experiment is divided into a hierarchical collection of experimental components (*expComponent* table), which can be further subdivided into experimental segments (*expSegment* table). An experimental component, for example, would be a recording session from a typical behavioral experiment. An experimental segment, therefore, would be an individual trial that was part of that session. However, in order to accommodate neuroanatomical, neurochemical, and neurophysiological data in

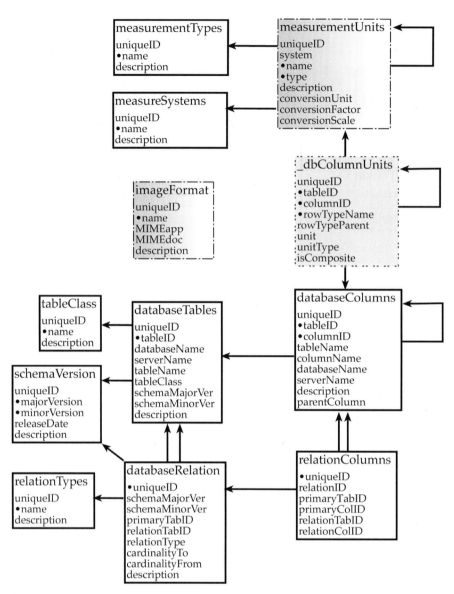

Figure 6 Schema for the metadata associated with all portions of the NeuroCore database. All tables depicted are sub-tables of the ***databaseMetadata*** table (Fig. 4). This information allows the complete structure and function of any NeuroCore database to be reconstructed. This includes information concerning all the tables and table columns, as well as descriptions and specifications of these columns. Also included is a description of all the relations between columns.

the same database, two parallel hierarchical descriptions have been implemented. First, a generic structure for any collection of data was implemented through the *expDataSet* and *expDataSegment* tables. These tables can store any number of unordered datasets and their respective data segments. For example, a neuroanatomical experiment may consist of a collection of stained histological sections from a specific brain where the researcher may be using multiple staining techniques (e.g., cell body stains, fiber stains, etc.). In this instance, each histological series using a specific stain can be considered an experimental data set; however, each of these data sets actually consists of a number of individual histological sections which can be considered the experimental data segments that constitute the various data sets.

Neurophysiological experiments, on the other hand, consist of an ordered collection of components where the data related to any one component may be dependent on previous components. Therefore, the *expSession* and *expTrial* tables have been implemented in order to define an ordered hierarchical description of an experiment. In addition to the standard experimental components and segments that can be defined, one can also group an ordered set of experimental segments, or trials, into experimental blocks through the *expBlock* table (each trial is related to a block through the *_blockTrial* table). Once the hierarchical description of an experiment has been defined, the experimental protocol associated with the experiment must then be defined.

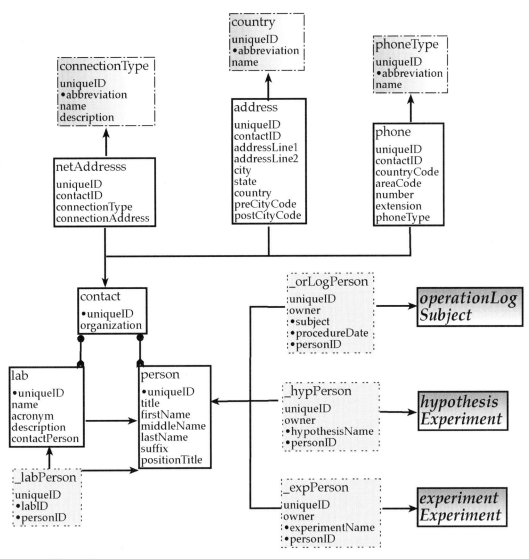

Figure 7 Schema for the contact portion of the NeuroCore core database. All tables depicted are sub-tables of the *contactDB* table (Fig. 4). All information regarding the researchers and their laboratories are stored in these tables.

FRAMEWORK FOR THE DESCRIPTION OF PROTOCOLS

Probably the most important aspect of storing an experiment in the NeuroCore database is the storage and description of the experimental protocols involved (Fig. 9). All protocols in the database are referenced through the *protocol* table and are identified by an ID and version number. In order to track these protocols, the complete ancestry (*protocolAncestry*) and history (*protDefaultHistory*) of the default version for a protocol are stored. This allows researchers to keep a history of the versions for a specific protocol with which they may be working. A protocol may also be named (*namedProtocol*) so that referencing a protocol is more natural.

Each protocol consists of a set of associated manipulations defined by the *_protocolManip* table where each manipulation is stored as a tuple in an extension table attached under the manipulation table (see Figure 14). Protocols stored in a NeuroCore database can consist of two very different types of collections of manipulations:

1. A set of ordered manipulations where each manipulation can be followed in a sequence by another manipulation without any regard to the exact timing of the specific manipulations. The ordering of these manipulations is handled by the *_protocolManip* table, where one is able to define where each manipulation starts in the sequence (manipStart) and end in the sequence (manipEnd).

2. A set of manipulations that occur at a specific time relative to the beginning of the experimental segment. This is very common in many behavioral and neurophysiological experiments. The timing of each manipulation relative to the beginning of the experimental segment is stored in the manipulation extension tables themselves.

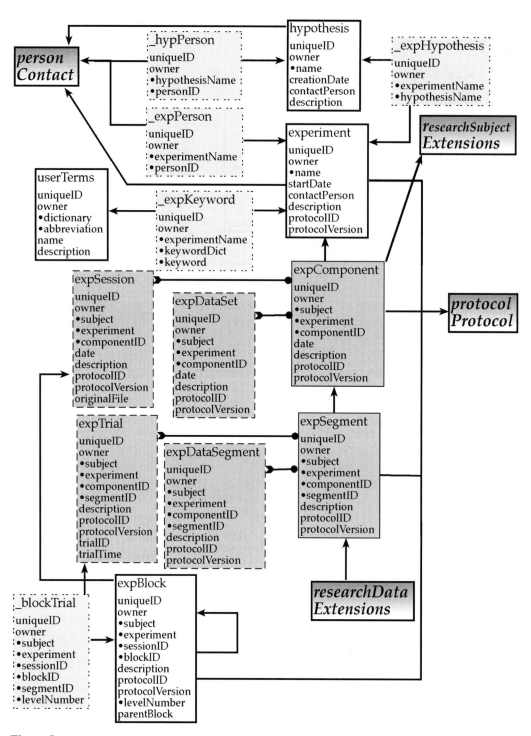

Figure 8 Experimental portion of the core schema. All tables depicted are sub-tables of the *experimentCore* table (Figure 4). Each experiment consists of a hierarchical collection of experimental components and segments which can be grouped into blocks.

The first collection of manipulations defined describes the standard process-oriented scientific data model that allows one to define a sequential collection of processes and the transitions between them. This protocol framework is the typical model used to describe experimental protocols for the genome databases developed by the Human Genome Project and other molecular biology databases (Chen and Markowitz, 1995; Lee *et al.*, 1993; Pratt and Cohen, 1992). However, this framework is not sufficient to describe the protocols involved in neuroscientific experiments. In order to be able to store neuroscientific protocols, provisions must also be made for the second class of protocols defined above, where the set of manipulations is specifically timed. One interesting aspect of neuroscientific experiments is that they can be described by both types of

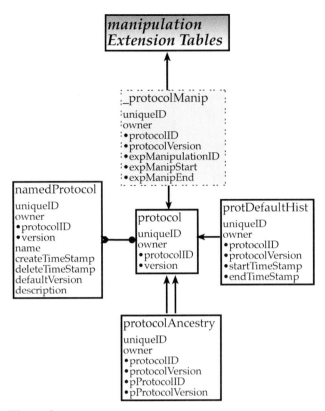

Figure 9 Section of NeuroCore database that provides the framework for the storage of protocols in the core database. All tables depicted are sub-tables of the *experimentCore* table (Figure 4). Each protocol consists of a set of manipulations that are grouped and/or ordered in time.

protocols described previously. By using the implemented framework, the NeuroCore core database is able to store both types of protocols together in a single framework.

As mentioned in the previous section, each experiment in NeuroCore is defined as a hierarchical collection of components and segments. In order to be able to store complete information regarding the experimental protocol, each entry in the experiment, component, block, or segment can have a protocol associated with it (segments are the only objects that are required to have a protocol). This allows for the researcher to store protocols at varying levels of the experimental hierarchy. This can be useful, for example, in defining the experimental conditions vs. the experimental manipulations used to collect the actual data. The experimental conditions, which may remain constant across a complete experiment, can be defined via a protocol referenced in the experiment table. This differs from the experimental manipulations, which can be referenced from the experimental component and segment tables.

SUBJECT SECTION

The last aspect of defining an experiment in a NeuroCore database is the definition of the research subjects associated with it (Fig. 10). Each research subject in the database is stored as a tuple in an extension of the *researchSubject* table. One standard extension to this table is the *animalSubject* table that defines animal research subjects of a particular species, as defined by the *animalSpecies* table. Each subject defined in NeuroCore can be associated with one or more experiments through the *_subjExperiment* table, allowing subjects to be represented in multiple experiments. A group of experimental subjects can also be classified as belonging to a research group (e.g., control animals, sham lesion animals, lidocaine infusion animals, etc., as defined in the *resGroupType* table), which is defined through the *researchGroup* table. Finally, the procedures performed on all research subjects are tracked by the *operationLog* table, which notes the procedures performed and the researchers performing those procedures through the *orProcedure, orProcedureType,* and *_orLogPerson* tables.

Reference Sections

One of the important aspects in the design of the NeuroCore database was inclusion of the framework necessary to describe neuroscientific experiments. One key component of this framework is the definition of the proper terminology. Two important classes of terminology are vital for the neurosciences, namely, the definition of neurochemical and neuroanatomical terminology. These two terminology classes are outlined in the following sections.

NEUROCHEMICAL REFERENCE SECTION

Many neuroscience experiments require the use of various chemicals, drugs, molecular markers, etc. To be able to reference such information, the necessary dictionary tables needed to be added to the NeuroCore framework (Fig. 11). The neurochemical dictionary consists of a dictionary of chemicals and chemical compounds stored in the *chemicalDict* table. All chemical compounds found in the dictionary are also related to the type of compound they are (e.g., protein, neurotransmitter, etc.) and stored in the *chemTypeDict* table, which is referenced through the *_chemDictType* table.

NEUROANATOMICAL REFERENCE SECTION

The neuroanatomical section of the core database contains information pertaining to anatomical terminology and specific anatomical objects (Fig. 12), as well as anatomical atlases (Fig. 13). All anatomical terminology is stored in the sub-tables of the *anatomicalTerms* table. Provisions have been made to store terminology regarding cell groups (*cellGroups*), fiber tracts (*fiberTracts*), ventricles (*ventricles*), and superficial structures (*superficial*), such as sulci and gyri, from various anatomical dictionaries and atlases defined in the *anatomicalDict* table. Terminology regarding cell types (e.g., Purkinje cell, granule cell) and cell processes (e.g., axon, soma)

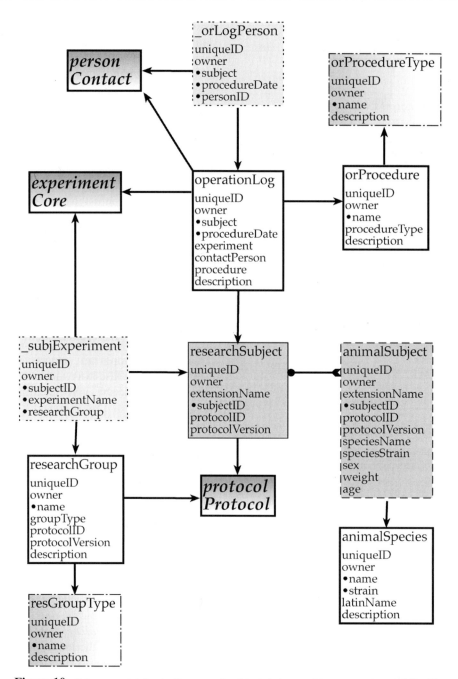

Figure 10 Schema pertaining to the research subjects being used in an experiment. All tables depicted are sub-tables of the *experimentCore* table (Figure 4). Each research subject in an experiment can belong to a specified research group.

can be defined in the *cellType* and *cellProcess* tables, respectively. Specific neuroanatomical information regarding individual cells and their anatomical locations can also be stored in the *cell* table. Each cell (i.e., single unit being recorded from, anatomical reconstruction of a single cell, etc.) also is located in a specific coordinate system (*coordSystem*).

In addition to these anatomical descriptors, researchers can store and access various data associated with published anatomical atlases. Neuroanatomical atlases (all located under the parent table *atlas*) can either be classified as two-dimensional, defined by the *atlas2D* table, or as three-dimensional, defined by the *atlas3D* table. All atlases are defined within a specific coordinate system (*coordSystem* table). Two-dimensional atlases can further be categorized as published atlases, via the *published2DAtlas* table, and histological series, via the *histoSeries2D* table. Each published atlas contains both atlas drawings (stored in the *atlasDrawing* table) and their associated atlas photographs. Histological series, on the other hand, consist just of a sequence of section images. All section images and atlas photographs

CHAPTER 3.2 Design Concepts for NeuroCore and NeuroScience Databases 147

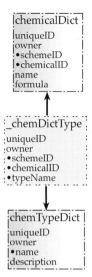

Figure 11 Neurochemistry portion of the NeuroCore database. All tables depicted are sub-tables of the *neuroChemCore* table (Figure 4). These tables define the chemical dictionary for all neurochemical referenced from within NeuroCore.

are stored in the *atlasPhoto* table. In addition, each atlas drawing has associated with it the various anatomical regions found in that particular section. These regions are defined through the *_drawingAnatTerm* table.

Extension Interface for the NeuroCore Database

The previous sections have described the essential components of the NeuroCore database that describe the framework for all neuroscientific experiments. However, this framework does not define the specific attributes regarding the research data, experimental protocols, and research subjects used in any neuroscience experiment. In order to completely store a particular experiment, the core database needs to be extended so that the details of a particular experiment can be stored. The database has been designed so that the core schema can be extended into five different areas (Fig. 14):

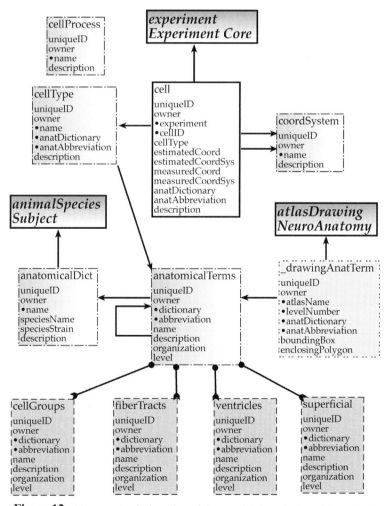

Figure 12 Neuroanatomical portion of the core database dealing with anatomical terms and objects. All tables depicted are sub-tables of the *neuroAnatCore* table (Fig. 4). All anatomical terms are stored with reference to an anatomical dictionary.

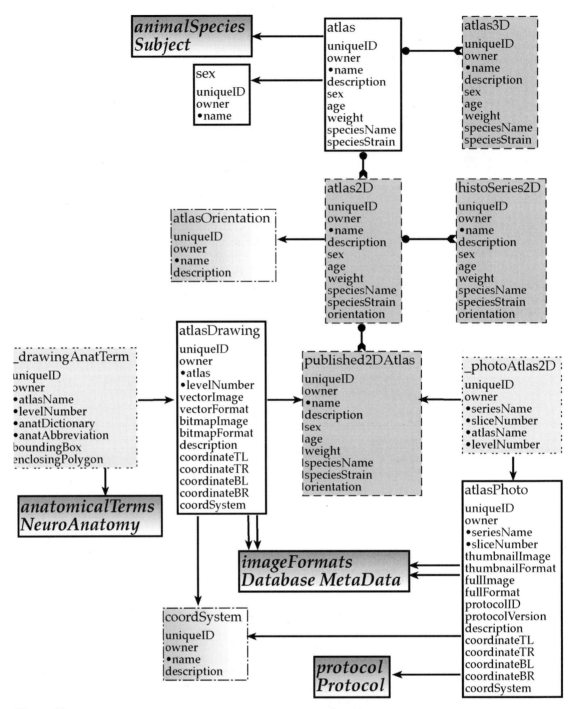

Figure 13 Neuroanatomical portion of the core database dealing with anatomical atlases. All tables depicted are sub-tables of the *neuroAnatCore* table (Fig. 4). Each atlas can reference the anatomical objects contained within it.

1. The research subject being used in an experiment through extensions to the *researchSubject* table. A standard *animalSubject* extension is already included.
2. The research data being collected in a particular experiment through extensions to the *researchData* table.
3. The actual manipulations which make up the protocols being used in a specific experiment through extensions to the *expManipulation* table.
4. The equipment being used in the experiment through extensions to the *expApparatus* table. The *_dataApparatus* and *_manipApparatus* tables relate this equipment to the specific data collected using the equipment.
5. The metadata (statistics, annotations, etc.) associated with each experiment through extensions to the *expMetadata* table. The *_dataMetaData* relates each piece of metadata to its original underlying data, thereby allowing researchers to generate new metadata on the original data sets.

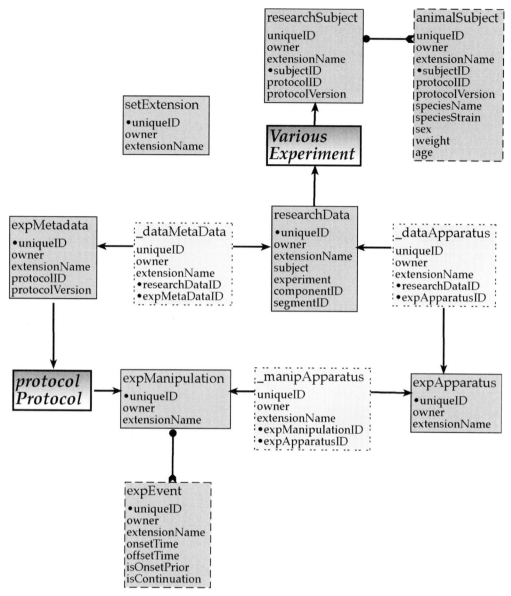

Figure 14 Areas of the core database where a user may extend the core database schema. All tables depicted are sub-tables of the *coreExtension* table (Fig. 4).

One very important standard extension included with NeuroCore is the *timeSeriesData* table. This table allows for the storage of time-series data through the use of a newly defined database object or database type. This special datatype is briefly described in the following section (a more complete description can be found in Appendix A2). A discussion of how a NeuroCore database may be extended for various experiments can be found in Chapter 3.1.

Neuroscience Time-Series Datatype

The Time-Series Datatype is constructed as a datablade in Informix by using Informix's opaque type. The opaque data type is a fundamental data type that a user can define. (A fundamental data type is atomic. It cannot be broken into smaller pieces, and it can serve as the building block for other data types.) When the user defines an opaque data type, the data type system of the database server is extended. The new opaque data type can be used in the same way as any built-in data type that the database server provides. To define the opaque data type to the database server, the user must provide the following information:

- A data structure that serves as the internal storage of the opaque data type
- Support functions that allow the database server to interact with this internal structure
- Optional additional routines that can be called by other support functions or by end users to operate on the opaque data type

For further information on implementation of the Time-Series datablade, please see Appendix A2.

Conclusion

The previous two chapters have provided an overview of the core functionality of the NeuroCore database. The core framework that has been introduced here is not able to store a complete neuroscientific experiment, as it lacks the extensions that allow one to fully describe the experimental preparation and the types of research data being collected. Case studies in extending the core for specific protocols were presented in Chapter 3.1. For the latest release information and further documentation of the NeuroCore database, please visit the University of Southern California Brain Project Web site at http://www-hbp.usc.edu/Projects/NeuroCore.htm.

References

Chen, I. A., and Markowitz, V. M. (1995). An overview of the object-protocol model (OPM) and the OPM data measurement tools. *Information Systems* **20(5)**, 393–417.

Lee, A. J., Rundensteiner, E. A., Thomas, S., and Lafortune, S. (1993). An information model for genome map representation and assembly. 2nd International Conference on Information and Knowledge Management.

Pratt, J. M., and Cohen, M. (1992). A process-oriented scientific database model. *ACM SIGMOD Record.* **21**, 3.

CHAPTER 3.3

User Interaction with NeuroCore

Edriss N. Merchant,[1] David A.T. King,[2] and Jeffrey S. Grethe[3]

[1]University of Southern California Brain Project, University of Southern California, Los Angeles, California
[2]Neuroscience Program, University of Southern California, Los Angeles, California
[3]Center for Cognitive Neuroscience, Dartmouth College, Hanover, New Hampshire

3.3.1 Introduction

For any database to be widely accepted and used it must be an experimental tool as well as a data storage, retrieval, and analysis system. To address this issue, we have begun development of an on-line notebook, a collection of Web-based tools for the researcher. The goal of the notebook is to have a laboratory-independent set of tools for viewing, recording, and comparing data across the country and around the world. To make data easily accessible to anyone with the proper access permissions, we are using the World Wide Web (WWW) for distribution.

The original on-line notebook was developed to investigate the use of the WWW as an interface to the NeuroCore database. This original interface was written using the Informix™ Web Datablade, which enables one to create Web applications that incorporate data retrieved dynamically from an Informix database. To produce a dynamic Web page, a Web browser passes an HTML form to Webdriver, Informix's interface to the Web server. The HTML form data are passed to the Webdaemon, which requests the proper application pages from the Informix database. These application pages are then parsed, and all SQL statements embedded in these pages are executed. A new HTML page is then dynamically constructed and the results are returned to the Webdaemon, which then forwards the page to the Web server through the Webdriver interface.

The original version of the on-line notebook contained various interfaces designed for specific types of queries. As an example of the functionality of these components, two of the query interfaces will be briefly presented. First, a generic query interface is shown.

Through this on-line interface (Fig. 1) any textual SQL query can be submitted to the NeuroCore database. The second interface consists of a graphical query based on an histological section contained in the database (Figs. 2- and 3). This notebook interface allows the user to search the data in the database using anatomical coordinates which are input graphically when the user selects the region or location of interest on the histological image. This original prototype version of the on-line notebook helped us choose the WWW as the delivery method for our interfaces to NeuroCore. In designing the current on-line notebook, several Web technologies have been considered, such as Java, Javascript, PERL, and the Informix Web Datablade. The prototype interface also gave us feedback from our users as to features they would like to have in future versions of the on-line notebook.

The major task in designing and building the current on-line notebook was to develop a collection of interfaces that a researcher could use in all aspects of his work. One can view the scientific process as a cycle of events (Fig. 4). Scientific experiments begin with knowledge from journal articles that is used to formulate new experiments. In order to allow researchers greater access to data in the future through on-line journals, a new Model for On-line Publications (MOP) must be developed. Once a new experiment has been formulated, the researcher must then collect new data, which will then be analyzed. This will require several tools. The first stage in using NeuroCore concerns the database itself: Can the current instance of the database store all the data that you wish to store? For example, a database used in previous experiments might allow a user to store data from single unit recordings and behavioral data. In your

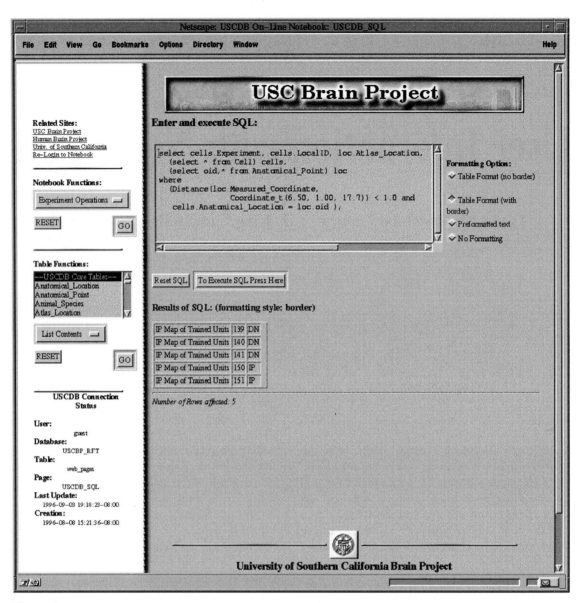

Figure 1 On-line SQL form allows users to interact with the Informix database over the Web using the standard SQL query language. The SQL query shown here searches for single unit recordings, stored in the database, that lie within a certain distance from an anatomical coordinate.

current experiment, however, you might wish to collect multiple unit recordings for which the database entries will contain slightly different information from the single unit entries already in your databases. To examine and modify the database, a schema browser (NeuroCore Schema Browser) will be necessary. Once a researcher has the correct extensions in place, a tool will be needed to enter the data being collected (Java Applet for Data Entry, or JADE). This data entry tool will need to be extensible so that different researchers can use the same foundation for their specific data entry needs. Once a researcher's data have been stored in the database, a couple of tools will be required to support the data management and analysis. First, a database browser (DB Browser) is needed to allow researchers to view and navigate the data stored in their databases. Second, an extensible analysis platform (DataMunch) must be available for users to analyze their data. The results from these analyses will then be used in producing a new journal article. For a database to be used widely by the community, it must be embedded at each stage of this life-cycle. To accomplish this, a suite of tools has been developed as part of an on-line notebook to support the various stages of this life-cycle. The five tools mentioned earlier will be discussed in the remainder of this chapter. Up-to-date information for all tools discussed in this chapter can be found online at http://www-hbp.usc.edu.

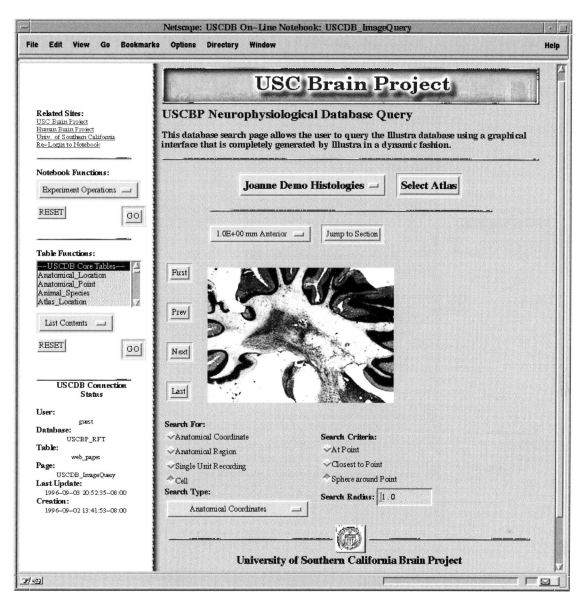

Figure 2 This atlas-based graphical query demonstrates Informix's ability to generate dynamic Web pages as well as the ability to store multimedia data such as images. Through this query, a user can select a search region by clicking directly on the anatomical image. Once the query is executed, the single units recorded in that search region are returned (Fig. 3).

3.3.2 NeuroCore Schema Browser

Introduction

The NeuroCore Schema Browser is a Web-based database client for the NeuroCore database. The goal is a fully functional database client that can be run from a Web browser and can access, retrieve, and manipulate both data and metadata on the USC Brain Project's NeuroCore database architecture. The schema browser allows a user to view the table structure of the Neuro-Core database in a hierarchical tree format. The browser interface contains four sections (Fig. 5). The menu allows users to login/logout and change databases. Tables of the NeuroCore database are displayed in the bottom left menu portion of the Schema Browser in a folder tree format, similar to a standard directory tree. By clicking to expand the folder, the child or extended tables are displayed for that particular table. When one clicks on the name of a table, the table information (such as column names, description, data type, etc.) is displayed in the top right portion of the Schema Browser.

Background/Evolution

The Schema Browser was initially developed as a joint venture between the University of Southern California (USC) Integrated Media Systems Center (IMSC) and the USC Brain Project. The requirements and design reports were developed as part of this collaboration, whereas the development of the final tool itself has been conducted by the USCBP.

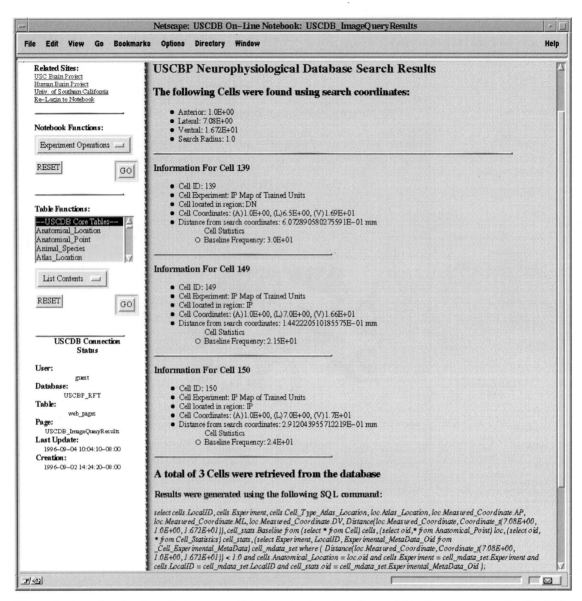

Figure 3 Textual results of graphical query provide information from the cells returned by the query generated in Fig. 2. The actual SQL query used to retrieve the data can be found at the bottom of the captured screen.

Implementation

The Schema Browser currently uses Java and JDBC to display the contents of the database hierarchy. The hierarchy and table information are dynamically loaded from the NeuroCore metadata tables, which the Web browser displays in a hierarchical folder tree format. The original version of the Schema Browser only allowed a user to view metadata associated with a NeuroCore database (i.e., the information regarding the tables and columns that comprise the database). The current release allows users to work directly with the database metadata stored in NeuroCore. This involves being able to view and extend the schema of the database. Because Informix is an object-relational database, this involves not only displaying the tables but also being able to view and manipulate the inheritance hierarchy.

Future Work

One of the most important aspects of the schema browser is the ability to modify the NeuroCore schema. Currently, users are only allowed to add new extension tables to the database through the interface. In future releases, to enter a new schema in the database (e.g., for a new experiment from a particular laboratory), the researcher will have the option of modifying an existing extension set or creating a new extension set. An extension set is a collection of table and datatype definitions that extend the NeuroCore database for a specific experiment or set of experiments. Modification of these extension sets will allow the researcher to save time by using pre-existing schema elements and will allow the researcher to build upon an existing schema. By creating a new set of database extensions, the researcher is able to

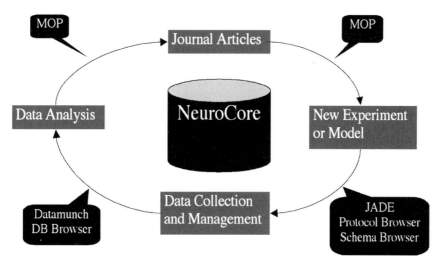

Figure 4 The scientific life-cycle. Scientific experiments begin with knowledge from journal articles that is used to formulate a new experiment. The researcher must then collect new data, which will then be analyzed. The results from these analyses will then be used to produce a new journal article. For a database to be used widely by the community, it must be embedded at each stage of this life-cycle. To accomplish this, various tools must be developed as part of an on-line notebook to support the various stages of this life-cycle.

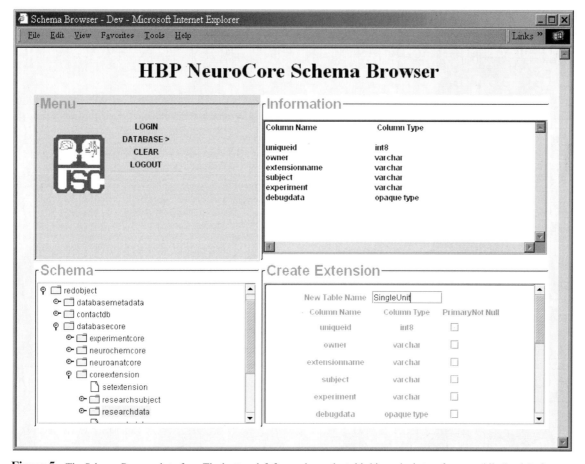

Figure 5 The Schema Browser interface. The bottom left frame shows the table hierarchy in tree format, while the right frame displays the description of the table chosen (top) as well as a form (bottom) to extend the database using the current table as a parent.

extend the NeuroCore schema to customize the database to meet the needs of his laboratory (see Chapter 3.1 for more information on how NeuroCore can be extended for various experiments). In addition to the ability to work with the database schema, the browser will also have to be able to work with objects and abstract data types created in an object-oriented fashion.

3.3.3 Java Applet for Data Entry (JADE)

Introduction

JADE is a Web-enabled tool that allows a neuroscience researcher to enter experiment data directly into a NeuroCore database. JADE is designed specifically for the NeuroCore database schema and offers the same flexibility and extendibility as the NeuroCore design. Like the NeuroCore database schema, JADE is designed to contain certain information elements that are common to all laboratory experiments, for example, all the descriptive information regarding an experiment (i.e., experiment name, experimenter, etc.). JADE can then be extended to meet the needs of a particular laboratory. Researchers can enter specific data regarding their particular experimental paradigms and conditions. One laboratory may be conducting an *in vitro* experiment where the application of various chemical agonists is required, whereas another laboratory may be recording neuronal unit data during a behavioral paradigm using various stimuli (see Chapter 3.1). It is therefore important that JADE be extendible in order to accommodate the needs of a particular laboratory similar to NeuroCore.

Figure 6 JADE three-panel interface: Menu (left), Body (upper right), and Messages (lower right). The Body displays the entry labels and entry fields that are dynamically retrieved from the laboratory database.

The interface for JADE (Fig. 6) consists of three different sections. The first section is the Menu, which contains the options that the user has to work with. The Body section contains the forms with which the user will most often interact; this section contains the labels and text fields for entering data. The third section of the JADE interface is the Messages section, which will send messages to users based upon their actions. JADE has been successfully used by researchers at USC to enter legacy data from traditional paper-based experimental notebooks.

Background/Evolution

The main design consideration for JADE was that the interface needed to have an extendible architecture. In the early development of JADE, interfaces for different laboratories were developed using a specific template for that particular laboratory. The templates were developed by identifying the specific information necessary to be entered for each experiment. The initial development was done for experiments where electrophysiological recordings were obtained (see Chapter 3.1). The information needed for JADE was extracted from descriptor files used by researchers to describe their data for a nongraphical data-entry tool that was being used to upload their data into a NeuroCore database.

The original JADE interface led a researcher through all the questions necessary to create the descriptor files they had been generating. After all the information was received, JADE produced SQL statements for entry into the database. The complete set of SQL statements was then transformed into a single transaction file so that this file could be sent to the database as a single processing block. However, one problem with this first version of JADE was that interfaces for a new laboratory experiment were not easily added to the base architecture. The new experiment would need to have a new template created, and JADE would have to be modified to accommodate this new template. A solution to this was to let JADE query the table definitions of the extension set that constituted a certain experiment. This would allow JADE to be able to produce dynamic interfaces based on the metadata describing the actual tables in an extended NeuroCore database; therefore, JADE must have connectivity to the NeuroCore database for data entry as well as information retrieval of database metadata for the generation of these dynamic interfaces. This more generalized version of JADE is currently under development and is nearing completion.

Implementation

In addition to the metadata tables in the NeuroCore database (see Chapter 3.2, Fig. 6) used to construct dynamic query interfaces, some JADE-specific tables were necessary. JADE would use these tables to build the user prompts accordingly. Because only the data-entry flow for each experiment varies, the core program retrieving and submitting data would remain unchanged. The critical step to create this generic JADE version was to identify the types of database access required by JADE during data entry. The procedure used by JADE to dynamically produce an interface depends on the type of data being entered. More specifically, whether the data are being entered into a single table or multiple tables and whether the data consist of a single entry or multiple entries. In order to accommodate these varying data-entry tasks, additional metadata tables specific to JADE had to be added to NeuroCore. In addition to the detailed description of where data must be entered throughout the database, the order in which data are entered is also extremely important. During the data-entry process, there are certain values that must be defined in describing the experiment before other information can be entered (e.g., to describe the protocol for individual experimental trials, one first must know how many trials there are). With the addition of JADE-specific metadata tables, NeuroCore and JADE have been able to produce dynamic data-entry forms.

Future Work

The current version of JADE has been used for data entry of legacy experiments in laboratories involved with the USC Brain Project. Once the initial design, implementation, and testing of JADE have been completed, further work on JADE has been planned to improve its capabilities. Based upon feedback from users entering data into their own laboratories' NeuroCore databases, a number of future enhancements have been identified:

- Streamline the automatically generated input forms to minimize the display of duplicate information.
- Enhance the user interface when entry fields reference values from other tables.
- Build an interface to populate the tables that JADE queries for building the prompts. Currently, the information is entered manually by a database administrator. A user interface would help administrators to view and update information for JADE.

3.3.4 Database Browser

Introduction

The Database Browser offers a neuroscience researcher the ability to view, browse, query, and display results of an experiment stored in the NeuroCore database. The interface was designed for the NeuroCore database schema and offers the same flexibility and extendibility as the NeuroCore design. By using the NeuroCore schema as a foundation, any database built

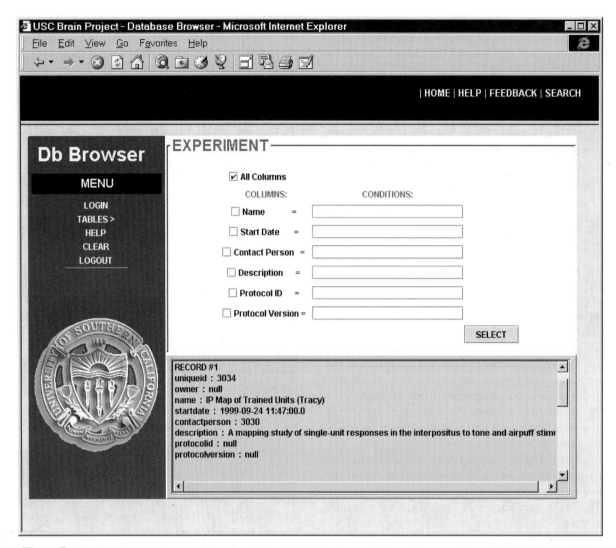

Figure 7 Database Browser three-panel interface: Menu (left), Body (upper right), and Messages (lower right). The Body displays the Columns and Search Criteria Text fields, while the Messages/Results section displays the results of the query performed.

using NeuroCore can use the Database Browser, as all NeuroCore databases contain the same core information and structure.

The user interface for the Database Browser consists of three sections (Fig. 7). The left-hand portion is the Menu for the Database Browser. The researcher has the options to Login, Access Table pages, get Help, Clear the contents of the other two sections, or Logout. The upper right-hand portion is the Body of the user interface, which changes based on the user's actions and is used to enter information or view the database contents. The lower right portion of the page, the Messages/Results section, is used to interact with the user. It will display information or instructions based upon the user's actions.

In order to facilitate searching, a user selects a particular table to search. The user selects the table name from the menu and is shown the columns for that table. The user is given a choice of checkboxes and text fields in two separate sections displayed side by side (Fig. 7). The section on the left contains a list of the columns that the user may select (with checkboxes) for the search results. For example, in the Experiment table, a user selecting the "Name" checkbox will be shown only the names of the experiments in the database. By the same method, a user choosing the "Name" and "Contact Person" checkboxes will be returned the name and contact person for all experiments stored in the database.

The section to the right contains a number of text boxes which correspond to a column of the table. These textboxes are used to allow a user to enter information or keywords for a search query. For example, using the same Experiment table, a user entering "Michael Arbib" in the text box corresponding to the Contact Person column will be shown the experiments in the database for which "Michael Arbib" is assigned as the Contact Person. The text box feature allows the user to narrow the search results on a table.

Another feature incorporated into the Database Browser is a specialized query selection. A particular laboratory may need a certain query performed

CHAPTER 3.3 User Interaction with NeuroCore

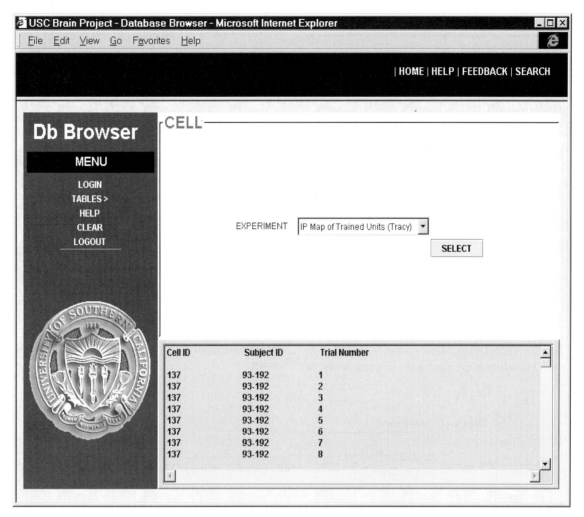

Figure 8 Specialized query result: This query returns the cell ID and corresponding trial and subject ID number for a recorded cell. The query is specialized for the needs of the USC Thompson Laboratory.

occasionally based upon the data stored in the database. For example, one query that has been incorporated is to identify the cells that are recorded during an experiment. The user who selects this query from the menu and selects an experiment from the database is then returned a listing of all the cells and their corresponding trial for that experiment (Fig. 8).

Implementation

The Database Browser was created initially as an interface so that researchers could verify their data were correctly stored within the NeuroCore database. A lot of work on an external interface had already been completed involving JADE so the Database Browser was built on this foundation. The Database Browser is a Java Applet that connects to the Informix database via JDBC.

Future Work

The initial development of the Database Browser has been completed, and further work has been identified to improve the tool. The program is being enhanced to dynamically display the table information and the necessary checkboxes and text input fields based on the information stored in the NeuroCore database metadata tables (see Chapter 3.1, Fig. 6). Currently, the Database Browser stores and retrieves the table information within its program. In order to add a new table for querying, the program must be changed to accommodate the new table. The same principle for storing the table information within the database and retrieving that information to build the interface that is incorporated in JADE needs to be applied to the Database Browser. This feature will improve the maintainability of the tool. Another feature that has been identified for future work is the dynamic retrieval of specialized queries for a particular laboratory. The query to retrieve cells corresponding to an experiment may not be used by all laboratories. In addition, new specialized queries may need to be added to the Database Browser. The queries would be stored within the database and could then be retrieved at a later time by the Database Browser when needed by a user.

3.3.5 DataMunch

Introduction

DataMunch is a free, open-source-code application that runs in the Matlab™ scripting language. These routines are used primarily to analyze two types of data:

- "Spike" data, that is, extracellular single or multiple-unit waveforms, spike times, or bin values obtained from single or multiple cells
- Continuously recorded behavioral data, such as (the magnitude of) eyeblink or limb movement

DataMunch (King and Tracy, 1999) was designed to analyze data from experiments using a classical eyeblink conditioning paradigm (see Chapter 3.1). The analysis program itself works in batch mode, allowing the user to define what data should be examined and what operations should be performed on the selected data.

A set of analysis records created from a collection of data is called an "analysis set." One is able to search each analysis set DataMunch creates with various tools in order to find collections of cells that match whatever combination of statistical criteria were defined. These selected data can then be analyzed and graphed in various ways:

- Plots of the distribution of spike-train onset times (or post-stimulus firing rates or response types, etc.) for all cells having strong correlations with behavioral onset times
- Histograms for each cell with this (or some other) combination of measurements
- Scatterplots of one statistic against another for all cells selected by the search parameters

In summary, DataMunch can take a set of data and produce a set of analysis records. It provides the user with a great deal of control over how data are analyzed as well as how they appear when plotted (see Fig. 9). A researcher can easily search through an analysis set with a convenient search tool to test hypotheses and, based on the results of this search, can create new plots that summarize part or all of the analysis set.

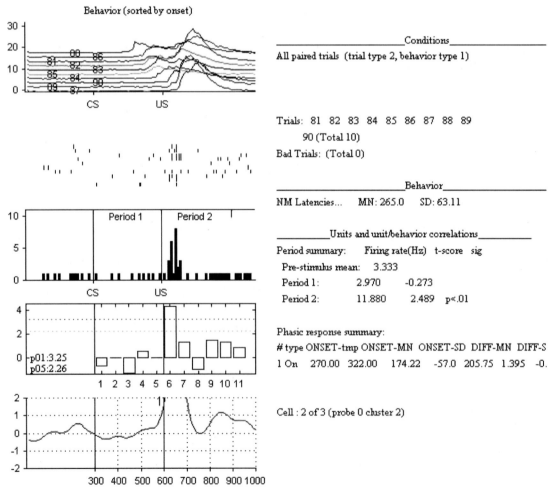

Figure 9 DataMunch: A typical example of output from an analysis of data from a classical conditioning experiment.

Background and Evolution

DataMunch is the data analysis package that members of the Thompson Laboratory have developed to analyze single- and multiple-cell spike data and behavioral (eyeblink) data. It was originally designed to recognize several popular file types, plot behavioral curves, spike rasters, summed histograms, and normalized firing rates for each cell. The program also calculates a set of statistics and displays them both within the plot and also in a separate spreadsheet suitable for further analysis. DataMunch was originally designed to work "locally," analyzing files on a single machine; however, current efforts are expanding DataMunch's capabilities so that it will be able to connect directly with a NeuroCore database and ultimately allow a convenient user interface through the Web. This functionality is being added using database and Web connectivity toolkits developed specifically for Matlab.

Future Work

There are two major efforts currently underway to connect DataMunch to the NeuroCore time-series database (for information on the time-series database, please see Chapter 3.1): data transfer between Matlab and DataMunch and an enhanced querying tool that interfaces DataMunch with the database. The querying tool will allow users to create advanced search statements by choosing variables such as response type (increases vs. decreases in firing rate), response timing (onset of neural responses), etc. For each variable, the user can enter a value or range of values to search for, and cells that fit this criteria are returned by the database for plotting and statistical analysis. Choices are also offered as to what kind of plots should be displayed and what should be done with the statistics returned.

3.3.6 Discussion

The development of the tools comprising the on-line notebook is an ongoing process. With each successive release, feedback from users highlights new functionality that can be incorporated within this framework. One such requested addition is the addition of a graphical way to reconstruct the protocol for a given experiment,

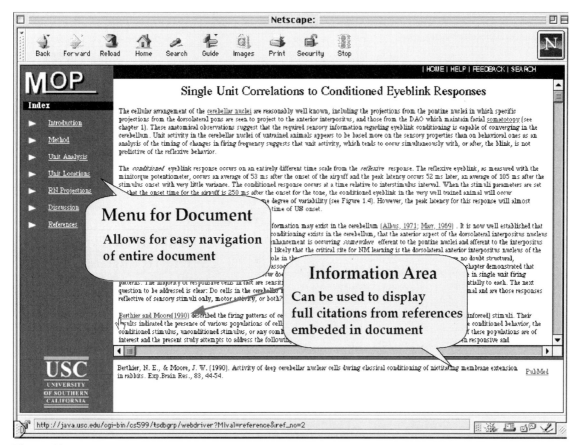

Figure 10 A prototype on-line journal interface. This interface uses features of HTML and Java to enhance the user interface for a journal article. Here, we have taken a chapter from a thesis that contains the data we have stored in the database that was developed (Tracy, 1995). This document has a navigation bar on the left for easy navigation and an information area on the bottom which, in this example, is giving the user information concerning a specific reference in the document. This allows the user to view select information without leaving his current view.

session, block, or trial. This Protocol Viewer would allow a user to graphically view protocols related to the data collected, to edit protocols, and to create new protocols. The Protocol Viewer is currently in the design stage of development; however, there is one aspect of a researcher's interaction with an on-line database that is of critical importance to the success of any scientific database, namely, the interaction between published journal articles, the data they contain, and the databases housing this data.

A scientist's life revolves around publication (Fig. 4). When designing a new experiment, many articles are read for background and information on experimental procedures. The scientist must then collect, store, and analyze all his data. These data will then be published as a journal article, which completes the cycle. Currently, the NeuroCore database and on-line notebook will allow researchers to collect and store, retrieve, and analyze their data. However, a critically important aspect of this cycle is the journal article itself. Science revolves around the publication and reading of peer-reviewed journal articles. With many journals going on-line, it is imperative that the databases we construct are able to interface with the on-line journal of the future. To this end, we have developed MOP (Model of On-line Publishing) as a prototype of such an interface. MOP highlights some important features that should be included in tomorrow's on-line journals.

The Concept of an Embedded Type

Each journal article contains many types of information (e.g., images, graphs, references). For each embedded type a set of functionality needs to be defined so that users can interact with these embedded objects. For example, a citation in the article, when selected, could display the full reference in another window without affecting the presentation of the article. A user could then interact with this reference independently of the article itself. Another example would be a graph containing various statistical data. In this instance, selecting the object would provide the user with a description of the graph object as well as specific menu options related to that graph object (Fig. 11). However, by activating the object (i.e., double-clicking the object), one could launch a separate window where users could interact directly with the graph and the data contained within it (Fig. 12).

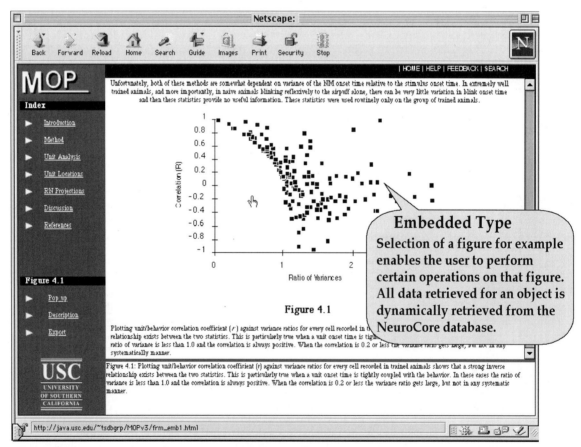

Figure 11 An embedded type in a document. In this example, the chart is an embedded type that can be operated on. In the lower left corner of the interface is a special contextual menu for this specific chart object. In the status window on the bottom of the interface is a complete description of the embedded object.

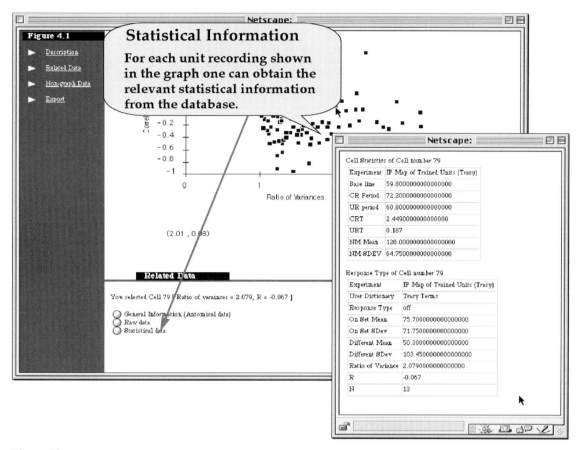

Figure 12 Once an embedded object has been activated, a new window with that object is launched. This allows access to other information stored in the database related to specific data within the object. In this example, a user has selected a specific data point and then has asked for more information concerning that data point from the database. In this instance, the user has retrieved statistical data concerning a specific cell found in the database.

A More Intuitive Interface Than What Is Found in Today's On-Line Journals

Most on-line journals do not take full advantage of the new Web-based medium they are using. Many still look and feel like a printed journal article, with the exception of some hypertext links to on-line reference information. To make an on-line journal more effective, one needs to take advantage of this new medium. For example, one could provide status windows and contextual menus to ease navigation of the document and references (Fig. 10).

The Ability To Drill Down into the Data That Generated a Specific Graph or Chart

With both journals and databases on-line it is now feasible to connect the summary charts and figures found in journal articles to the actual data that generated them (Fig. 12). This will allows users to re-analyze the underlying data themselves, examine the data to find data points that were clipped or not shown in the figure, and to find other data related to individual data points.

References

King, D. A. T., and Tracy, J. (1999) DataMunch Web Site. http://www.novl.indiana.edu/~dmunch.

Tracy, J. (1995). Brain and Behavior Correlates in Classical Conditioning of the Rabbit Eyeblink Response, Ph.D. thesis. University of Southern California, Los Angeles.

PART 4

Atlas-Based Databases

Part 4

Non-Book Databases

CHAPTER 4.1

Interactive Brain Maps and Atlases

Larry W. Swanson
*The Neuroscience Program and University of Southern California Brain Project,
University of Southern California, Los Angeles, California*

Abstract

The emergence of powerful and practical computer graphics applications in the last decade is revolutionizing the production and use of brain maps for summarizing the results of experimental neuroanatomy. Perhaps the most far-reaching consequence of electronic or digital brain maps will be their use in databases of graphically presented structural information that may be queried via the World Wide Web. The design of such databases is a major goal of the USC Brain Project, and they will be most powerful if results are presented on, or indexed in a systematic way to, standard or reference atlases of the brain. In practice, however, direct comparisons of histological structural data (as opposed to tomographic scans) obtained from different brains are problematic because of possible differences in plane of section, shrinkage or swelling, and nonlinear warping that has a different pattern for each histological section as it is mounted on glass for microscopic observation. Computer graphics applications also can be used to create three-dimensional models of the brain, as well as two-dimensional flatmaps, and they also may serve as templates for electronic databases.

4.1.1 General Features of Maps

In principle, there is little if any difference between maps of the Earth's surface, which we are all familiar with, and maps of the nervous system—and of the brain, in particular (see Robinson and Petchenik, 1976; Swanson, 1992). First and foremost, maps are *representations* of some part of the environment on a flat surface; that is, maps are an abstraction or simplification of an object (such as the Earth), typically from three dimensions to two. Traditional maps provide three obvious advantages. First, three-dimensional objects can be represented in two dimensions for publication. Second, large or small objects can be scaled to convenient sizes. And, third, the essential features of a complex object can be represented in a simplified, abstract way. The most obvious disadvantages of maps are that by their very nature they cannot reproduce all of the features of the mapped object and they impart misinformation if errors are made during the process of abstraction. In the end, the usefulness of any map is a function of its accuracy, clarity, and ability to display a particular type of information. It should be obvious that there is no one "correct" way to produce a map; after all, over 200 different projections of the Earth's surface onto a flat plane have been devised over the years, and none is ideal (Snyder, 1993). This is a consequence of the fact that in transferring a pattern from any curved surface (except a cone) onto a flat surface, only one of the three important features of distance, area, and shape can be maintained accurately.

While this is all well and good, it is obvious that mapping a complex three-dimensional solid such as the brain is immensely more difficult than mapping the surface of an approximate sphere (the Earth) onto paper. One might think that mapping the brain is more like mapping the Earth considered as a solid sphere, but even this is a gross oversimplification because the Earth is formed essentially of concentric layers, whereas the central nervous system (CNS—the brain and spinal cord) is essentially a hollow tube whose walls contain irregularly stacked cell groups that are differentiated to wildly different extents along the longitudinal axis (see Alvarez-Bolado and Swanson, 1996).

The history of attempts to map the CNS is very long indeed—it can be traced back some 2500 years—and many highly ingenious procedures have been developed (Swanson, 1999). Nevertheless, the widespread availability of personal computers in the 1990s has stimulated a radically new approach to cartography, in general, and to brain mapping in particular. The purpose of this chapter is to explore some of the new ways that brain maps can be produced and used in electronic formats and to introduce the reader to specific developments in the USC Brain Project. However, to place these developments in context, it is important first to review certain general problems faced by brain mappers: the overall structure of the brain and methods used to determine the architecture of brain circuits.

4.1.2 Overall Structure of the Brain

As mentioned above, from a topological perspective the vertebrate CNS has a rather simple shape and location: a dorsal midline tube that is closed at both ends, and that starts in the head (as the brain) and ends in the abdomen (as the spinal cord). However, the actual geometry of the adult CNS is very complex, mainly because: (1) the longitudinal axis of the tube is highly curved; (2) the walls of the tube are highly differentiated in each "segment" of the tube (endbrain, interbrain, midbrain, hindbrain, and spinal cord); and (3) connections between cell groups stacked within the walls are exceedingly complex, showing massive divergence and convergence. To gain some appreciation for the task faced by brain mappers, it is helpful to consider some of the major physical features of the mammalian brain, and for this we shall compare certain features of this organ in the rat and human.

Organization and Number of Parts

As a general organizing principle, it is useful to start by pointing out that a great deal of embryological and neuroanatomical work over the last century (well reviewed in Nieuwenhuys et al., 1997) suggests that the brains of all mammals share the same basic architectural plan, with secondary variations on a common theme being characteristic of individual species (and tertiary variations being characteristic of individual members of a species). It goes without saying that brain circuits are formed by individual nerve cells or neurons (the neuron doctrine), and that in general information is received by a neuron's cell body and dendrites and then is transmitted to other cells by its axon (the theory of functional polarity). What, then, about a comparison of the human and rat brain (for further references, see Swanson, 1995)?

Based on round numbers (usually order of magnitude estimates), the following comparisons are useful. To begin with, the human brain weighs about 750 times as much as the rat brain (about 1500 g vs. 2 g; with overall dimensions of about $15 \times 15 \times 10$ cm vs. $2 \times 1 \times 1$ cm). Within those volumes, the human brain contains on the order of 10^{11} neurons (and 10^{12} glial or support cells), whereas the rat brain contains on the order of 10^3 fewer cells of either type (10^8 neurons and 10^9 glial cells).

For mapping purposes, it is also important to appreciate the dimensions of individual neurons: in humans, the cell body (the soma) ranges between 5 to 50 μm in diameter, whereas this range is 5 to 25 μm in rats. Of perhaps more interest, the length of individual axons in humans can reach on the order of a meter and on the order of 0.1 m in rats, whereas figures for the diameter of axons range between 0.1 and 15 μm in humans and 0.05 to 5μm in rats. Thus, there is a scaling factor between nerve cell diameter and axon length of 10^7 in humans and 10^6 in rats. Obviously, this places severe limitations on brain mappers in terms of accurate depiction of results—this is simply not possible when dealing with pathways between different cell groups. These results must be presented schematically.

Luckily for brain mappers, neurons tend to cluster into more or less distinct cell groups, which have been referred to variously as centers, nuclei, cortical areas, and so on. There appear to be on the order of 500 such major centers (which themselves may be subdivided) in the mammalian brain (typically, each center is found on the right and on the left side), and on average each cell group may contain on the order of five different cell (neuronal) types, based primarily on connections and secondarily on neurochemistry and physiology (see Swanson, 1996). Simple multiplication indicates that there are on the order of 2500 major types of neurons in the mammalian brain.

One reason that the circuits formed by these neurons are so complex is that parent axons typically branch (collateralize) extensively, and it seems reasonable based on a broad survey of the literature to assume that, on average, the axon of each type of neuron branches to innervate on the order of 10 different cell groups. If there are 2500 different cell types, this would imply that there are on the order of 25,000 different macroconnections or macropathways making up the circuitry of the brain. However, the true complexity of brain circuitry is fully appreciated only by realizing that each macroprojection typically branches prolifically within its target cell group, so that it has been estimated that the human brain contains on the order of 10^{14} synapses (functional contacts between an axon terminal and another cell, usually a neuron, in the brain), and the rat brain 10^{11}. The fact that the physiological effectiveness ("strength"), and total number, of individual synapses in a terminal field (the microcircuitry) may depend on their previous history of activity—the biological foundation of learning and memory—only adds another level of complexity to the challenge faced by brain mappers.

4.1.3 Experimental Circuit-Tracing Methods

The structural complexity just reviewed has forced neuroscientists to develop over the last century experimental methods for dissecting and characterizing the structural organization and chemical content of neural circuits. Very briefly, contemporary neuroscientists rely on combining two classes of methods for characterizing the structural organization of connections between cell groups, and use brain maps to summarize the results for publication (for references to these and other neuroanatomical methods see Dashti *et al.*, 1997; Swanson, 1998–1999, 1999). One class involves the physiological transport of injected markers (tracers) up and/or down the interior of the axon. These tract-tracing methods rely on the physiological transport of markers from axon terminals to neuronal cell bodies of origin (retrograde tracing), or on transport from cell bodies to axon terminals (anterograde tracing).

The other class of methods involves performing chemical reactions on histological sections of brain tissue (histochemistry), which allows the cellular localization of molecules. The most widely used methods at this time involve *immunohistochemical* localization of neurotransmitter-related molecules within specific circuits and *hybridization histochemical* localization of messenger RNAs (mRNAs) encoding molecules of interest. Immunohistochemistry involves localizing specific molecules (antigens) in tissue sections with antibodies that have been tagged with markers that can be seen under the microscope, whereas hybridization histochemistry involves localizing nucleic acid molecules (especially mRNAs) in tissue sections with complementary strands of nucleic acids that have been tagged with markers that can be seen under the microscope.

As we shall now discuss, one major challenge to brain mappers is how to represent histochemical staining patterns, and the results of axonal transport experiments, on schematic representations of histological sections—bearing in mind the physical characteristics outlined above.

4.1.4 Atlases: Slice-Based Sampling and Standard Brains

The traditional method of representing information about brain structure on a series of slices through the organ actually dates back to Vesalius's revolutionary work in 1543 (Fig. 1a) and has been refined progressively ever since (Swanson, 1999). Because histological sections for experimental neuroanatomy are relatively thin (e.g., 30 μm thick) to increase microscopic resolution, serial sections are virtually never illustrated in publications (a rat brain cut in the transverse plane has almost 700 frozen sections 30 μm thick). Instead, evenly spaced (e.g., 1-in-10) series of sections might be illustrated, or, alternatively, unevenly spaced series of sections might be used more efficiently if data are clustered in certain regions of the brain (Fig. 1b). The neuroanatomist chooses a sampling method that best represents the data. It should be noted in passing that whereas confocal microscopy now can be used to reconstruct three-dimensional datasets from thick histological sections, the volume of tissue involved is always very tiny compared to the volume of the entire brain.

Vesalius produced a series of maps (an atlas) of progressively deeper horizontal slices through the human brain, and over the years atlases of slices cut in one or another of the three standard anatomical planes (horizontal, sagittal, and frontal or transverse) in many different species have been prepared. Nevertheless, it is worth stating the obvious: any particular brain (or brain block) can be sliced physically only in one plane; therefore a decision has to be made as to the most appropriate or useful plane in which to section a particular brain or brain block. Approaches based on using multiple brains, each cut in a different plane of section, to reconstruct three-dimensional datasets or atlases will be discussed below.

Many factors go into determining the most appropriate plane of section for a particular brain or brain feature. However, the factor that concerns us here is the ability to compare easily the results of different experiments, that is, results obtained from different animals. Direct comparisons are always easiest to make when patterns of data are viewed in the same plane of section using a standard format. For example, in comparing the distribution pattern of two neurotransmitter receptors in a complex region such as the visual cortex, it is better to compare the patterns in adjacent sections cut in the same plane rather than in one frontal section and one horizontal section. These considerations assume even more importance when graphical databases are a goal. In principle, an electronic database of graphical neuroanatomical information would be most efficient and accurate if all results were plotted in a standardized way on a standard brain with standardized nomenclature, because under these circumstances queries could be stated precisely, and the results of searches would be unambiguous.

As discussed elsewhere we are far from this state of affairs at the present time (Swanson 1992, 1998–1999). In a nutshell, we are still profoundly ignorant about the fundamental organizing principles of brain architecture and circuitry because of the bewildering complexity outlined above. As a result, there is controversy, confusion, and lack of detailed knowledge about virtually all aspects

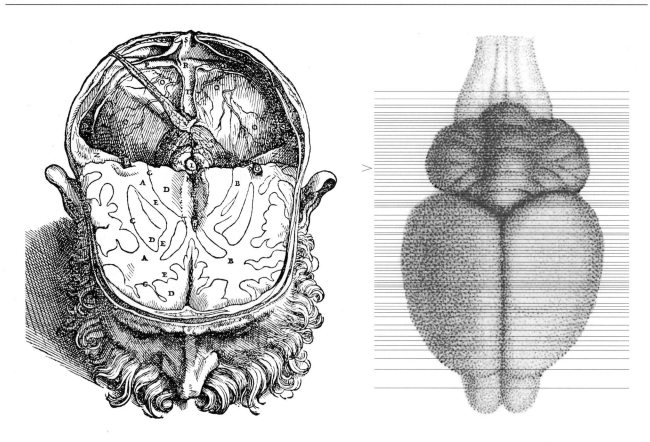

Figure 1 (**Left**) A drawing of the partly dissected human head from Vesalius's 1543 book, *The Fabric of the Human Body*. It shows a horizontal section through the rostral half of the cerebral hemispheres (indicated by A–E) after removal of the skull cap, with the caudal half of the hemispheres cut away to show the brainstem (M indicates the tectum) and cerebellum (indicated by O). (**Right**) A dorsal view of the rat brain (the drawing is from Leuret and Gratiolet, 1857) with a superimposed set of lines indicating the 73 transverse atlas levels of *Brain Maps* (Swanson, 1992, 1998–1999). The arrowhead along the left margin of atlas level lines indicates level 51, which is shown in Fig. 2. (From Swanson, L. W. (1998–1999). *Brain Maps: Structure of the Rat Brain*. Second rev. ed., Elsevier, Amsterdam. With permission. From Swanson, L. W. (1992). *Brain Maps: Structure of the Rat Brain*, Elsevier, Amsterdam. With permission.)

of brain morphology, and at a very basic level, as far as databases are concerned, one result is an unsystematic, incomplete, and contradictory nomenclature. Because of all this, a good case can be made for avoiding attempts to force the use of standard interpretations (e.g., nomenclature) of brain structure—they can only reinforce current misconceptions and retard work designed to understand the true structure of the brain (Swanson, 1992).

Having said this, it is nevertheless essential to have a standard (or perhaps better "reference") brain(s) with accompanying graphical atlas interpretations that can be used for *indexing* any and all results in a particular species (with a particular age and sex). That is, it should be possible, or results should be presented in such a way that it is possible, to say precisely how (or actually to illustrate graphically how) a given interpretation of brain structure is related to the interpretation of brain structure in a standard or reference brain and atlas. This is very easy to do when true synonyms for structures are involved (over 800 are provided in the second edition of our atlas, *Brain Maps*; Swanson 1998–1999) and progressively more difficult when there are differing opinions about the placement of borders or even about the existence of particular structures. Nevertheless, it should be obvious that when anatomical descriptors are used, their precise meaning should be made clear—which all too often has not been the case. The neuroanatomical literature is replete with ambiguities about location and structural details, which led Cajal (see Swanson and Swanson, 1995) a number of years ago to point out that "...there can never be enough [drawings], particularly in anatomy, where one could argue that drawings are more important than text...they are documents of inestimable value that future generations may refer to with advantage in the never-ending battle of opinion and theory."

This of course raises the question as to the best way to present graphically a standard or reference atlas of the brain. Basically, 500 years of experience has led to the current standard of representing the outlines of cell groups and major fiber tracts on atlas drawings of the brain. This is somewhat equivalent to drawing the outlines of the continents, as well as of the various countries, on maps of the world. Once these basic templates have been prepared, they can be used to plot an infinite variety of other data, such as transportation systems, population distributions, weather patterns, and so on. And now that it is obvious that electronic databases (both graphics and text) will be established, it will be necessary to develop a thesaurus of neuroanatomical nomenclature. The same "search word" can have very different meanings to different authors, and the same structure can have many different names. The amount of scholarship required to do a thorough retrospective analysis of neuroanatomical nomenclature usage is probably not feasible—there are hundreds of thousands, if not millions, of references in many different languages spanning about 2500 years. In the long run, the simplest solution is to insist from now on that anyone using neuroanatomical nomenclature define precisely what the terms mean (surprisingly, this is very rare at the moment—the dawn of the 21st century).

Normal Brain Structure (Cell Groups and Fiber Tracts)

As just mentioned, the bare essentials for a good traditional brain atlas include clear indications of how the major cell groups and fiber tracts are distributed within a series of histological sections. At the gross anatomical level, this amounts to showing the spatial distribution of the main centers or nodes in brain circuitry (e.g., cell groups are analogous to major cities on a continent map), and the main fiber tracts (bundles of conducting axons) between them (e.g., fiber tracts are analogous to major highways on a continent map).

The traditional, best, and most convenient way to display the organization of cell groups in a histological section of the CNS is with a Nissl stain, which relies on basic aniline dyes interacting with nucleic acids in the section. The result is a very clear picture of the distribution pattern, size, orientation, shape, and staining intensity of neurons (as well as glial, endothelial, and mast cells) in the brain; the creation of such pictures is known as cytoarchitectonics. In contrast, fiber tracts in the brain are typically revealed with a myelin stain, and the formal study of their spatial distribution is known as myeloarchitectonics. Interestingly, a good indication of major fiber tract distribution in frozen sections can often be gained simply by utilizing dark-field illumination. For an introduction to the literature on these approaches see Swanson (1998–1999, 1999).

A Typical Atlas Level (*Interactive Brain Maps*)

The traditional approach just described for illustrating the disposition of major cell groups and fiber tracts was used for our atlas of the adult male rat brain (Swanson, 1992), which was prepared in the transverse (frontal) plane because experience has shown that in this species a series of transverse sections is most commonly useful for mapping datasets that extend through much of the brain (instead of being confined to one cell group or another). Of the 556 serial sections through the brain, 73 were illustrated in detail; virtually every known cell group is represented on at least two of these levels.

One unusual feature of this atlas at the time of publication in 1992 was that the brain maps were drawn with a computer graphics application (Adobe Illustrator) rather than with pen and ink (Fig. 2); traditional photographs of histological sections were scanned and used as templates for tracing on the computer monitor, aided by a microscope (with the corresponding histological sections) placed next to the monitor. In retrospect, this was a fortunate choice of methods because: (1) the vector-based drawings are much smoother than pen-and-ink drawings, and they can be modified much more conveniently; (2) the vector graphics files can be scaled virtually infinitely without loss of resolution; (3) as we shall return to later in this chapter, the electronic format of the maps lends them immediately to use on the World Wide Web, especially as templates in a standardized atlas for databases of graphical neuroanatomical information (Dashti *et al.*, 1997); and (4) it has become clear that the electronic format allows one to develop a new generation of atlases that are interactive rather than static (printed).

The atlas drawings from the first edition of *Brain Maps: Structure of the Rat Brain* (Swanson, 1992) were soon made available on floppy discs (Swanson, 1993), and the files simply consisted of the brain-section drawings (which are of the right side of the brain and include a gross outline of the brain and ventricles, cell group outlines, fiber tract outlines, abbreviations, and a mask so that overlapping stacks of sections can be used). The second edition of *Brain Maps* (Swanson 1998–1999) included a double CD-ROM with a much more advanced version of the 73 atlas-level templates. The major technical advance that led to the new electronic atlas format was the introduction of a layer manager palette to computer graphics illustration applications, and its use prompted us to refer to the new atlas as *Interactive Brain Maps*. In essence, contemporary illustration applications (based on vector graphics) allow a great deal of flexibility in modifying, viewing or hiding, and printing or not printing various components of a file containing an atlas level, and one can do so essentially in real time.

The layer manager palette allows one to create an essentially infinite set of transparent (or translucent), perfectly aligned overlays for a map template and to arrange and view them in any order. In other words, a layer manager is simply a list of overlays. The stacking order can be changed in any way desired, and individual layers can be named, hidden or shown, and printed or not printed.

In *Interactive Brain Maps*, different layers are used to display the following features of each atlas level: (1) a bilateral drawing of the brain, (2) a unilateral photograph of the Nissl-stained histological section used to prepare the drawing, (3) a grid of physical coordinates,

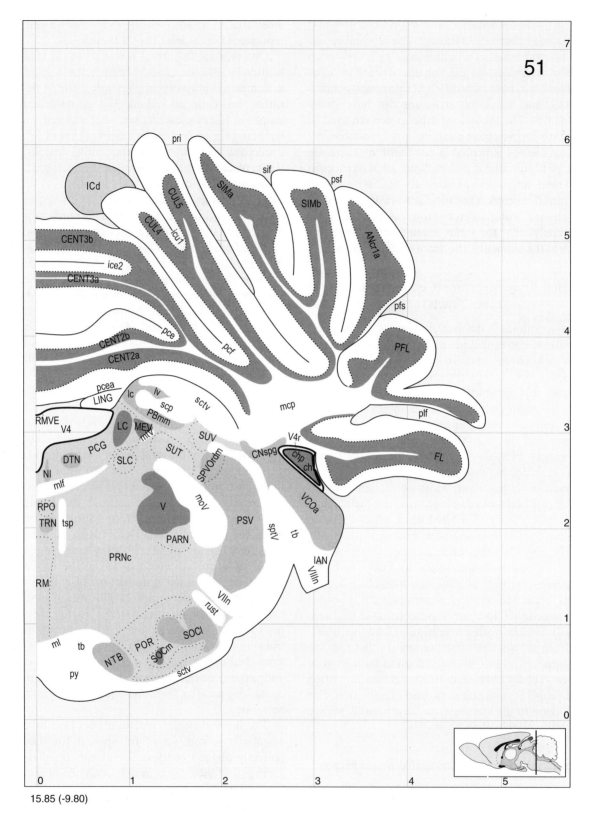

Figure 2 A map of atlas level 51 from the first edition of *Brain Maps* (Swanson, 1992). The drawing was made by tracing a photomicrograph of the corresponding Nissl-stained histological section using a computer graphics application and checking all details by viewing the section under a microscope at the side of the monitor. The position of the section/level is indicated in the schematic sagittal section in the lower right corner of the figure and by the arrowhead in Fig. 1 (right). The grid behind the map is a scale in millimeters; the number in the lower left is the distance along the rostrocaudal (z) axis. Thus, the brain is placed in a Cartesian coordinate system, and every location has an x, y, and z coordinate (in addition to being in a named structure). All structures have been arranged in a hierarchical nomenclature table, with references to the primary literature. (From Swanson, L. W. (1992). *Brain Maps: Structure of the Rat Brain*, Elsevier, Amsterdam. With permission from Elsevier Science.)

(4) a grid of stereotaxic coordinates and a database fiducial (see below), (5) a list of abbreviations, and (6) a mask for stacking levels. A Web site for the second edition of *Brain Maps*, including *Interactive Brain Maps*, may be found at http://www.elsevier.com:80/homepage/sah/brainmaps/. It contains sample files that can be downloaded for manipulation and printing. Use of these interactive template files on a computer highlights immediately the advantages over static printed maps.

4.1.5 Transferring Data from Experimental Brain to Standard (Atlas Reference) Brain

The time-honored method of mapping neuroanatomical data essentially involves drawing the cell bodies and fiber tracts as they appear on an individual histological section and then drawing the data (axonal or histochemical labeling, for example) on that map—using a camera lucida or equivalent approach. These drawings or maps are an abstraction of data, and the data are represented in a drawing of the histological section that contains them. However, when histological analysis is involved, no two brains are cut in exactly the same plane, and each section from every brain is distorted in a different way. These problems can be approached more or less successfully in the following ways, with the goal in mind of comparing the results of different experiments as plotted on a standard or reference series of templates. The results of such comparisons are, of course, qualitative rather than quantitative, but they are very useful for comparing general distribution patterns. The only way of comparing distribution patterns quantitatively is to carry out multiple staining methods on individual sections so that the patterns are compared directly in the same section.

Linear Rescaling

Different histological procedures result in more or less shrinkage or expansion of the brain and/or tissue sections. For example, our atlas brain was embedded in celloidin for sectioning, and as a result underwent considerable shrinkage (on the average, about 38% dorsoventrally, 21% rostrocaudally, and 28% mediolaterally). To a first approximation it seems reasonable to assume that a great deal of this distortion may be corrected for by linear scaling along each of the three cardinal axes (which can be accomplished trivially in computer graphics applications; Fig. 3a). However, it is not possible to be sure how much nonlinear distortion is present, due, for example, to mechanical factors such as the presence of fiber tracts running through gray matter (the physical texture of white and gray matter is quite different). In *Brain Maps*, such linear rescaling was used to produce a stereotaxic coordinate system from the physical coordinate system. This approximation was based on a com-

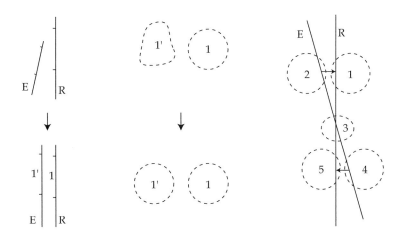

a. Align, linear rescale   b. nonlinear warping   c. data transfer, different plane

Figure 3 Key factors related to comparing spatial features in a histological section from an atlas or reference brain (R) to those in an experimental brain (E). **(a)** Assuming the two brains are cut in the same plane, the simplest first step would involve correcting for shrinkage or expansion by linear rescaling and then aligning the two sections. **(b)** The second step would involve removing as much nonlinear warping of the experimental section as possible. Here structure 1 is distorted in the experimental section (it has shape 1′). The location of structure 1 (and 1′) is indicated in (a). **(c)** If, or really because, the experimental brain section (E) is cut in a different plane than the reference or atlas brain section (R), errors in the assignment of spatial location to data occur. For example, data in structure 2 of the experimental brain section will be mapped onto structure 1 of the atlas or reference brain level, and data in structure 4 of the experimental section will be mapped onto structure 5. Accurate data transfer occurs only at one point in structure 3, and transfer errors increase as the distance from this point increases (and as the angle between the two sections increases).

parison with the atlas of Paxinos and Watson (1986), where frozen sections were used, and there was relatively little shrinkage distortion. In the case of the brain used for our atlas, nonlinear distortions are at least an order of magnitude less than the linear shrinkage.

Nonlinear Warping

Because histological sections are so thin (e.g., 30 μm) relative to the dimensions of the section itself (about 5000 × 6000 μm for a typical rat brain section) and because brain tissue is so fragile to begin with, these sections undergo considerable nonlinear, uncontrollable distortion or warping when mounted on glass slides for observation under the microscope (Fig. 3b). Mounting such a tissue section is a little like trying to lay a giant crêpe down flat on a plate—assuming one could lay the same crêpe down over and over, its exact shape would be different each time because it is so thin and pliable. If an experimental brain could be cut in exactly the same plane of section as a reference brain, then a wide variety of warping algorithms (see Chapter 4.6 and Toga, 1999) could be applied to remove distortions in sections of the experimental brain, relative to those of the reference brain (after the application of linear rescaling, if necessary). However, it is incumbent upon the user to quantitate the extent of distortion correction, and no one has remotely approached the highly accurate removal of warping from tissue sections relative to atlas levels (in terms of cell group and fiber tract borders). And, of course, warping cannot deal with common defects in tissue sections such as folds, tears, bubbles, and holes.

Correcting for Plane of Section Differences

Another very serious problem is that every brain processed for histology is cut in (at least a slightly) different plane of section because of technical factors involved in mounting the brain on a microtome stage for cutting. The results of this difference in plane of section for direct mapping of experimental data onto standard atlas levels is illustrated in Fig. 3c: data are not placed in the spatially correct location, and errors involved become greater (1) the greater the difference in plane of section, and (2) the farther away one is from the point of intersection of the two sections (Swanson, 1998–1999).

Conceptually at least, there are two obvious solutions to this problem. First, the neuroanatomist can sit at the microscope for many hours correcting the plane of section difference mentally, using a method that has been described in some detail (Swanson, 1998–1999). This method requires a great deal of patience and knowledge, is exceptionally time consuming, and obviously is qualitative in nature; however, it is the only method that has been shown empirically to work satisfactorily at the present time for large-scale brain mapping of data onto standard atlas levels.

The second approach would involve resectioning a three-dimensional computer graphics model of the brain in the plane of the experimental brain, subjecting the experimental section to linear scaling and warping, and then transferring the data to the template from the three-dimensional model, ideally using image analysis methods. However, to be maximally useful for comparison and databasing purposes, a continuous, three-dimensional model of the data itself would have to be constructed, and then the data model would have to be resliced in a standard or reference plane. It does not seem likely that this goal will be accomplished fully in the foreseeable future (see below), although a very good start at laying the groundwork for this approach has been made (Chapter 4.6).

4.1.6 Toward Textual and Graphical Databases on the Web

Because of the explosion in neuroanatomical data beginning around 1970, there is no doubt that electronic databases available over the World Wide Web would be of great value to experimental and computational scientists alike. Obviously, there are two broad classes of relevant databases: textual and graphical. Approaches to textual databases (and knowledge bases) of neuroanatomical information are dealt with in Chapter 6.2, where NeuroScholar™ is considered in detail (also see http://neuroscholar.usc.edu/). The development of our database and query manager for atlas-based neuroanatomical data, NeuARt, is described in Chapter 4.3, where the need for both textual and spatial query managers (Dashti *et al.*, 1997; Shahabi *et al.*, 1999) is emphasized. The goal is to integrate seamlessly these textual and graphical databases of neuroanatomical information. At the simplest level of integration, the *names* of structures in the atlas ("places") templates of the graphics database can be used as *keywords* for querying the textual database.

4.1.7 Three-Dimensional Computer Graphics Models of the Brain

Excellent three-dimensional drawings of brain structure also date back to the sixteenth-century work of Vesalius (and the long-unpublished drawings of Leonardo). Particularly good modern examples can be found in books by Krieg (1955, 1966), who used a systematic method based on reconstructing series of histological sections from rat and human brains, and by Nieuwenhuys and colleagues (1988, 1997), who by and large used more traditional artistic methods. In addition, many histological atlases of the brain have used sections cut in the three standard planes (for the rat, see Paxinos and Watson, 1986, 1998; Kruger *et al.*, 1995).

Today, sophisticated computer graphics methods can be used to design, construct, and display digital three-dimensional models of the brain. There are two basic approaches to three-dimensional computer graphics modeling: voxel-based and vector-based. "Voxels" are "volume elements" (rather than "pixels" or "picture elements"), and their use was spurred by the development of computerized tomography (e.g., CAT, PET, and fMRI scans). The great advantages of this approach are that a "solid" three-dimensional image of the living brain (or other organ) can be obtained, and when this image is resectioned electronically in any plane the slices are virtually perfectly aligned (see Toga and Mazziota, 1996). The major disadvantages of this approach are low resolution and contrast relative to histological sections prepared for microscopic examination (see above and Fig. 4).

The vector-based approach depends on reconstructing surfaces from series of cross-sections with vector graphics—that is, building computer graphics models of the brain from drawings such as those prepared for our atlas, *Brain Maps* (Swanson, 1993, 1998–1999). These highly simplified models have the advantages of relatively small file sizes and essentially infinite scalability (see Chapter 4.6). One of their major disadvantages is that they are reconstructed from histological sections. This is a problem for two major reasons. First, it is not possible to align a series of histological sections of the brain absolutely correctly because there are no invariant fiducial structures (Weninger *et al.*, 1998), although accuracy may be increased significantly by comparing section alignment with an MRI scan of the head from which the brain was removed. And, second, as mentioned above, each histological section undergoes unique nonlinear distortion when mounted on glass; therefore, the outlines (or "surface") of a brain structure reconstructed from a series of histological sections will have more or less "noise" (deviation from true position and shape) based on a number of factors, including the skill of the histologist who mounted the sections on glass. These considerations have led to the conclusion that the templates or maps generated for *Brain Maps* (Swanson, 1993, 1998–1999) must be redrawn and redesigned for use in constructing a three-dimensional model of the rat brain. In addition to smoothing outlines in the maps to eliminate warping, many more sections of the brain used to prepare *Brain Maps* must be drawn (only 73 of 556 have thus far been drawn completely), and complete outlines of all individual structures (blobs or paths) must be provided.

At least four major uses of three-dimensional computer graphics models of the brain come readily to mind. First, as mentioned above, they could be sliced in the same plane of section as any experimental brain, for easier mapping of data onto computer graphics templates. Obviously, once a brain cut in a particular plane of section (say transverse) is rendered as a three-dimensional model, the latter can then be resliced in the other two cardinal planes of section (say horizontal and sagittal). Second, the delineation of surface features required for the construction of three-dimensional models lends itself immediately to quantitative morphology (for example, measurements of volume, surface area, shape, and distance). In principle, vector graphics surfaces can be analyzed mathematically, in terms of both their geometry and topology. Third, three-dimensional models of the brain can be used to illustrate clearly the physical location and shape of brain structures and pathways (as is so helpful in textbooks). And, fourth, it will be possible to animate three-dimensional models of the brain and nervous system as a whole. For example, "physical models" of brain circuitry could be animated in terms of

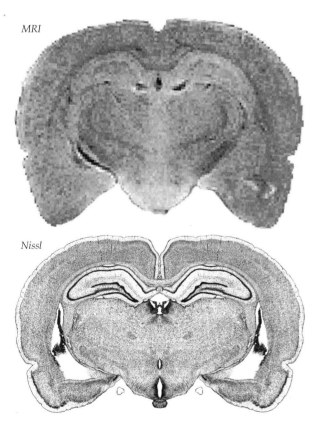

Figure 4 A comparison of the resolution obtained in photos of rat brain sections from MRI **(top)** and Nissl-stained histological sections (bottom). The MRI voxels were 100 μm^3 and data for the whole brain took about 24 hours to collect. The dataset was obtained by Russell Jacobs and Eric Ahrens at Caltech. The Nissl-stained section is from *Brain Maps* (Swanson, 1998–1999) and was used to prepare atlas level 32. Needless to say, the histological section can be viewed under the light microscope, which has a resolution on the order of 1 μm. The MRI image is from a male rat that was the same size as the rat used to prepare *Brain Maps*, and the section shown here is approximately the same as level 32 in *Brain Maps*. In comparing the two images, the most obvious difference is that the MRI does not have cellular resolution; cells in the rat brain range between 5 and 25 μm in diameter. (Both photos from Swanson, L. W. (1998–1999). *Brain Maps: Structure of the Rat Brain*. Second rev. ed. Elsevier, Amsterdam. With permission from Elsevier Science.)

dynamic patterns of action potentials or information flow within specific pathways between specific cell groups, or more accurately, between specific cell types. It remains to be determined how far this approach must be refined before it can escape the realm of a gross cartoon that ignores the true subtleties of neuronal information processing.

4.1.8 Two-Dimensional Flatmaps: Schematic Circuit Diagrams and Distribution Patterns

Although they contain very significant distortions, maps are used much more widely than globes because they are so convenient. There has been relatively little work toward a systematic flatmap of the brain because this would involve flattening a highly compartmentalized solid object rather than just the surface of a sphere. The best examples so far have been unilateral flatmaps of the amphibian brain (Herrick, 1948) and mammalian brain (Nauta and Karten, 1970), and a bilateral flatmap of the rat (Swanson, 1992) and human (Swanson, 1995) CNS (Fig. 5) based on a fatemap of the embryonic neural plate, which is a flat sheet topologically (see above).

Our flatmap has a number of obvious uses, especially for comparing patterns of gene expression and neurochemical distribution, and for illustrating schematically the organization of various circuits. These applications are facilitated greatly by use of the flatmap as a template in an electronic database. For example, different pathways or sets of pathways can be stored in transparent overlays (layers) and can then be displayed in any desired combination, just as for the data layers over atlas levels considered above. In fact, to aid in the design of rat brain circuits, all of the major pathways and cranial nerves have been placed in a standard way over the flatmap, with different functional systems in different layers (Swanson, 1998–1999). Obviously, flatmap layers containing various expression patterns, circuit elements, and other information could be stored in a database for retrieval when needed.

4.1.9 The Future: Atlases as Expandable Databases and Models

It is becoming obvious very quickly that the traditional role of brain maps and atlases is changing in a fundamental way. Instead of being static images of the

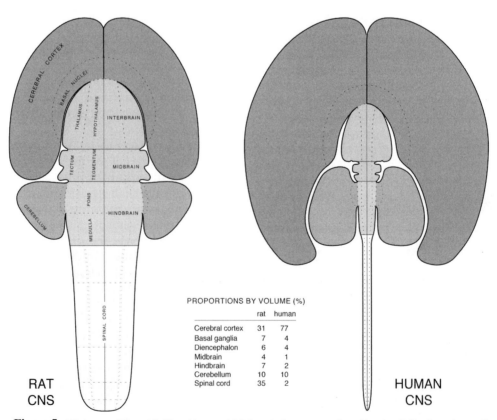

Figure 5 Flatmaps of the rat (left) and human (right) central nervous system. For simplicity, the volume of structures in the actual brain is proportional to their area in the schematic flatmap. The table in the center of the figure compares the size of major central nervous system divisions between rat and human. See text for further details. (From Swanson, L. W. (1995). *Trends Neurosci.*, **18**, 471–474. with permission from Elsevier Science.) (See color plates.)

brain—whether slices or volumes—they will be dynamic computer graphics models that serve multiple, expandable purposes, including the framework for databases. This approach is already practical for much simpler applications such as Geographical Information Systems (Chapter 4.2), and the computer-aided design of buildings, cars, and airplanes. Functioning, useful applications to biology, including neuroscience, will be next!

The development of optimal digital brain models and databases will utilize all of the topics covered in this book. First, there is the construction of an interactive "physical" model of the brain using three-dimensional computer graphics software. Because of the brain's structural complexity, these models (say, for different species, sexes, and ages) will probably never be completed, but instead will become progressively more detailed. As more is learned about the brains of various species, controversies about nomenclature, connections, and so on will gradually be resolved. This has already happened for most of the rest of the body—there is very little controversy about the structural organization and nomenclature of the skeletal, muscular, cardiovascular, and digestive systems, for example—and the same will inevitably apply to the nervous system in due time.

A second step will involve building a whole variety of neuroscience databases, implementing ways to federate them seamlessly, designing powerful ways to search for information within the federation, and then tying the database federation to the three-dimensional computer graphics models of the brain just mentioned. And, as a third phase, one might envision creating dynamic computer graphics models of brain structure and function based on knowledge extracted from the databases. In essence then, the computer graphics models could become knowledge bases of brain structure and function. As such, they could be used to test hypotheses generated from existing knowledge and to suggest new hypotheses that could be tested experimentally. How far will it be possible to develop the idea of a computer graphics virtual brain whose parts are fully documented by links to databases, knowledge bases, and modeling/simulation tools?

References

Alvarez-Bolado, G., and Swanson, L. W. (1996). *Developmental Brain Maps: Structure of the Embryonic Rat Brain.* Elsevier, Amsterdam.

Dashti, A. E., Ghandeharizadeh, S., Stone, J., Swanson, L. W., and Thompson, R. H. (1997). Database challenges and solutions in neuroscientific applications. *Neuroimage* **5**, 97–115.

Herrick, C. J. (1948). *The Brain of the Tiger Salamander.* University of Chicago Press, Chicago.

Krieg, W. J. S. (1966). *Functional Neuroanatomy,* Third ed., revised and enriched. Brain Books, Evanston, IL.

Krieg, W. J. S. (1955). *Brain Mechanisms in Diachrome.* Second ed. Brain Books, Evanson, IL.

Kruger, L., Saporta, S., and Swanson, L. W. (1995). *Photographic Atlas of the Rat Brain: The Cell and Fiber Architecture Illustrated in Three Planes with Stereotaxic Coordinates.* Cambridge University Press, New York.

Leuret, F., and Gratiolet, P. (1857). *Anatomie comparée du système nerveux.* Bailliére, Paris.

Nauta, W. J. H., and Karten, H. J. (1970). A general profile of the vertebrate brain, with sidelights on the ancestry of cerebral cortex. In *The Neurosciences: Second Study Program* (Schmitt, F. O., Ed.). Rockefeller University Press, New York, pp. 7–26.

Nieuwenhuys, R., ten Donkellar, H. J., and Nicholson, C. Eds. (1997). *The Central Nervous System of Vertebrates.* Springer-Verlag, Berlin.

Nieuwenhuys, R., Voogd, J., and Chr. van Huijzen (1988). *The Human Nervous System: A Synopsis and Atlas.* Springer-Verlag, Berlin.

Paxinos, G., and Watson, C. (1998). *The Rat Brain in Stereotaxic Coordinates.* Fourth ed. Academic Press, San Diego, CA.

Paxinos, G., and Watson, C. (1986). *The Rat Brain in Stereotaxic Coordinates.* Second ed. Academic Press, Sydney.

Robinson, A. H., and Retchenik, B. B. (1976). *The Nature of Maps. Essays Toward Understanding Maps and Mapping.* University of Chicago Press, Chicago.

Shahabi, C., Dashti, A. E., Burns, G., Ghandeharizadeh, S., Jiang, N., and Swanson, L. W. (1999). Visualization of spatial neuroanatomical data. In *Visual Information and Information Systems* (Huijsmans, D. P., and Smoulders, A. W. M., Eds.). Springer, Berlin, pp. 801–808.

Snyder, J. P. (1993). *Flattening the Earth: Two Thousand Years of Map Projections.* University of Chicago Press, Chicago.

Swanson, L. W. (2000). A history of neuroanatomical mapping. In *Brain Mapping: The Applications* (Toga, A. W., and Mazziotta, J. C., Eds.). Academic Press, San Diego, CA, pp. 77–109.

Swanson, L. W. (1998–1999). *Brain Maps: Structure of the Rat Brain. A Laboratory Guide with Printed and Electronic Templates for Data, Models, and Schematics,* Second rev. ed. Elsevier, Amsterdam.

Swanson, L. W. (1996). Histochemical contributions to the understanding of neuronal phenotypes and information flow through neural circuits: the polytransmitter hypothesis. In *Molecular Mechanisms of Neuronal Communication* (Fuxe, K., Hökfelt, T., Olson, L., Ottoson, D., Dahlström, A., and Björklund, A., Eds.). Pergamon Press, Elmsford, NY, pp. 15–27

Swanson, L. W. (1995). Mapping the human brain: past, present, and future. *Trends Neurosci.* **18**, 471–474.

Swanson, L. W. (1993). *Brain Maps: Computer Graphics Files* Elsevier, Amsterdam.

Swanson, L. W. (1992). *Brain Maps: Structure of the Rat Brain,* Elsevier, Amsterdam.

Swanson, N., and Swanson, L. W. (Translators). (1995). *Santiago Ramón y Cajal: Histology of the Nervous System in Man and Vertebrates.* Vol. I. Oxford University Press, New York, p. xiv.

Toga, A. W. (Ed.). (1999). *Brain Warping,* Academic Press, San Diego, CA.

Toga, A. W., and Mazziota, J. C. (Eds.). (1996). *Brain Mapping: The Methods.* Academic Press, San Diego, CA.

Weninger, W. J., Meng, S., Streicher, J., and Müller, G. B. (1998). A new episcopic method for rapid 3–D reconstruction: applications in anatomy and embryology. *Anat. Embryol.* **197**, 341—348.

Vesalius, A. (1543). *De Humani Corporis Fabrica Libri Septem,* Oporinus, Basel. For an English translation of his work on the brain, see Singer, C. (1952). *Vesalius on the Human Brain: Introduction, Translation of Text, Translation of Descriptions of Figures, Notes to the Translations, Figures.* Oxford University Press, London.

CHAPTER 4.2

Perspective: Geographical Information Systems

Cyrus Shahabi[1] and Shuping Jia[1,2]
[1]*University of Southern California Brain Project and Computer Science Department*
University of Southern California, Los Angeles, California
[2]*Intuit Inc., San Diego, California*

Abstract

The goal of neuroscience research is to understand the structure and function of brains in general and the human brain in particular. We use a brain atlas as a standard for each species to consolidate data from different researchers. In order to achieve this objective, a spatial structure of the brain atlas as well as a standard coordinate system need to be defined. Similar challenges have been addressed in the area of the Geographical Information System (GIS); however, because of the differences between brain atlases and geographical atlases, some of the solutions proposed by GIS cannot be directly applied to neuroscience. In this chapter, we discuss some of the distinctions between brain and geographical atlases and describe our approach to tackle the challenges resulting from these differences.

4.2.1 Introduction

The emergence of neuroinformatics as a discipline has prompted the need for a standardization and coordination of neuroanatomical terminology and coordinate systems. These are cornerstones of effective information sharing among scientists and applications. At present, brain atlases provide the main practical standardized global maps of neural tissue. In this regard, the USC Brain Project uses the Swanson atlas (Swanson, 1992, 1993) as a standard to consolidate neuroscience data on the rat, which includes neuroanatomical and neurochemical data, time-series data, and publications.

As a direct result of the interconnection and consolidation of neural data, many neuroinformatic navigation scenarios will become feasible. For example, a neuroscientist can start data navigation from a repository of digital publications, select a paper, and then request to zoom into the Swanson atlas to see the corresponding brain section discussed in the experimental section of the paper. Alternatively, one might start from navigating the Swanson atlas and then request to view all the publications available linking certain research protocols to a specific brain region. Similar navigation can be initiated from a time-series data repository or a database containing neurochemical experiments; however, our concern is much more general, and the approaches we develop can apply to other atlases of the rat brain—such as the Paxinos-Watson atlas (Paxinos and Watson, 1986) or to comprehensive atlases of the brains of other species. While noting this generality, we nonetheless focus on the rat brain and the Swanson atlas for this chapter.

In order to employ the Swanson atlas as a standard to consolidate data from different applications, one should be able to relate each piece of data to a region on the brain atlas. This can be achieved in two steps:

1. Construct a spatial structure for the brain atlas.
2. Identify a standard coordinate in order to "map" data to the defined spatial structure of the atlas.

Similar issues have been addressed by the Geographical Information Systems (GIS) to map data on geographical atlases (Bekkers, 1998; Goodchild *et al.*, 1996; Guttman, 1984; Poage, 1998); however, due to inherited differences between brain atlases and geographical atlases, the solutions proposed by GIS are not directly useful for neuroscience applications. In particular, GIS usually replaces the circular globe by a series of flat maps where the relationship between these "sheets" is one either of adjacency or overlap. By contrast, the brain is an inherently three-dimensional structure so that the notion of adjacency of sheets or "levels" (as termed in the Swanson atlas) is different, as will be logically characterized in this chapter.

In this chapter, we first provide an overview of the state of the art in GIS. Subsequently, we discuss what GIS has to teach neuroinformatics, both indicating what we have done at USCBP, as well as the promises and pitfalls. Finally, we conclude by summarizing the main GIS techniques that NeuARt has used and those that look promising for the future extensions. At the end, we also provide an itemized list of URLs where one can find useful GIS material.

4.2.2 Overview of GIS

There are several different definitions for a Geographical Information System (GIS), depending on the type of user and the application domain (Medeiros and Pires, 1994). We start with the definition provided by Goodchild (1991) as "a digital information system whose records are somehow geographically referenced." From a database perspective, GIS manages *georeferenced* data. Georeferenced data consists of data elements whose locations are spatially referenced to the Earth (Carter, 1989). Three different data types are usually managed by GIS: (1) conventional record-based data consisting of alphanumeric attributes, (2) image (or *raster*) data, and (3) spatial (or *vector*) data. We focus our attention on raster and spatial data.

Raster Data

Raster data are simply a bitmap that, if displayed, illustrate the image of an area. Raster data are tightly related to what is called *field* view (Medeiros and Pires, 1994) or *space* view (Guting, 1994). This view sees the world as a continuous surface (layer) and is used to describe the entire space and every point in it; however, if the space consists of several objects, there exists no information on the location of these objects and their spatial relationships. For that, we need the vector or spatial data (see the following subsection). Raster data are useful to represent continuous phenomena such as atmospheric pressure.

Spatial Data

Spatial data are the data representing objects in space with identity, well-defined extents, locations, and relationships (Guting, 1994). The database system managing spatial data is called a spatial database system, which is typically an extension of a regular database system with additional capabilities to handle spatial data such as spatial data models, query languages, spatial index structures (e.g., R-Trees; see Guttman, 1984)), and support for spatial joins. Spatial objects are typically modeled as points, lines (or curves), and polygons. (Note that several other constructs such as networks and partitions are also discussed in the literature [for example, see Guting, 1994], but their descriptions are beyond the scope of this chapter.) A point represents only the location of an object, but not its extent or shape (e.g., a Zip Code area may be modeled by the latitude and longitude of its center). A line (can be represented by the coordinates of its two endpoints or can be a curve consisting of multiple line segments or a curve modeled as a spline curve represented by its parameters, or...) represents moving through space or connections in space (e.g., roads, rivers, cables). A polygon (or region) represents an object with extent and shape (e.g., city, Zip Code area). Fig. 1 depicts one approach to store different spatial data types as conventional records in a database (ARC/INFO, 1991). However, with recent advances in object-relational database management systems (OR-DBMS), a database can be extended to support spatial data types as first-class residents of the database system. Examples are Oracle 8i's spatial cartridge, the NCR TOR spatial UDT/UDF, and Informix Geodetic spatial datablade. These systems have data types such as points or polygons predefined and hence can be used by the user applications and/or SQL3 query language directly. Moreover, most of the commercial systems define special cases of polygon objects such as circles and rectangles as additional spatial data types.

Besides the representation illustrated in Fig. 1, there is an alternative way to represent polygons and arcs within ARC/INFO. This representation relies on the following three major concepts: (1) connectivity, (2) area, and (3) contiguity. We consider each concept in turn.

1. *Connectivity:* The x,y pairs along each arc define the shape of the arc. The endpoints of an arc are called nodes. Each arc can only have two nodes. Arcs can meet only at their endpoints. By tracking all the arcs that meet

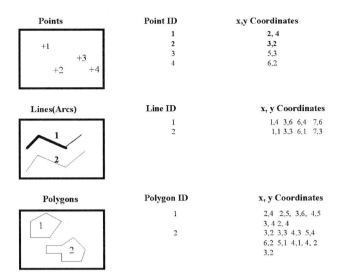

Figure 1 Multiple data types using the vector data structure in ARC/INFO. (From ARC/INFO (1991). *User Guide*.)

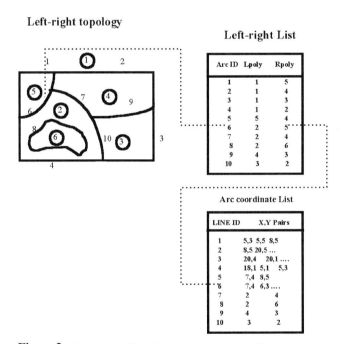

Figure 2 Contiguity. Circled numbers are polygon IDs; regular ones are arc IDs. (From ARC/INFO (1991). *User Guide*.)

at any endpoints, we can identify which arcs are connected.

2. *Area definition:* Polygons can be represented as a set of arcs instead of a closed loop of *x,y* coordinates. A list of the arcs that make up a polygon is stored and used to construct the polygon when necessary.

3. *Contiguity:* Two polygons sharing a common arc are adjacent. If a standard direction is imposed on every arc, then for each arc one can list the polygons on its left and right (see Fig. 2).

Spatial Relationships

One of the major reasons for storing data in spatial format is for later querying and accessing this data through spatial queries. These queries can be submitted either through a graphical user interface (for example, by interacting with a map-based GUI) or through utilization of SQL3 queries. The spatial queries typically look for specific spatial relationships between a query object and the objects in the database. Theses spatial relationships can be categorized into three types (Guting, 1994):

1. *Topological relationships:* These represent those spatial relationships that are invariant under topological transformations (i.e., rotation, scale, and translation). Between two polygon objects (or connected areas with no holes) there exist six of such relations (Egenhofer and Herrings, 1990): disjoint, overlap (or intersect), touch (or meet), inside, cover, and equal. The same concept can be extended for other spatial data types. For example, two points can only be disjoint or equal. Almost all of the commercial spatial database systems (e.g., extended OR-DBMSs) support all these topological relationships.

2. *Direction relationships:* Here, we are interested in finding the objects that satisfy a certain location in space with respect to a query object. These include relations such as above, below, north of, southwest of. There are two approaches to support direction relationships: project-based (Frank, 1992) and cone-based (Pequet and Ci-Xiang, 1994). With the former, a plane is partitioned into subpartitions, and the direction relation between two objects is identified by the subpartitions they occupy. With the latter, the plane is divided into five partitions that define the primitive direction relations (see Fig. 3).

3. *Metric relationships:* These represent the relationships that can be quantified such as "distance < 1000."

Now, a sample window query might be "Find all the theaters within 2 miles of my house." To support this query, the system needs to examine for "inside" relationships between coordinates of theaters (i.e., latitude and longitude) in the database and the circle defined by the user, with its center being the user's home address and the radius being the 2-mile distance.

Fig. 4 illustrates a sample GUI to query for average households color-coded within different states. The user can zoom in and out or change directions by clicking on the menu buttons.

Layers, Data Features or Coverages

The common requirement to access data on the basis of one or more classes of phenomena has resulted in several GISs employing organizational schemes in which all data of a particular level of classification such as roads, rivers, or vegetation types are grouped into so-called "*feature planes*", "*layers*", or "*coverages*" (ARC/INFO, 1991; Burrough and McDonnell, 1998; Jones,

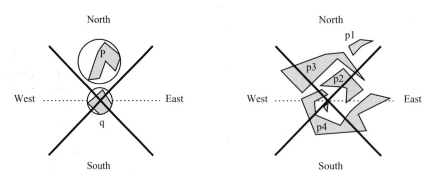

Figure 3 Cone-based direction relationships when objects spanning one or more direction(s).

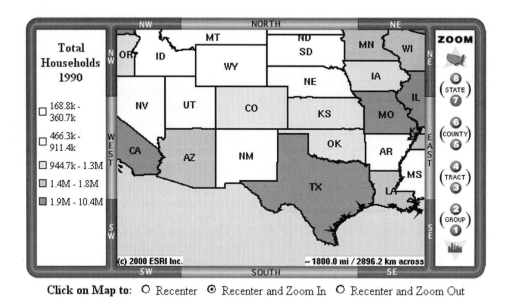

Figure 4 A GIS GUI for submitting spatial queries. (From http://www.esri.com/data/online/tiger/index.html.)

1997). In order to avoid confusion between this concept and the concept of "layers" used in the Swanson atlas, hitherto we will use the term *coverage* in this chapter.

Coverage represents the main method for storing vector data. A coverage is a digital version of a single map sheet and generally describes one type of the map spatial data, such as streets, parcels, soil units, wells, or forest stands. A coverage contains both the location data and thematic attributes for spatial data types in a given area. Coverage logically organizes spatial data into layers or themes of information. A base map can be organized into layers such as streams, soils, wells, administrative boundaries, and so on. A coverage consists of topologically linked spatial data and their associated descriptive data.

A coverage is stored as a set of spatial data types such as points, arcs (or lines), and polygons. The combination of spatial data present in a coverage depends upon the geographical phenomena to be represented. One approach to store information about a coverage in a database is as follows (ARC/INFO, 1991). Each arc record contains the arc's ID, location, and shape information defined as a series of x,y coordinates. Furthermore, for each arc record we can store the IDs of the polygons on its left and right, assuming that the coverage contains polygon features and a direction can be imposed on each arc. Subsequently, each polygon record contains a list of all the arcs defining the boundary of the polygon. No coordinates are explicitly stored for polygons, which means the polygons are stored topologically (Fig. 5 illustrates this representation).

The power of a coverage lies in the relation between the spatial data and tabular (descriptive) data. The important characteristics of this relation are:

1. A one-to-one relationship exists between data types that appear in the coverage and corresponding records representing the data type (e.g., x,y coordinates of a point data type).
2. The link between the data types within a coverage and their corresponding records is maintained through unique identifiers assigned to each data element.

Fig. 6 depicts several coverages, each of which is composed of numerous spatial data, superimposed on a single map sheet. The full system is available for interaction and download on the Web at the following URL: http://www.esri.com/data/online/tiger/index.html.

Polygon-arc topology

Polygon-arc List

POLY ID	LINE ID
2	4,6,7,10,0,8
3	3,10,9
4	7,5,2,9
5	1,5,6
6	8

Arc coordinate List

LINE ID	X,Y Pairs
1	5,3 5,5 8,5
2	8,5 20,5 ...
3	20,4 20,1 ....
4	18,1 5,1 5,3
5	7,4 8,5
6	7,4 6,3 ....
7	2 4
8	2 6
9	4 3
10	3 2

Figure 5 Polygon-arc topology. Circled numbers are polygon IDs; regular ones are arc IDs. (From ARC/INFO (1991). *User Guide*.)

The figure illustrates four coverages consisting of line data types—roads, railroads, transportation, and utility lines—and three coverages consisting of area data types:

1. Statistical boundaries, such as census tracts and blocks
2. Local government boundaries, such as places and counties
3. Administrative boundaries, such as congressional and school districts

The figure also shows three coverages of point data types:

1. Point landmarks, such as schools and churches
2. Area landmarks, such as parks and cemeteries
3. Key geographic locations, such as apartment buildings and factories

Registration and Rectification

Thus far, we used the term "coordinates" as a standard way to represent a point. These coordinates with GIS are the geographic coordinates of the point or its latitude and longitude. The latitude and longitude values for any point on the Earth are fixed for a given referencing system. In GIS, rectification and registration are

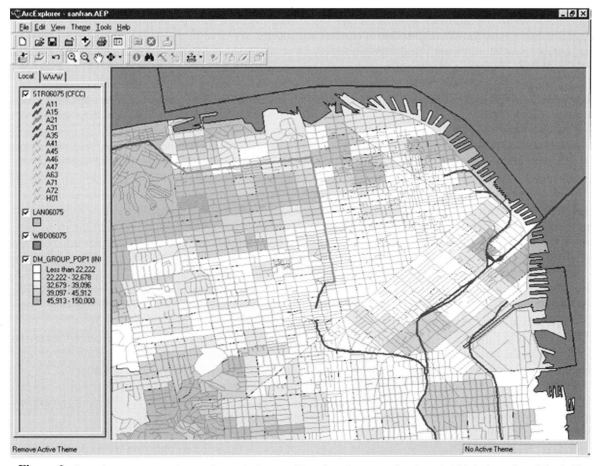

Figure 6 Several coverages superimposed on a single map. (From http://mapserver2.esri.com/cgi-bin/usdemog? m=3 & cd=5.)

based on a family of mathematical tools that are used to modify the spatial arrangement of objects in a data set into some other spatial arrangement (Goodchild *et al.*, 1996). Their purpose is to modify these geometrical relationships without substantively changing the contents of the data itself. Rectification involves manipulating the raw data set so that the spatial arrangement of objects in the data corresponds to a specific geocoding system. For example, a rough sketch of an undeveloped lot, made on-site by a landscape architect, typically does not show objects in their true spatial relationships. Once the sketch has been brought back into the office, the sketch can be re-drafted (perhaps as an overlay on a reproduction of an existing map that shows legal boundaries and easements) so it conforms to a useful coordinate system. Registration is similar to rectification; however, it is not based on an absolute georeferencing scheme. Both sketches from a field team and uncontrolled aerial photography have distorted views of the Earth's surface. The registration process involves changing one of the views of surface spatial relationships to agree with the other, regardless of any particular geodetic referencing system.

Generally, we do not have an exact solution to the problem of rectification and registration, either because we do not know the details of the map projection in the datasets, or because the data do not conform to a standard projection or georeferencing system. An example of the latter situation may be found in an oblique aerial photograph, in which the changes in scale across the image are due to the particular configuration of platform altitude, camera system alignment, and topography. A common approach, based on statistical operations, is called a *ground control point rectification/registration*. In this technique, locations of objects in the source data (often based on the arbitrary reference grid of raster row and column) are compared to their locations on a trustworthy map. A statistical model is then used to convert the data into a new set, with the desired geometrical characteristics.

An alternative procedure, often called a *rubber sheeting operation*, is as follows. Suppose we start with an uncontrolled off-nadir air photograph. This image will have many kinds of distortion due to aircraft motion, perspective problems contributed by the curvature of the Earth, scale changes due to topography, distortions in the camera lens, and so on. Next, we place the film over a "correct" map of the Earth, selected based on the desired map projection and scale. Subsequently, we identify a location on the map that is easy to distinguish on the image and run a pin vertically through the location on the images, down to the corresponding location on the map. This fixes the first location on the film to the corresponding location on the map. Finally, we identify another unambiguous location, place a pin through the photograph at this point, and run the pin down to the corresponding point on the map. The identified image points that have been attached to the map are the ground control points or tie points, and the number of points required depends on the geometric properties of both the desired map projection as well as the images. This rectification/registration process involves building a numerical coordinate transformation between the original image coordinates and the rectified (or "true") coordinates. Frequently, these transformations are based on polynomials, whose coefficients are computed by regression on the coordinates.

4.2.3 Atlas-Based Neuroscientific Data

The Swanson atlas was prepared from a rat brain sectioned in the coronal (frontal or traverse) plane. From 556 serial sections, 73 levels were chosen and illustrated as representative of the entire rat brain. Each brain is divided into regions and then further subdivided into subcortical nuclei and cortical areas. Neuroanatomical descriptions may use standard three-dimensional coordinates relative to a prominent landmark, but they are much more likely to be described by their approximate, relative position to the borders and small-scale detail of the structures that are contained in the region of interest. The plane of section of every experimental brain is different, so some coordinate transformation (or "re-slicing") must be applied to the data in order to map from an experimental brain section onto an atlas drawing (see Chapter 4.1 of this volume for more information).

Each level of the Swanson atlas captures the major cell groups of the rat brain, along with their subdivisions. The atlas gross structure hierarchy consists of: gross segments, segments, region, cell groups and fiber-tracts, and parts, from most general to most specific. For example, the septal region consists of the medial septal nucleus, the lateral septal nucleus, and the septohippocampal nucleus. These structures are related to each other in the same level by subclassing and/or adjacency relationships. Adjacency relationships describe which structures have common boundaries (Swanson, 1992, 1993). It is also important to note that structures in the atlas typically exist in more than one adjacent level (for example, the lateral septal nucleus exists on 12 levels) and that the same structure is typically bounded on more than one side. Because they are complex shapes, each structure can have few or many boundary structures (adjacency relations).

Several types of data can be placed over the atlas levels in order to summarize the results of experimental work in a systematic way (see Chapter 4.3 for examples and figures). Each experiment consists of several tract fibers, conceptualized as a layer superimposed on top of its corresponding atlas level.

In essence, the collection of atlas levels and the data layers is intended to illustrate the major cell groups and fiber tracts of the rat brain that can be identified with the

Nissl stain, which shows the distribution of cells in a histological section. It is analogous to a conventional geographical atlas that summarizes the distribution of water and land masses, along with the borders of countries at a particular time in history. Such maps are quite useful because they can be used as templates for plotting many other types of information: highway systems, population densities, energy production and distribution systems, and so on. However, with the passage of time, political boundaries change, and the amount of information available to plot continues to increase. Thus, the atlas templates (or levels) themselves will change over time. For the remainder of this section, we discuss an approach to represent the atlas levels and data layers within a spatial database system, in order to utilize the available GIS tools as much as possible for their management.

4.2.4 Raster Data

Originally, both the atlas levels and the experimental data layers were only two stacks of two-dimensional images. We still do store the atlas levels as raster data in our database. The database is managed by Informix Universal Server v9.2, which is an OR-DBMS. The levels are stored as BLOBs (Binary Large Objects) within the relations. Other attributes include the level ID and image size. The purpose of storing the levels as images, in addition to their spatial structures (see below), is solely for the purpose of quick visualization. That is, the Java applet that serves as the GUI (see Chapter 4.3 for details) can quickly retrieve the entire image corresponding to an atlas level in one shot and display it to the user. The data layers, however, are not stored as raster data anymore.

Spatial Data

Spatial representations of both the Swanson atlas and experimental data have to be stored into the database. Originally, the Swanson atlas consists of a set of 73 electronic drawings in Adobe Illustrator. The curves and lines of the drawings delineate the brain structures, but the spatial structure of many of the constituent spline curves do not fully enclose their respective nuclei in a topologically consistent manner. Some regions lie in areas without complete boundaries, so the exact location of their borders remains unclear. To solve this problem, the Spatial Index Manager was developed to impose a spatial structure onto the atlas drawings, by using a combination of automation and expert user intervention with the GIS geographical topological mapping program (see Shahabi *et al.*, 1999, and Chapter 4.3). This process has converted the atlas drawings into "intelligent templates" in which every point "knows" both the spatial extent and the name of the region that contains it. This "knowledge"—a Spatial Indexing Scheme—is then inherited by any regional data registered against the atlas and thus supports spatial queries anchored by references to particular brain regions, spatial features, or three-dimensional coordinates. In practice, we used the spatial data types defined by the Geodetic datablade v9.14 (an Informix spatial extension) to store atlas spatial data as polygons, arcs, and points.

As for data layers, however, the best spatial representation would be spline curves. Unfortunately, such a spatial data type is not supported by Geodetic; therefore, as a temporary solution, we have them stored as polygons. This representation is, of course, not an optimal one as some spatial relationships between an atlas structure and an experiment (e.g., overlap relation) would not return accurate results. This is because the two corresponding polygons might, say, overlap but the contained spline curves and the structure might not. We could have stored the experiments as a set of points, but this would result in a very low performance for spatial queries.

Spatial Relationships

Once both atlas levels and data layers are spatially stored in the database, all the spatial relationships supported by Geodetic are at our disposal. This includes all the topological relationships discussed above, as well as a limited number of direction and metric relationships. Consequently, one can ask, for example, for all the experiments that had impacts on a specific structure on a specific level of the atlas. This query can be translated to a spatial query that looks for "overlap relationship" between the polygon representing the structure and all the experiments across multiple layers. Now, suppose we georeferenced (or should we say "*neuroreference*") all the neuroinformatics publications with the corresponding coordinates of each structure to which they refer. As a result, we can search for all publications corresponding to a specific structure of the atlas. Note that interacting with a GUI, and not necessarily describing the query in any natural language, can form these queries. Currently, NeuARt's GUI allows the user to select a point, a rectangle, or a circle on a specific atlas level and ask for relevant materials (for now, only experimental data are neuroreferenced).

Coverages

At the first glance, one may be tempted to use the concept of GIS "coverages" to represent atlas "levels;" however, GIS coverages are used to conceptualize a collection of related phenomena—for example, a series of roads or rivers or a specific vegetation type—and there is no tight relationship between two coverages. Basically, there are inherit differences in consistency between GIS coverages when they overlap as opposed to consistency between brain boundaries as we move from level to level through the atlas of the three-dimensional structure of brain. That is, with brain atlases, brain structures exist over many levels. For example, the CA1 field of the hippocampus extends from level 28 to 40 of the Swanson atlas. This is analogous to having a river starting at one coverage and end on another.

On the other hand, it seems more appropriate to apply the concept of coverage to experimental data layers. Consequently, multiple experimental results obtained for an identical level can be considered as multiple coverages superimposed on top of that level. This is identical to representing different coverages for roads and rivers all mapped on top of, say, the California map.

As discussed above, works in GIS and spatial databases on spatial indexing and querying are largely based on the assumption that data are represented in two dimensions via latitude and longitude and can be represented as points, arcs, or polygons. On the other hand, brain regions are "volumes" and have complex shapes (e.g., CA1 field of hippocampus). Extending the works in GIS and spatial databases to support queries across levels on such complex shapes poses several research topics.

Registration and Rectification

In order to register data from other applications onto the atlas, all data have to be registered onto the same atlas. In the neuroanatomy domain, the registration task is more challenging than for GIS. First, there is no standard coordinate system in neuroanatomy corresponding to latitude/longitude in the geographical system. Second, there are no consistent naming standards; the same region in the same species may have different names in different atlases. Finally, the brain varies from one rat to another, not like the single Earth that is not changing rapidly.

To integrate data from different applications into the Swanson atlas, data should be rectified to the atlas and certain registration has to be done. That is, all data need to agree on some unique but shared characteristics to consolidate the information. The registration and rectification methods and theories in GIS, which we discussed above, are applicable here. Obviously, the main problem that arises here is that data source and target site might not agree on a unique frame of reference. Or, even if they do agree, they might not contain the required information for registration. For example, the selected experimental data of a certain query might be from different database and might use a different atlas to reference a brain region than what have been used by the Swanson atlas (Swanson, 1992, 1993).

For neuroanatomy domain, consider the registration of brain-slice images against a standard brain atlas so that one can ask a query to view all the images available about a particular region as seen in the atlas. The challenge is that the brain structure is three dimensional, while a standard brain atlas (e.g., Swanson's atlas) is a series of two-dimensional images. For three-dimensional brain structure projection into two-dimensional images, there is no mature or standard algorithm similar to the geo-map projection algorithms (e.g., azimuth projection, conic projection, cylindrical projection, etc.). To match a two-dimensional image with one of the standard atlas levels, it should be sliced in an identical plane; however, a rat brain is very small, and the odds of slicing a particular brain in the same plane as the atlas sections are very small. The process of superimposing data on the atlas requires a high level of expertise and patience for several reasons. Even so, this cutting and fixing procedures still may cause unpredictable nonlinear distortions of the tissues. Another challenge is the inter-individual variation and time-series variation of the brain structure. These variations make it very difficult to find "ground control points" like in geographical system to execute registration/translation. The approach that USC Brain Project has taken to this problem is to match a given section against many different slices through a three-dimensional brain estimated from the Swanson atlas. Taking advantage of this approach, their tool can be considered as a translation function (either rectification or registration) with two-dimensional rat brain images sliced in different planes as input and the corresponding level(s) and section(s) in the Swanson atlas as output (see Chapter 4.4 for more information).

Another challenge in registration is the uncertainty when dealing with multiple independent translation functions. For example, with neuroscience applications, there exists another standard rat brain atlas, the Paxinos-Watson atlas (Paxinos and Watson, 1986). Not only does this atlas consist of different number of levels, but also each level is based on slicing a different rat brain in a different plane and is partitioned into different sections with different labels. The registration of experimental data on one atlas can be translated to that of another atlas given that a translation function exists. However, the result of this translation and hence the registration might not be exact and certain. That is, a section in the Swanson atlas can correspond to one or more sections of the Paxinos atlas with some probability. The rectification and translation functions need to be extended in order to take into account this uncertainty and provide a method of representing the "spatial" uncertainty to the user. This extension imposes new research challenges that are currently under investigation at USCBP.

4.2.5 Conclusion and Web Resources

We described the primitive data types, operations, and features required by any GIS application. We demonstrated how these features can be supported through a spatial database management system and mentioned several commercial products and the ways they support these features. Subsequently, we explained both similarities and distinctions of a neuroinformatic atlas-based application to typical GIS applications. Consequently, we provided an overview of the GIS features helped in our implementation of NeuARt (see Chapter 4.3) as well as discussing those unique features required by our neuroinformatic atlas-based application that cannot be immediately supported by conventional GIS tools. Specifically, with NeuARt we store atlas layers as both image (raster) types and as collections of spatial objects. The experiments are stored only spatially and data belonging to the same experiments are grouped together as layers similar to the concept of coverage with GIS. Because we utilized a commercial database system with spatial capabilities, most of spatial relationships become readily available to us. Therefore, our Java-based GUI can support spatial queries such as overlap, contain, or point queries by defining a point (by mouse click) or a rectangle or a circle. Finally, neuroreferencing objects such as publications, relevant time-series data, and experimental data remains a challenge due to the three-dimensional and complex structure of the brain.

Below, we provide a list of useful Web sources for GIS and spatial database-related materials.

1. NCR TOR Database: http://www3.ncr.com/product/teradata/object/index.html
2. Informix Geodetic Spatial Datablade: http://www.informix.com/informix/products/options/udo/datablade/dbmodule/informix5.htm
3. Oracle 8 Spatial Cartridge: http://technet.oracle.com/products/oracle8/info/sdods/xsdo7ds.htm
4. Professor Hanan Samet's page at UMD provides spatial Java plug-ins, codes, and index structures: http://www.cs.umd.edu/~brabec/quadtree/index.html
5. Professor Ralph Hartmut Guting's page at Praktische Informatil IV provides a nice overview and survey on spatial models and query languages: http://www.informatik.fernuni-hagen.de/import/pi4/gueting/home.html
6. GIS WWW Resource List: http://www.geo.ed.ac.uk/home/giswww.html
7. ESRI—The GIS Software Leader. This site features free GIS software, online mapping, and GIS training, demos, data, product: http://www.esri.com/

References

American National Standard for Information Systems. (1986). *Database Language SQL* American National Standards Institute, New York.

ARC/INFO (1991). *User Guide, Data Model, Concepts, and Key Terms.* ESRI.

Bekkers, J. (1998). Datamining with GIS, an (im)possibility, in *ESRI International User Conference Proceedings.*

Carter, J. (1989). Defining the Geographical Information System. In *Fundamentals of Geographical Information Systems: A Compendium.* American Society for Photogrammetry and Remote Sensing, pp. 3–8.

Dashti, A. E., Ghandeharizadeh, S., Stone, J., Swanson, L. W., and Thompson, R. H. (1997). Database challenges and solutions in neuroscientific applications. *NeuroImage J.* **5(2)**, 97–115.

Egenhofer, M., and Herring, J. (1990). A mathematical framework for the definition of topological relationships, in *Proceedings of the 4th International Symposium on Spatial Data Handling*, Zurich, pp. 803–813.

Frank, A. U. (1992). Qualitative spatial reasoning about distances and directions in geographic space. *J. Visual Languages Computing.* **3**, 343–371.

Goodchild, M. (1991). Integrating GIS and environmental modeling at global scales. *Proc. GIS/LIS '91.* **1**, 40–48.

Goodchild, M., Steyaert, Louis, and Maidment (1996). *GIS and Environmental Modeling: Progress and Research Issues.* Wiley. New York.

Guting, R. H. (1994). An introduction to spatial database systems. *VLDB J.* [special issue], **3(4)**, 357–399.

Guttman, A. (1984). R-trees: a dynamic index structures for spatial searching. In *Proceedings of ACM-SIGMOD.*

Jones C. B. (1997). *Geographical Information Systems and Computer Cartography.* Addison-Wesley, Reading, MA, p. 30.

Medeiros, C. B., and Pires F. (1994). Databases for GIS. *ACM SIGMOD Record.* **23**, 107–115.

Paxinos, G., and Watson C. (1986). *The Rat Brain in Stereotaxic Coordinates.* 2nd ed. Academic Press, New York.

Pequet, D., and Ci-Xiang, Z. (1992). An algorithm to determine directional relationship between arbitrary shaped polygons in the plane. *J. Pattern Recognition.* **20(1)**, 65–74.

Poage, J. F. (1998). Analyzing spatially enabled health care data using the Dartmouth atlas data viewer, in *ESRI International User Conference Proceedings.*

Shahabi, C. et al. (1999). Visualization of spatial neuroanatomical data. In *Proceedings of Visual '99.* Amsterdam, Netherlands, Springer, pp. 201–208.

Swanson, L. W (1993). *Brain Maps: Computer Graphics Files.* Elsevier. Amsterdam.

Swanson, L. W. (1992). *Brain Maps: Structures of the Rat Brain.* Elsevier, Amsterdam.

Thompson, P. M., Schwartz, C., and Toga, A. W. (1996). High-resolution random mesh algorithms for creating a probabilistic 3D surface atlas of the human brain. *NeuroImage* **3**, 19–34.

CHAPTER 4.3

The Neuroanatomical Rat Brain Viewer (NeuARt)

Ali Esmail Dashti,[1] Gully A. P. C. Burns,[2] Shahram Ghandeharizadeh,[3] Shuping Jia,[4] Cyrus Shahabi,[3] Donna M. Simmons,[2] James Stone,[5] and Larry Swanson[6]

[1]*Computer Engineering Department, College of Engineering and Petroleum, Kuwait University, Al-Khaldya, Kuwait*
[2]*Department of Biological Sciences, University of Southern California, Los Angeles, California*
[3]*Department of Computer Science, University of Southern California, Los Angeles, California*
[4]*Intuit Inc., San Diego, California*
[5]*Neuroscience Program, University of California, Davis, California*
[6]*The Neuroscience Program and University of Southern California Brain Project, University of Southern California, Los Angeles, California.*

4.3.1 Introduction

A neuroanatomical atlas is an idealized three-dimensional map of brain tissue. Like any map, it can be used to navigate (by identifying the position of apparatus within the brain with stereotaxic coordinates within experiments), and it may be used to provide a standard "geographical" frame of reference (for data from neuroscientific experiments). This chapter describes an ongoing informatics-based project to simplify and optimize the process of superimposing neuroanatomical data onto the plates of a brain atlas through the use of multimedia databasing and network technology. The NeuroAnatomical Rat Brain Viewer (NeuARt) is based on neuroanatomical extensions to the NeuroCore database system, allowing neuroscientists to browse, compare, and query the complex spatially distributed patterns of label obtained from different experiments at their desktops through an network connection.

Consider the initial, primary form of neuroanatomical data as a large number of sections mounted onto slides providing a stacked three-dimensional map of different staining in the tissue at intervals through the brain. The physical characteristics of the physical preparation of these slides affect the viewing conditions of the data, so that one could almost think of the primary data as an analog data resource that only can be interfaced with through the use of a microscope. The primary data contains the full wealth of information that is present in the stained tissue and subsequent stages progressively simplify this information.

Neuroanatomists often depict their data by faithfully representing individual cases with photographs or *camera-lucida* drawings or by performing detailed examination of the tissue with computational tools (with products such as NeuroLucida from MicroBrightField, Inc., or the SEM Align tool developed as part of the Human Brain Project; Harris, 1999). Such representations are usually exquisitely detailed and very compelling and are becoming increasingly powerful when combined with high-powered computational analysis. However, their complexity makes them difficult to generalize from and computational methods to automate the process of transforming such results into a standard frame of reference (such as an atlas) cannot perform at the required level of accuracy required by expert anatomists.

Over a period of roughly 20 years, workers within the Swanson laboratory have amassed a wealth of original research data from anterograde tract-tracing experiments, retrograde labeling experiments, *in situ* hybridization and immunohistochemical staining (e.g., Risold and Swanson, 1997; Swanson, 1987; Swanson and Cowan, 1975;). The neuroanatomical data that will be considered are *Phaseolus vulgaris leuco-agglutinin* (PHAL) tract-tracing data (Gerfen and Sawchenko, 1984). These are by no means representative of all different types of neuroanatomical data, but they provide a starting point from which we intend to generalize.

Before workers in the Swanson laboratory publish their PHAL data, they transfer it to atlas plates of the rat brain by examining the tissue microscopically and drawing individual PHAL-stained fibers by hand onto a digital version of the atlas (Swanson 1998). This procedure requires a high level of expertise and patience because the orientation of the plane of section of the tissue cannot correspond exactly to that of the atlas and the cutting and fixing procedures cause unpredictable distortions of the tissue. The experimenter is forced to perform some degree of subjective interpretation while generating the drawings. If sufficient care is taken over this procedure, the end product is a highly detailed and accurate representation of the labeling pattern in the histological slide, but the procedure is extremely time consuming (Dashti *et al.*, 1997).

Fig. 1 illustrates an example of this sort of data. The left-hand drawing shows atlas level 14 of the Swanson 1998 atlas with the distribution of calcitonin gene-related peptide (CGRP) drawn in the left-hand hemisphere (lines = fibers, circles = pericellular baskets; see Risold and Swanson, 1997) and PHAL fiber labeling from injections into the posterior part of the anterior hypothalamic nucleus drawn in the right (AHNp; see Risold *et al.*, 1994). The space limitations of this document only permit a small "magnified view" of the high level of spatial detail of these data. The right-hand side of Fig. 1 shows the flatmap representation of the output connections of AHNp with a small "magnified view" of the detailed organization of this connection system. At present, the work in the Swanson laboratory emphasizes the use of the PHAL tract-tracing technique (Gerfen and Sawchenko, 1984; see also Chapter 6.3 concerning the NeuroScholar project), but any sort of anatomical data could be represented in this way (i.e., as overlays on atlas plates).

Before the NeuARt project was instigated, data were shared and examined in the Swanson laboratory by importing data by examining copies of the Adobe Illustrator files that make up the atlas with imported layers representing data from different experiments. When comparing data from two or three experiments, this was satisfactory, but when we wished to look for patterns in the data over large numbers of experiments this process became extremely cumbersome, requiring in excess of 30 to 40 layers to a single file. Reorganizing the data was impossible without laboriously copying layers between files, and the process of scanning data across atlas levels required the opening of several of these large Illustrator files and switching between them.

We reported much of the conceptual background and the proposed functionality of the NeuARt application in a paper published in the journal "*Neuroimage*" in May 1997 (Dashti *et al.*, 1997) and a more recent description of the architecture of all aspects of the system appears in the Neuroinformatics Workbench Manual (see the Website for the USC Brain Project for the most recent version). The current NeuARt application consists of a graphical user interface and neuroanatomical extension tables and relations for the NeuroCore Informix database. These tables hold details of the manipulations and protocols used in anatomical experiments and the drawings and photographs of the data, combined with index information describing how these data should be overlaid onto the atlas.

4.3.2 The NeuARt System

The basic functionality of the NeuARt viewer is to allow users to query a database of results from

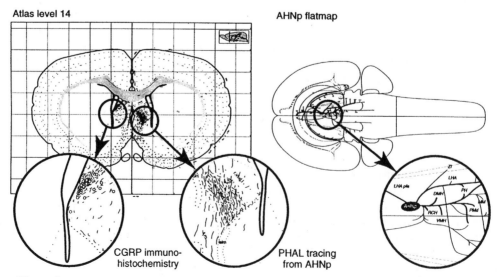

Figure 1 Illustration of detailed anatomical data from the Swanson laboratory; see text for description.

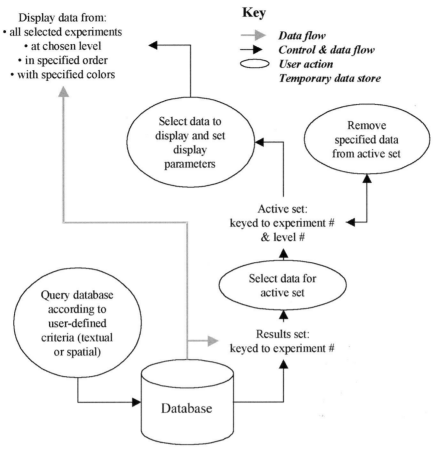

Figure 2 High-level flow diagram depicting the software requirement specification for the NeuARt system.

neuroanatomical experiments and to place results of queries into an *active set*. The display of the active set over atlas plates may be manipulated dynamically so that each individual result set may be displayed, hidden, recolored, reordered, or discarded. Users may browse data at different levels throughout the active set. The high-level organization of the system that underlies this functionality is shown in Fig. 2.

Fig. 2 illustrates the use of the system. At any time, users may perform any of the tasks shown in ellipses, assuming that the current state of the system provides those tasks with the necessary input data from either the database or temporary data stores within the application. This diagram illustrates the "two pass paradigm" method of gaining access to the database. The database serves data at two stages, first as low-bandwidth input to the results set as a simple identifying index, so users may choose to select it to add to the active set, or as the full data as input to the display. Segmenting the load on the database into these two sections means that the system does not serve full database sets unless the user actually wants to look at them. In the following sections we will discuss the schema of the Neuroanatomical database extensions to the NeuroCore system that we use to store the data and the programming design of the user interface itself.

Neuroanatomical Extensions for the NeuroCore Database

There are many kinds of neuroanatomical data. In fact, there are as many as there are different experimental procedures. An increasingly diverse number of sophisticated staining methods are available to neuroanatomical researchers. Data can be viewed in different ways and at different scales with light, confocal, electron, or magnetic resonance imaging microscopes. Neuroanatomical methodology is becoming increasingly more diverse, meaning that the schema used to capture the data must be extensible to be able to accommodate for new types of data.

The organization of the database underlying NeuArt is an extension to the NeuroCore system, meaning that not only is the basic design extensible, but non-neuroanatomical aspects of the data use the same schema as other NeuroCore databases. This creates a broad common substrate which makes the process of designing common tools more straightforward. The database can accommodate this variability by extending existing relations or adding new relations as the complexity of the data grows. Within this document, we will refer to the combination of the core section of the database (NeuroCore) and the neuroanatomical extensions (AnatExt) as the Neuroanatomical Repository of Data (Anat-RED).

The format of the data that are currently being stored in the Anat-RED are based on the methodology of a single laboratory, that of L. W. Swanson at the University of Southern California. Comparing data from different laboratories can be extremely difficult, if the underlying nomenclature of each laboratory is different, if different techniques are used, or if the data are presented in a different computational format. By presenting a way of viewing the data of a well-respected research laboratory with reference to a standard atlas (which, incidentally, was produced by the same group), we seek to illustrate the potential of this approach by example. If other laboratories find these methods useful, then NeuARt could provide a means to compare data between laboratories.

The central challenge presented to the designers of Anat-RED is to provide a unified format that captures the data of different laboratories but still represents the neuroanatomical data at a suitable level of complexity to allow the use of multiple atlases, nomenclatures, and imaging systems. The Swanson (1998) atlas has an extensive glossary, which will soon be incorporated into RED's lookup tables to provide a way of cross-referencing the nomenclature with structures that are defined as shapes in the atlas. At the present time, the neuroanatomical data under investigation are taken from adult rats. The general framework of the Anat-RED could be applied to any species or age, as long as a brain atlas for that animal could be incorporated into the database.

The tables of the NeuroCore section of the Anat-RED provide the general framework for neuroscientific experimental data. The extensions we describe here (AnatExt) provide the support to represent specific neuroanatomical research data. There are two classes of extensions: *neuroanatomical manipulations* (AnatManip) and *neuroanatomical research data* (AnatData). The AnatManip extension defines the possible manipulation techniques associated with anatomical protocols. The AnatData extension defines the spatial data associated with the anatomical data.

We have defined five classes of anatomical manipulations (Fig. 3):

1. *Chemical treatment:* These database classes describe drug treatments, specific diets, or experimental manipulations that have been performed on an animal prior to experimentation.
2. *Tracing:* These classes describe the procedures involved in performing a tract-tracing experiment; for example, the data in this section would describe the injection parameters and anesthetic regimen employed in a PHAL injection.
3. *Fixation:* These classes are used to describe the methods employed to fix the brain tissue. This would include a rigorous description of the perfusion methodology.
4. *Sectioning:* The classes of this section would be used to describe the precise nature of the sectioning techniques employed in the experiment, including a description of the type of cutting performed (e.g., freezing microtome, cryostat, etc.).
5. *Staining:* At the present time there are three separate subclasses to denote three different types of staining that are used to reveal the neuroanatomical properties of the tissue. They are listed below:

a. *Histochemical:* These are general histochemical stains such as Nissl or Golgi staining.
b. *Immunohistochemical:* These are immunohistochemical stains where the stained particles are attached to antibodies that recognize specific proteins.
c. *In situ hybridization:* These data refer to staining techniques that stain specific mRNA or DNA fragments.

A protocol can be made up of several manipulations, and these manipulation techniques are defined in isolation from specific protocols. Protocols can be associated with research data at a number of levels: at the level of individual experiments, sessions, segments, or subjects. On the other hand, manipulations are not associated directly with any of the research data.

The current implementation of AnatData assumes that anatomical data consist mostly of spatial data. At its most basic form, it consists of some sort of digital extraction of data from an actual experimental brain. This extraction could be either in two- or three-dimensional form, depending on the size of the brain, the experiment type, and the subject animal. The two-dimensional extraction methods usually constitute the slicing of the research brain and then using either a scanning or photographic technique to digitize the data. The three-dimensional extraction methods are based on MRI scanning technology. Our Anatomical-Research Data extensions provide the necessary framework to capture all of the above data types; however, because our lab generates mostly two-dimensional data, the focus of the AnatData extensions is based on two-dimensional data (Fig. 4).

User Interface Applet

We interpret the spatial structure of neuroanatomical data through visualization. Analysis of patterns of cellular properties in brain tissue with quantitative statistical methods can be used to make objective interpretations (Roland and Zilles, 1994; Zilles, 1985), but the use of these analyses is limited by the complexity of the tissue and the inherent difficulties of obtaining quantitative neuroanatomical data, and it is a well-known truism that the best pattern-recognition system available is currently the human visual system. Simple visualization is therefore unlikely to be superseded by quantitative statistical analyses among the majority of experimental

Chapter 4.3 The Neuroanatomical Rat Brain Viewer (NeuARt)

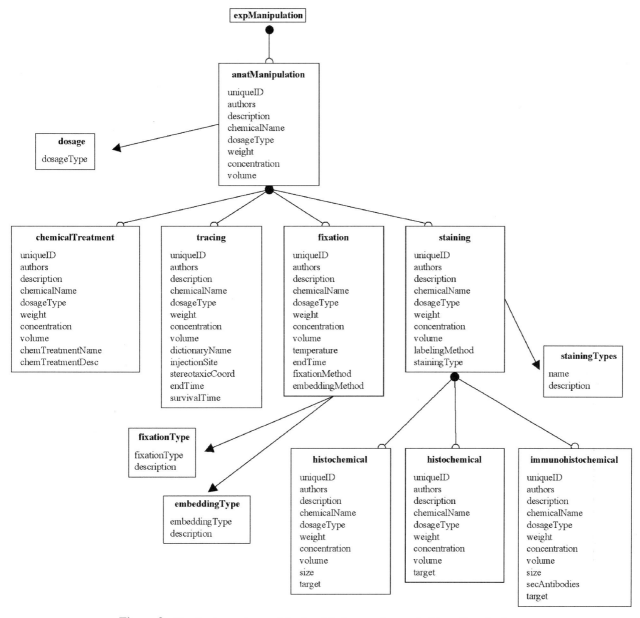

Figure 3 The database schema of the AnatManip extensions to the NeuroCore database.

workers. In any case, a visual representation of the data can be regarded as a standard requirement of any neuroanatomical storage system.

The architecture of the user interface consists of three types of modules: user-interface managers (UIM), a user-interface server (UIS), and a data server (DS). The UIM modules provide the direct interface for the user and communicate only with the UIS module, which coordinates their interactions. The UIS delivers queries to, and receives results from, the database via the data server. The specification of the database management system is contained with the DS and so can be changed, simply by replacing the DS module. This modular design will facilitate future modifications and extensions (e.g., modifying or adding UIM modules or changing the database implementation). The Swanson laboratory has a database of roughly 700 drawings from 43 separate studies in the form of layers in Adobe Illustrator.

The user interface consists of seven components: the Display Manager, the Query Manager, the Results Manager, the Active Set Manager, the Level Manager, the Viewer Manager, and the Inquire Manager. The Display Manager is where the data are visualized and may be considered the central focus of NeuARt. There are four types of interaction among the managers and the underlying data.

The first type of interaction (shown in Fig. 5A) involves the Display Manager, the Query Manager, the database (in this case, Informix Universal Server), and the Results Manager. In general, the user defines the textual and spatial query using the Query Manager and the Display Manager, respectively, and then submits this

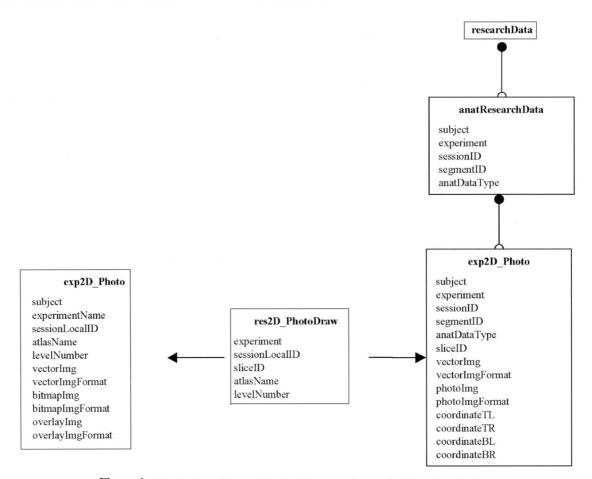

Figure 4 The database schema of the AnatData extensions to the NeuroCore database.

combined query to the database. The database returns the result to the Results Manager.

The second type of interaction (shown in Fig. 5B) involves the Results Manager, the database, the Level Manager, the Active Set Manager, and the Display Manager. After receiving the textual result of the query, the user may select a number of the experiments to be returned in their entirety for more elaborate study. This is sometime referred to as a two-pass paradigm (see above). The database returns these results into the Level Manager, the Active Set Manager, and the Display Manager.

The third type of interaction (shown in Fig. 5C) involves the Viewer Manager and the Display Manager. The user may customize aspects of the Display Manager using the Viewer Manager. The fourth type of interaction (shown in Fig. 5D) involves the Display Manager and the Inquire Manager. Using the inquire tool on the Display Manager, the user may query about the atlas structures. We have designed the underlying architecture of NeuArt to be as modular as possible. This was done in order to allow the various managers, each of which can be considered as a module, to be upgraded or replaced as new functionality is desired or as new data types become available.

The overall functionality is divided between user interface and data access, each of which is represented by an individual server. The user-interface server (UI-server) handles all communication between the managers described in the previous section. Because no manager communicates to another except via the UI-server, we were able to define clearly the interfaces between the managers and the UI-server. As long as a replacement manager conforms to these interface requirements, the system will still function correctly; however, if a new type of manager is implemented, then it is necessary to make changes to the UI-server (note that only one module has to be updated). Similar reasoning applies to the separation between the UI-server and the data server (D-server). To access data, the managers have to make the request to the UI-server, which alone communicates with the D-server. Currently, the data server is configured to communicate with an Informix database (see Fig. 13.3a). Although we have a number of future enhancements planned for NeuARt, we feel that the best enhancements will

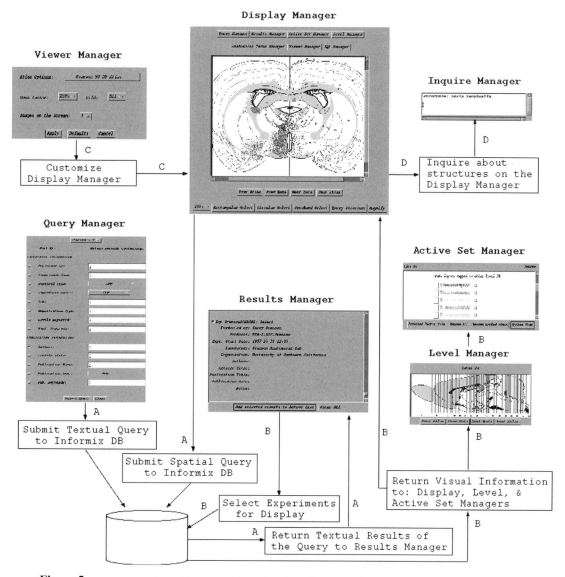

Figure 5 The organization of the NeuARt user interface, illustrating the global organization of the system.

probably result from user's comments, requests, and complaints.

DISPLAY MANAGER

The Display Manager is the focal point of most of the user interactions with the system (Fig. 7). These interactions take several forms, and the Display Manager can perform several tasks:

1. It can control the display of other managers.
2. It can control which atlas level is displayed.
3. It can enhance the ability of the user to specify spatial queries of the database.

The top two rows of buttons control the visibility of the other managers. So as not to clutter the user's view of NeuArt, the other managers will run even though they are not visible. Activating the appropriate button on the top two rows will make the corresponding manager vis-

ible. If the manager is obscured by another window, it will be raised to the top.

The third row of buttons control which atlas level is displayed. These buttons are of two types. The first type (Prev Atlas/Next Atlas) will switch the display to the previous or next atlas level as appropriate. If data layers are mapped to the newly displayed atlas level, their display will be controlled by the settings of the Active Set Manager, which we describe below. The second type of buttons (Prev Data/Next Data) will switch the display to the previous or next atlas level where data layers can be found in the active set. Any intervening atlas layers that do not contain data are skipped.

The final row of buttons enables the user to specify queries based on the spatial nature of the atlas or the data. The user may choose one of three selection tools (rectangular, elliptical, or freehand) with which to select an area of interest on the atlas. By such a selection, the

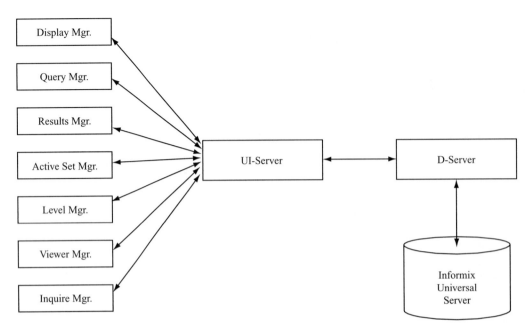

Figure 6 Relationship between NeuARt's managers, servers, and the underlying database.

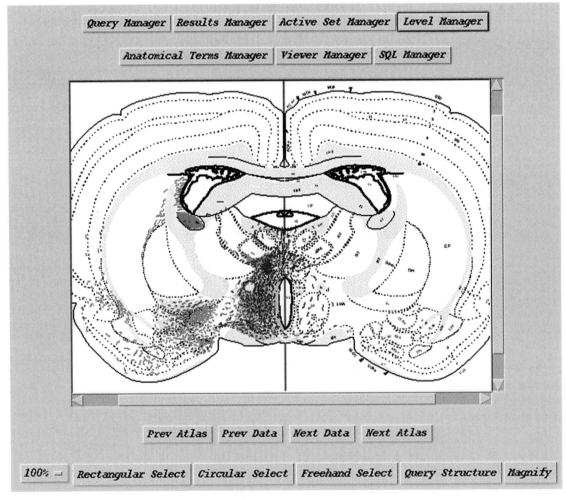

Figure 7 A screenshot of NeuARt's Display Manager.

user restricts the query so that only results that contain some data within the selected area are returned. An additional button is used to make name queries of the atlas. By the user's selecting the "Query Structure" button and then clicking on the atlas, the database will return the name of the anatomical structure in which the user clicked.

QUERY MANAGER

The purpose of the Query Manager (Fig. 8) is to enable the user to specify textual conditions that the results of the query must satisfy. Fields to specify various attributes of data from either published or unpublished sources, as well as from a variety of experimental protocols, laboratories, authors, etc., are provided. For cases in which the user will select from a limited number of cases, the user is presented with the possible choices and may select one. The operation of the Query Manager is that a blank field enforces no restriction upon the query; however, a non-blank field must be matched by the returned data. Multiple non-blank fields are combined by logical AND to determine the semantics of the query. To make multiple queries easier to perform, the Query Manager maintains a history of previously issued queries. This history may be used to fill in the fields as in a prior query; the user may then do any desired editing before issuing another query. The user submits a query to the database by activating the "Submit Query" button at the bottom of the Query Manager.

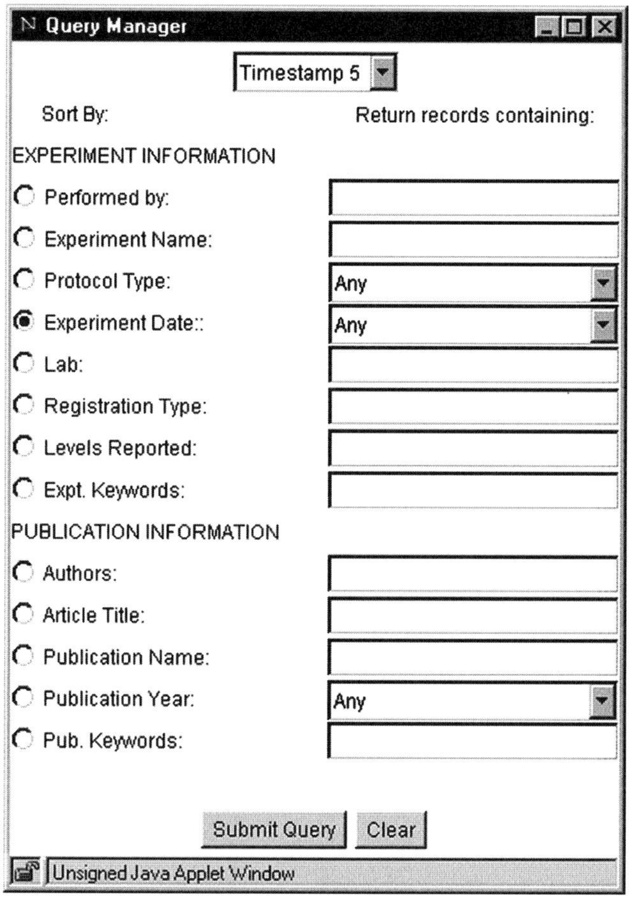

AN ADDITIONAL, RELATED APPLICATION: THE SPATIAL INDEX MANAGER (SIM)

The spatial organization of the regions in the Swanson brain atlas and of experimental data that can be overlaid onto them is valuable information that we wish to use within NeuARt. To this end, we have devised and implemented an application called the *Spatial Index Manager* (SIM) that permits polygons to be drawn over the atlas plates using a combination of automation and expert user intervention with a topological mapping program (Fig. 9). This process converts the atlas drawings into "intelligent templates" in which every point "knows" both the spatial extent and the name of the region that contains it. This "knowledge" is then inherited by any regional data registered against the atlas in the Neuroanatomical database and may thus support spatial queries anchored by references to particular brain regions, spatial features, or three-dimensional coordinates. The current version of SIM is implemented in the Java language. Similar to NeuARt's data server, it communicates to the Informix Universal Server via RMI. It stores the identified topological structures in Informix's spatial datablade format. Two major functions of SIM are

1. *Freehand drawing:* This function allows users to identify objects by freehand drawing polygons around them, labeling them, and storing them into the database. Through this function we can impose topological structures on both the atlas and the experimental data.

2. *Fill function and automatic boundary generation:* This function can semi-automatically identify objects with closed structures and store them as polygons into the database system. This is achieved by filling a selected closed structure with a certain color and then automatically detecting the convex hull of the colored region. Another function of this module is to check whether a freehand drawn polygon is closed or not.

SPATIAL QUERIES

The user may spatially query the atlas structures and/or query the combination of the atlas structures and the overlay data using the Display Manager. To specify spatial queries, the user may use SQM, which is designed in order to support spatial queries on both the atlas images and experimental data. SQM extends the NeuArt user interface permitting the user to: (1) point at a structure and see the name and corresponding information about the structure (including the list of publications with experiments on that structure), and (2) select an area (as a rectangle or a circle) and find all the experiments that are contained in, contain, or overlap with the

Figure 8 A screenshot of the Query Manager taken from NeuARt running as an applet over the World Wide Web.

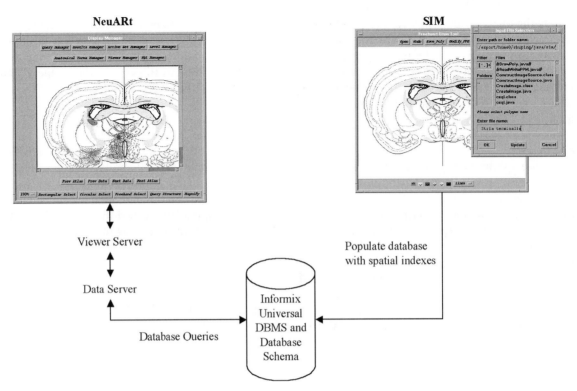

Figure 9 The computational architecture of the NeuARt and SIM systems.

selected area. SQM achieves its spatial query functionality by utilizing the Java 2D API on the user interface side and the Informix spatial datablade on the database server side. In addition, it utilizes topological information generated by the Spatial Index Manager (see above) for both atlas images and experimental data. The Query Manager utilizes the spatial description as an additional constraint to the textual attributes that it contains. After submitting a query, with the Query Manager and the spatial query tools of the Display Manager, the database returns results to the Results Manager (via the viewer server).

RESULTS MANAGER

A description of all results matching a user's query are returned to the Results Manager (Fig. 10). While scrolling through these descriptions, the user selects results of interest by clicking on the "Select" checkbox for each result. When the user is satisfied that the selections are finished, clicking on the "Add selected results to Active List" button will inform the Active Set Manager (see below) that new data are being mapped to atlas layers. Should the user decide instead that the results are completely inappropriate, the user may select the "Clear All" button. This will result in the Results Manager being emptied of all results, and the user may initiate a completely new query.

ACTIVE SET MANAGER

The Active Set Manager enables the user to control the presentation of the data layers on an atlas level (Fig. 11). Its operation is tied closely to the Level Manager. Upon selection of a new data atlas level in the Level Manager, the Active Set Manager scrolls to show the data layers mapped to that level, and the background changes to white to indicate that it is selected. In the Active Set Manager, all of the other atlas levels containing data are also present and may be accessed by use of the scroll bars. The background for a non-selected atlas level will remain gray to indicate its status.

The entries in the Active Set Manager are the identification strings that were used to select these experiments in the Results Manager, providing a reminder of the data source. Each ID is displayed in a different color from other IDs which is the same color used to display the data layer in the Display Manager. In an atlas level with many data layers mapped to it, confusion will result if all data layers are present at all times. To aid the user in evaluating the different data layers, the Active Set Manager enables the user to change the order of display of the data layers, as well as to specify whether a data layer is displayed at all. Furthermore, the user may remove entries in the Active Set Manager by using the "Remove ALL" button and the "Remove Marked Items" button in conjunction with the checkboxes to the right of the IDs of the experimental data. The "Remove ALL" button removes all data layers mapped to the indicated atlas level, and the "Remove Marked Items" button removes only those items with selected checkboxes.

The display order of the data layers is the order in which they are listed in the Active Set Manager. The

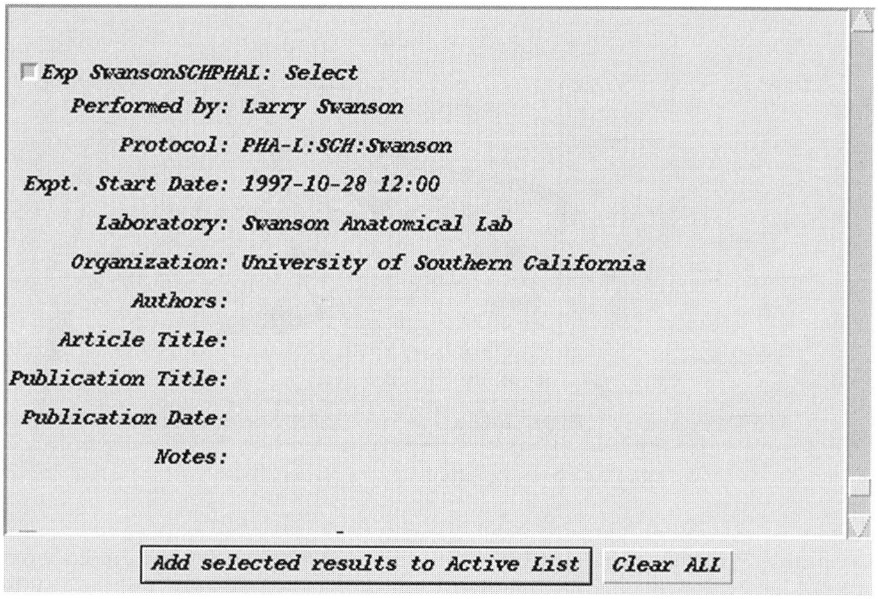

Figure 10 A screenshot of the Results Manager window, illustrating the results of a single experiment from the Swanson laboratory.

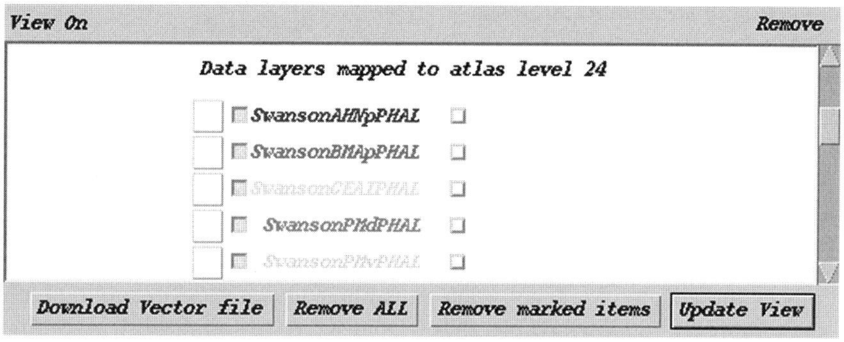

Figure 11 A screenshot of the Active Set Manager window, illustrating several datasets concurrently present on level 24 of the atlas.

uppermost-listed data layer is displayed on top of all data layers listed beneath it, and so on down the list. To change the display order, the user clicks on the square pad to the left of the ID and then moves the mouse cursor to the desired position while holding down the mouse button. Upon release of the mouse button, the list of IDs is rearranged to reflect the new display order.

In certain circumstances too many data layers may be mapped to the atlas level for the data to all be displayed simultaneously and still be understood. The user may deal with this situation by selectively displaying the data layers. This selectivity is specified by the checkbox to the left of the ID. If the checkbox is unchecked, the data layer will not be displayed, regardless of its display order. Should the user so choose, an Adobe Illustrator-compatible vector file may be downloaded to the user's system from the database, showing the selected atlas level with the data layers mapped to it. This is accomplished by clicking on the "Download Vector" file button. If the user wishes to use the atlas level in a publication, the quality of the bitmapped images displayed in the Display Manager will probably not be of sufficiently high quality.

LEVEL MANAGER

It is the responsibility of the Level Manager to provide the user with feedback about which atlas levels have data mapped to them and which atlas level is currently displayed in the Display Manager and to enable the user to move from one atlas level to another. As shown in Fig. 12, the Level Manager shows a mid-saggital view of the brain with vertical lines superimposed on it. The position of the lines correspond to the position of the atlas levels. The vertical lines are drawn in one of three colors. A light gray line indicates an atlas level that has no experimental data mapped to it. Lines drawn in blue indicate that the corresponding atlas level has one or more data layers mapped to it. A red line indicates the current atlas level shown in the Display Manager. The position of the red line moves as the user changes which atlas level is currently displayed. The Level Manager has four buttons that operate in the same manner as the corresponding

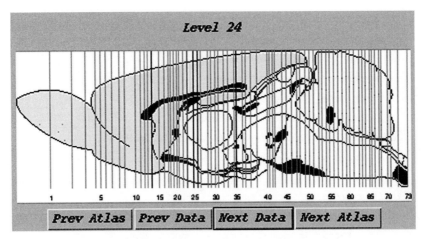

Figure 12 A screenshot of the Level Manager.

buttons in the Display Manager to change atlas levels. At the top of the Level Manager is a display indicating the number of the currently displayed atlas level.

INQUIRY MANAGER

Currently, the sole function of the Inquiry Manager is to display information returned as a result of a "structural query" in the Display Manager Fig. 13). The information returned is selectable text and may be pasted into the Query Manager if so desired in support of query specification.

VIEWER MANAGER

The purpose of the Viewer Manager (Fig. 14) is to control certain functions of the Display Manager, such as the number of Display Manager windows visible, the image magnification, the presence or absence of the physical-coordinates grid, and the chosen atlas. All of these choices may be specified by selecting relevant items from the lists shown, and the desired change will be brought about by clicking on the "Apply" button.

4.3.3 Discussion

This chapter has presented an introduction to a neuroanatomical informatics tool built on the NeuroCore architecture developed elsewhere within the USCBP. This tool provides a way of performing essential everyday visualization tasks within an anatomical laboratory without concomitant informational problems that we have found obstructive in the past. NeuARt is still a fragile prototype, requiring additional development in order to make it robust enough for general use, but it remains a coherent practical design and implementation that fulfills its original objectives well.

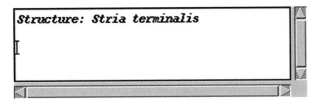

Figure 13 A screenshot of the Inquiry Manager.

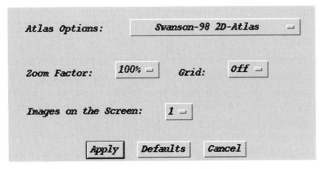

Figure 14 A screenshot of the Viewer Manager.

References

Dashti, A., Ghandeharizadeh, S., Stone, J., Swanson, L. W., and Thompson, R. (1997). Database challenges and solutions in neuroscientific applications. *Neuroimage* **5(2)**, 97–115.

Gerfen, C. R., and Sawchenko, P. E. (1984). An anterograde neuroanatomical tracing method that shows detailed morphology of neurons, their axons and terminals—immunohistochemical localization of an axonally transported plant lectin. *Phaseolus vulgaris* leukoagglutinin (PHA-L). *Brain Res.* **290**, 219–238.

Harris, K. (1999). Structure, development, and plasticity of dendritic spines. *Curr. Opin. Neurobiol.* **9**, 343–348.

Risold, P., Canteras, N., and Swanson, L. W. (1994). Organization of projections from the anterior hypothalamic nucleus: a *Phaseolus vulgaris*-leucoagglutinin study in the rat. *J. Comp. Neurol.* **348(1)**, 1–40.

Risold, P., and Swanson, L. W. (1997). Connections of the rat lateral septal complex. *Brain Res. Brain Res. Rev.* **24(2–3)**, 115–95.

Roland, P., and Zilles, K. (1994). Brain atlases—a new research tool. *Trends Neurosci.* **17**, 458–467.

Swanson, L. W. (1998). *Brain Maps: Structure of the Rat Brain*. Elsevier Science, Amsterdam.

Swanson, L. W. (1987). The hypothalamus. In *Handbook of Chemical Neuroanatomy.* Vol. 5. *Integrated Systems of the CNS* (A. Bjorkland, T. Hokfelt, and Swanson, L. W., Eds.). Elsevier Science, Amsterdam, pp. 1–124.

Swanson, L. W., and Cowan, W. M. (1975). Hippocampo-hypothalamic connections: origin in subicular cortex, not ammon's horn. *Science* **189(4199)**, 303–4.

Zilles, K. (1985). *The Cortex of the Rat: A Stereotaxic Atlas.* Springer-Verlag, Berlin.

CHAPTER 4.4

Neuro Slicer: A Tool of Registering 2-D Slice Data to 3-D Surface Atlases

Bijan Timsari, Richard M. Leahy, Jean-Marie Bouteiller, and Michel Baudry
University of Southern California Brain Project, University of Southern California, Los Angeles, California

Abstract

In this article, we present a novel method for accurate registration of autoradiographs of rat brain to a standard atlas. This method minimizes the errors produced by nonlinear deformations of experimental samples and decreases the distortions caused by possible differences in the orientation and level of the sections between the experimental specimen and the reference brain. We have created a three-dimensional (3-D) surface atlas based on the two-dimensional (2-D) cryosection images used as reference maps of the rat brain. The surface can be resliced into new sections with a resolution high enough to cover a wide range of sectioning angles for experimental brains. Shape matching and transformation from sample data to the atlas is performed by using the Thin Plate spline model.

4.4.1 Introduction

Today, many clinical applications rely on medical imaging. Disease diagnosis, therapy planning, and surgical operations are often supported by several imaging modalities providing complementary information. There are many instances in which it would be desirable to integrate the information obtained from different studies on the same subject or similar studies on different subjects. Medical image matching is a difficult task, owing to the distinct physical realities represented by the imaging modalities, the differences in patient positioning, and the varying image acquisition parameters. Image registration is the fundamental task required for comparing two or more images acquired at different times, from different sensors, or from different subjects. What is usually referred to as *registration* is the process of spatially aligning two images by transforming the coordinates of each point in one image into the coordinates of the (physically) corresponding point in the other one. The determination of the optimal registration method depends on the types of variations between the images. In the literature, a broad range of methods for image registration has been proposed for various types of data and problems. This broad spectrum of methodologies makes it difficult to classify and compare techniques, as each technique is often designed for a specific application and not necessarily for specific types of problems or data. In recent years, a few surveys on registration methods in general image processing systems (Brown, 1992) and those with application in medical imaging (Maintz and Viergever, 1998; Maurer and Fitzpatrick, 1993; van den Elsen, 1994) have been published. More focused surveys with special attention to brain imaging have also been prepared (Thompson and Toga, 1999).

In general, the type of transformation used to register images is one of the best ways to categorize the methodology and select a technique for a particular application. In this context, a transformation is defined as a mapping between two images both spatially and with respect to intensity, which may either act on the entire image as a whole (global mapping) or on the regions of the image based on their spatial locations (local mapping). The required characteristics of the transformation class used in the image registration paradigm are dictated by the specifics of the problem at hand. In situations where the relative size or shape of image components is to be retained, a rigid body transformation (a combination of rotation, translation, and scale change) is all that is required. Additionally, if shearing (a non-orthonormal rescaling operation) also needs to be accounted for, an affine transformation must be used. Affine transformations preserve parallelism between lines, but not necessarily their lengths. In more general cases, where the straightness of the lines is guaranteed to be preserved but parallelism between them is in general not preserved, a projective (or perspective) transformation must be used. Projective mappings account for distortions due to the projection of the objects at varying distances from the sensor onto the image plane. The most general class of transformations is elastic mappings. Elastic mappings are appropriate for problems that map straight lines into curves. Most elastic transformations try to conform to a set of physical constraints such as continuity, smoothness, and minimal energy state.

While the characteristics of the appropriate transformation class for a particular image registration problem are determined by the functional specifications of the mechanism responsible for image variations, the dimensionality of the transformation function is determined by the dimensions of the input and output image datasets. Image registration may be performed in any dimension, whether all dimensions are spatial or time is an added dimension. One-dimensional registration problems are exclusively used for performing temporal match on a time series of spatially consistent images. Two-dimensional methods may apply to intrinsically 2-D images, projection images, or separate slices from tomographic data. On the other hand, 3-D methods apply to tomographic images as a volumetric dataset not as a set of individual slices. In both case the matching procedure may include time as an extra dimension. Compared with 3-D to 3-D registration methods, 2-D to 2-D registration is far less complex, but it relies on an implicit assumption that the images to be matched are made exactly in the same plane with respect to a reference coordinate system. To meet this assumption, special provisions are required to ensure correct positioning of the object with respect to the imaging sensors. Even with the complicated imaging protocols these requirements demand, for some applications such as tomographic medical imaging the accuracy is rarely sufficient for 2-D to 2-D methods; therefore, most of the algorithms proposed for registration of tomographic datasets are 3-D to 3-D. In the literature, the class of 2-D to 3-D registration techniques is generally reserved for applications concerning direct alignment of spatial data to projective data (e.g., a pre-operative CT image to an intra-operative X-ray image) or the alignment of a single tomographic slice to spatial data (Maintz and Viergever, 1998).

In this article, we will study the problem of registering autoradiographic slice images of rat brain to a cryosection imaging based atlas. To solve this problem, we constructed a 3-D surface atlas from the available reference slice images and then resliced the reconstructed surface to generate new reference sections. Final registration is performed using a 2-D to 2-D technique; however, because the overall matching is between a 2-D image and a 3-D surface, this method can be categorized as a 2-D to 3-D method. Before we get into the detailed description of the USC Brain Project histological image registration tools we will quickly review different classes of image registration techniques that are currently used for medical images in general and brain images in particular to provide the reader with a general understanding of available methodologies along with their specific strong and weak points.

4.4.2 Classification Criteria

Registration methods are developed for a wide variety of applications and based on considerably different assumptions. Various discerning criteria have been proposed for classification of these methods (Maintz and Viergever, 1998). We have chosen a subject-based classification scheme to categorize registration methods into two classes: intrasubject and intersubject. The scope of our classification is restricted to methods that register brain images, and it covers all different brain imaging modalities. To keep our study brief, for each category we have selected for review only those methods that represent the dominant strategies in medical image registration (i.e., model-based approaches and intensity-based approaches).

As a common characteristic, all model-based algorithms are based on identifiable features of a geometric model of the brain. These explicitly defined features of the model correspond to anatomically significant elements of the brain, such as functionally important surfaces, curves, and point landmarks. Anatomical elements are matched with their counterparts, and their correspondences guides the transformation of one brain to another.

Intensity-based approaches use some statistical criteria to match the regional intensity patterns of the two images. Typically, the matching problem is cast as a problem in optimization theory for finding a transformation function that maximizes some mathematical measure of similarity between the deforming image and target. Measures of intensity similarity can include squared differences in pixel intensities, regional correlation, or mutual information.

4.4.3 Intrasubject Image Matching

When the goal of the registration task is to match the images acquired from a single subject at different times and under various conditions it is referred to as *intrasubject registration*. Such matches can occur within a given modality or between modalities. The most straightforward case is when the datasets are obtained using the same modality and the matching is accomplished by a linear transformation amounting to rotation and translation of the dataset. Another component may possibly be required for scaling to allow for the adjustment of size while maintaining shape as invariant. The result is a nine-parameter fit (rotation, translation, and scaling along each of the three spatial axes).

The problem of intrasubject, cross-modality registration is more difficult because of the modality-specific inconsistencies between images, such as incomplete coverage, intensity distortions, geometric distortions, and lack of contrast. One of the most common applications of intrasubject registration used in almost any type of diagnostic and interventional procedure is to match anatomical structures of the brain obtained from magnetic resonance imaging (MRI) with the physiological activities of each brain structure acquired from positron emission tomography (PET).

Model-Based Strategies

POINT PAIRS MATCHING

In the literature, a variety of techniques for mapping one brain to another using correspondences between homologous point landmarks have been described. The diversity of methods originates from the level of user interaction in identification of landmarks and the approach chosen for matching corresponding pairs of points. A well-known technique to solve 3-D rigid point to point match is using singular value decomposition (SVD) to find a transformation that minimizes the root mean square distance between homologous points in two datasets. Evans *et al.* (1991) apply this method to PET and MR brain images, while Hill *et al.* (1991) use it to register computerized tomography (CT) and MR skull base images. For the images that are misaligned by small rigid or affine transformations, cross-correlation can be used as a match metric to give a measure of the degree of similarity between two images. Maguire *et al.* (1986, 1991) used user-identified anatomical landmarks and external markers to find the affine transformation between tomographic brain images by optimizing the cross-correlation in areas around the landmarks.

Another frequently used approach is based on moment matching. From the moments of inertia, the principal axes of the object can be derived. Translation is found based on the position of the centroid of landmarks after equating the first-order moments. Rotation is calculated by aligning the landmark population's principal axes of inertia or equating the second-order moments. Calculating scaling factors is also possible from the relative magnitude of the principle axes (Alpert *et al.*, 1990; Bajcsy and Kovacic, 1989; Kovacic *et al.*, 1989).

CURVE MATCHING

Curves, both in 2-D and 3-D, have been used as the basic feature for developing several image-matching algorithms. Matching curves without well-defined starting and ending points is the major bottleneck in most of these algorithms. Balter *et al.* (1992) first determine the overlap between corresponding curves by searching for the best fit of the local curvatures along the curves. From the overlapping segments, a number of point pairs are generated that are used in a direct point to point matching algorithm. Gueziec and Ayache (1992) make a table for all the curves in one image, with an index on their curvature and torsion parameters. Votes for particular transformations are generated by scanning the curvature and torsion parameters in the second image. From these votes a single rigid transformation is generated.

SURFACE MATCHING

A number of methods employ parametric correspondence between brain surface models extracted from tomographic images for matching them. The most popular method in this category is the "head-hat" surface matching technique (Pelizzari *et al.*, 1989) which tries to find a rigid (optionally affine) transformation that optimally fits a set of points (*hat*) extracted from the contours in one image to the surface model (*head*) extracted from another image. The *hat* fits the *head* when the mean squared distance between the *hat* points and the *head* surface is minimized.

Intensity-Based Strategies

Intensity-based approaches try to match the regional intensity patterns of the two images by comparing some statistical measure of similarity of the voxels. There are two distinct classes of approaches in this category. The first approach is to reduce image intensity content to a representative set of scalars and orientations from which the center of gravity and principal axes of the image will be calculated. Registration is then performed by aligning the center of gravity and the principal orientations (Alpert *et al.*, 1990; Banerjee and Toga, 1994). The second is to use the full image content throughout the registration process. Theoretically, these methods are more appealing because they use all of the available information for calculating the transformation function. The best example of this class of methods is registration based on minimization of variance of intensity ratios (Woods *et al.*, 1992, 1993). In this methodology, the rotation and translation parameters of the transformation function are adjusted iteratively by minimizing the variance of the ratios between intensity values of all voxels in one image and a single voxel intensity value in the other.

4.4.4 Intersubject Image Matching

Intersubject registration is the task of matching two images belonging to different subjects. Despite the variations in size, orientation, topology, and geometric complexity of cortical and subcortical structures, the assumption underlying intersubject matching methods is that, at a certain level of representation, the topological structure of the brain is invariant among normal subjects (Rademacher et al., 1993). The quantitative comparison of brain architecture across different subjects requires a structural framework in which individual brain maps can be integrated. "*Atlas*" is the general term used to refer to such a framework derived from either a single representative brain or an average anatomic template. The standardized 3-D coordinate system of the atlas (*stereotaxic* space) supplies a quantitative spatial reference system in which the variability of anatomical features from different individuals can be compared (Evans et al., 1996) and multiple brain maps from different modalities can be correlated.

Brain Atlases

Classic atlases were developed based on a detailed representation of the individual anatomy of one brain (or a few brains) in a standardized 3-D coordinate system (Matsui and Hirano, 1978; Ono et al., 1990; Paxinos and Watson, 1986; Schaltenbrand and Bailey, 1959; Schaltenbrand and Warren, 1977; Swanson, 1992; Talairach and Tournoux, 1988). Using these atlases generally involves employing simple proportional scaling systems to stretch or contract a given subject's brain to match the atlas. The classic 3-D neuroanatomic human atlases have proven to be useful for providing reference information required for stereotaxic surgical procedures (Kikinis et al., 1996). However, as expected, such atlases are less accurate at brain sites with more intersubject variability and regions far from the landmarks of the reference system. To reflect the complex structural variability of individual brains the construction of brain atlases to represent large populations of subjects has become the focus of intensive research in recent years (Evans et al., 1992; Friston et al., 1995; Steinmetz et al., 1989; van Buren and Maccubbin, 1962; Vries and McLinden, 1988).

CRYOSECTION IMAGING-BASED ATLASES

The purpose of cryosectioning is to collect high spatial imagery of the whole brain. Cryomacrotome sectioning of the frozen brain produces whole brain sections for staining and forms the basis of a reconstructed 3-D digital dataset. Frozen brains are usually sectioned in the coronal plane using a hardened steel knife, then the sections are photographed using digital imaging techniques. Several digital atlases have been developed using this method (Bohm et al., 1983; Greitz et al., 1991; Swanson 1992; Toga et al., 1995).

MAGNETIC RESONANCE IMAGING-BASED ATLASES

In recent years, development of a true 3-D atlas of the human brain has been made possible because of the growth in brain imaging technologies and the advances in development of 3-D segmentation methods. Consequently, for the first time in medical imaging history, anatomical details can be appreciated in 3-D instead of being restricted to 2-D images of 3-D structures. The major defect in MRI-based atlases is still the relatively low resolution and lack of anatomic contrast in important subregions. Where resolution is not a factor for development of an atlas, as in construction of population-based average atlases, MRI has proved to be very efficient (Andreasen et al., 1994; Evans et al., 1992). Average intensity atlases are useful for automated registration and mapping of MRI and fMRI datasets into stereotaxic space (Evans et al., 1994).

MULTI-MODALITY ATLASES

In 1993, the U.S. National Library of Medicine funded a team of researchers under the Visible Human Project to produce thousands of razor-thin slices of two cadavers (male and female) which were photographed to capture detailed images of the structures in the body. The outcome was over 15 gigabytes of image data, including CAT scans, MRI, and cryosection images. Combining the strengths of each imaging modality, the Visible Human dataset, although not a brain atlas, has the quality and accessibility to be used as a test platform for developing methods and standards (Spritzer et al., 1996).

PROBABILISTIC ATLASES

Due to the large variation in the anatomy of brain in different individuals, an atlas based on a single subject's anatomy cannot be a true representative of the whole population. To generate anatomical templates whose data can be extendable to a population, probabilistic atlases were introduced. These atlases contain precise statistical information on positional variability of important functional and structural organizations of the brain (Matsui and Hirano, 1978).

Model-Based Strategies

The first stage of any intersubject registration method is the transformation of individual data to the space occupied by the atlas. A stereotaxic coordinate system has been used conventionally as a framework for representing quantitative information about 3-D spatial relationships in the brain. The most widely used stereotaxic system introduced by Talairach (Talairach and Szikla,

1967; Talairach and Tournoux, 1988) is defined by a set of piece-wise affine transformations applied to 12 rectangular regions of brain, specified by drawing lines from the anterior and posterior commissures to the extrema of the cortex. Despite the widespread usage of Talairach stereotaxic system as a standard coordinate system for researchers to compare and contrast their results (Fox *et al.*, 1985; Friston *et al.*, 1991), the fact remains that linear transformations (rotation, scaling, translation) by themselves are not sufficient to remove the large intersubject structural variation in the anatomy of brain. Elastic transformations are required to adapt the shape of an atlas to reflect the anatomy of an individual subject.

Among model-based techniques, deformable models offer the unique characteristic of accommodating the significant variability of biological structures over time and across different individuals by combining geometry, physics, and approximation theory. Geometry serves to represent object shape, physics imposes constraints on how the shape may vary over space and time, and approximation theory provides the formal mechanism for finding an optimal model fitting the measured data. For broader shape coverage, deformable models usually employ geometric representations that involve many degrees of freedom, such as splines. The freedom of model is limited by physical principles that impose intuitively meaningful behavior on the geometric substrate. Physically, a deformable model can be viewed as an elastic body subject to a set of constraints which is responding naturally to a system of applied forces. An energy function defined in terms of the model's geometric degrees of freedom quantifies the deformation.

Point Pairs Matching

A global transformation derived from matching pairs of corresponding points on two images cannot account for local geometric deformations. In case of an approximating transformation, the local distortions are spread throughout the image, while for interpolating transformations high-order polynomials will usually be required, and these behave erratically. Piece-wise interpolation techniques based on point matching can account for deformations which vary across different regions of the image. In this methodology, a spatial mapping transformation is specified for each coordinate which interpolates between the matched coordinate values. Various choices of interpolating functions have been used for medical image matching, including thin-plate spline (Bookstein, 1989), 3-D volume spline (Davis *et al.*, 1997), elastic body spline (Davis *et al.*, 1997), and multiquadric and shifted log interpolants (Frank, 1979; Hardy, 1990; Ruprecht and Muller, 1995).

Curve Matching

Choosing the best feature space to use for matching will significantly improve the registration result. Structural similarity and as a result anatomical accuracy can be increased if matching is based on features extracted from intrinsic structures of the images. Edges, contours, and boundaries are frequently used as a feature space because they represent the intrinsic structure of an image. In recent years, several warping transformations based on lines and curves with salient features have been introduced (Declerck *et al.*, 1995; Joshi *et al.*, 1995). Curve-matching algorithms try to match curves with their counterparts in the target image using geometric features such as torsion, curvature, and local Frenet frames as guiding parameters (Gourdon and Ayache, 1993; Kishon *et al.*, 1991; Vaillant and Davatzikos, 1997).

Surface Matching

A natural extension of matching techniques based on curve and line feature spaces is surface-based warping algorithms. Such algorithms warp one image to another by elastically deforming a surface extracted from the source dataset to its homologous surface in the target domain. The success of these methods depends on the accuracy of the surfaces they match. Several algorithms have been proposed for extracting brain surfaces from 3-D image volumes. The analytical form of the surface is calculated either directly by the extraction algorithm or in a later processing through a flattening procedure. Extraction algorithms that start with a parametric model for the surface and try to deform it to match a target boundary end up with an analytic form (MacDonald *et al.*, 1993; Vaillant and Davatzikos, 1997; Xu and Prince, 1997), while others create a segmented volume whose boundary defines the surface (Sandor and Leahy, 1995). The latter approach generally produces a more detailed and accurate description for the surface which may justify the cost of the additional flattening step required for obtaining its analytic description (Drury *et al.*, 1996). From the extracted surfaces represented in analytical form, warping functions can be computed to map surfaces from the atlas to the target image. To guide the mapping from atlas to target anatomically significant features of both objects are detected and forced to match (Collins *et al.*, 1996; Thompson and Toga, 1996). Deformable models that can accurately locate image features such as snakes and their extension have been successfully used for this purpose (Sandor and Leahy, 1995).

Intensity-Based Strategies

In many applications, it is desired to map data from a variety of modalities into an atlas coordinate space without requiring structural correspondence at a local level. With the development of average-intensity MRI atlases in the Talairach coordinate system, automated image registration algorithms have emerged for this purpose. These algorithms try to align new datasets with the atlas by maximizing a measure of intensity similarity, such as

3-D cross-correlation (Collins et al., 1994, 1995) or ratio image uniformity (Woods et al., 1992).

On the other hand, when the goal is perfect intersubject registration of brain data, the atlas must be deformed in an extremely local level to fit the shape of the target anatomy. This requires the warping transformation to be of high spatial dimension. Local deformations do not necessarily preserve the connectivity and topology of transformed objects. To ensure that the atlas will not break into parts under such complex transformations it is typically considered to be embedded in a 3-D deformable medium which can be either an elastic material (Bajcsy and Kovacic, 1989; Broit, 1981; Miller et al., 1993) or a viscous fluid (Christensen et al., 1994). All intensity-based approaches basically try to redistribute the image intensity in the atlas to make it more similar to the target. Different algorithms vary depending on how the similarity function is defined and maximized.

ELASTIC MEDIUM APPROACHES

In this approach, a brain atlas modeled as an elastic medium is subject to two systems of opposing forces, external forces trying to deform the atlas and internal restoring forces generated by the elasticity of the material. The displacement field at equilibrium state is described by the Navier-Stokes equation:

$$\mu \nabla^2 U(x) + (\lambda + \mu)\nabla(\nabla \cdot U(x)) + F(x) = 0 \quad \forall x \in R \tag{1}$$

where R is discrete lattice representation of the atlas,

$$\nabla \cdot U(x) = \sum_i \partial u_i / \partial x_i \tag{2}$$

shows the cubical dilation of the medium; ∇^2 is the Laplacian operator; λ and μ define the elastic properties of the body; $U = U(x_1, x_2, x_3)$ is the displacement field; and $F(x)$ is the external force. This system of partial differential equations can be solved iteratively on the defined lattice and interpolated trilinearly to obtain a continuous displacement field.

VISCOUS FLUID APPROACHES

Linear elastic models are valid only for small deformations. Restoring forces are proportional to the deformed distance and as a result the displacement field resulting from these forces is limited to be low in magnitude, leaving the atlas incompletely warped onto the target. Viscous fluid based models were developed to overcome this limitation of elastic models. The displacement field obtained from the solution to the elastic equilibrium equation is used as an initial value for the viscous fluid differential equation:

$$\alpha \nabla^2 V(x,t) + \beta \nabla(\nabla^t V(x,t)) = F(x) \tag{3}$$

where α and β are viscosity constants; $F(x)$ is the external deforming force (per unit volume); and $V(x,t)$ is the instantaneous velocity of the deforming atlas at location x at time t. The velocity, $V(x,t)$, is related to the displacement field, $U(x,t)$, by the equation:

$$V(x,t) = \frac{\partial U(x,t)}{\partial t} + V(x,t)\nabla \cdot U(x,t) \tag{4}$$

As the atlas deforms to match the target over time, the external force goes to zero, which causes the velocity to go to zero, and final match is obtained.

4.4.5 Neuro Slicer: USCBP Histological Registration Tool

Many biological measurements of the brain depend upon the qualitative evaluation of radiolabelled samples to provide information about protein metabolism, gene expression, and nucleic acid sequences. Quantitative analysis of autoradiographic data helps neuroscientists to determine the functional properties of a particular region in response to a given treatment or to obtain some knowledge about the mechanisms for expressing certain responses. Furthermore, integrating and comparing data from multiple subjects provides the researchers an understanding of the functional behavior and structural organization of the brain.

Most autoradiographic image data used in histological studies, though information rich and high resolution, present non-linear deformations due to specimen sectioning. These deformations together with natural morphological differences between individual brains make the task of comparing results from different studies more challenging. For a meaningful interpretation, when analyzing autoradiographic images, the data collected from multiple subjects should be compared with respect to some standardized base of reference. This requires an atlas of brain anatomy with the resolution of an autoradiogram, which considering the current state of imaging technology is limited to cryosection-imaging based atlases.

Ideally, the data in each experimental brain section must be mapped to the corresponding homologous section in the atlas, and all the quantitative measurements have to be made with respect to the atlas. Selection of homologous sections is traditionally done by an expert based on visual judgement. However, in the very likely situation when the orientation or position of the plane of sectioning of the experimental brain does not match that of the reference brain, this method is potentially error prone.

As an alternative, if instead of a series of 2-D cryosection images of brain, a 3-D atlas is used, then theoretically it would be possible to find homologous atlas sections for any given experimental brain section without any error. Based on this, we have developed a method for non-rigid registration of autoradiographic images of rat brain using a 3-D surface atlas. Our approach is categorized in the class of model-driven, feature-point-based, 2-D to 3-D registration methods and comprises the following steps.

Reconstruction of Surface Atlas

Compared to 2-D atlases, a full 3-D atlas contains a spatially continuous description of the physical entity it is representing. Due to the limited resolution of current non-invasive 3-D medical imaging modalities which is far less than the resolution required for histological studies, 2-D cryosection imaging-based atlases are commonly used for such studies. Swanson's rat brain atlas (Swanson, 1992), one of the most comprehensive atlases of this type for rat brain, consists of 73 coronal (frontal or transverse) sections in which major cell groups and fiber tracts along with their subdivisions are labeled. In the structural hierarchy of this atlas gross segments, regions, cell groups, and fiber tracts are identified by their boundaries. Tissue labeling and region identification were both performed manually section by section. This has naturally caused loss of data continuity along the dorso-ventral axis. A sample section is displayed in Fig. 1. To restore the lost continuity, one can interpolate between the data in available sections. Depending on the desired accuracy, different interpolating functions can be used for this purpose.

In the digital version of Swanson's atlas, image data are conveniently stored in two formats: raster and vector. The raster file format is best for representing image intensity value at each voxel and is used when interpolating the cross-sectional density maps to create 3-D volume data. Vector format, on the other hand, is for representing curves and contours. All the information required to reconstruct surfaces by interpolating between curves is abstracted in this format. Although, for a quantitative study, the full 3-D form of both the volume and the surface are required, for the purpose of visualization one may choose only the surface atlas to take advantage of the speed and capabilities of the available surface rendering tools.

Reconstructing the surface of an object from a series of planar contours corresponding to the intersection of its boundary with a set of parallel planes is of major interest in computer-aided design of mechanical and architectural structures. Several general purpose and many custom-made programs have been developed for this purpose. We have used Microstation, one of the available commercial packages for solid modeling, drafting, and visualization of CAD products, for creating the 3-D surface of the labeled structures in Swanson's rat brain atlas.

The line drawings, marking the boundary contours of different structures in each one of the 73 sections of the atlas, were imported into this program in vector format. A bicubic B-spline model was then used to fit a parametric surface to a selected subset of these line drawings. The parametric model of the reconstructed surfaces can be easily used for calculating the cross-sectional profile of the surface within any arbitrary plane. Fig. 2 shows a sample surface reconstructed by this method. This surface represents the left half of rat brain cortex cut at the top to display the internal structures of the brain.

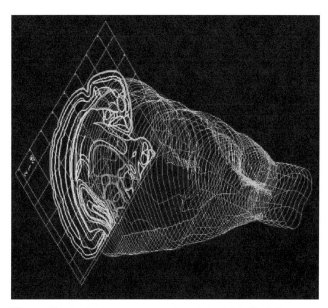

Figure 1 A sample section from Swanson's rat brain atlas. The contours represent the boundaries of the structures in the left hemisphere of rat brain.

Figure 2 A wire-frame representation of the left half of the reconstructed rat brain cortex. The top part of the brain is removed to display internal structures.

Matching Experimental Data to Atlas

In histological studies, when the autoradiogram of an experimental specimen is to be compared with a standard atlas, classically, an expert selects a particular level of the atlas as the closest match for the experimental image. In this selection he uses his knowledge and experience to look for similar shape patterns between two images. The shape of an object is described by words or quantities that do not change when the object is moved, rotated, enlarged, or reduced. Mathematically, shape is defined as the geometric properties of a configuration of points that are invariant to changes in translation, rotation, and scale (Kendall, 1984). By quantifying shapes and defining a measure for shape distance one can automate the process of atlas matching.

Mathematically, shapes are defined as equivalence classes of discrete point sets, called *landmarks*, under the operation of Euclidean similarity transforms (Bookstein, 1991). Landmarks are labeled points on a biological image located according to some rule. Generally, on a biological image, landmarks are selected as a set of anatomically significant points that are easily identified across all images. The similarity between two landmark shapes is measured in terms of their Procrustes distance. Informally, the squared procrustes distance between two sets, A and B, of landmark points is the minimum sum of squared Euclidean distances between the landmarks of A and the corresponding landmarks in point sets C as C ranges over the whole set of shapes equivalent to B (Bookstein, 1991). To measure the shape distance between two point sets, first their centroids have to be translated to the origin of their common coordinate system. Then their centroid size, the square root of the summed squared distances between landmarks and their centroid, is scaled to unity. Finally, these scaled and translated images are superimposed at their centroid and one of them is rotated with respect to the other to minimize the sum of squares of the residual distances between matched landmarks. The squared Procrustes distance between the shapes is the sum of squares of those residuals at the minimum state (Bookstein, 1991).

The only nontrivial part of this algorithm is calculating the rotation required to superimpose the two images. Let X_1 and X_2 be $k \times 3$ matrices for the coordinates of k landmarks after the centering and re-scaling has been done. It can be shown (Rohlf, 1990) that if the singular-value decomposition of $X_1^t X_2$ is UDV^t with all elements of D positive, then the rotation required to superimpose X_2 upon X_1 is the matrix VU^t.

To avoid the computational cost of singular-value decomposition, one can use other methods to approximate Procrustes distances. We can use the notation of complex algebra to represent the two sets of landmark points by two complex vectors $Z_j = (Z_{1j}, Z_{2j}, \ldots, Z_{kj})^t$, $j = 1, 2$. After the centering and re-scaling steps, we have:

$$\sum_i Z_{ij} = 0 \quad (j = 1, 2) \tag{5}$$

$$\sum_i Z_{ij} Z_{ij}^* = 1 \quad (j = 1, 2) \tag{6}$$

where * denotes complex conjugate. The superposition of the second image on the first is given approximately by:

$$Z_2 \rightarrow (\sum_i Z_{i1} Z_{i2}^*) Z_2 \tag{7}$$

and the squared Procrustes distance between the two shapes is approximately (Bookstein, 1997):

$$D_p^2(Z_1, Z_2) = 1 - \left| \sum_i Z_{i1} Z_{i2}^* \right|^2 \tag{8}$$

Within this framework, matching an experimental image to the atlas is equivalent to finding a profile from the atlas with a configuration of landmarks similar to the experimental image. This is essentially a search for a particular pattern of 2-D feature points in a 3-D feature spaces. There is an inherent ambiguity involved in this process because the mapping between 3-D vectors and their projections in 2-D space is many to one. To remove this ambiguity one has to limit the search only to 2-D projection spaces. We have accomplished this with the following method.

A certain number of anatomically significant landmark points are selected and their locations are marked with small circles on the brain sections where they appear. These circles will form continuous cylindrical tubes after 2-D to 3-D interpolation is applied to consecutive sections of the brain. Fig. 3 shows the reconstructed landmark tubes in the portion of the brain that contains the hypocampus.

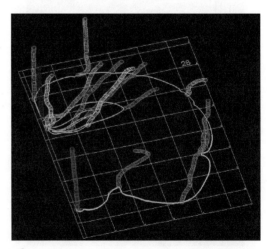

Figure 3 Landmarks marked with small circles on different sections of the brain form cylindrical tubes when reconstructed using a 2-D to 3-D interpolation technique.

Having a 3-D atlas with embedded labeled landmarks provides us the possibility of generating arbitrarily oriented reference sections with pre-labeled landmark points on them. By re-sectioning the 3-D atlas with a cutting plane with adjustable position and varying orientation, a set of 2-D profiles is formed. The collection of all profiles forms the search space for the matching algorithm. In order to get the desired spatial resolution and to cover the possible range of variation in the cutting angle of the experimental brain, re-sectioning is performed by making cuts in the coronal plane as well as planes rotating within the range of $-10°$ to $+10°$ (with increments of $0.5°$) around medio-lateral and dorso-ventral axes. A 5–micron distance between two consecutive coronal planes appears to give sufficient spatial resolution for our purpose.

The same set of landmark points is correspondingly marked on experimental images. To match an experimental image to the atlas, the Procrustes distance between its landmark points and the profiles in the pool of atlas sections is calculated. The minimum distance determines the profile of the standard brain that is most similar to the experimental image. To speed up the search for the closest atlas profile, matching is broken down into several stages. The vicinity of the experimental section in the brain is determined using a search over the entire brain with relatively widely spaced profiles with large angular differences. Then, using higher resolution profiles, the margins are narrowed repeatedly until the best match is found.

Warping Experimental Image

In general, when the mapping between two images is guided by the specification of the correspondence between landmark points various deformation functions can be found that are consistent with the displacements assigned at landmark points. The ambiguity lies in the construction of the non-rigid part of the transformation after the correspondences between the first three pairs of landmarks are used for calculation of the rigid portion of the mapping. One way to resolve this ambiguity is to require the deformation field to have minimum irregularity or roughness by adding a regularizing functional to the mapping (Tikhonov and Arsenin, 1977). A regularizer function penalizes large values of derivatives and thus smooths the deformation field, but at the same time it should create warping transformations that satisfy displacement values at landmark points and do not vary erratically elsewhere.

Thin-plate splines are functions which minimize the penalty:

$$\left(\frac{\partial^2 Z}{\partial x^2}\right)^2 + 2\left(\frac{\partial^2 Z}{\partial xy}\right)^2 + \left(\frac{\partial^2 Z}{\partial y^2}\right)^2 \qquad (9)$$

known as the bending energy of function,

$$Z(x,y) = -U(r) = -r^2 \log r^2 \qquad (10)$$

over the class of all functions Z taking the specified displacement values at landmark points. A thin-plate spline function has a quadratic form in r, the distance $(x^2 + y^2)^{1/2}$ from the origin of the Cartesian coordinate system, and is the fundamental solution of the biharmonic equation:

$$\Delta^2 U = \left(\frac{\partial^2}{\partial x^2} + \frac{\partial^2}{\partial y^2}\right)^2 U = 0 \qquad (11)$$

the equation for the shape of a thin steel plate lofted as a function $Z(x,y)$ above the plane (x,y) (Bookstein, 1997). The mapping that extends the displacement field from a set of pairs of landmark points to the entire 2-D space is given by:

$$\begin{pmatrix} x' \\ y' \end{pmatrix} = A\begin{pmatrix} x \\ y \end{pmatrix} + T\begin{pmatrix} Z_x(x,y) \\ Z_y(x,y) \end{pmatrix} \qquad (12)$$

where A represents a 2×2 linear transformation matrix, and T is the translation vector. The vector $Z = (Z_x, Z_y)$ represents the nonlinear part of the warping transformation defined by:

$$Z_\eta(x,y) = \sum_{i=1}^{N} \omega_{\eta i} U(|(x,y)^t - (x_i, y_i)^t|) \qquad (13)$$
$$\eta \in \{1, 2\}$$

in which U is the thin-plate spline function. The unknown coefficients $\omega_{\eta i}$ along with the elements of matrix A and translation vector T are determined by evaluating the equation at the given landmark points $(x_i, y_i)^t$ and solving the resulting linear system of equations.

Results

We have created a 3-D atlas for the rat brain based on a series of maps originally generated by Swanson. We used vector representation of a subset of line drawings in the atlas, corresponding to the boundaries of different structures in the brain, to extract the location of their control points. These data were then used to describe the outer boundaries of the structures with a bicubic B-spline parametric surface model. Using the parametric model for the surface, we can easily calculate the cross-sectional profile of the surface within any arbitrary plane.

To describe the rat brain-shape information in terms of the configuration of a finite number of discrete points, we have identified a set of anatomically significant landmark points and manually marked them on each section of the atlas. When reconstructed, these landmarks form continuous tubes. The assembly of tubes constitutes a 3-D object with a shape equivalent to the atlas. A pool of profiles for the standard rat brain is generated by reslicing the 3-D atlas shape-equivalent object. To cover

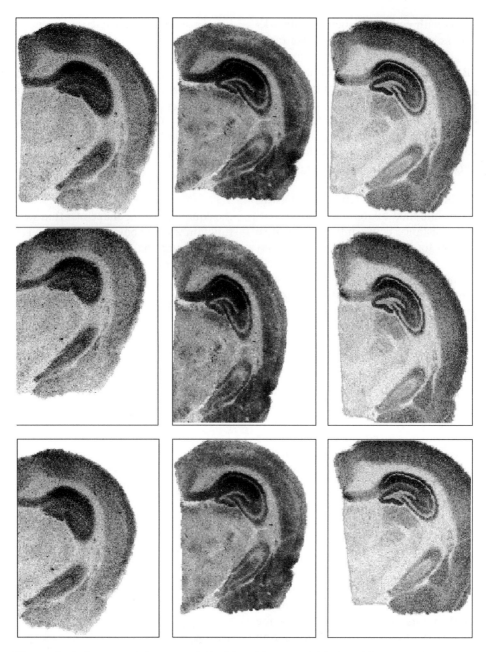

Figure 4 Performance of the proposed algorithm. (First row) Original experimental section. (Second row) Experimental section mapped to the best matching atlas section selected by an expert. (Third row) Experimental section mapped to the best matching atlas section found based on the proposed algorithm.

possible deviations in the cutting angle of experimental brain from that of standard brain, re-slicing is performed in steps of 5 microns and each step generates sections in the coronal plane as well as planes rotating within the range of -10 to $+10°$ (with increments of $1°$) around medio-lateral and dorso-ventral axes. The same set of landmark points is correspondingly marked and labeled on experimental images. To match an experimental image to the atlas, the Procrustes distance between its landmark points and the profiles in the pool of atlas sections is calculated. The minimum distance determines the profile of the standard brain that is most similar to experimental image.

A series of experiments was carried out to evaluate the performance of the algorithm. In each experiment, a sample section of a rat brain was mapped to the atlas using two different methods:

1. By following the classic method and using the original 2-D atlas, an expert was asked to select a section from the atlas which he found most similar to the experimental section.
2. By following the described algorithm, predefined landmark points were labeled on the experimental section and then the procrustes distance from the sections in the searching pool was calculated. The

best match in this case was the one having the shortest distance.
3. In both cases, the experimental section was mapped to its best match using thin-plate spline transformation.

All experiments showed that by using the second method we can gain significant improvements in the matching results. Fig. 4 show three sample results of these experiments. The first row shows the original image of experimental brain sections, the middle row is the result of warping the experimental images to atlas sections selected by an expert using thin-plate spline, and the last row shows the result of our registration/warping tool.

References

Alpert, N. M., Bradshaw, J. F., Kennedy, D., and Correia, J. A. (1990). The principal axis transformation a method for image registration. *J. Nucl. Med.* **31**, 1717–1722.

Amit, Y. (1997). Graphical shape templates for automatic anatomy detection with applications to MRI brain scans. *IEEE Trans. Med. Imag.* **16(1)**, 28–40

Andreasen, N. C., Arndt, S., Swayze, V., Cizadlo, T., Flaum, M., O'Leary, D., Ehrhardt, J. C., and Yuh, W. T. C. (1994). Thalamic abnormalities in schizophrenia visualized through magnetic resonance image averaging. *Science* **266**, 294–298.

Bajcsy, R., and Kovacic, S. (1989). Multiresolution elastic matching. *Computer Vision, Graphics, Image Processing* **1(46)**, 1–21.

Balter, J. M., Pelizzari, C. A., and Chen, G. T. Y. (1992). Correlation of projection radiographs in radiation therapy using open curve segments and points. *Med. Phys.* **19(2)**, 329–334.

Banerjee, P. K., and Toga, A. W. (1994). Image alignment by integrated rotational and translational transformation matrix. *Phys. Med. Biol.* **39**, 1969–1988.

Bohm, C., Greitz, T., Kingsley, D., Berggren, B. M., and Olsson, L. (1983). Adjustable computerized brain atlas for transmission and emission. *Tomography, Am. J. Neuroradiol.* **4**, 731–733.

Bookstein, F. L. (1997). Landmark methods for forms without landmarks, morphometrics of group differences in outline shape. *Med. Image Anal.* **1**, 225–243.

Bookstein, F. L. (1991). *Morphometric Tools for Landmark Data, Geometry, and Biology.* Cambridge University Press, New York.

Bookstein, F. L. (1989). Principal warps, thin-plate splines and the decomposition of deformations. *IEEE Trans. Pattern Analysis Machine Intelligence* **11(6)**, 567–585.

Broit, C. (1981). Optimal Registration of Deformed Images, Ph.D. thesis. University of Pennsylvania, Philadelphia.

Brown, L. G. (1992). A survey of image registration techniques. *ACM Computing Surveys.* **24(4)**, 325–376.

Christensen, G. E., Rabbitt, R. D., and Miller, M. I. (1994). 3-D brain mapping using a deformable neuroanatomy. *Phys. Med. Biol.* **39**, 609–618.

Collins, D. L., Neelin, P., Peters, T. M., and Evans, A. C. (1994). Automatic 3D intersubject registration of MR volumetric data into standardized Talairach space. *J. Comp. Assisted Tomography.* **18(2)**, 192–205.

Collins, D. L. Holmes, C. J., Peters, T. M., and Evans, A. C. (1995). Automatic 3D model based neuroanatomical segmentation. *Human Brain Mapping* **3**, 190–208.

Collins, D. L., Le Goualher, G., Venugopal, R., Caramanos, A., Evans, A. C., and Barillot, C. (1996). Cortical constraints for non-linear cortical registration. In *Visualization in Biomedical Computing* (Hohne, K. H. and Kikinis, R., Eds.). Lecture Notes in Computer Science. Vol. 1131. Springer-Verlag, Berlin, pp. 307–316.

Davatzikos, C., and Prince, J. L. (1996). Convexity analysis of active contour problems. In, *Proceedings of CVPR*. June 17–20, San Francisco, CA.

Davis, M. H., Khotanzad, A., Flaming, D. P., and Harms, S. E. (1997). A physics-based coordinate transformation for 3D image matching. *IEEE Trans. Med. Imaging* **16(3)**, 317–328.

Declerck, J., Subsol G., Thirion, J. P., and Ayache, N. (1995). *Automatic Retrieval of Anatomical Structurers in 3D Medical Images.* INRIA Technical Report **2485**, 153–162.

Drury H. A., van Essen, D. C., Anderson, C. H., Lee, C. W., Coogan, T. A., and Lewis, J. W. (1996). Computerized mapping of the cerebral cortex, a multiresolution flattening method and a surface-based coordinate system. *J. Cog. Neurosci.* **8(1)**, 1–28.

Evans, A. C., Collins, D. L., and Holmes, C. J. (1996). Computational approaches to quantifying human neuroanatomic variability. In *Brain Mapping, The Methods* (Toga, A. W., and Mazziotta, J. C., Eds.). New York Academic Press, New York.

Evans, A. C., Collins, D. L., Neelin, P., MacDonald, D., Kamber, M., Marrett, T. S. (1994). Three-dimensional correlative imaging, applications in human brain mapping. In *Functional Neuroimaging, Technical Foundations.* (Thatcher, R. W., Hallett, M., Zeffiro, T., John, E. R., and Huerta, M., Eds.), Academic Press, Orlando, FL, pp. 145–162.

Evans, A. C., Collins, D. L., and Milner, B. (1992). An MRI-based stereotactic brain atlas from 300 young normal subjects. In *Proceedings of the 22nd Annual Symposium of the Society of Neuroscience* Anaheim, CA, p. 408.

Evans, A. C., Marrett, S., Torrescorzo, J., Ku, S., and Collins, L. (1991). MRI-PET correlation in three dimensions using a volume of interest (VOI). atlas. *J. Cereb. Blood Flow Metab.* **11**, A69–A78.

Fox, P., Perlmutter, J., and Raichle, M. (1985). A stereotactic method of anatomical localization for PET. *J. Comput. Assisted Tomogr.* **9**, 141–153.

Franke, R. (1979). *A Critical Comparison of Some Methods for Interpolation of Scattered Data.* Naval Postgraduate School Technical Report NPS-53-79-003.

Friston, K. J., Ashburner, J., Frith, C. D., Pline, J. B., Heather, J. D., and Frackowiak, R. S. (1995). Spatial registration and normalization of images. *Human Brain Mapping* **2**, 165–189.

Friston, K. J., Frith, C. D., Liddle, P. F, and Frackowiak, R. S. (1991). Plastic transformation of PET images. *J. Comput. Assisted Tomogr* **15(4)**, 634–639.

Gambo-Aldeco, A., and Fellingham L. (1986). Correlation of 3D surfaces from multiple modalities in medical imaging. In *Medical Imaging, Processing, and Display and Picture Archiving and Communication Systems for Medical Applications.* SPIE Medicine XIV/ PACS IV, February, Newport Beach, CA.

Gourdon, A., and Ayache, N. (1994). *Registration of a Curve on a Surface using Differential Properties.* INRIA Internal Report No. **2145**, 187–192.

Greitz, T., Bohm, C., Holte, S., and Eriksson, L. (1991). A Computerized Brain Atlas, Construction, Anatomical Content and Application. *J. Comp. Assisted Tomogr* **15(1)**, 26–38

Gueziec, A., and Ayache, N. (1992). Smoothing and matching of 3D space curves. In *Proceedings of the European Conference on Computer Vision.* [AU21]pp 620–629.

Haar, B. M. *et al.* (1993). Higher order differential structure of images. In *Proceedings of the 13th International Conference on Information Processing in Medical Images* (Barrett, H. H., and Gmitro, A. F., Eds.), 77–93.

Haar, B. M., Florack, L. M. J., Koenderink, J. J., and Viergever, M. A. (1991). Scale-space, its natural operators and differential invariants. In *Proceedings of the 12th International Conference on Information Processing in Medical Images* (Colchester, A. C. F., and Hawkes, D. J., Eds.), **511**, pp. 239–253.

Hardy R. L. (1990). Theory and applications of the multiquadric-biharmonic method, 20 years of discovery 1968–1988. *Computers Math. Appl.* **19**, 163–208.

Hill, D. L. G., Hawkes, D. J., Crossman, J. E., Gleeson, M. J., Cox, T. C. S. et al. (1991). Registration of MR and CT images for skull base surgery using point-like anatomical features. *Br. J. Radiol.* **64(767)**, 1030–1035.

Hohne, K. H., Bomans, M., Riemer, M., Schubert, R., Tiede, U., and Lierse, W. (1992). A 3D anatomical atlas based on a volume model. *IEEE Comput. Graphics Appl.* **12**, 72–78.

Joshi, S. C., Miller, M. I., Chiristensen, G. E., Banerjee, A., Coogan, T. A., and Grenander, U. (1995). Hierarchical brain mapping via a generalized dirichlet solution for mapping brain manifolds, vision geometry IV. *Proc. SPIE Conf. Optical Sci. Engin. Instr.* **2573**, 278–289

Kendall, D. G. (1984). Shape-manifolds, Procrustean metrics, and complex projective spaces. *Bull. London Math. Soc.* **16**, 81–121.

Kikinis, R., Shenton, M. E., Iosifescu, D. V., McCarley, R. W., Saiviroonporn, P., Hokama, H. H., Robatino, A., Metcalf, D., Wible, C. G., Portas, C. M., Donnino, R., and Jolesz, F. (1996). A digital brain atlas for surgical planning. model-driven segmentation, and teaching. *IEEE Trans. Visualization Comp. Graphics.* **2(3)**, 232–241

Kishon, E., Hastie, T., and Wolfson, H. (1991). 3-D curve matching using splines. *J. Robotic Syst.* **6(8)**, 723–743.

Kovacic, S., Gee, J. C., Ching, W. S. L., Reivich, M., and Bajcsy, R. (1989). Three-dimensional Registration of PET and CT images. In *Proc. Ann. Int. Conf. IEEE Eng. Med. Biol. Soc.* Vol 11. IEEE Computer Society Press, Los Alamitos, CA, pp. 548–549.

MacDonald, D., Avis, D., and Evans, A. C. (1993). Automatic parameterization of human cortical surfaces. *Annual Symp. Info. Proc. in Med. Imag. (IPMI)*.

Maguire, G. Q., Noz, M., Rusinek, H., Jaeger, J., Kramer, E. L., Sanger, J. J., and Smith, G. (1991). Graphics applied to medical image registration. *IEEE Comp. Graphics Appl.* **11**, 20–28.

Maguire, G. Q., Noz, M. E., Lee, E. M., and Schimpf, J. H. (1986). Correlation methods for tomographic images using two and three dimensional techniques. In *Information Processing in Medical Imaging* (Bacharach, S. L., Ed.). Martinus Nijhoff, Dordrecht, pp. 266–279.

Maintz, J. B. A., and Viergever, M. A. (1998). *A Survey of Medical Image Registration Medical Image Analysis* **2(1)**, 1–36.

Maurer, C. R., and Fitzpatrick, J. M. (1993). A review of medical image registration. In *Interactive Image-Guided Neurosurgery* (Maciunas, R. J., Ed.). American Association of Neurological Surgeons, Parkridge, IL, pp. 17–44.

Matsui T., and Hirano, A. (1978). *An Atlas of the Human Brain for Computerized Tomography*. Igako-Shoin.

Mazziotta, J. C., Toga, A. W., Evans, A. C., Fox, P., and Lancaster, J. (1995). A probabilistic atlas of the human brain, theory and rationale for its development. *NeuroImage* **2**, 89–101

Meyer, C. R., Leichtman, G. S., Brunberg, J. A., Wahl, R. L., and Quint, L. E. (1995). Simultaneous usage of homologous points, lines and planes for optimal 3-D linear registration of multimodality imaging data. *IEEE Trans. Med. Imaging* **14**, 1–11.

Miller, M. I., Christensen, G. E., Amit, Y., and Grenander, U. (1993). Mathematical textbook of deformable neuroanatomies. *Proc. Nat. Acad. Sci.* **90(24)**, 11944–11948.

Nowinski, W. L., Fang, A., Nguyen, B. T., Raphel, J. K., Jagannathan, L., Raghavan, R., Bryan, R. N., and Miller, G. (1997). Multiple brain atlas database and atlas-based neuroimaging system. *J. Image Guided Surgery* **2(1)**, 42–66.

Ono M., Kubik S., and Abernathey C. D. (1990). *Atlas of the Cerebral Sulci*. Thieme Medical Publishers, Stuttgart.

Payne, B. A., and Toga, A. W. (1990). Surface mapping brain function on 3D models. *IEEE Comp. Graphics Appl.* **10**, 33–41.

Paxinos, G., and Watson, C. (1986). *The Rat Brain in Stereotaxic Coordinates*. Academic Press, San Diego, CA.

Pelizzari, C., Chen, G., Sperling, D., Weichselbaum, R., and Chen, C. (1989). Accurate three-dimensional registration of CT, PET, and/or MRI images of the brain. *J. Comp. Assisted Tomogr* **13**, 20–26.

Rademacher, J., Caviness, V. S., Steinmetz, H., and Galaburda, A. M. (1993). Topographic variation of the human primary cortices, implications for neuroimaging, brain mapping and neurobiology. *Cerebral Cortex* **3(4)**, 313–329.

Rohlf, F. J., and Slice, D. (1990). Extensions of the Procrustes method for the optimal superposition of landmarks. *Systematic Zool.* **39**, 40–59

Ruprecht, D., and Muller, H. (1995). A framework for scattered data interpolation. In *Visualization in Scientific Computing* (Goebel, M., Muller, H., and Urban, B., Eds.). Springer-Verlag, Vienna.

Sandor, S. R., and Leahy, R. M. (1995). Towards automated labeling of the cerebral cortex using a deformable atlas. In *Info. Proc. in Med. Imag.* [AU30] (Bizais, Y., Barillot, C., and Di Poala, R., Eds.), June. pp. 127–138.

Schaltenbrand, G., and Bailey. P. (1959). *Introduction to Stereotaxis with an Atlas of Human Brain*. Stuttgart, New York.

Schaltenbrand, G., and Wahren, W. (1977). *Atlas for Stereotaxy of the Human Brain*. Georg Thieme Verlag, Stuttgart.

Sokolnikoff, I. S. (1956). *Mathematical Theory of Elasticity*. McGraw-Hill, New York.

Spritzer, V., Ackerman M. J., Scherzinger, A. L., and Whitlock, D. (1996). The visible human male, a technical report. *J. Am. Med. Informatics Assoc.* **3(2)**, 118–130.

Steinmetz, H., Furst, G., and Freund, H. J. (1989). Cerebral cortical localization, application and validation of the proportional grid system in magnetic resonance imaging. *J. Comp. Assisted Tomogr.* **13**, 10–19.

Strother, S. C., Anderson, J. R., Xu, X. L., Liow, J. S., Bonar, D. C., and Rottenberg, D. A. (1994). Quantitative comparisons of image registration techniques based on high-resolution MRI of the brain. *J. Comp. Assisted Tomogr.* **18**, 954–62.

Swanson, L. W. (1992). *Brain Maps*, Structure of the Rat Brain. Elsevier Science, New York.

Szeliski, R., and Lavallee, S. (1994). Matching 3-D anatomical surfaces with non-rigid volumetric deformations. In *Applications of Computer Vision in Medical Image Processing*. Spring Symposium Series, AAAI, Stanford, CA.

Talairach, J., and Szikla, G. (1967). *Atlas d'Anatomie Sterotaxique du Telencephale*, Etudes Anatomo-Radiologiques. Masson & Cie, Paris.

Talairach, J., and Tournoux, P. (1993). *Referentially Oriented Cerebral MRI Anatomy. Atlas of Stereotaxic Anatomical Correlations for Gray and White Matter*. Georg Thieme Verlag, Stuttgart.

Talairach, J., and Tournoux, P. (1988). *Co-Planar Stereotaxic Atlas of the Human Brain*. Georg Thieme Verlag, Stuttgart.

Thompson, P. M., and Toga, A. W. (1996). A surface-based technique for warping three-dimensional images of the brain. *IEEE Trans. Med. Imaging* **15(4)**, 1–16

Thompson, P. M., and Toga, A. W. (1999). Anatomically Driven Strategies for High-Dimensional Brain Image Warping and Pathology Detection. In *Brain Warping* (Toga, A. W., Ed.). Academic Press, New York.

Thurfjell, L., Bohm, C., Greitz, T., and Eriksson, L., (1993). Transformations and slgorithms in a vomputerized brain atlas. *IEEE Trans. Nucl. Sci.* **40(4, pt. 1)**, 1167–91

Tikhonov, A. N., and Arsenin, V. A. (1977). *Solutions of Ill-Posed Problems*. Winston and Sons, Washington, D.C.

Toga, A. W., Santori, E. M., Hazani, R., and Ambach, K. (1995). A 3D digital map of rat brain. *Brain Res. Bull.* **38(1)**, 77–85.

Toga, A. W., and Thompson, P. M. (1999). Multimodal Brain Atlases. In *Advances in Biomedical Image Databases* (Wong, S., Ed.). Kluwer Academic Press, New York.

Vaillant, M., and Davatzikos, C. (1997). Mapping the cerebral sulci, application to morphological analysis of the cortex and to non-rigid registration annual sump. *Info. Proc. Med. Imaging (IPMI)*.

Vaillant, M., and Davatzikos, C. (1996/7), Finding parametric representation of the cortical sulci using an active contour model. *Med. Image Anal.* **1(4)**, 295–315.

van Buren, J., and Maccubbin, D. (1962). An outline atlas of the human basal ganglia with estimation of anatomical variants. *J. Neurosurg.* **19**, 811–839.

van den Elsen, P. A. (1994). Retrospective fusion of CT and MR brain images using mathematical operators. In *Applications of Computer Vision in Medical Image Processing*. Spring Symposium Series, AAAI. Stanford University, Stanford, CA, pp. 30–33.

van den Elsen, P. A. (1992). Multimodality Matching of Brain Images, PhD thesis. Utrecht University, The Netherlands.

van den Elsen, P. A., Pol, E. J. D., and Viergever, M. A. (1993). Medical image matching a review with classification. *IEEE Eng. Med. Biol.* **12**, 26–39.

Vries, J., and McLinden, S. (1988). Computerized three-dimensional stereotactic atlases. In *Modern Stereotactic Neurosurgery*. Little Brown & Company, Boston.

Vvedensky D. D., and Holloway, S. (1992). *Graphics and Animation in Surface Science*. Adam Hilger, UK.

Woods, R. P., Mazziotta, J. C., and Cherry, S. (1993). MRI-PET registration with automated algorithm. *J. Comp. Assisted Tomogr.* **17(4)**, 536–546.

Woods, R. P., Cherry, S. R., and Mazziotta, J. C. (1992). Rapid automated algorithm for aligning and reslicing PET images. *J. Comp. Assisted Tomogr.* **16(4)**, 620–633.

Xu, C., and Prince, J. L. (1997). Gradient vector flow, a new external force for snakes. In *IEEE Proc. CVPR*. Computer Society Press[AU35], Los Alamitos, pp. 66–71.

CHAPTER 4.5

An Atlas-Based Database of Neurochemical Data

Rabi Simantov,[1] Jean-Marie Bouteiller,[2] and Michel Baudry[2]
[1]*Molecular Genetics Department, Weizmann Institute of Science, Rehovot, Israel*
[2]*University of Southern California Brain Project, University of Southern California, Los Angeles, California*

Abstract

The aim of this chapter is to outline the approaches and tools used in our laboratory for building a useful neurochemical database. The data currently collected by a typical neurochemical laboratory are very diverse and heterogeneous and include molecular, neurochemical, and anatomical information. We have therefore selected a few prototype studies in the area of excitatory synaptic neurotransmission and learning models to illustrate the architecture of the object-oriented database being used to construct the neurochemical database within the framework of the NeuroCore database schema for neuroscience data of the USC Brain Project (USCBP).

4.5.1 Synaptic Neurotransmission: Molecular and Functional Aspects

Inter-neuronal communication in the central as well as peripheral nervous systems depends on a complex sequence of neurochemical events taking place at the synaptic level. The molecules mediating synaptic transmission consist of about nine small specialized molecules, amino acids, derivatives of amino acids, or other chemical structures (such as amines) and a large number (50 to 70) of neuroactive peptides. In addition, receptors, ion channels, G-proteins, transporters (carriers), cytoskeletal proteins, and various enzymes play a major role in this process. A major goal of modern neurochemistry is to provide a detailed mapping and a complete understanding of all these elements and to integrate them into synaptic models that not only can account for experimental observations but can also direct research to test new hypotheses. Furthermore, such detailed mapping could then be used to generate an atlas-based database of neurochemical features for all neuronal cell types and pathways. Another unique feature of the brain is the fact that many more genes are expressed in brain cells than in any other cell types, and the identification and localization of brain-specific genes is of primary importance for understanding brain function. This task will become more important as the genomes of complex organisms will become fully available in the near future.

A question that has fascinated neuroscientists for decades concerns the understanding of the effects of past experience on the characteristics of synaptic transmission, as it is widely believed that information in the central nervous system (CNS) is stored as distributed modifications of synaptic efficacy in complex neuronal networks. It is now well established that the underlying mechanisms for such modifications involve alterations in the expression of various genes, as well as structural reorganization. Two simple forms of synaptic modifications—long-term potentiation (LTP) and long-term depression (LTD), representing a long-lasting increase or decrease in synaptic efficacy resulting from the activation of monosynaptic excitatory connections by specific pattern of electrical activity, have been extensively studied, as it is widely assumed that they represent cellular mechanisms of information storage (Baudry and Davis, 1991, 1994, 1996). Although very few studies have addressed the issues of the duration of these types of synaptic modifications in intact animals, LTP has been shown to last for weeks and possibly longer in hippocampus (Staubli and Lynch, 1987). In turn, these types of synaptic modifications have been incorporated as learning rules in neural networks and the behavior of these networks in various learning tasks investigated (Lynch and Baudry, 1988). Thus not only do these neurochemical databases need to incorporate widely disparate types of information, but they should also be easily linked with brain atlases, as well as with time-series databases and models of neuronal networks.

4.5.2 Roles of Glutamatergic Synapses in LTP and LTD

Excitatory synapses use mainly glutamate as a neurotransmitter, and both LTP and LTD have been shown to occur at glutamatergic synapses and to involve glutamate receptors. There are four families of glutamate receptors: AMPA (α-amino-3–hydroxy-5–methyl-4–isoxazole proprionic acid), NMDA (N-methyl-D-aspartate), kainate, and metabotropic glutamate receptors (Hollman and Heinemann, 1994). AMPA and kainate receptors are relatively simple transmitter-gated channels, while NMDA receptors are both voltage-and glutamate-gated ion channels. Furthermore, the NMDA receptor/channel is a calcium channel. The glutamate metabotropic receptors belong to the family of G-protein coupled receptors and their activation leads to changes in second messenger systems (Conn and Pin, 1997). A number of studies have shown that LTP and some forms of LTD (but not all) at glutamatergic synapses depend on NMDA receptor activation and are associated with alterations in the properties of AMPA receptors (Baudry and Davis, 1991, 1994, 1996). In contrast, LTD in cerebellum has been shown to involve both the AMPA and the glutamate metabotropic receptors (Linden and Connor, 1993). It is thus important to have detailed knowledge concerning the relative location and numbers of AMPA, NMDA, and metabotropic receptors at different synapses. Furthermore, experimental evidence has accumulated indicating that different synapses exhibit various forms of plasticity with different time-courses, mechanisms, and functional relevance for behavioral learning. An important challenge for the future will be to understand the relationship between the cellular features of synaptic modifications and the behavioral features of learning and memory processes. Of particular interest are the links between the time courses of the establishment of different forms of plasticity and those of the behavioral phenomena to which they could be related. As pointed out by several authors (Goelet et al., 1986; McGaugh, 1989), there are more than two time courses for memory, and the distinction between short-term and long-term memory does not seem to capture all the temporal features of memory processes. At a more theoretical level, it remains to be demonstrated that time courses or other features of synaptic modifications are necessarily isomorphic to features of memory processes. It is our assumption that the further development of an integrated NeuroCore database environment in which neurochemistry, neurophysiology, and behavioral studies are brought together in conjunction with network models will provide critical answers to these questions (see Chapter 2.3 and others in this book).

In particular, two subfields within the hippocampus have very peculiar features: the mossy fiber terminal zone in CA3 and the stratum lacunosum moleculare in CA1. The first one exhibits an NMDA receptor-independent form of LTP, and there is considerable discrepancy as to the nature of this form of LTP. The latter receives a distinct projection from the entorhinal cortex via the perforant path, and little is known concerning the mechanisms and functions of synaptic plasticity at the synapses generated by these fibers. Furthermore, considerable evidence indicates that LTP at the mossy fiber/CA3 synapses exhibit features different from those found at other glutamatergic synapses. Such peculiar features should provide unique opportunities to apply an integrated approach of the type developed by the USCBP.

4.5.3 Glutamate Receptor Regulation and Synaptic Plasticity

Much has been learned over the last 15 years concerning the characteristics and regulation of ionotropic receptors. At the molecular level, molecular biologists have cloned families of genes coding for subunits of the three families of ionotropic glutamate receptors, the NMDA receptors, the AMPA receptors, and the kainate receptors, as well as for members of the glutamate metabotropic family (Hollman and Heinemann, 1994). Changes in the expression of these genes under different experimental conditions have been documented (Dingeldine et al., 1999). Increased attention has been devoted to the study of post-translational regulation of glutamate receptors, as this level of regulation is well fitted to produce rapid modifications of synaptic efficacy at glutamatergic synapses. In particular, phosphorylation of AMPA and NMDA receptors by several protein kinases, including protein kinase A, protein kinase C, calcium/calmodulin protein kinase, and tyrosine kinases has been shown to produce functional alterations of the receptors (Bi et al., 1998). Partial proteolysis of both AMPA and NMDA receptor subunits as a result of the activation of the calcium-dependent protease, calpain, has been shown to generate new species of the receptors lacking a small fragment in the C-terminal domains of several subunits of the receptors (Bi *et al.*, 2000). As the C-terminal domain exhibits multiple consensus sites for phosphorylation, it is likely that phosphorylation and proteolysis interact to regulate the properties of AMPA and NMDA receptors. In addition, the phospholipid environment of the AMPA receptors appears to be critical for regulating the binding and functional properties of the receptors (Massicotte and Baudry, 1991). The C-terminal domain of receptor subunits has also recently been shown to be critically involved in the targeting of receptors to the appropriate neuronal domains, and proteins interacting with the C-terminal domains of NMDA and AMPA receptors have been identified (Ziff, 1997). However, little is known concerning the mechanisms regulating the insertion and the internalization of the receptors. Likewise, very little information is currently available regarding receptor turnover rates and their modifications under various experimental conditions (Luscher *et al.*, 1999; Shi *et al.*, 1999). Both NMDA and AMPA receptors are involved in various pathological conditions. In particular, epilepsy is characterized by increased neuronal excitability that could be due to modifications of the properties of AMPA and/or NMDA receptors (Mody and Staley, 1994). Finally, both types of receptors have been associated with the phenomenon of excitotoxicity (Choi and Rothman, 1990). Thus, although a rich literature is available concerning the characteristics and regulation of the properties of glutamate receptors and their participation in synaptic plasticity processes, much more needs to be understood in order to integrate these data in neurochemical databases and in models of synaptic plasticity in neural networks.

4.5.4 How To Build a Useful Neurochemical Database

The exceptional diversity and heterogeneity in neurochemical information being collected nowadays, in the same or different laboratories, require development of unique tools for storage, manipulation, and comparative analysis of the data. Neurochemical studies involve anatomical mapping of various proteins and receptors and of the activity of various types of enzymes; they also involve collecting information about the functions and neurochemical properties of different ion channels and data about the effect of drugs, neurotoxins, hormones, nutrients, etc. on the brain. In the framework of the USC Brain Project, we have developed an object-oriented database in which a wide variety of neurochemical data can be easily stored, retrieved, and analyzed.

The database design process is represented in Fig. 1. The first step of the database design consisted in

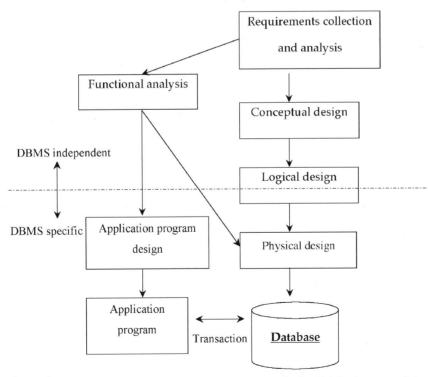

Figure 1 Database design process. In order to design an effective high-level conceptual data model, several steps need to be followed. The step in which the collaboration between neuroscientists and computer scientists is the tightest is requirement collection and analysis. The result of this step is a concise definition of users' requirements. The conceptual design allows a more abstract description of the data types, relationships, and constraints. The next step, the logical design, is the actual implementation of the database using a database management system (DBMS). The last step characterizes the specification of internal storage structures and file organization.

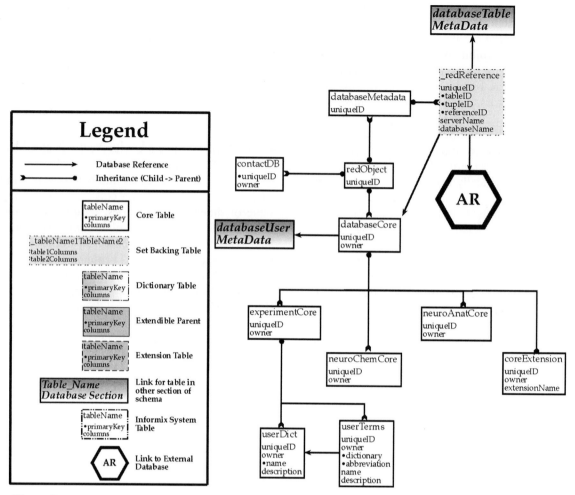

Figure 2 Top-level schema of the Core database, NeuroCore, and NeuroChem. This figure represents the top-level schema of the Core database tables and the parent tables of all children tables created within NeuroCore and NeuroChem. It shows the different modules of NeuroCore (databaseMetadata: necessary information regarding tables, columns, and relations within the database; contactDB: information regarding researchers and laboratories; databaseCore: parent for all experiment related tables in the database, its direct children are experimentCore, neuroChemCore, neuroAnatCore and coreExtension). This figure illustrates the tight interaction that exists at the higher hierarchy level between NeuroCore and NeuroChem.

analyzing the requirements for data collection and analysis. During this step, the database designers interviewed prospective database users (in our case, neurobiologists) to understand and document their requirements. The result of this step was a concisely written set of users' requirements. These requirements should be specified in as detailed and complete a form as possible. In parallel with specifying the data requirements, it was useful to specify the known functional requirements of the application. These consist of user-defined operations (or transactions) that will be applied to the database, and they include both retrievals and updates (functional requirements).

Once all the requirements were collected and analyzed, the next step consisted of creating a conceptual schema for the database, using a high-level conceptual data model. The NeuroCore database architecture, discussed in other chapters of this book, has been used as a framework in which NeuroChem has been developed in order to store neurochemical data. This step is called *conceptual database design*. The conceptual schema is a concise description of the data requirements of the users and includes detailed descriptions of the data types, relationships, and constraints related to the data. Because these concepts did not include any implementation details, they were usually easier to understand and could be used to communicate with non-technical users. The high-level conceptual schema could also be used as a reference to ensure that all users' data requirements were met and that the requirements did not include any conflicts.

The next step in our database design was the actual implementation of the database, using a commercial DBMS (this step is called *data model mapping*). Finally, the last step was the physical database design phase, during which the internal storage structures and file organization for the database were specified. In parallel with these activities, application programs had to be

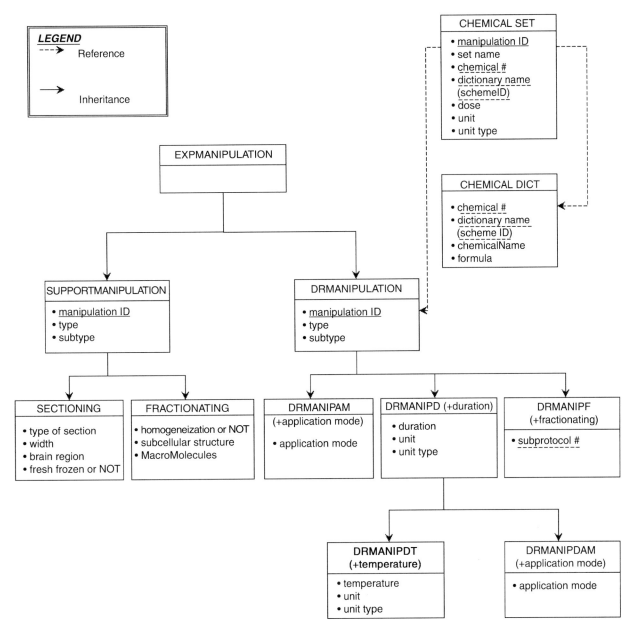

Figure 3 NeuroChem architecture. NeuroChem is built using Informix, a database management system (DBMS) that allows direct reference between tables and inheritance. Attributes that refer to other objects located in other tables are called *references*. They represent relationships among tables. The *Sectioning* and *Fractioning* tables are subclasses from the superclass *supportManipulation*. They inherit the same attributes as their superclass, as well as having attributes of their own. The architecture described here takes full advantage of the inheritance principle. The tables *DrManipAM*, *DrManipD*, and *DrManipF*, for example, have the same attributes as their parent table, *DrManipulation*, but add attributes related to their specificity (i.e., respectively, application mode, duration, and fractioning).

designed and implemented as database transactions-corresponding to the high-level transaction specifications.

The first architecture of the database is represented in Figs. 2 and 3. The database management system that was chosen is Informix. The Informix Universal Server is an object-relational database management system (ORDBMS) that combines both relational database and object-oriented programming technologies. Informix's relational database system supports standard SQL data types, functions, and queries. Additional features extend the standard relational database model with object-oriented features and include the ability to add new user-defined data types and functions, built-in support for large objects as datatypes (e.g., images, audio, video, etc.), and support of object-oriented concepts such as inheritance and function overloading.

4.5.5 Incorporating Neurochemical Data into the NeuroCore Repository of Empirical Data

A close collaboration between neuroscientists and computer scientists has been necessary. In order to define the data that populate the database, neurochemical publications are used. Parameters of experiments and the corresponding results are entered into the database. This primary set of data will allow us to test, validate, and improve the quality of the database design and architecture. Three sets of experiments included in the coming two sections will be presented to illustrate the effort in this area. These experiments are representative of the types of studies conducted in our laboratories.

Role of Proteases of the Calpain Family in LTP

We recently showed that *in situ* activation of calpain, a calcium-activated neuronal protease, produced marked alterations in the immunoreactivity of several subunits of AMPA receptors, in particular, GluR1 subunits, as well as in NR2 subunits of NMDA receptors (Bi *et al.*, 1996, 1997). Calpain was found to have a selective proteolytic activity on the C-terminal domains of GluR1, GluR2/3, and NR2. To further characterize calpain-mediated GluR1 proteolysis, we studied the effect of incubating frozen-thawed rat brain in the presence or absence of calcium on GluR1 immunoreactivity, examined with antibodies directed towards the N- or the C-terminal domains of the subunits (Fig. 4). Analysis with C-terminal antibodies (Figure 4D–F) showed that incubation with calcium (Figure 4D) decreased GluR1 immunoreactivity in several hippocampal regions, including the stratum lacunosum-moleculare of the CA1 region, stratum moleculare of the dentate gyrus, and dendritic fields in the cortex. No such effect was observed when GluR1 was detected with the N-terminal antibodies (Figure 4A,B). These effects were blocked by calpain inhibitors, indicating that calpain activation produces partial proteolysis of GluR1 subunits (Bi *et al.*, 1996, 1997). These results are of particular interest considering the evidence implicating calpains, calcium and glutamate receptors in mechanisms of synaptic plasticity and neurodegenerative processes (Lynch *et al.*, 1986; Bi *et al.*, 2000).

To illustrate the population of the database by experimental data, Table 1 shows the overall architecture of the database, illustrated with reference to a specific experiment.

Glutamate Transporters Modulate Glutamate Neurotransmission: Regulation by Kainic Acid in Rat Brain and Hippocampal Organotypic Cultures

Upon its release and interaction with pre- and postsynaptic receptors, glutamate is removed by an active

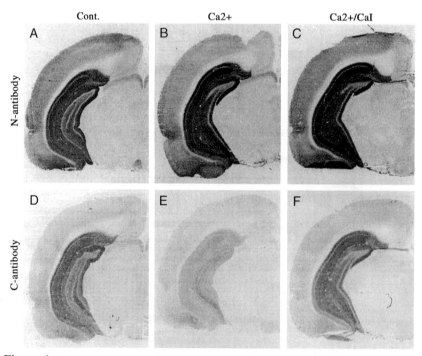

Figure 4 Effect of calcium on GluR1 immunoreactivity. Frozen-thawed rat brain sections were incubated in the absence or presence of calcium at 37°C. After incubation sections were processed for immunohistochemistry with antibodies directed to the N- or the C-terminal of the GluR1 subunits of AMPA receptors. Analysis with the C-terminal antibodies (C and D) showed that incubation with calcium (D) decreased GluR1 immunoreactivity in several hippocampal regions. No such effect was observed when GluR1 was detected with the N-terminal antibodies (A and B). These effects were blocked by calpain inhibitors, indicating that calpain produces partial proteolysis of GluR1 subunits.

Table 1 Study: Effect of Calcium Treatment of Frozen-Thawed Rat Brain Sections on GluR1 Subunits of AMPA Receptors

Antibodies directed towards N-terminal domain	Antibodies directed towards C-terminal domain
No effect	Incubation with calcium induces a decrease in GluR1 immunoreactivity in several hippocampal regions
	Effect blocked by calpain inhibitors
	⇒ Effect produces partial proteolysis of GluR1 sub-units

uptake system made up of specific glutamate transporters. During the last few years it has been found that neurons as well as glial cells possess glutamate transporters, indicating therefore that both cell types are involved in controlling the extracellular level of this major excitatory neurotransmitter. Five glutamate transporters have been cloned so far, and they have been named *excitatory amino acid transporters 1–5* (EAAT1–5) (Gegelashvili and Schousboe, 1997). Of this group, EAAT1 and EAAT2 were found mainly in astrocytes, EAAT3 in neurons distributed in many brain regions, EAAT4 in cerebellar Purkinje cells, and EAAT5 in the retina. Keeping in mind that the rate and efficiency of glutamate reuptake have various implications, we analyzed the expression of glutamate transporters upon neuronal activation, using kainic acid as a prototype inducer. Two experimental conditions have been used: (1) following systemic injection of kainic acid, and (2) following *in vitro* treatment of hippocampal slices maintained in cultures.

Expression of the glial (EAAT2) and neuronal (EAAT3) transporters was determined at the protein and mRNA levels, using specific antibodies and oligonucleotide probes, respectively (Simantov et al., 1999a). The immunocytochemical analysis indicated that treatment of young adult rats with kainic acid decreased EAAT3–immunoreactivity in stratum lacunosum moleculare of the hippocampus within 4 hours, prior to any evidence of neuronal death. Upon pyramidal cell death (5 days after kainate treatment) EAAT3 expression in CA1, CA2, and stratum lacunosum moleculare was practically abolished. The fast effect of kainic acid on the expression of EAAT3 was confirmed by *in situ* hybridization; kainic acid treatment decreased EAAT3–mRNA levels in CA1 and CA3 regions of the hippocampus within 4– to 8 hours. Kainate had an opposite and more widespread effect on EAAT2 expression; both the immunocytochemistry and *in situ* hybridization showed a modest but significant increase in the level of EAAT2 in several hippocampal and dendate regions. Developmental analysis of EAAT2 and EAAT3 expression by *in situ* hybridization indicated that the fast regulation of the transporters upon kainic acid application was absent in rats younger than 21 days (Simantov et al., 1999a). This result is in line with previous studies indicating the resistance of young rats to kainic acid neurotoxicity (Tremblay et al., 1984). Hippocampal organotypic cultures, which lack a major input of fibers from the entorhinal cortex through the perforant path, were used to further analyze the kainate effect. Treatment with kainic acid slowly decreased ^{3}H-D-aspartate uptake, an effect reaching a maximum decrease at 17 to 48 hours (Fig. 5). These studies indicated that one of the earliest effects of kainic acid in hippocampus consists of a down-regulation of the neuronal transporter EAAT3 in restricted regions, concurrent with a modest increase in the glial transporter EAAT2. These and additional studies suggest that glutamate transporters play an important role in kainate-induced neurotoxicity (Simantov et al., 1999). The implications of these observations regarding the role of glutamate transporters in modulating neuronal plasticity, seizure, and LTP/LTD are yet to be determined.

In an attempt to further elucidate the role of different glutamate transporters in hippocampal subfields involved in LTP and LTD, expression of the neuronal or glial transporters in hippocampal slice cultures was inhibited with selective antisense oligonucleotides, and kainate toxicity was determined. These experiments confirmed earlier studies (Bruce et al., 1995) that kainate is more toxic in CA3 than in CA1. More interestingly,

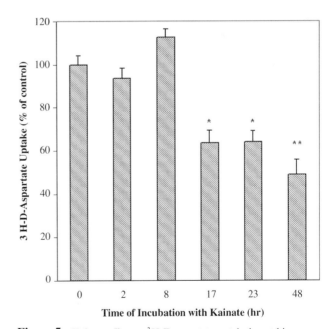

Figure 5 Kainate effect on ^{3}H-D-aspartate uptake by rat hippocampal slice culture. Hippocampal slices were prepared and maintained in culture as described in Simantov et al. (1999a). Uptake of ^{3}[H]-D-aspartate (200 nM final concentration) was determined at 0– to 48 hours after treatment with 10 (μM kainate. Data are means SD from 4– to 6 experiments, each assayed in quadruplicates. * and ** indicate statistically different from control, $p < 0.01$ and 0.001, respectively.

treatment with an antisense oligonucleotide to EAAT3 increased kainate toxicity in CA3, but had an opposite effect on the CA1 layer (Simantov et al., 1999b). Antisense to EAAT2 increased kainate toxicity in both CA1 and CA3. NMDA was more cytotoxic in the CA1 than CA3 cell layer, and the antisense oligonucleotides to EAAT3 and EAAT2 increased NMDA toxicity in both CA1 and CA3 subfields. These and additional experiments (Simantov et al., 1999b) indicate that glutamate transporters play a key role in regulating glutamate concentration in the vicinity of the pyramidal cell layers. EAAT3, the neuronal transporter, has a unique role in CA3.

The set of experiments described above is far more complex than the first one we examined; however, the flexibility of the architecture adopted still allows the storage of these experiments. Table 2 shows the overall architecture of the experiment as stored in the database. In the next section are the details of one of the protocols used for the experiment described above. Furthermore, the tables populated with data specific to the protocol used here illustrate the database population.

Overall architecture of the experiment as stored in the database:

EXPERIMENT: EFFECT OF KAINIC ACID AND PENTOBARBITAL TREATMENTS ON THE IMMUNOHISTOCHEMICAL DISTRIBUTION OF GLUTAMATE TRANSPORTERS IN RAT HIPPOCAMPUS

Subject (hypothesis): Short- and long-term treatments with kainic acid alter the expression of glutamate transporters.

EXP ID: 0001

Date: 5.5.1997

Treatments: 15 Sprague-Dawley rats (males, 3 months old) were divided into three groups of five rats each and treated intraperitoneally (ip) as follows (1–3):

1. Control group
2. 4–hour treatment
3. 5–day treatment

The control group received 1 ml 0.15 M NaCl (saline). Groups 2 and 3 were treated with 10 mg/Kg kainic acid, prepared in saline.

Tissue Preparation (4–10):

4. Rats were anesthesized with sodium pentobarbital (150 mg/kg).
5. They were then perfused intracardially with phosphate buffered saline (PBS).
6. Perfusion was followed by 300–400 ml of 4% paraformaldehyde in 0.1 M phosphate buffer (PB, pH 7.4).
7. After perfusion, the brain was removed
8. The brain was then washed with PB.
9. The brain was transferred to PB containing 30% sucrose.
10. Sections (40 μm thick) of the brain were cut, using a microtome.

Immunostaining (11–24):

11. Three brain sections from the hippocampal region of each rat were preincubated for 1 hour in 10 mM Tris-buffered saline (TBS, pH 7.4) containing 0.1% Triton X-100 and 4% normal goat serum.
12. The sections were then rinsed for 30 minutes in TBS.

The sections were then incubated for 48 hour in TBS containing 2% normal goat serum, 0.1% Triton X-100, and one of the following three antibodies:

13. Anti-EAAT1 (0.2 μg/ml)
14. Anti-EAAT2 (0.17 μg/ml)
15. Anti-EAAT3 (0.06 μg/ml)

Sections were then:

16. Rinsed (30 minutes, TBS)

Table 2 Study: Analysis of Expression of Glutamate Transporters Following Kainic Acid-Induced Seizure Activity

Two experimental conditions	Treatment with kainic acid	Earlier effects of kainic acid on hippocampus are
1 Following systematic injections of kainic acid	4 hours: Decrease in EAAT3 immunoreactivity in stratum lacunosum moleculare of hippocampus	Down-regulation of EAAT3
2 Following *in vitro* treatment of hippocampal slices maintained in culture	5 days: EAAT3 expression in CA1, CA2, and stratum lacunosum moleculare practically abolished	Modest increase in glial transporter EAAT2
	In situ hybridization: Same effect on EAAT3 mRNA levels in CA1 and CA3 within 4 to 8 hours	Decrease in-H-D-aspartate uptake
	In situ hybridization + immunocytochemistry: Increase in EAAT2 levels	

CHAPTER 4.5 An Atlas-Based Database of Neurochemical Data

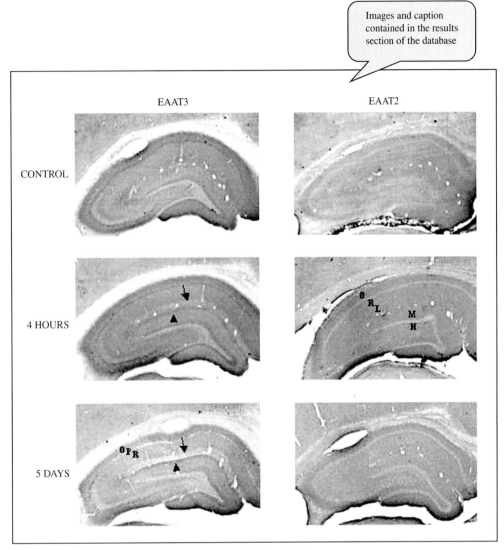

Images and caption contained in the results section of the database

Figure 6 Immunocytochemical analysis of the neuronal (EAAT3) and glial (EAAT2) transporters in kainite-treated rats. Sections from control and kainite-treated rats (4 hours and 5 days) were analyzed. O: Stratum oriens, P: pyramidal cell layer, R: stratum radiatum, L: stratum lacunosum moleculare of the hippocampus, M: molecular layer of the dentate gyrus, H: hilus. Arrow and arrowhead point to significant changes in immunoreactivity in stratum lacumosum moleculare of the hippocampus and outer molecular layer of the dentate gyrus, respectively.

17. Incubated for 60 min in TBS containing 2% normal goat serum, 0.1% Triton X-100, and goat anti-rabbit antibodies (1:200 dilution)
18/19. Rinsed *twice* with TBS
20. Stained with Vectastain ABC Kit (Vector Laboratories)
21. Dried
22. Washed in H_2O
23. Dehydrated with alcohol
24. Mounted.

Imaging and Storing the Data: Sections were scanned and data stored in PC 307 C:/Windows/RS/GlutTr/May 5 as Neuro1-5.tif, glt1a-e.tif, or glast1-5.tif.

4.5.6 Available Resources

NeuroChem: Current Status

The first version of NeuroChem is implemented as an extension of NeuroCore. This version has been implemented based on the needs expressed for the proper storage and retrieval of information regarding the experiments described above and other experiments currently being run in Dr. Baudry's laboratory (University of Southern California). It is in actuality being populated. So far, only protocol-related and chemicals-related data have been entered. The data obtained from the experiments (brain slices, arrays of data, etc.) have not been stored yet. Even though NeuroChem has been really

Table 3 DRUGMANIPULATION

uniqueid	owner	extensionname	manipulationid	type	subtype
			18	ICC	Rinsing
			19	ICC	Rinsing
			20	ICC	Staining
			21	ICC	Dry
			22	ICC	Washing
			23	ICC	Dehydration

Table 4 DRMANIPD

uniqueid	owner	extensionname	manipulationid	type	subtype	duration	durationUnit	dUnitType
			11	ICC	Preincubation	1	HR	Duration
			12	ICC	Rinsing	30	Min	Duration
			13	ICC	Incubation	48	HR	Duration
			14	ICC	Incubation	48	HR	Duration
			15	ICC	Incubation	48	HR	Duration
			16	ICC	Rinsing	30	Min	Duration
			17	ICC	Incubation	60	Min	Duration

Table 5 CHEMICALSET

uniqueid	owner	extensionname	manipulationid	setname	schemeid	chemicalid	dose	doseunit	dunittype
			11	TBS 7.4	IND	7	10	mM	conc.
			11	Triton X100	IND	8	0.1	%	conc.
			11	Normal goat Serum	IND	9	4	%	conc.
			13	N G S	IND	9	2	%	conc.
			13	TBS	IND	7	–	–	–
			13	Triton X100	IND	8	0.1	%	conc.
			13	Anti EAAT1	IND	10	0.2	microg	conc.
			14	NGS	IND	9	2	%	conc.
			14	TBS	IND	7	–	–	–
			14	Triton X100	IND	8	0.1	%	conc.
			14	Anti EAAT2	IND	11	0.17	microg	conc.
			15	NGS	IND	9	2	%	conc.
			15	TBS	IND	7	–	–	–
			15	Triton X100	IND	8	0.1	%	conc.
			15	AntiEAAT3	IND	12	0.06	microg	conc.
			12	TBS	IND	7	–	–	–
			16	TBS	IND	7	–	–	–
			17	TBS	IND	7	–	–	–
			17	NGS	IND	9	2	%	conc.
			17	Triton X100	IND	8	0.1	%	conc.
			17	Goat anti-rabit serum	IND	13	1/200	–	dilution
			18	TBS	IND	7	–	–	–
			19	TBS	IND	7	–	–	–
			20	Vectastain ABC kit	IND	14	–	–	–
			22	H_2O	IND	15	–	–	–
			23	Alcohol	IND	16	–	–	–

stable so far and responded in a reliable manner to the queries that have been run throughout its development, the plan for the near future is to test the reliability and consistency of the links within the tables of NeuroChem and between the NeuroCore schema and NeuroChem. For this purpose, collaborations will be established between Dr. Baudry's group and Jade developers. These collaborations should also be of benefit to

Jade as far as testing its ability to deal with data it was not originally designed for.

For the latest release information and further documentation of the NeuroChem database, please visit the University of Southern California Brain Project Website at http://www-hbp.usc.edu/Projects/neurochem.htm.

Integration with Three-Dimensional Registration and NeuArt

Many neuroanatomical studies of brain tissue sections depend on the qualitative evaluation and quantitative analysis of radiolabeled samples obtained from experimental animals. Such a tight interaction between neuroanatomy and neurochemistry illustrates the need of integrated computational tools that help neuroscientists with regards to each aspect of their work. The integration of computational sets of software tools and databases has been an increasing field of study (Peterson, 1995; Dashti et al., 1997; Bloom, 1996) in the past few years to provide a means for conveying complete descriptions of experiments, a platform for virtual experiments; and an interface where modeling and experimental results could be exchanged.

The integration of NeuroChem with three-dimensional registration and NeuArt is possible via the NeuroCore framework (Diallo et al., 1999) and represents a real opportunity for neuroscientists to retrieve data related to neurochemistry, neuroanatomy, and electrophysiology. For example, it is possible for a neuroscientist to query NeuroChem on the various experiments realized on Sprague-Dawley rats treated with kainic acid, to retrieve these experiments, to observe images of brain sections, to compare them with other brain sections issued from an atlas, and to be able to determine in a precise manner the location and angle of the actual experimental brain sections as compared to the sections present in the atlas. Such integration should create a means for conveying complete descriptions of experiments and a set of computational tools for complete analysis.

4.5.7 Conclusion

Considering the immense amount and heterogeneity of the neurochemical information being collected at the present time, it is becoming necessary to develop new methods for accumulating, handling, and querying the data. Indeed, effort is being made by several research centers to construct databases and repositories to address this issue (Sheperd et al., 1998). In this chapter, we have summarized our effort within the framework of the USC Brain Project and have provided step-by-step examples of two prototypes of neurochemical studies. The architecture of the object-oriented database presented herein provides the groundwork for the construction of a global neurochemical database, in which a wide variety of neurochemical data from different laboratories can be easily stored, available to the entire scientific community, and productively analyzed. The object-oriented architecture and the use of a core database such as NeuroCore also provide a solid basis for the creation of an integrated computer-aided neuroscience.

References

Baudry, M., and Davis, J. L., Eds. (1996). *Long-term Potentiation*. Vol. 3. MIT Press, Cambridge, MA.

Baudry, M., and Davis, J. L. (1994). *Long-Term Potentiation*. Vol 2. MIT Press Cambridge, MA.

Baudry, M., and Davis, J. L. (1991). *Long-Term Potentiation*. Vol 1. MIT Press. Cambridge, MA.

Bi, X., Bi, R., and Baudry, M. (2000). Calpain-mediated truncation of glutamate ionotropic receptors. In *Calpain Methods and Protocols*. Humana Press, Ottawa.

Bi, X., Standley, S., and Baudry, M. (1998). Post-translational regulation of ionotropic glutamate receptors and synaptic plasticity. *Int. Rev. Neurobiol* **42**, 227–284.

Bi, X., Chen, J., Dang, S., Wenthold, R. J., Tocco, G., and Baudry, M. (1997). Characterization of calpain-mediated proteolysis of GluR1 subunits of α-amino-3-hydroxy-5-methylisoxazole-4-propionate receptors in rat brain. *J. Neurochem.* **68**, 1484–1494.

Bi, X., Chang, V., Molnar, E., McIlhinney, R. A., and Baudry, M. (1996). The C-terminal domain of glutamate receptor subunit 1 is a target for calpain-mediated proteolysis. *Neuroscience* **73**, 903–906.

Bloom, F. E. (1996). The multidimensional database and neuroinformatics requirements for molecular and cellular neuroscience. *Neuroimage* 4, S12–3.

Bruce, A. J., Sakhi, S., Schreiber, S. S., and Baudry, M. (1995). Development of kainic acid and N-methyl-D-aspartic acid toxicity in organotypic hippocampal cultures. *Exp. Neurol.* **132**, 209–219.

Choi, D. W., and Rothman, S. M. (1990). The role of glutamate neurotoxicity in hypoxic-ischemic neuronal death. *Annu. Rev. Neurosci.* **13**, 161–182.

Conn, P. J., and Pin, J. P. (1997). Pharmacology and functions of metabotropic glutamate receptors. *Annu. Rev. Pharmacol. Toxicol.* **37**, 205–37.

Dashti, A. E., Ghandeharizadeh, S., Stone, J., Swanson, L. W., and Thompson, R. H. (1997). Database challenges and solutions in neuroscientific applications. *Neuroimage* **5(2)**, 97–115.

Diallo, B., Travere, J. M., and Mazoyer, B. (1999). A review of database management systems suitable for neuroimaging. *Methods Inf. Med.* **38(2)**, 132–9.

Dingledine, R., Borges, K., Bowie, D., and Traynelis, S. F. (1999). The glutamate receptor ion channels. *Pharmacol. Rev.* **51**, 7–61.

Gegelashvili, G., and Schousboe, A. (1997). High affinity glutamate transporters, regulation of expression and activity. *Mol. Pharmacol.* **52**, 6–15.

Goelet, P., Castellucci, V. F., Schcher, S., and Kandel, E. R. (1986). The long and the short of long-term memory a molecular framework. *Nature* 322, 419–422.

Hollmann, M., and Heinemann, S. (1994). Cloned glutamate receptors. *Annu. Rev. Neurosci.* **17**, 31–108.

Linden, D. J., and Connor, J. A. (1993). Cellular mechanisms of long-term depression in the cerebellum. *Curr. Opin. Neurobiol.* **3**, 401–406.

Luscher, C., Xia, H., Beattie, E. C., Carroll, R. C., Zatrow, M. V., Malenka, R. C., and Nicoll, R. A. (1999). Role of AMPA receptor cycling in synaptic transmission and plasticity. *Neuron* **24**, 649–658.

Lynch, G., and Baudry, M. (1988). *Structure-Function Relationships in the Organization of Memory*, Perseptives in Memory Research, The MIT Press, Cambridge, MA, pp. 23–91.

Lynch, G., Larson, J., and Baudry, M. (1986). Proteases, neuronal stability, and brain aging, an hypothesis. In *Treatment Development Strategies for Alzheimer's Disease* (T. Crook, R. Bartus, S. Ferris, and S. Gershon, Eds.). Mark Powley Assoc., Madison, CN, pp. 127–163.

Massicotte, G., and Baudry, M. (1991). Triggers and substrates of hippocampal synaptic plasticity. *Neuro. Biobehav. Rev.* **15**, 415–423.

McGaugh, J. L. (1989). Involvement of hormonal and neuromodulatory systems in the regulation of memory storage. *Ann. Rev. Neurosci.* **12**, 255–287.

Mody, I., and Staley, K. J. (1994). Cell properties in the epileptic hippocampus. *Hippocampus* **4(3)**, 275–280.

Peterson, B. E. (1995). Exploration and virtual experimentation in a local neuroscience database. *J. Neurosci. Methods* **63(1–2)**, 159–74.

Ramez, E., and Navathe, S. B. (1994). *Fundamentals of Database Systems.* Second ed., Addison-Wesley, Boston, MA.

Shepherd, G. M., Mirsky, J. S., Healy, M. D., Singer, M. S., Skoufos, E., Hines, M. S., Nadkarni, P. M., and Miller, P. L. (1998). The Human Brain Project, neuroinformatics tools for integrating, searching and modeling multidisciplinary neuroscience data. *Trends Neurosci.* **21**, 460–468.

Shi, S. H., Hayashi, Y., Petralia, R. S., Zaman, S. H., Wenthold, R. J., Svoboda, K., and Malinow, R. (1999). Rapid spine delivery and redistribution of AMPA receptors after synaptic NMDA receptor activation. *Science.* **284**, 1811–1816.

Simantov, R., Crispino, M., Hoe, W., Broutman, G., Tocco, G., Rothstein, J. D., and Baudry, M. (1999a). Changes in expression of neuronal and glial glutamate transporters in rat hippocampus following kainate-induced seizure activity. *Mol. Brain Res.* **65**, 112–123.

Simantov, R., Liu, W., Broutman, G., and Baudry, M. (1999b). Antisense knockdown of glutamate transporters alters the subfield selectivity of kainate-induced cell death in rat hippocampal slice cultures. *J. Neurochem.* **73**, 1828–1835.

Staubli, U., and Lynch, G. (1987). Stable hippocampal long-term potentiation elicited by "theta" pattern stimulation. *Brain Res.* **435**, 1–2.

Tremblay, E., Nitecka, L., Berger, M. L., and Ben-Ari, Y. (1984). Maturation of kainic acid seizure-brain damage syndrome in the rat. I. Clinical, electrographic and metabolic observations. *Neuroscience,* **13**, 1051–1072.

USCBP Report 97–01 (J. S. Grethe, J. M. Mureika, and Mi. A. Arbib).

Ziff, E. B. (1997). Enlightening the postsynaptic density. *Neuron.* **19**, 1163–1174.

PART 5

Data Management

CHAPTER 5.1

Federating Neuroscience Databases

Wen-Hsiang Kevin Liao and Dennis McLeod
University of Southern California Brain Project
University of Southern California, Los Angeles, California

Abstract

There are three key aspects of sharing and interconnection in a federated database environment: information discovery, semantic heterogeneity resolution, and system-level interconnection. Although the focus here is on the system-level interconnection process of the three key aspects described above, we also provide brief discussions on the other two aspects and summarize our approaches to those two issues. Our approach to support information sharing among databases is based on the import/export paradigm. In this paradigm, a component database of a federation decides the portion of its database to be exported and offers the methods to others on how the information can be shared. Users of other component databases who are interested in using the information can import it using one of the sharing methods provided by that component database. Information sharing is thus reached based upon the agreements (contracts) between each pair of component databases in which sharing is needed. A set of sharing primitives/tools is designed to support these sharing activities in a uniform manner. The sharing primitives/tools let users focus on how their information could be shared efficiently rather than on the underlying implementation. This enables users of a component database to utilize not only the information in their own component database but also the information in other component databases of the same federation.

5.1.1 Introduction

With the rapid growth of computer communication networks over the last decade, a vast amount of diverse information has become available on these networks. Users often find it desirable to share and exchange information that resides either within the same organization or beyond organizational boundaries. A federated database system is a collection of autonomous, heterogeneous, and cooperating component database systems. Component database systems unite into a loosely coupled federation in order to achieve a common goal: to share and exchange information by cooperating with other components in the federation without compromising the autonomy of each component database system. Such environments are increasingly common in various application domains, including office information systems, computer-integrated manufacturing, scientific databases, etc. Fig. 1 shows a generic view of a typical federated database system. Each component database participates in the federation through the sharing network. Component databases could reside within the intranet of an organization or belong to several organizations which are interconnected through the internet. The sharing network provides services to individual components for exchanging information across database boundaries.

There are three key aspects of sharing and interconnection in a federated database environment. These may

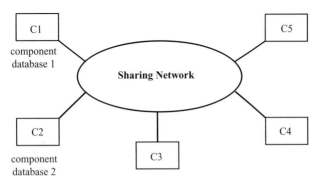

Figure 1 Generic view of a federated database system.

be viewed in the context of a given component database (C), whose user intends to import information from and/or export information to other component databases.

1. *Information discovery:* The information discovery process pertains to finding out what information can be shared in the first place. From the viewpoint of a user of component database C, the main concern is to discover and identify the location and content of relevant, non-local information units in the federation.

2. *Semantic heterogeneity resolution:* Various kinds of diversity may exist in the autonomous component databases in a federation. The similarities and differences between the information of component C and relevant non-local information need to be resolved so they can be integrated and shared.

3. *System-level interconnection:* Based upon existing networking technologies, the mechanisms and their implementation must support actual sharing and transmission of information to and from C and other components in the federation.

The focus of this chapter is on the system-level interconnection process of the three key aspects described above. In addition, we also provide brief discussions on the other two aspects and summarize our approaches to those two issues. The remainder of the chapter is organized as follows. Section 5.1.2 briefly describes the dynamic classificational ontology approach to information discovery problem. Section 5.1.3 outlines the spectrum of semantic heterogeneity and introduces the approach of meta-data implantation and stepwise evolution. Section 5.1.4 discusses the issues related to system-level interconnection. Section 5.1.5 describes the characteristics of sharing patterns. Section 5.1.6 lays out the system architecture of our approach to providing sharing primitives/tools. We conclude the chapter with a sharing scenario using neuroscience databases in Section 5.1.7.

5.1.2 Information Discovery

The information discovery process involves finding relevant information units from the vast amount of data located in the component databases of a federation. Information discovery in a federated database environment faces a number of challenging problems due to its distinct characteristics: the volume of shared information is large, the unit of information is substantially structured, and the number of participants is large. A scalable and practical mechanism for information discovery should accommodate these characteristics to facilitate information sharing in the federated database environment. A common approach to mediating information discovery in such an environment is to adopt a common ontology (a collection of concepts and their relationships that are used for describing information units) as the basis for mutual understanding among participants. However, the approach does not scale well and it is difficult to build and maintain a common ontology as the number of participants becomes large. The concept and mechanism of dynamic classificational ontology (DCO; see Kahng and McLeod, 1996, 1998) was proposed to address the problems of common ontology and to illustrate how DCO can facilitate information-sharing activities.

The DCO approach is based on the observations that it is practical to minimize the size and complexity of the ontology in order to keep the system scalable, it is extremely difficult to reach a total agreement on an ontology when the number of participants is more than a handful, and it is beneficial to allow the ontology to change dynamically as the system evolves. In order to reduce the size of the common ontology, a DCO contains a small amount of high-level meta-knowledge on information exported from information providers. Specifically, it contains a collection of concepts and their relationships to be used for classification of exported information. The approach relies on classification because it is an effective scheme for organizing a large amount of information.

A DCO consists of a base ontology and a derived ontology. The base ontology contains an application-specific collection of concepts and their relationships that are agreed upon among participants and used by them for high-level description and classification of shared information units. The derived ontology contains information on additional relationships among some concepts in the base ontology, which are derived from the base ontology and the collection of exported information. Such relationships typically involve inter-related concepts that are useful for describing and classifying other information (for example, relationships among the collection of research subjects). The advantage of this approach is that participants are not required to agree on those relationships in advance; rather, those relationships are allowed to change as the usage of involved concepts changes. At the cost of information providers' cooperative efforts, this approach supports effective information sharing in the federated database environment.

5.1.3 Semantic Heterogeneity Resolution

The semantic heterogeneity resolution process resolves the similarities and difference of information in a component database and the information being imported from other components in the federation. Semantic heterogeneity in federated database systems is primarily caused by design autonomy of component database systems. Component database systems may employ different design principles for modeling the same or related data, thus resulting in semantic heterogeneity. The heterogeneity in a federation may be at various levels of abstraction. Components may use different database model in modeling data. Even if they use the same database model, they may come up with different conceptual schemas for the same or related data. Even object representation and low-level data formats differ from component to component employing the same database model. Finally, they may use different tools to manage and provide an interface for the same or related data.

The approach of meta-data implantation and stepwise evolution (Aslan and McLeod, 1999) to resolve semantic heterogeneity within the federated database environment involves a partial database-integration scheme in which remote and local (meta-)data are integrated in a stepwise manner over time. The meta-data implantation and stepwise evolution techniques can be used to inter-relate database elements in different databases and to resolve conflicts on the structure and semantics of database elements (classes, attributes, and individual instances). The approach employs a semantically rich canonical data model and an incremental integration and semantic heterogeneity resolution scheme. Relationships between local and remote information units are determined whenever enough knowledge about their semantics is acquired. The folding problem—folding remote meta-data (conceptual schema) into the local meta-data to create a common platform through which information sharing and exchange becomes possible—is solved by implanting remote database elements into the local database, a process that imports remote database elements into the local database environment, hypothesizes the relevance of local and remote classes, and customizes the organization of remote meta-data.

The advantages of this approach include the fact that it requires minimum global knowledge and effort from federation users both before and during interoperation. Global structures which should be maintained in the federation are minimum. Inter-relationships between schema elements in different components are highly dynamic. It recognizes the fact that knowledge required to relate inter-database elements may not be available, derivable from within the federation, or obtainable from users prior to interoperation.

5.1.4 System-Level Interconnection

The system-level interconnection process supports the actual sharing and transmission of information units among component databases in a federation. Considering the federation environment described above, each component database may be controlled by different database management systems and reside in different parts of the network. The sharing and transmission of information among component databases are not directly supported by most existing database management systems, and new mechanisms and tools to support the exchange of information are needed. In our approach, we adopt the import/export paradigm to facilitate information sharing among component databases. In this paradigm, the user of a component database decides the portion of the database to be exported and offers the methods to others as to how the information can be shared. Any user of a component database who is interested in using the information exported by a particular component database can then pick a portion of exported information and choose one of the sharing methods provided by that component database. Information sharing is thus achieved based upon the agreements (contracts) reached between each pair of component databases requiring sharing. A set of sharing primitives/tools is designed to support these sharing activities in a uniform manner. The sharing primitives/tools allow users to focus on how their information could be shared efficiently rather than on the underlying implementation. This enables users of component databases to utilize not only the information in their own component databases but also the information in other component databases of the same federation.

Import/Export Based Sharing Paradigm

Our approach to supporting information sharing between databases is based on the import/export paradigm (Fig. 2). The users of a database can acquire information units from other databases and use them to answer queries as if the imported information units are part of their own database. The goal is to allow the incorporation of objects in remote databases into the user's local database. In so doing, remote objects should appear transparent to users of the local database. This means that the user utilizes the remote data the same way she/he manipulates the local data.

The import/export approach is distinguished from most distributed query processing approaches, which send subqueries to be executed on remote environments and present the combined results back to its users, in that queries are executed under the user's local environment on the information units carried over the network from various sources. In our approach, almost all acquired information units are located locally in a consistent

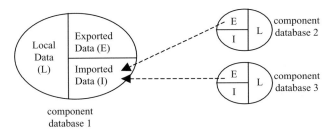

Figure 2 Import/export-based sharing paradigm.

organization and thus allow users to treat both its own data and imported data in a uniform way.

Using the import/export paradigm, we can demonstrate how personnel information from several component databases can be used to answer a query. As shown in Fig. 3, BMW, Atlas, and Berger's lab are three autonomous databases of USCBP and each of them maintains personnel information of its group. The staff table in the local database contains information on those staff who work for USCBP but do not belong to any of the three groups. Under the import/export paradigm, the personnel information of each group can be imported using various sharing mechanisms. These imported personnel information and the local staff table can then be combined as a single USCBP personnel table which contains information on all USCBP personnel. It is then possible to efficiently answer a query such as "show me the names and phone numbers of all USCBP personnel."

5.1.5 Characteristics of Sharing Patterns

Several key characteristics of shared data may affect the performance of the actual sharing between two databases within our federation framework (Alonso and Barbara 1989). The number of imported information units and the size of each imported information unit determine the additional local space needed to hold all these imported information units. The frequency of updating data at the source may affect how often the changes have to be propagated to its importers. To determine the best way of information sharing is nontrivial. We believe that the information exporters know best the characteristics of data and performance and they can recommend sharing methods that are most suitable for information importers and can be afforded by the exporters (owners of the data) (Fig. 4). The importers would select the data of interest and then select one of the recommended sharing methods that would work best for the applications with the lowest overhead (Fig. 5).

Several sharing mechanisms have been designed to facilitate the retrieval of data and increase the availability of data. These mechanisms include direct link, copy, caching, time-bounded copy, and remote query execution. A set of primitives/tools has been designed to support all these sharing methods within our database federation framework. These sharing methods can be summarized as follows:

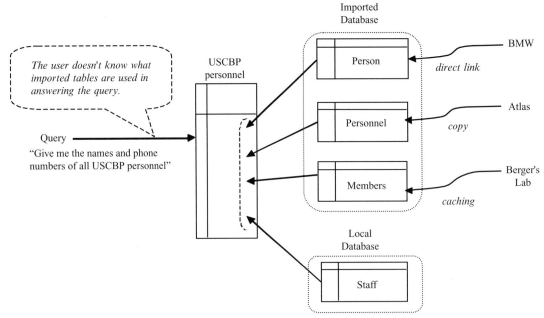

Figure 3 Information sharing example using import/export paradigm.

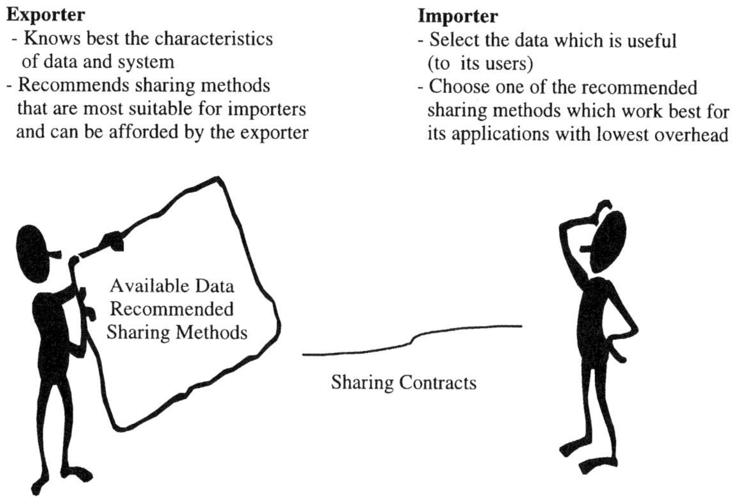

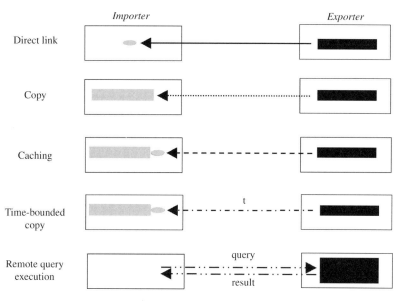

Figure 4 Sharing scenario.

Figure 5 Sharing methods.

1. *Direct link:* The direct-link method imports only the references to the sources of imported information, and each subsequent access will result in fetching data from their source databases. It is most suitable for frequently changed source data as it guarantees that the most current data are retrieved. It is expensive in that data must be retrieved each time it is accessed and data might not be available due to network problems or poor system performance on the exporter side.

2. *Copy:* The copy method duplicates all the data of interest on the importer database. While it has the best performance because copies of data are stored locally at the importer side and there is no need to retrieve data through a network, the imported data might become stale due to missed updates of data on the exporter database. Larger space is also required locally to store copied data. The copy mechanism is most suitable for importing those data which are rarely changed once they are published.

3. *Caching:* The caching sharing mechanism is a compromise between the direct-link and copy mechanisms. The importers keep both references to data sources as well as local copies, a procedure that provides better access performance than direct-link as the importers can use local copies directly if they are still up to date. It is also possible to use local copies without checking whether they are still current, if the network connection is not available. However, it also introduces additional overhead to the exporters as they are responsible for propagating updates to the importers' databases. The caching sharing method is most useful in sharing data that are updated irregularly.

4. *Time-bounded copy:* The time-bounded copy sharing mechanism is a special case of the caching method. It targets those data that are updated periodically. Local copies will be refreshed after a specified period of time based on the characteristics of their data sources. While it has the same benefits as those of the caching method, it also relieves some of the burden of the data exporters as the importers are responsible for retrieving data for every specified time period, but the time-bounded sharing mechanism is only suitable for those data that are updated regularly.

5. *Remote-query execution:* The remote-query execution method is designed to address those occasions when the set of data to be shared is so enormous that it is not practical to even import the references of data from an exporter's database. In such circumstances, it is more practical to execute a query request on the exporter's side rather than bring all the data to the importer's side.

5.1.6 System Architecture for Sharing Primitives/Tools

The implementation of sharing primitives/tools is based on the architecture as shown in Fig. 6. Each database component contains a set of data to be exported and a set of data imported from others. The import agent and export agent are responsible for negotiating the sharing agreement and transmission of data among database components. Sharing contracts, agreements between an exporter and importer on a set of data, are stored in each database as auxiliary data. Other types of data include:

1. *Local data:* Local data are all the information owned by and under the management of a component database.

2. *Exported data:* Exported data are that portion of local data intended to be shared with other component databases. Each exported table is associated with a list of available sharing methods recommended by the component database.

3. *Imported data:* Imported data contain information imported from other component databases.

4. *Imported contracts:* Imported contracts keep track of information on imported objects, including the provider, sharing method, and other parameters.

5. *Exported contracts:* Exported contracts keep track of information on exported objects, including the consumer, sharing method, and other parameters.

6. *Data export agent:* The Java-based export agent handles all the communication between the local database and remote databases. It is responsible for serving the requests for information sharing from other databases. These requests include building sharing contracts, transporting exported data, and propagating update on exported data using the caching method.

7. *Data import agent:* The Java-based import agent communicates with export agents of those databases from which the data are imported. It makes initial requests for sharing remote information and retrieves the data set once a sharing agreement is reached. It is also responsible for receiving updated data from remote databases and propagating the data to local copies.

There are also two sets of functions that allow users to specify which part of a database should be exported and from which database the data should be imported:

1. *Database exporting functions:* These functions provide users the capability to export part of their database and specify its supporting sharing mechanisms. They also allow users to browse all exported information

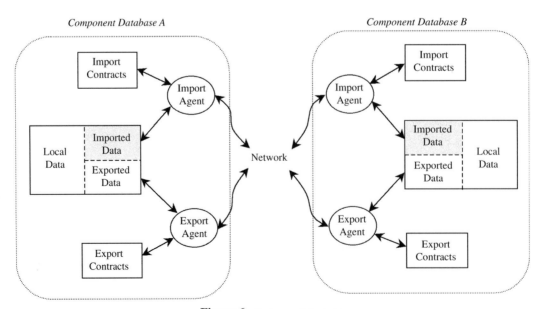

Figure 6 System architecture.

and modify sharing mechanisms as the sharing environment evolves with time.

2. *Database importing functions:* End users use these functions to acquire information on data being exported from other databases. After the set of relevant data is selected, these functions transform the request and direct the import agent to reach agreement with remote databases. An inventory of all imported data and their sharing mechanisms is also maintained in the user's local database environment.

5.1.7 Federating Neuroscience Databases

The sharing primitives/tools and ontological conversions (conversions from one concept/inter-relationship/terminology framework to another) can support inter-database referencing at various levels. Developing and deploying such techniques will allow referencing database subsets across database boundaries. It is essential to construct ontologies to describe the key concepts, inter-relationships, and terminology for databases based upon NeuroCore (experimental data), NeuARt (atlas-based data), and NeuroScholar (connectivity data), as well as models stored in BMW (Brain Models on the Web) and data summarized in NSW (Neural Summaries on the Web). The sharing primitives/tools can then be used to support labeled interconnection between "information units" in the various databases. It will thus be possible to link, for example, an experiment with the protocol it employs with atlas structures that it involves. Fig. 7 shows the "USCBP federation information flow", the architecture of our approach to neuroscience information sharing. Here, we see direct interconnection among structured databases—those managed by NeuroCore or which are NeuroCore compatible. For example, experimental data units in the Thompson lab may refer to experimental data units in the Baudry lab. Given the rich meta-data in structured databases, interconnections can be established between database items and item collections—employing derived data specification techniques (e.g., SQL3 views). Sharing of atlas-based data, modeling, and other information is done via the use of specific interconnection by hand or via an ontology. Here, we employ a formal framework for specific kinds of inter-relationships (e.g., between experimental protocols and models). The role of ontologies in interconnection involves the mapping of concepts ("nouns," such as experimental protocols) and interrelationship labels ("verbs," such as supports) from one database to another. Our approach is to devise, implement, and deploy a sharing mechanism based upon these techniques. Our work draws extensively upon the results we have obtained to date at USCBP (Aslan and McLeod, 1999; Hammer and McLeod, 1999; Kahng and McLeod, 1998).

The sharing primitives/tools can be used further to devise, implement, and deploy a personal neuroscience information manager to allow a researcher to maintain a personal "view" of the USCBP federation, in particular tracking relevant subsets of information and their interconnections. Such personal neuroscience databases may contain references to other databases—specifically, subsets of those databases, items in them, etc. It will also allow "local" annotation of data, so that an investigator may bring together, inter-relate, and document working data. This system will also allow researchers to add their own "local" data or interconnections, information that is "private" to the researcher. Fig. 8 demonstrates the functional use of this personal neuroscience information

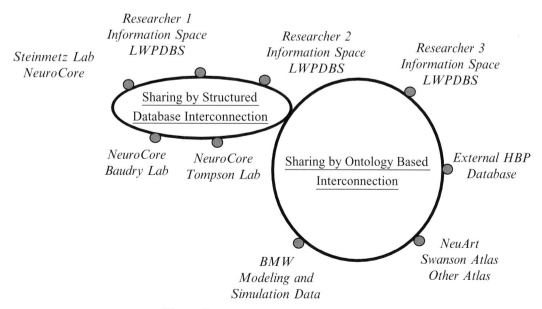

Figure 7 USCBP federation information flow.

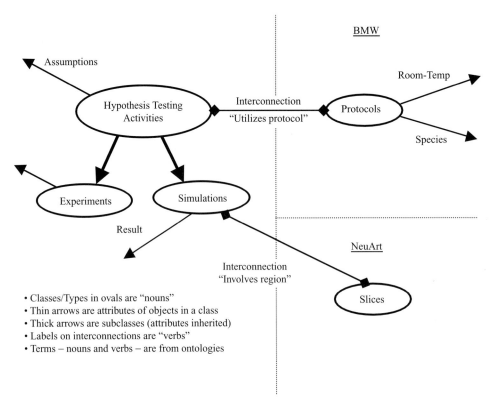

Figure 8 Illustration of LWPDBS.

manager,—which may also be referred to as a lightweight (running on a personal machine, non-expert useable) personal database system (LWPDBS). The figure shows a simple graphical view of a portion of a researcher's information space. Here, on the left side, we see some local information classes/types, such as Experiments and Simulations, along with their superclass—Hypothesis Testing Activities. Each class represents a kind of information unit (object) and specifies the attributes that apply to members of those classes (e.g., the attribute Assumptions of Hypothesis Testing Activities). On the right side, we see two remote classes of information (from two other databases), which are interconnected to members of "local" classes. Note that the connections are labeled with a term (verb) from an ontology that covers this area of neuroscience. This LWPDBS will allow a researcher to maintain a "guide" to other databases and to locally maintain inter-connections and annotations of the researcher. In effect, the researcher is provided with a customized view of a subset of the neuroscience database federation, along with researcher-specific augmentations. Key challenges in realizing the LWPDBS and the interconnection mechanism underlying it include:

1. Developing an effective and efficient representation of interconnections
2. Managing the ownership and dynamics of interconnections
3. Allowing interconnections to be created, modified, traversed, and searched
4. Achieving an effective functional researcher interface for local data creation, manipulation, and export

Our approach will be to focus upon these key challenges, while adopting commercial product solutions to the underlying general database management issues. This is most consistent with our use to date of the object-relational database management system, Informix. Our plan is to use "small" versions of Informix or Oracle (both are object relational) to support the LWPDBS.

References

Alonso, R. and Barbara, D. (1989). Negotiating data access in federated database systems. *ICDE*, 56–65.

Aslan, G. (1998). Semantic Heterogeneity Resolution in Federated Databases by Meta-Data Implantation and Stepwise Evolution, Ph.D. dissertation. University of Southern California, Los Angeles.

Aslan, G., and McLeod, D. (1999). Semantic heterogeneity resolution in federated databases by meta-data implantation and stepwise evolution. *Int. J. Very Large Databases*, in press.

Ceri, S., and Pelagatti, G. (1984). *Distributed Databases*, Principles and Systems. McGraw-Hill, New York.

Clifford, N. (1992). The Prospero file system, a global file system based on the virtual system model. *J. Comp. Syst.* **5(4)**, 407–432.

Elmasri, R., and Navathe, S. (1994). *Fundamentals of Database Systems*. Second ed. Benjamin/Cummings, Menlo Park, CA.

Fang, D., Ghandeharizadel, S., and McLeod, D. (1996). An experimental object-based sharing system for networked databases. *VLDB J.* **5**, 151–165.

Garcia-Molina, H. *et al.* (1995). Integrating and accessing heterogeneous information sources in TSIMMIS. In *Proceedings of AAAI. Symposium on Information Gathering*. Standford, CA, pp. 61–64.

Hammer, J., and McLeod, D. (1999). On the resolution of representational diversity in multidatabase systems. In *Management of Heterogeneous and Autonomous Database Systems* (Elmargamid, A., Rusinkiewicz, M., and Sheth, A., Eds.). Morgan Kaufman, San Mateo, CA.

Heimbigner, D., and McLeod, D. (1985). A federated architecture for information management. *ACM Trans. Office Inf. Syst.* **3(3)**, 253–278.

Hull, R., and Zhou, G. (1996). A framework for supporting data integration using the materialized and virtual approaches. In *Proceedings of the SIGMOD Conference*, Montreal, pp. 481–492.

Informix, Inc. (1997a). *Informix Guide to SQL*. Tutorial Version 9.1, Informix Press.

Informix, Inc. (1997b). *An Introduction to Informix Universal Server Extended Features*. Informix Press.

Kahng, J., and McLeod, D. (1998). Dynamic classificational ontologies, mediation of information sharing in cooperative federated database systems. In *Cooperative Information Systems*, Trends and Directions (Papazaglou, M., Ed.). Academic Press, New York.

Kahng, J., and McLeod, D. (1996). Dynamic classificational ontologies for discovery in cooperative federated databases. In *Proceedings of the First International Conference on Cooperative Information Systems*. Brussels, Belgium, pp. 26–36.

Kili, E. *et al.* (1995). *Experiences in Using CORBA for a Multidatabase Implementation*, DEXA Workshop 1995, London, pp. 223–230.

Litwin, W., Mark, L., and Roussopoulos, N. (1990). Interoperability of multiple autonomous databases. *ACM Computing Surv.* **22(3)**, 267–293.

Lyngbaek, P., and McLeod, D. (1990). A personal data manager. In *Proceedings of the International Conference on Very Large Databases*, pp. 14–25.

Sheth, A., and Larson, J. (1990). Federated database systems for managing distributed, heterogeneous, and autonomous databases. *ACM Computing Surv.* **22(3).**

Stonebraker, M. *et al.* (1996). Mariposa, a wide-area distributed database system. *VLDB J.* **5**, 48–63

Kim, W., Ed. (1995). *Modern Database systems*, The Object Model, Interoperability and Beyond. ACM Press, New York.

Kim, W. *et al.* (1993). On resolving schematic heterogeneity in multidatabase systems. *Distributed Parallel Databases* **1(3)**, 251–279.

Wiederhold, G. (1994). Interoperation, mediation, and ontologies. In *Proceedings of the International Symposium on Fifth-Generation Computer Systems*, December 1994. Tokyo, Japan, pp. 33–84.

Wiederhold, G. (1992). Mediators in the architecture of future information systems. *IEEE Computer* **25(3)**, 38–49.

CHAPTER 5.2

Dynamic Classification Ontologies[1]

Jonghyun Kahng and Dennis McLeod
Computer Science Department, University of Southern California, Los Angeles, California

Abstract

A cooperative federated database system (CFDBS) is an information sharing environment in which units of information to be shared are substantially structured, and participants are actively involved in information sharing activities. In this chapter, we focus on the problem of building a common ontology for the purpose of information sharing in the CFDBS context. We introduce the concept and mechanism of the dynamic classificational ontology (DCO), which is a collection of concepts and inter-relationships to describe and classify information units exported by participating information providers; a DCO contains top-level knowledge about exported information units, along with knowledge for classification. By contrast with fixed hierarchical classifications, the DCO builds domain-specific, dynamically changing classification schemes. Information providers contribute to the DCO when information units are exported, and the current knowledge in the DCO is in turn utilized to assist information sharing activities. We will show that, at the cost of information providers' cooperative efforts, this approach supports effective information sharing in the CFDBS environment.

5.2.1 Introduction

With the rapid growth of computer communication networks over the last decade, a vast amount of information of diverse structure and modality has become available on the networks. We consider this environment from the viewpoint of a *Cooperative Federated Database System* (CFDBS; see Heimbigner and McLeod, 1985); here units of information to be shared are substantially structured and participants are actively involved in information sharing activities.

A CFDBS consists of a number of autonomous *Information Frameworks* (IFs) and *Information Repositories* (IRs), as well as one or more *Dynamic Classificational Ontologies* (DCOs) (Fig. 1). IRs are major sources of information, while IFs are principally portals to IRs and other IFs (because an IR is a special kind of IF, the IF in the following will stand for both IFs and IRs unless otherwise mentioned). A DCO is a common ontology (a collection of concepts and their relationships to describe information units) which serves the basis of mutual "understanding" among participating IFs. Information is shared via mediators provided to support *import* (folding remote information into local environments), *export* (registering information to share), *discovery* (searching for relevant information), and *browsing* (navigating through information sources). Participants in the CFDBS communicate through an agreed-upon common data model and language.

Information sharing in the CFDBS faces a number of challenging problems due to the large volume of information and the rich structure of information units. Our approach is to dynamically build a common ontology, which is used by participants to describe and interpret

---

[1]This research has been funded in part by NIMH under grant no. 5P01MD/DA52194-02, and in part by the Integrated Media Systems Center, a National Science Foundation Engineering Research Center with additional support from the Annenberg Center for Communication at the University of Southern California and the California Trade and Commerce Agency.

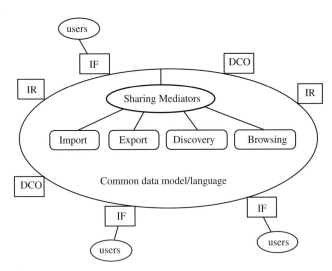

Figure 1 Top level view of a cooperative federated database system (CFDBS).

information that they share. An emphasis of our approach is based on the observation that it is extremely difficult to reach a total agreement on an ontology if the number of participants is large, and that the ontology should be allowed to change dynamically as the CFDBS evolves.

In this chapter, we present the concept and mechanism of the DCO, which addresses problems of the common ontology, and we illustrate how the DCO facilitates information-sharing activities. In order to reduce the size of the DCO, it contains a small amount of high-level meta-knowledge on exported information. Specifically, it contains a collection of concepts and their relationships that is to be used for classification of exported information. We rely on classification because it is an effective scheme for organizing a large amount of information. Further, some relationships in the DCO are not pre-determined but are computed based on exported information. Such relationships typically involve inter-related concepts which are useful for describing or classifying other concepts (for example, relationships among a set of subjects). An advantage of this approach is that participants are not required to agree on those relationships in advance. Another advantage is that those relationships are allowed to change as the usage of involved concepts changes.

The remainder of this chapter is organized as follows. Section 5.2.2 describes the spectrum of heterogeneity present in the CFDBS environment. Section 5.2.3 examines issues associated with the common ontology for resolution of semantic heterogeneity, and reviews related research. Section 5.2.4 discusses the role of classification as an information organization scheme and discusses the representation of classification. Section 5.2.5 describes the DCO in detail, while Section 5.2.6 shows how the knowledge in the DCO is utilized in the mediation of information-sharing activities, in particular, export and discovery. Section 5.2.7 concludes this chapter.

5.2.2 Heterogeneity

A key aspect of the CFDBS is heterogeneity of information at two levels of abstraction:

1. *Data model heterogeneity:* Information systems may use different collections of structures, constraints, and operations (i.e., different data models) to describe and manipulate data. For example, an information system may use a DBMS that supports object-based data modeling and an OSQL; another may store data in a collection of HTML documents and access them through http; and yet another may use a UNIX file system with various file management tools.

2. *Semantic heterogeneity:* Information systems may agree on a data model, but they may have independent specifications of data. This exhibits a wide spectrum of heterogeneity because most data models offer many different ways to describe the same or similar information.

Data Model Heterogeneity

Apparently, there are two alternatives to resolve data model heterogeneity. The first is to translate between every distinct pair of data models; the second is to adopt a common data model and translate between each data model and the common one. In the CFDBS, in which the number of distinct data models is expected to be large, the second alternative is more cost effective and scalable. This follows from a simple calculation of the number of necessary translations: $O(n^2)$ vs. $O(n)$, where n is the number of distinct data models. The price to pay for this alternative is that participating information systems should agree on a common data model; however, the cost of adopting a common data model can be well justified by its benefits. In fact, nearly all proposed systems for database interoperation assume some common data model (Arens *et al.*, 1993; Garcia-Molina *et al.*, 1995; Hammer and McLeod, 1993; Levy *et al.*, 1996; Mena *et al.*, 1996; Sciore *et al.*, 1992).

There is a tradeoff in choosing a common data model. Simple data models reduce the degree of potential semantic heterogeneity and the maintenance cost, but they limit the capability of information sharing; the opposite applies to semantically rich data models. A good example can be drawn from the recently exploding World Wide Web (WWW). Although it provides a great opportunity for people around the world to initiate information sharing, it comes with some intrinsic drawbacks. First of all, its data model is too simple to effectively describe diverse information. The simplicity, of course, represents both sides of a coin. It is the simplicity in part that has made the WWW so rapidly accepted in the Internet community. The simplicity would be acceptable as long as the information to be shared remains simple. This, however, is not the case, because people are now becoming more and more ambitious about sharing diverse information, both structured and unstructured, using WWW.

In the CFDBS, it is essential to adopt a data model that is more expressive than simple hypertext or flat files/tables, for example, because the CFDBS is intended to share information with diverse structures. An advantage of semantically rich models is that no information is lost when translation is done from less expressive models. It is also essential to have operations (query and manipulation languages) that are expressive enough to meet various demands and yet primitive enough to understand easily. An object-based data model might be a good choice because it is easy to understand (compared to richer models such as those of the KL-ONE family, which are more popular in the AI community), it allows effective data modeling in various application domains, and translation from other popular models is reasonably feasible. However, the query language for object-based data models needs improvement. Currently, variations of OSQL are the most prevalent languages for object-based models, but they fail to take advantage of object-oriented concepts such as inheritance. This is mainly because OSQL has its origin in SQL for relational models. Another drawback of OSQL is that users need quite a bit of training before effectively using it. A language that allows users to navigate through databases comfortably without formulating complicated queries would make information sharing in the CFDBS much more effective.

As object-based models, relational models, and some extensions of them have been widely adopted as a common data model in federated database environments, taxonomies of semantic heterogeneity allowed in such models have been extensively studied for the last decade (Kent, 1989; Kim et al., 1993; McLeod, 1992; Sheth and Kashyap, 1992). For the purpose of illustration, two university databases described in an object-based data model are shown in Fig. 2. We summarize incompatibilities between objects resulting from the semantic heterogeneity:

1. *Category:* Two objects from different information sources are under compatible categories if they represent the same or similar real-world entities. Specifically, they may have equivalence, subconcept/superconcept, or partially overlapping relationships. For example, Employees in A and People in B are equivalent because they both represent employees of the universities; Persons in A is a superconcept of People in B because the former represents a more general category of human beings than the latter; Students in A and People in B may be partially overlapping because some students may be employees, as well. On the other hand, Courses in A and People in B are under incompatible categories.

2. *Structure:* Two objects of a compatible category may have different structures. For example, Employees in A and People in B have quite different structures: People has an attribute "birthday," but Employees does not; the attribute "phone-nos" of Employees is equivalent to a combination of the attributes "work-phone" and "home-phone" of People. Another common example of structural incompatibility is that an attribute in one database is an object in another.

3. *Unit:* Two objects under a compatible category with a compatible structure may use different units. Salary in A and B gives an example of this incompatibility, given that the former is measured in dollars whereas the latter is in francs. Quality grades are another frequently encountered example. A grade may be measured on the scale of A, B, C, etc., or it may be measured on the scale of 1 to 10, for instance.

Other incompatibilities orthogonal to the above ones include:

4. *Terminology:* Synonyms and homonyms cause terminological incompatibilities. The attributes "SSN" of Employees and "ID" of People are an example of synonyms.

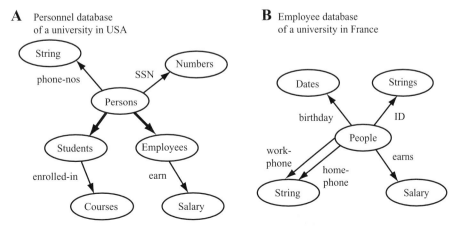

Figure 2 Examples of semantic heterogeneity.

5. *Universe of discourse:* The semantics of data are often hidden in the context. For example, the currencies used in A and B are presumably dollar and franc, respectively, considering their locations.

Resolution of semantic heterogeneity is at the center of interoperation in the CFDBS. Because of the difficulty of the problem, however, decades of research have been able to provide only primitive solutions to the problem, and there is little consensus on how to get beyond them. Among the first three incompatibility problems, we focus on the first one because locating relevant (i.e., categorically compatible) objects alone, setting aside their structural and unitary compatibilities, is a challenge in the CFDBS environment and because resolution of the first should precede that of the others.

A common approach to semantic heterogeneity resolution is to adopt a common ontology as a basis for mutual understanding. This introduces another level of agreement among participants in addition to an agreed-upon common data model. The remainder of this chapter is focused on this approach.

5.2.3 Common Ontology

An ontology is a collection of concepts and interconnections to describe information units. In particular, the common ontology in the CFDBS is to describe information exported from information sources. Fig. 3 shows a generic architecture for information sharing in the CFDBS. Information to be exported is first extracted from information sources and then translated from local data models to a common data model. Semantic heterogeneity among the exported information is resolved by mapping it into a common ontology. In other words, the common ontology is used to describe the exported information. Information sharing is facilitated by mediators provided for export, import, discovery, etc. There are several issues in working with a common ontology:

1. *Contents:* A common ontology could be as simple as a collection of concepts whose relationships are unspecified, or as complicated as a complete collection of concepts and their relationships that is enough to unambiguously describe all the exported information (like an integrated schema, for instance). Since neither of these two extremes are practical, most proposed systems adopt a common ontology that lies between the two. The contents of the common ontology strongly depends on the kinds of semantic heterogeneity that are to be resolved.

2. *Mapping:* Exported information needs to be mapped to (or described by) the common ontology. This process is typically the most labor intensive and time consuming one and is primarily carried out by domain experts. Thus, semi-automatic tools to assist this process would be very useful.

3. *Relevance:* Similarities and differences between two information units from different information sources or relevance of exported information to a given request must be determined at some point during the information-sharing activities.

4. *Maintenance:* Building a common ontology in the first place before any information sharing occurs is a challenging problem. Further, it is very helpful to allow evolution of the common ontology.

The problem of the common ontology has been addressed either implicitly or explicitly in several different contexts, including database interoperation, information retrieval, and Internet resource discovery.

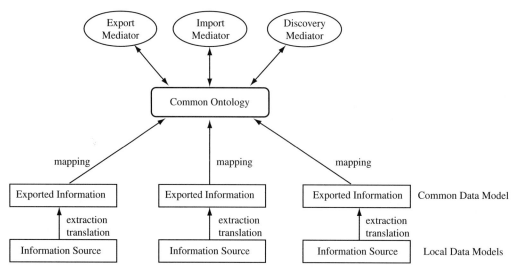

Figure 3 Information sharing in the CFDBS.

Database Interoperation

Early studies in database interoperation paid attention to tightly coupled federated database systems in which the common ontology is an integrated database schema (Batini *et al.*, 1986; Sheth and Larson, 1990). The focus of these systems is to build a database schema that supersedes all component database schemas and to define mappings between the integrated schema and component schemas. A number of techniques have been proposed for this purpose. Extensions of existing relational or object-based data models to improve capability of removing the ambiguity resulting from semantic mismatches among information units from different information sources have also been proposed. Some AI-oriented systems (Arens *et al.*, 1993; Levy *et al.*, 1996) use richer data models (those of the KL-ONE family) and focus on efficient query processing. This tightly coupled approach is not suitable for large systems such as the CFDBS. First, it is very difficult to construct an integrated schema if there are more than a few information sources. Second, a complete resolution needs to take care of detailed semantic conflicts, which could very well result in undesired complications. Third, evolution of the system is made difficult because every change in individual information sources must be reflected into the integrated schema.

Because of these difficulties, a more practical approach for the CFDBS is loosely coupled federated systems (Heimbigner and McLeod, 1985; Sheth and Larson, 1990; Wiederhold, 1992), in which the common ontology provides partial information about participating information sources. The choice of the common ontology in this approach strongly influences the functionality and capability of the system. Examples of proposed common ontologies include: a set of meta-attributes (Sciore *et al.*, 1992), a network of terms (Fankhauser and Neuhold, 1992), concept hierarchies (Yu *et al.*, 1991), summary schema hierarchies (Bright *et al.*, 1994), a set of canonical terms and transformer functions (Mena *et al.*, 1996), and a collection of concepts and relationship descriptors (Hammer and McLeod, 1993). Most of these systems emphasize relevance computation (or query processing) with a given common ontology and mappings; others are concerned with mappings/relationships to a given common ontology. A common drawback of these systems is that they do not deal with the problem of building and evolving the common ontology; the common ontology is defined in advance and more or less fixed.

Information Retrieval

Traditional information retrieval systems are concerned with instance-level (vs. type/class-level) information, as the type of information to be shared is documents with well-known properties such as title, authors, subjects, etc. As in database interoperation, common ontologies play an important role in these systems. In particular, the focus is on measuring the relevance of two documents or relevance of documents to a given request.

The simplest approach is to rely on keyword matching. That is, keywords are extracted from each document either manually or automatically, and two documents are compared based on the extracted keywords (Salton, 1989). The common ontology in this case is implicitly all words in a natural language with relationships among words nearly ignored. This can be improved by introducing synonyms or by replacing extracted keywords with their stems, but it is still too primitive to be useful in a more cooperative environments such as the CFDBS.

Another common approach is to take a collection of pre-classified subjects (a common ontology; see Samme and Ralston, 1982) and to assign a few of them to each document. While the pre-defined classification does include relationships between subjects, it has several undesirable features. First, it is hierarchical for the most part. That is, it contains only subsumption relations between subjects. Although cross-references between related subjects are often a part of the classification, they are not enough to represent overlapping relationships among subjects. Second, it tends to be static. Revision of the classification requires much time and effort. Consequently, it fails to accommodate dynamically changing usage of subjects. Third, it is typically huge and difficult to understand because it usually covers all disciplines and because it contains many artificial terms that are not commonly used in documents. These features make it difficult to apply this approach to the CFDBS environment.

An active area of research in information retrieval is to build term relationships from existing documents. Developed techniques include thesaurus-group generation (Chamis, 1988; Slaton, 1989), concept networks for concept retrieval (vs. keyword retrieval; see Chen and Lynch, 1992), and latent semantic indexing by singular value decomposition (Deerwester *et al.*, 1990). They basically rely on statistical analysis of term occurrence patterns in documents. This research is related to our approach, although our mechanism, assumptions, and context are quite different.

Internet Resource Discovery

It is interesting to observe that Internet resource discovery tools have followed in the footsteps of information retrieval systems. A number of systems based on keywords have been developed and are in use today. As expected, however, searching is not as efficient as desired due to their limitations; the precision of search results is so low that users need to spend much time to sort out retrieved information. To remedy such problems, some recent systems took the approach of classification. Yahoo

(2000), for example, takes a hierarchical classification of subjects and classifies URL objects by those subjects, which is reminiscent of the subject classification used in many library systems. Another one is Harvest (Bowman *et al.*, 1994), in which each broker specializes in a certain category such as technical reports, PC software, etc. It effectively divides the WWW space into several categories, and searching is carried out under each category. This is useful when users know which category of objects is relevant to their interests. In summary, a common ontology plays an important role in information sharing in the CFDBS environment. Many proposed systems adopt a common ontology as the basis for information sharing, but methodologies to construct and evolve the common ontology require more investigation.

5.2.4 Classification

Studies in cognitive science have shown that classification is the most basic scheme that humans use for organizing information and making inferences (Cagne *et al.*, 1993). Categories of objects have features that help identification of the categories, and objects are recognized by associating them with categories. For example, some children distinguish cats from dogs by the feature that cats have whiskers. In principle, all objects in the universe could be placed into a single classification tree; however, that is not the way humans picture the universe. Instead, there are categories at a certain level of generality (basic-level categories) on which people agree the most. When people were asked to list all the features of objects in categories such as trees, fish, birds, chairs, and cars, there was a high level of agreement among people with respect to the common features of those objects. Agreement is less prominent for superordinate categories such as plants, animals, furniture, and vehicles, as well as for subordinate categories such as robin, chicken, sedans, trucks, etc. Further studies showed that basic-level categories are the first ones learned by small children. An implication of these results is that classification is an effective method to organize a large amount of information, and it is natural to classify objects in two steps; the first is to classify objects into basic-level categories, and the second is to further classify objects in individual categories as necessary.

Following these research findings, our approach to constructing a common ontology is based on classification. That is, the common ontology will contain interrelated concepts that are just enough to classify exported information. In particular, the classification is organized around basic-level categories that are specific to the application domain. If independent information systems are in similar application domains, they are likely to agree on basic-level categories, regardless of their underlying data models and physical data structures. For example, most university databases will include information about courses, students, faculty, staff, and libraries at the top level. The agreement on basic-level categories would be very helpful for information sharing in the environment of a large-scale CFDBS.

Classification, in fact, has played an important role in information management. Most of the popular information management systems such as relational/object-based database management systems (DBMSs), the WWW, and hierarchical file systems provide constructs for classification. Some of them facilitate the two-step classification scheme that was mentioned above. To explore the representation of classification in the common ontology, we will examine each data model in turn.

In relational and object-based DBMSs, objects to be modeled are first classified into tables or classes. In the latter, objects in a class can be further classified into subclasses, resulting in class hierarchies. Classification mechanisms directly supported by the systems stop here. But, for a large amount of information, it is useful to classify objects in individual tables or classes. The systems provide an indirect mechanism for that; objects in a table or a class are implicitly classified by their attribute values. Fig. 4a shows a fragment of a library system. In this example, publications are broken down into three subclasses, where all four classes can be regarded as basic-level categories. In addition to the class hierarchy, publications are implicitly classified by subjects; that is, they can be grouped by the same or related subjects. Likewise, journals are classified by affiliated organizations as well as subjects.

Hierarchical file systems are supported in virtually all modern operating systems. A primary use of such file systems has traditionally been the management of personal files, such as documents and programs. Since computer communication networks became highly available, the file systems have been used as the primary storage of information by Internet resource-sharing tools such as WWW, Gopher, and anonymous ftp; they are now an important information organization tool. In hierarchical file systems, a directory can be used as a class of objects, and files in the directory can be regarded as objects in that class. If the number of files in a directory becomes large, they can be broken into subdirectories, resulting in a finer classification of objects. Fig. 4b shows a top-level directory structure of a typical anonymous ftp site. Many anonymous ftp sites have similar directory names and structure in the first one or two levels of directory trees; those directories tend to represent basic-level categories.

The WWW is an interesting invention for various reasons. It is basically a network of URL objects. A main strength of the WWW is that it supports diverse kinds of URL objects including HTML documents, images, video objects, and audio objects. It also provides gateways to Gopher, network news, and anonymous ftp sites. Its capability to organize information, on the other hand, is primitive. It is even more primitive than hierarchical file systems in the sense that it does not

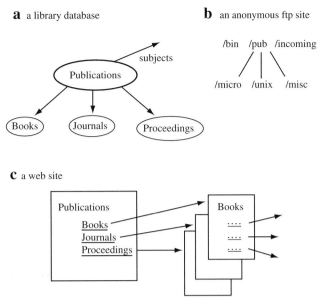

Figure 4 Classifications.

support any second-order modeling primitive that can be used for classification. That is, there is no notion of "type", "class", or "directory" as a collection of similar objects. Consequently, classification of objects is totally up to the person who manages information (see Fig. 4c, for example).

From these observations, relational and object-based data models are good candidates for the representation of classification; however, note that tables or class hierarchies with attributes (commonly referred to as a *database schema*) are insufficient to describe the two-step classification. For example, publications are implicitly classified by their subjects (Figure 4a), but that classification would not be very useful if the subjects are not well understood. That is, understanding relationships among subjects would be necessary to make that classification meaningful. The common ontology in our approach addresses this problem.

5.2.5 Dynamic Classificational Ontology

Information sharing in the CFDBS is centered around common ontologies, termed *Dynamic Classificational Ontologies* (DCOs). The CFDBS environment is characterized by a large amount of information with diverse semantic heterogeneity. Resolving semantic heterogeneity and setting up an interoperative environment are therefore extremely difficult and typically costly tasks. To assist in these tasks, the DCO keeps a small amount of meta-information on concepts (information units) exported from IFs; a DCO maintains a common ontology to describe and classify exported concepts.

If the number of participating IFs is more than a handful, it is very difficult to draw a total agreement on a common ontology. Among other difficulties, it is common in the CFDBS environment that certain concepts, such as "interoperability", are loosely defined but frequently used. Moreover, usage of concepts will keep changing as the CFDBS evolves. It would therefore be impractical to make precise definitions of all concepts in advance and enforce them. To address these problems, the DCO dynamically develops a common ontology that accommodates different understanding of concepts and their relationships in different IFs. Further, evolution of the common ontology is based on the input from individual IFs; the common ontology is maintained by their collaboration. This reduces the central coordination and the cost of setting up a cooperative environment.

A DCO consists of a *base ontology* and a *derived ontology*. The base ontology contains an ontology to describe and classify concepts exported by IFs, and the derived ontology contains an additional ontology to help classification of exported concepts in finer grains. The former is typically static and maintained by a DCO administrator; the latter is dynamic and computed based on the base ontology and the population of exported concepts. Fig. 5 shows the flow of information among the DCO and IFs in a CFDBS. As we see here, export is a part of a learning cycle: it adds knowledge to the DCO while being guided by the knowledge in the current DCO.

Classificational Object-Based Data Model

Knowledge in the DCO is represented by the *Classificational Object-based Data Model* (CODM). The basic unit of information in this model is the *concept*; concepts are grouped into *classes*. A class may have one or more *properties*. A collection of classes and their

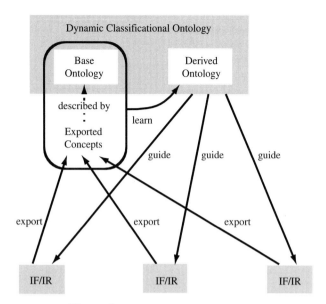

Figure 5 Information flow in a CFDBS.

properties is termed a *schema*. Looking ahead, Fig. 6a shows an example of the schema. The CODM supports generalization/specialization and inheritance of properties from superclasses to subclasses (Cardenas and McLeod, 1990).

In addition, the CODM supports conceptual relationships between concepts, and concept operators, which are useful for dynamic classification. A concept is essentially a representative of a set of real-world entities. Two concepts are *disjoint* if the two sets of entities that they represent are disjoint. One concept is a *superconcept/subconcept* of the other if the set of entities represented by the former is a superset/subset of that represented by the latter; otherwise, two concepts are *overlapping*. A concept operator takes one or more concepts and produces a new concept. There are three concept operators:

1. *Conceptual union* (*OR*): The concept (A *OR* B) represents a set of entities that is the union of the two sets represented by concepts A and B.
2. *Conceptual intersection* (*AND*): The concept (A *AND* B) represents a set of entities that is the intersection of the two sets represented by concepts A and B.
3. *Conceptual negation* (*NOT*): The concept (*NOT* A) represents a set of entities that is the complement of the set represented by A.

A concept is a *composite concept* if it can be decomposed into other concepts and concepts operators; otherwise, it is a *simple concept*.

A property is a mapping from a class to another class (a *value class*); that is, a property assigns to each concept of a given class a *value* which is composed from concepts of the value class. A property is *single valued, multi-valued, or composite valued* if the property value takes a simple concept, a set of simple concepts, or a composite concept, respectively. The first two are common in object-based data models, but the third is unique to the CODM.

The composite-valued property is introduced in the CODM because the multi-valued property generates ambiguities in some cases. To illustrate this point, consider the following examples, where pairs of subjects are given to describe some research articles:

1. {AI, knowledge representation}: If the article is about knowledge representation techniques in AI, it probably means (AI *AND* knowledge representation); that is, it covers the overlapping area of AI and knowledge representation.
2. {object-based data model, relational data-model}: If the article introduces and compares the two models, then (object-based data model *OR* relational data model) might be a better expression.
3. {database, network}: If the article is a survey of database technology, and a part of the article covers network-related materials(database *AND* network) *OR* (database *AND* (*NOT* network)) might well represent its intention, meaning that both network-related and not network-related materials are covered in the article.

The ambiguities are present because subjects are not independent of each other. In general, a property may be defined as composite valued when concepts of its value class are inter-related with each other.

Base Ontology

A base ontology consists of a schema in the CODM and concepts of selected classes. Fig. 6a shows an example of the schema. Classes in the base ontology represent (basic-level) categories of concepts, and properties represent relationships between such categories. Every concept in the base ontology has two required properties: *owner* is the owner of the concept (see below), and *time-of-entry* is the time when the concept was recorded. Fig. 6b describes a concept DBNETLAB (for convenience, a concept will be identified by a textual string) of the class Research-Lab, where its property names and values are given (e.g., owner: usc-database-if indicates that the value of the property owner is the concept usc-database-if).

The owner of a concept is either the DCO or an IF. Concepts owned by the DCO (e.g., concepts of Organization, Subject, Hobby) are a part of the base ontology. Concepts of other classes (e.g., Research-Lab, Text, Time) are exported by IFs and owned by them. Only the owner of a concept may remove or change it. Concepts owned by the DCO along with the schema are used to describe exported concepts. For example, Fig. 6b shows that the concept of the class Research-Lab was exported by the IF usc-database-if, and the value of its property institutions was chosen from the concepts of the class Organization which are owned by the DCO.

The ownership of concepts depends, in part, on how much information is going to be managed by the DCO.

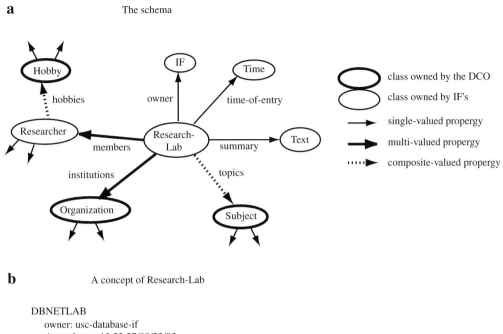

Figure 6 A base ontology.

For instance, Organization could be owned by IFs or by the DCO. If the DCO is planned to rigorously follow up information about organizations in any detail, the latter might be a good choice. In this case, the base ontology should include most of known organizations so that exported concepts may refer to them. On the other hand, Organization could be owned by IFs if, for instance, only the names of organizations are to be kept in the DCO.

Another (more critical) reason for the DCO to own and manage some concepts is to assist classification. In Fig. 6, research labs are implicitly classified by their topics (as well as by any other properties). This classification could be ambiguous, however, because subjects are interrelated with each other and the relationships tend to be non-objective. It is therefore necessary to understand conceptual relationships among subjects in order to make the classification useful. To deal with this problem, *topics* is defined as a composite-valued property, and the concepts of Subject are owned by the DCO in our example. Conceptual relationships among subjects are determined by statistical analyses, which will be discussed next.

Derived Ontology

A derived ontology records information about conceptual relationships. Specifically, if concepts of a class are owned by the DCO, and the class is the value class of a composite-valued property, conceptual relationships among those concepts enter the derived ontology. We will first discuss how composite concepts can be interpreted as they are the main source of information, and then describe how to derive conceptual relationships from them.

INTERPRETATIONS OF COMPOSITE CONCEPTS

Suppose that simple concepts C1, C2,..., CN of a class are owned by the DCO. A composite concept C composed from these concepts can be interpreted in two different ways:

1. *Open interpretation:* C can be interpreted as (C AND (CS1 OR (NOT CS1)) AND (CS2 OR (NOT CS2)) AND....AND (CSm OR (NOT CSm))), where CS1, CS2,..., CSm are simple concepts that do not appear in C. For example, (database AND network) means (database AND network AND (artificial-intelligence OR (NOT artificial-intelligence)) AND (operating-systems OR (NOT operating-systems))...). In other words, a composite concept may or may not be related to the concepts that are not explicitly mentioned in it.

2. *Closed interpretation:* C can be interpreted as (C AND (NOT CS1) AND (NOT CS2) AND...AND (NOT CSm)), where CS1, CS2,..., CSm are simple concepts that do not appear in C. For example, (database AND

network) means (database *AND* network *AND* (*NOT* artificial-intelligence) *AND* (*NOT* operating-systems) ...). In this case, a composite concept is not related to the concepts that are not explicitly mentioned in it.

A composite concept given by a user might require different interpretations depending on his/her intention. If a user is searching for some information, and the composite concept is provided as a specification of desired information, the open interpretation is probably a better one. That is, the user may not care whether or not the information that he or she wants is also related to other information. On the other hand, if a user is asked to describe some information by a composite concept as precisely as possible, the closed interpretation may be closer to his or her intention. This is because the user would try not to leave out any relevant concepts.

The open interpretation is safer, while the closed one is more informative. If users are not forced to adhere to either of the two interpretations, it is most likely that they will produce composite concepts whose interpretation falls somewhere between the two. The DCO takes the closed interpretation for composite concepts given by exporters. We will later show how the DCO can help them progressively formulate informative composite concepts that are subject to the closed interpretation.

CONCEPTUAL RELATIONSHIPS

The population of exported concepts is the basis for the derivation of conceptual relationships. We first define the *frequency* of composite concepts: For a composite concept C of a class Q, and a composite-valued property p whose value class is Q, the *frequency* of C is the number of superconcepts of C among the values of p. For the example in Fig. 6a, if some values of topics are (database *AND* network), (database *OR* network), or ((database *OR* information retrieval) *AND* network), each of them counts toward the frequency of a concept (database *AND* network) as all of them are its superconcepts.

As in mining association rules (Agrawal *et al.*, 1993), we introduce a variable to indicate the significance of statistical data: A *minimal support* is the frequency such that any frequency below it is considered as statistically insignificant.

Thus, if the minimal support is 10 and the frequency of (database *AND* complexity theory) is less than 10, then there is not enough data to determine whether database and complexity-theory are related.

We introduce another variable, the *tolerance factor*, to indicate confidence of derived conceptual relationships. Conceptual relationships are defined with the tolerance factor: Suppose that concepts C1, C2, and (C1 *AND* C2) have frequencies f1, f2, and f3, respectively (see Fig. 7a), and t is the *tolerance factor*. When f3 is larger than the minimal support:

1. C1 and C2 are disjoint concepts within a tolerance factor t, if both f3/f1 and f3/f2 are smaller than t.
2. C1 is a subconcept of C2 within t (C1 < C2) or C2 is a superconcept of C1 within t (C2 > C1), if f3/f1 is larger than (1 - t).
3. C1 and C2 are equivalent concepts within t (C1 = C2), if C1 is a subconcept of C2 within t and vice versa.
4. C1 and C2 are overlapping concepts within t otherwise.

When f3 is smaller than the minimal support, C1 and C2 are considered as disjoint concepts.

Conceptual relationships between C1 and C2 are summarized in the table in Fig. 7a. The tolerance factor represents statistical variations. There are two main causes for such variations; the first is simply that IFs may make mistakes at the time of export, and the second is that different IFs may have somewhat different understandings of involved concepts.

Conceptual relationships among two or more concepts can be best illustrated by diagrams, as in Fig. 7b; this figure shows some relationships among the concepts of Subject. Numbers in the figure indicate frequencies of the concepts; there are currently 400, 51, and 2 research labs whose topics include (computer-science *AND* (*NOT* database)), (computer-science *AND* database), ((*NOT* computer-science) *AND* database), respectively, and so on. Assuming a tolerance factor of 10%, the figure shows that database is a subconcept of computer-science, database and complexity-theory are disjoint concepts, and database and network are overlapping concepts. Fig. 7(c) shows how conceptual relationships may evolve. In the example, computer-science began as a part of mathematics and has grown out of it so that the two are now more or less separate disciplines.

The derived ontology is not fixed but dynamically changes as the population of exported concepts grows. Compared to using fixed conceptual relationships, such as pre-defined hierarchical classifications, this approach has several advantages: the derived ontology can be progressively built up, it may change as usage of concepts changes, and it will shape up in such a way as to reflect domain-specific usage of concepts. It is important to note that the aim of dynamically building the derived ontology is not to derive exact conceptual relationships, but to evolve a collection of reasonably agreeable conceptual relationships.

5.2.6 Mediators for Information Sharing

We will discuss in this section how the knowledge in the DCO is used by mediators provided for information sharing. First of all, the derived ontology depends heavily on information provided by IFs, and it might be unrealistic to expect IFs to provide precise descriptions of information to export from the beginning. We will show how the DCO can help them progressively formulate their descriptions. Discussed next will be discovery:

	$f_3/f_1 < t$	$t < f_3/f_1 < 1-t$	$f_3/f_1 > 1-t$
$f_3/f_2 < t$	disjoint	overlapping	$C_1 < C_2$
$t < f_3/f_2 < 1-t$	overlapping	overlapping	$C_1 < C_2$
$f_3/f_2 > 1-t$	$C_1 > C_2$	$C_1 > C_2$	$C_1 = C_2$

a Definitions

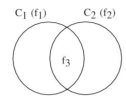

b Examples

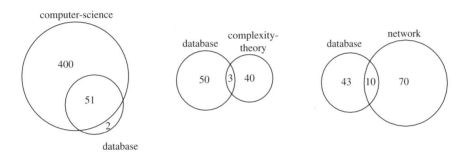

c Evolution

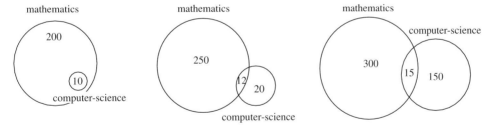

Figure 7 Conceptual relationships.

in the presence of abundant information, one of critical problems is to sort out information that is relevant. It is therefore essential to measure the relevance of available information to a given discovery request. We will show how to do that with the help of the DCO.

of the value class (e.g., institutions: University of Southern California). In particular, if the property is composite valued (e.g., topics, hobbies), the IF is allowed and encouraged to refine the value as precisely as possible with the help of the DCO.

Export Mediator

An IF exports a concept by submitting an entry using the schema of the base ontology as a template. The entry should include the name of the class to which the concept belongs, along with values for its properties; Fig. 8 shows an example. If the value of a property is concepts owned by the IF, it is accepted as entered (e.g., members: McLeod, McNeill). On the other hand, if it is owned by the DCO, it should be composed from existing concepts

CLASS: Research-Lab
owner: usc-bp-if
time-of-entry: 09:12:35/09/15/95
summary: The lab has been developing neuroscience databases that contain information about related literature and experimental data.
topics: (discovery AND scientific-db)
institutions: University of Southern California
members: McLeod, McNeill

Figure 8 An entry for export.

The export mediator utilizes the knowledge in the DCO to help IFs formulate the description of concepts that they export, especially when the description involves composite-valued properties and concepts owned by the DCO. It applies the following strategies to achieve this goal with minimal interaction with IFs.

STRATEGY 1

For a composite concept (as the value of a composite-valued property) given by the IF, concepts that are overlapping with it are retrieved from the DCO and presented to the IF so that the composite concept may be modified, restricted, or extended with them.

This is a rather straightforward strategy of utilizing conceptual relationships. Lines 1 through 3 in Fig. 9 show that application, heterogeneous-db, data-model, language, etc. turned out to be overlapping with the given composite concept (discovery *AND* scientific-db), and the IF added three of them to the composite concept.

STRATEGY 2

Among the overlapping concepts found by Strategy 1, only the concepts that are not subconcepts of others are presented to the IF to enable the IF to refine the composite concept from the top level to lower ones in progressive steps. This strategy reduces the number of related concepts to present to the IF so that it is not overwhelmed by a large number of concepts to choose from. Concepts that are left out will be further explored later only if their superconcepts are determined to be relevant by the IF. Lines 3 and 4 in Fig. 9 show that subconcepts of data-model (i.e., object-based-model and relational-model) are presented to the IF in the second round because data-model was selected by the IF in the previous round.

STRATEGY 3

If two concepts A and B given by the IF are disjoint, the IF is asked to choose either (A *AND* B) or (A *OR* B). It is useful to distinguish between (A *AND* B) and (A *OR* B) in order to improve the accuracy of the derived ontology. The list of concepts given by the IF is by default regarded as a conceptual intersection of those concepts. If A and B are disjoint, then (A *AND* B) is a non-existing composite concept (i.e., its frequency is insignificant), and it may not be what the IF intended. Lines 5 through 7 demonstrate this strategy. In this example, object-based-model and relational-model are assumed to be disjoint in the current DCO. If the IF insists that the previously given input is correct, it provides a basis for the composite concept (object-based-model *AND* relational-model) to develop and for the conceptual relationship between object-based-model and relational-model to change in the DCO.

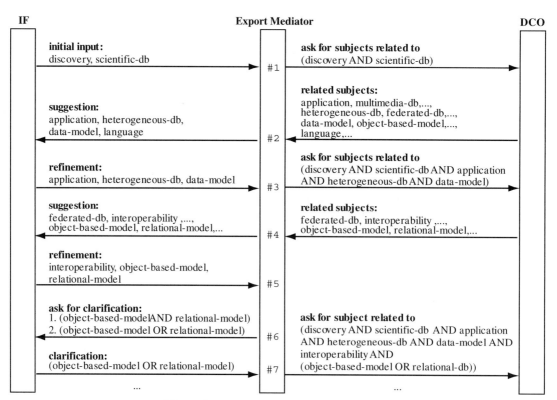

Figure 9 Description of topics of a research lab.

Discovery Mediator

An IF submits a discovery request using the schema of the base ontology as a template, as for export. The request includes the class name of the concepts in which the IF is interested. It may also specify some or all property values of the concepts (see Fig. 10). As in export, the discovery request can be first validated and refined with the help of the DCO. This will make the discovery request precise so that the precision and recall of retrieved results will be high.

Once the request for discovery is constructed, the next step is to retrieve relevant ones from exported concepts. Critical in this step is to measure relevance of exported concepts to the discovery request. For that purpose, we introduce a *relevance factor* (RF), which measures the relevance using the knowledge in the DCO. The RF is first computed for each property, and the final RF is the product of all those relevance factors. Retrieved concepts will be listed with corresponding relevance factors in the decreasing order of the RF. In the following definition of the RF, we will use examples of the discovery request shown in Fig. 10 and the exported concept DBNETLAB shown in Fig. 6b.

For the property whose value is not specified in the discovery request such as owner and institutions:

$$RF = 1$$

For a single-valued or multi-valued property:

$$RF = \frac{\# \mid D \cap O \mid}{\# \mid D \cup O \mid}$$

where D is the set of the property values in the discovery request, O is that of the exported concept, and $|S|$ is the cardinality of the set S. For example, the RF for members is

$$RF = \frac{\# \mid \{McLeod, Smith\} \cap \{Aslan, Kahng, Liao, McLeod\} \mid}{\# \mid \{McLeod, Smith\} \cup \{Aslan, Kahng, Liao, McLeod\} \mid}$$

For a composite-valued property,

$$RF = \frac{\# \mid D \text{ AND } O \mid}{\# \mid D \text{ OR } O \mid}$$

where D is the property value in the discovery request, O is that of the exported concept, and $|C|$ is the frequency of the concept C. For example, assuming conceptual relationships given in Fig. 11, the RF for topics is

CLASS: Research-Lab
owner: *
time-of-entry: *
summary: *
topics: ((database OR network) AND information-retrieval)
institutions: *
members: McLeod, Smith

Figure 10 A discovery request.

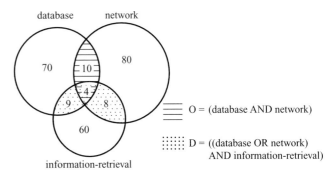

Figure 11 Conceptual relationships among concepts of Subject.

$$RF = \frac{4}{4 + 10 + 9 + 8} = \frac{4}{31}$$

The final RF for the above examples is

$$RF = \frac{1}{5} \times \frac{4}{31} = \frac{4}{155}$$

To illustrate the advantages of the RF over relevance measurements in conventional information retrieval systems, suppose that $D = (A \text{ AND } B)$ and $O = A$ for some composite-valued property in the above notation, and compare the RF with Radecki's coefficient, which measures similarity between two Boolean expressions (Radecki, 1982). It takes the same form as the RF for composite-valued properties, but the interpretation of $\#|C|$ is different: C is first converted into a disjunctive normal form $C = (C_1 \text{ OR } C_2 \text{ OR } ... C_n)$ such that each C_i contains all the terms that appear in C, and then $\# \mid C \mid = n$. For our example, Radecki's coefficient is

$$\frac{\# \mid D \text{ AND } O \mid}{\# \mid D \text{ OR } O \mid} = \frac{\# \mid A \text{ AND } B \mid}{\# \mid A \mid}$$
$$= \frac{\# \mid A \text{ AND } B \mid}{\# \mid (A \text{ AND } B) \text{ OR } (A \text{ AND } (\text{NOT } B)) \mid}$$
$$= \frac{1}{2}$$

regardless of how the two concepts A and B are related. In contrast, the RF is

1. 1, if A is a subconcept of B (i.e., (A *AND* B) = A).
2. 0, if A and B are disjoint concepts (i.e., (A *AND* B) does not exist).
3. Between 0 and 1, depending on how much A and B overlap with each other.

That is, the RF results in more meaningful relevance measurements by taking advantage of the knowledge in the DCO on generality of concepts as well as relationships among them.

5.2.7 Conclusions

We introduced the dynamic classificational ontology (DCO) for mediation of information sharing in the cooperative federated database system environment. In the presence of a large amount of heterogeneous information, it is beneficial to reduce the central coordination and distribute the maintenance cost. To this end, the DCO keeps only a small amount of meta-information, specifically a common ontology to describe and classify information exported by participating IFs, and it is maintained by their cooperative efforts. The classificational object-based data model was introduced as a model that facilitates two-step classification; concepts are classified into classes of basic-level categories, and concepts of each class can be further classified by their property values. While relying on partial agreement among participating IFs, the DCO is progressively established and dynamically adapts to changing usages of concepts.

We have developed an experimental prototype of the DCO, and applied it to document search problems in Medline (a medical information retrieval system provided by the Norris Medical Library at USC) in the context of the USC Brain Project (Arbib et al., 2000). We have developed a datamining algorithm that is advantageous for library systems with deep hierarchies of terms such as Medline. Preliminary results indicate that the precision and recall of document searches in Medline can be significantly improved by the interactive query refinement and the relevance measurement that are supported by the DCO. We are currently extending the system to support browsing and pattern discovery based on the knowledge accumulated in the DCO.

References

Arbib, M. A. et al. (2000). USC Brain Project, http://www-hbp.usc.edu/

Arens, Y., Chin, Y., Chee, C.-H., and Knoblock, C. A. (1993). Retrieving and integrating data from multiple information sources. *Int. J. Intelligent Cooperative Inf. Syst.* **2(2)**, 127–158.

Agrawal, R., Imielinski, T., and Swami, A. (1993). Data mining: a performance perspective. *IEEE Trans. Knowledge Data Eng.* **5(6)**, 914–925.

Batini, C., Lenzerini, M., and Navathe, S. B. (1986). A comparative analysis of methodologies for database schema integration. *ACM Computing Survey* **18(4)**, 323–364, December 1986.

Bowman, C. M., Danzig, D. B., Hardy, D. R., Manber, U., and Schwartz, M. F. (1994). The Harvest Information Discovery and Access System. In *Proceedings of the Second International World Wide Web Conference,* October, Chicago, IL, pp. 763–771.

Bright, M. W., Hurson, A. R., and Pakzad, S. (1994). Automated resolution of semantic heterogeneity in multidatabases. *ACM Trans. Database Systems* **19(2)**, 212–253.

Cagne, E. D., Yekovich, C. W., and Yekovich, F. R. (1993). *The Cognitive Psychology of School Learning.* Harper-Collins, New York.

Cardenas, A. F., and McLeod, D. (1990). *Research Foundations in Object-Oriented and Semantic Database Systems.* Prentice Hall, Englewood Cliffs, NJ.

Chamis, A. Y. (1988). Selection of online databases using switching vocabularies. *J. Am. Soc. Inf. Sci.* **39(3)**, 217–218.

Chen, H., and Lynch, K. L. (1992). Automatic construction of networks of concepts characterizing document databases. *IEEE Trans. Syst. Man Cybernetics.* **22(5)**, 885–902.

Deerwester, S., Dumais, S. T., Furnas, G. W., Landauer, T. K., and Harshman, R. (1990). Indexing by latent semantic analysis. *J. Am. Soc. Inf. Sci.* **41(6)**, 391–407.

Fankhauser, P., and Neuhold, E. J. (1992). Knowledge based integration of heterogeneous databases. In *Proceedings of the IFIP WG2.6 Database Semantics Conference on Interoperable Database Systems (DS-5).* November, Lorne, Victoria, Australia, pp. 155–175.

Garcia-Molina, H., Hammer, J., Ireland, K., Papakonstantinou, Y., Ullman, J., and Widom, J. (1995). Integrating and accessing heterogeneous information sources in TSIMMIS. In *Proceedings of the AAAI Symposium on Information Gathering*, March, Stanford, CA, pp. 61–64.

Hammer, J., and McLeod, D. (1993). An approach to resolving semantic heterogeneity in a federation of autonomous, heterogeneous database systems. *Int. J. Intelligent Cooperative Inf. Syst.* **2(1)**, 51–83.

Heimbigner, D., and McLeod, D. (1985). A federated architecture for information management. *ACM Trans. Office Inf. Syst.* **3(3)**, 253–278.

Kent, W. (1989). The many forms of a single fact. In *Proceedings of the IEEE COMPCON Spring '89.* February-March, San Francisco, CA. pp. 438–443.

Kim, W., Choi, I., Gala, S., and Scheevel, M. (1993). On resolving schematic heterogeneity in multidatabase systems. *Distributed Parallel Databases* **1(3)**, 251–279.

Levy, A. Y., Rajaraman, A., and Ordille, J. J. (1996). Querying heterogeneous information sources using source descriptions. In *Proceedings of the International Conference on Very Large Data Bases.* Bombay, India. pp. 251–262.

McLeod, D. (1992). The remote-exchange approach to semantic heterogeneity in federated database systems. In *Proceedings of the Second Far-East Workshop on Future Database Systems.* April, Kyoto, Japan, pp. 38–43.

Mena, E., Kashyap, V., Sheth, A., and Illarramendi, A. (1996). OBSERVER. an approach for query processing in global information systems based on interoperation across pre-existing ontologies. In *Proceedings of the First IFCIS International Conference on Cooperative Information Systems.* June, Brussels, Belgium, pp. 14–25.

Radecki, T. (1982). Similarity measures for boolean search request formulation. *J. Am. Soc. Inf. Retrieval* **33(1)**, 8–17.

Salton, G. (1989). *Automatic Text Processing,* Addison-Wesley, Reading, MA.

Sammet, J. E., and Ralston, A. (1982). The new (1982). computing reviews classification systemsfinal version. *Commun. ACM.* **25(1)**, 13–25.

Sciore, E., Seigel, M., and Rosenthal, A. (1992). Context interchange using meta-attributes. In *Proceedings of the First International Conference on Information and Knowledge Management.* Baltimore, MD, pp. 377–386.

Sheth, A., and Kashyap, V. (1992). So far (schematically). yet so near (semantically). In *Proceedings of the IFIP WG2.6 Database Semantics Conference on Interoperable Database Systems (DS-5).* November, Lorne, Victoria, Australia, pp. 283–312.

Sheth, A., and Larson, J. (1990). Federated database systems for managing distributed, heterogeneous, and autonomous databases. *ACM Computing Surveys* **22(3)**, 183–236.

Wiederhold, G. (1992). Mediators in the architecture of future information systems. *IEEE Computer* **25(3)**, 38–49.

Yahoo (2000). http://www.yahoo.com/.

Yu, C., Sun, W., Dao, S., and Keirsey, D. (1991). Determining relationships among attributes for interoperability of multi-database systems. In *Proceedings of the First International Workshop on Interoperability in Multidatabase Systems.* April, Kyoto, Japan, pp. 251–257.

CHAPTER 5.3

Annotator: Annotation Technology for the WWW

Ilia Ovsiannikov and Michael Arbib
University of Southern California Brain Project
University of Southern California, Los Angeles, California

Abstract

Annotation technology is a theoretical analysis of annotations. Grounded in studies of paper-based, handwritten notes, it applies the findings to electronic media to produce a set of practical recommendations for annotation software development. In the theoretical section, behavioral, ergonomical, functional, computational, and other aspects are discussed. The implementation part presents Annotator, a system for online annotations on arbitrary hypertext documents which is based on the concepts of annotation technology.

5.3.1 Introduction

There is no doubt that you have read books and taken notes. The custom of annotating is very common, especially during studying, reviewing, proofing, or research. The reader can spend a lot of time and effort on it, and the next time one takes those documents in one's hands, the annotations will readily help refresh one's memory, locate a paragraph of interest, provide proofing instructions, or do one of the many other important things that make annotations indispensable in everyday life.

Consider the following paradigm as a concrete example. In the academic environment, researchers and students annotate papers as a matter of daily routine. Later they write reports, manuscripts for which the annotations are reused. It is the nature of this activity to collect shreds of informations from a large variety of printed publications. Many a time the writer has to go over stacks of annotated papers in search of the particular excerpts of interest. This task can be very demanding and time consuming. This chapter presents the concept of *annotation technology* (AT), which explores various aspects of electronic annotations and offers solutions that can help this process.

Electronic annotations have their roots in handwritten notes; however, advances in personal computing have allowed us to take a step ahead. With the coming of electronic annotations, the student in the above-mentioned paradigm receives a variety of benefits. Annotations and original documents are now just one click away. The student can also easily keep track of the many old papers that would otherwise be half forgotten. Electronic annotations can be searched, shared, and used to construct comprehensive summaries or bibliographies. Applications of electronic annotations are by no means limited to the academic environment. Overall, insights offered by AT can leverage performance and decrease human effort in activities ranging from personal notetaking to office work and publishing.

5.3.2 Overview of Existing Annotation Software

Before we start with the discussion of annotation technology, let us take a brief look at existing annotation software systems. These systems can be roughly divided into two major categories: personal and collaborative. Just as the name suggests, personal systems are intended primarily for individual use. Microsoft Word (Microsoft, 1998) is one example of such software. MS Word allows the adding of typed, handwritten, or voice comments to documents. When the document is opened again in a next session these comments can be examined, edited, or deleted. In this paradigm, annotations become an integral part of the document and cannot be shared among many users.

Annotation sharing is one of the key features that distinguishes collaborative systems from personal ones. Third Voice (1999) represents an example of such a system that has come into existence quite recently. An add-on to Netscape Communicator 4.x, it allows attaching notes to selected parts of hypertext pages. If a note is marked "public," it will also show up on the many screens of other users that come to this page. These users can respond to that note to begin a discussion and thus work together, or collaborate, on some problem.

Multimedia-enabled annotation systems represent a special class of software by themselves. Some of these systems, such as XLibris (Price et al., 1998), focus on the problems of integrating handwritten notes into electronic documents. XLibris accepts pen input that is superimposed on the view of a document. The user can choose between pens of different colors, a marker, and an eraser. Later on, annotation clippings can be reviewed and searched in the Readers Notebook. XLibris also permits creation of various user-defined links that can be traversed to jump to other document parts.

Other multimedia systems such as DIANE (Benz et al., 1997) concentrate on supporting annotations on video, sound recordings, and all other kinds of media. Thanks to a dynamic download of viewer code, DIANE supports annotations on arbitrary types of media. In general, systems of this class often serve one purpose, namely to index the datastream by attaching to it human- or machine-readable labels. This index can be utilized to tag or search data that otherwise are difficult to index.

One more software system that must be mentioned here is EndNotes (Niles, 1998). Strictly speaking, this is not an annotation system per se, but it will be relevant to our upcoming discussion of annotation technology. EndNotes is a personal bibliography database that contains records of various kinds of publications that the user chooses to retain. It is used extensively during composition to find and properly format literature references. For our purposes, EndNote's search represents the feature of interest. We can view the bibliography database as a database containing stand-alone notes along with referencing information. A query into such a database will effectively search annotations. This is not the kind of search available for us when we deal with ordinary handwritten notes, nor is it characteristic of all systems of electronic annotations, but it is central to the design for annotation technology that we present here. For a more thorough review of existing annotation software, please refer to Ovsiannikov et al. (1999) which discusses more than a dozen systems, offers a feature-by-feature comparison, and also looks at the implementation side.

5.3.3 Annotation Technology: An Integrative Approach

Annotation technology (AT) is a theoretical analysis of annotations aimed at producing a set of recommendations for the design of electronic annotation software and is the basis for the Annotator system we have developed as part of the USC Brain Project. In this chapter, AT is presented from an integrative point of view. We start by sketching out behavioral aspects of handwritten, paper-based annotations and then make a transition to ergonomic, functional, and computational aspects of electronic ones.

In the beginning, let us introduce some new terminology that will be used throughout the rest of the chapter. Here we should mention that electronic annotations have a concrete foundation in ordinary handwritten, paper-based ones and, perhaps, should be thought of as their extension. Thus we feel it appropriate to accompany this part of the article with examples from paper documents (see Fig. 1). This figure shows the words "frame of reference" selected by means of circling. This is an example of a text *atom*, which is a contiguous portion of text. In general, documents often contain information expressed in forms other than text. For instance, it can be figures, tables, or even, in the case of electronic documents, sound or video. Hence, in order to be more general, we can extend the definition of atom to cover all other forms of presentation as follows: an *atom* is any selected piece of information, which is contiguous with respect to its form of presentation.

For the purpose of annotation atoms sometimes are found aggregated in a collection, as shown in Figs. 2 and 3. Here one can see that the equations containing the first and the second derivative are both selected by means of

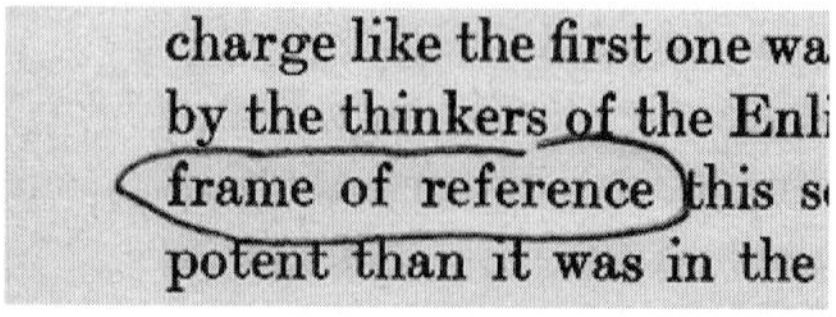

Figure 1 Atom created by circling several words.

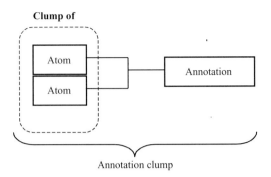

Figure 2 An annotation attached to a clump of atoms forms an annotation clump.

it $\dot{x}(t) = 0$ and that $\ddot{x}(t) = 0$. Using (1)

$- k(x - x_0)$

given force u satisfies

they do not change, due to the equilibrium state.

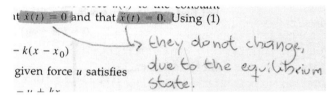

Figure 3 Annotated clump of atoms that were created by highlighting.

highlighting. This allows us to speak of them as atoms. Next, we can say that the two atoms are interrelated in at least two ways: schematically, by the hand-drawn connecting link, and semantically, by the reader's comment referring to both of them. This example illustrates the important concept of a *clump*, which is a set of semantically related atoms linked to the same comment.

Often, the reader not only marks up document portions to create a clump but also attaches to the clump some *annotation*. Let us define an *annotation* to be a datum added by a third party to the original document. From the presentational aspect, this datum can be a handwritten note, a symbol, a drawing, or even a multimedia clip. Fig. 3 shows an example of a handwritten annotation. Notice that the annotation text "they do not change..." connects to its clump by means of an arrow. This arrow represents an example of a *link*. A *link* is something that connects clumps of atoms and annotations together into an *annotation clump* or just *clump*. In Fig. 3, the annotated clump consists of two highlighted equalities and the handwritten text.

Handwritten Annotations

To get a good grip on the new terminology, let us consider some more examples of handwritten annotations. These figures were handpicked from a large number of books that went through students' hands. Annotations from such a source tend to be mostly academic and study oriented but do an excellent job when it comes to demonstrating the multiplicity of annotation form and content.

doubt; they make this decisio
but in search of security. The st
orary scene who is not concern
ut with man's *soul* considers this

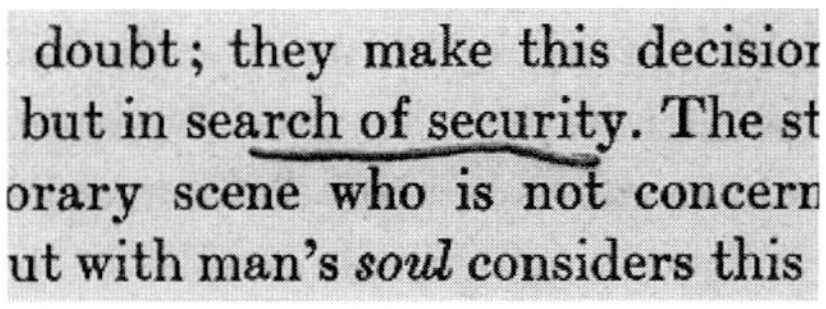

Figure 4 Small atom created by underscoring.

Consider Fig. 1, which shows the words "frame of reference" circled up into a text atom. The reader wanted to select this particular notion out of the document. Because this atom consists of only three words, selecting it by circling was convenient as the method of choice; however, circling is just one out of many methods for selecting short phrases, individual words, or parts thereof. Fig. 3 shows two atoms that are highlighted with a marker, and Fig. 4 presents an example of using an underscore. These selection methods are very popular and convenient as well, but the list is not over. Short atoms can also be enclosed in a box (which is something in between underscoring and circling), struck out, or bracketed or put in parentheses.

What if the text to be selected is longer than just a few words? In such a case, one can continue using the methods outlined above, but the bigger the atom gets, the more handwriting work must be done. When its length is on the order of several lines, *range* selections become the method of choice. Fig. 5 shows an example of a range, which is the bracket located at the right margin of the document. Imagine the atom in Fig. 5 being underscored and you will see that range is a very compact and convenient method. Not only is too much underscoring difficult to carry out, but it also can easily make the document illegible.

m like an Islamic motif. And it was,
an Week (lettering like *Newsweek*), a
e paper I had bought in a kiosk. The
it was protesting against what had
on, all the Islamic bans on alcohol,
fashions, music, mixed bathing,
over illustration showed a twisted
n replaced by iron bars. The family
om—father, mother, two children—
ut where their faces should have been

endered to the saviour and avenger.
, individualism—in the great city the
shed again. That seemed to be the

extremely important!!

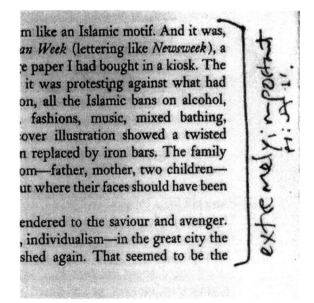

Figure 5 A larger selection uses a bracket to create a closed range atom.

The range in Fig. 5 is called a *closed* range, as it clearly marks its boundaries by means of the angled ends. Other types of text ranges include open, detailed, and hybrid. An *open* range is one that does not set a clear-cut boundary on either side, usually by omitting the angled ends. Such a selection is quite unprecise but occurs commonly. A *detailed* range always marks both of its ends explicitly and precisely. An example of a *hybrid* range is shown in Fig. 6. Here, its starting boundary, "The psychoanalyst," is detailed, as it is specified exactly to within the word. On the other hand, its ending is open, because the markings show that it is located somewhere in the line starting with "systems."

Ranges also differ with respect to the choice of the delineating bracket. Some readers prefer vertical lines or square brackets, while others opt for curly brackets or even tall parentheses. The choice is often a matter of habit, but sometimes it carries additional semantics. Here we will present only one example of markings loaded with semantics, as a longer discussion would be beyond the scope of this chapter. As a rule, there is no second interpretation as to the meaning of text that is crossed out. In editing and proofing, this means that the text atom must be deleted from the document.

Step the atom length up even more and chances are it will not fit a single page. One way to select such big portions of text is to allow ranges to continue onto the next page or pages. Just as in the case with regular ranges, all the range types and markings mentioned above are available here, as well.

Throughout this discussion, we have talked about selecting text by the character or line. These are just two of the four primary *levels of granularity* of selection. On the next level, *page*, the annotator selects atoms by the page. There is no doubt you have worked on this level many a time. One example could be marking pages in the corner or employing a bookmark. The highest level of granularity is the level of the document. We know we are working at this level whenever we affix a post-it note to the first page of a paper or jot a summary in the free space next to the title or even stack up the printed contents of our bookshelf into a "clump of books," so to speak. So far in our atom typology we have concentrated on text atoms only. This type is very widely used and is quite important in everyday work. When it comes to annotating pictures, a different atom type is used. For the reasons of space we will limit our discussion to text only.

that we have to give up the concern for the soul if we
do not accept the tenets of religion. The psychoanalyst is in a position to study the human reality
behind religion as well as behind nonreligious symbol
systems. He finds that the question is not whether man

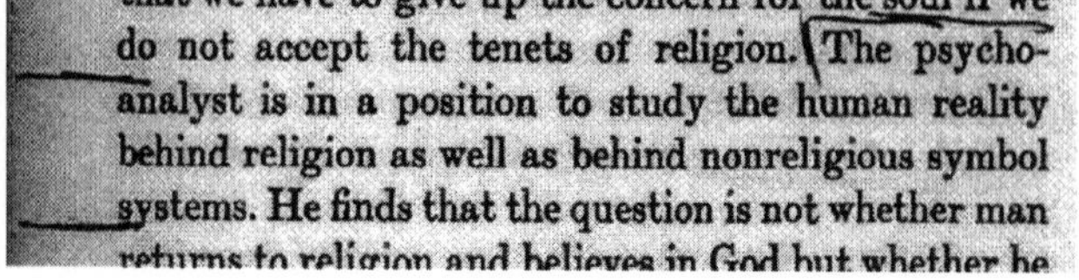

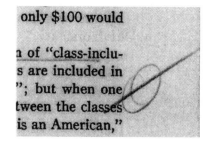

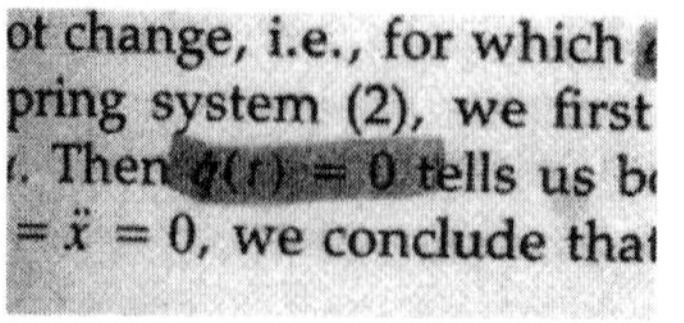

Figure 6 One boundary of every hybrid range is specified precisely while the other is not.

Let us continue on with our discussion of handwritten notes and take a closer look at annotations. Just like atoms, handwritten annotations can be of text or picture type. Fig. 3 shows an example of a text annotation that is attached to a clump of atoms by means of an arrow. By looking at this figure, one can see that the meaning of the attached note is pretty much clear to anyone who is familiar with that document. This is an example of a *common* annotation, as opposed to idiomatic or idiosyncratic annotations. Fig. 7 shows a very different annotation that resembles the "empty set" symbol in form. This is an example of a *pictorial* annotation, which can also be considered *idiosyncratic*, as it looks like it is meaningful only to the person who jotted it down. In between these two poles you can find *idiomatic* annotations, which are characteristic of people of a certain profession (e.g., proofreaders).

What unites annotations in Figs. 3 and Fig. 5 is that they are both *explicitly* marked. Consider Fig. 8, which shows a single atom created by means of a marker pen and has no note attached to it. Does that render it useless? Not at all. Even when atoms or clumps of atoms have no accompanying notes, the atom markings serve as annotations themselves. In particular, in Fig. 8 the atom's yellow color is equivalent to writing "this is important" next to the selection. Such apparently missing annotations occur very commonly and are called *implicit*. Generally speaking, annotators very often skip putting down notes of particular functions or semantics that occur too often. Such "compression" is performed in order to save time during writing and effort during reading and to avoid clutter.

The implicit annotation in Fig. 8 is conveyed by means of a special color. Are there any other forms of presentation? It turns out that there is a variety of atom markings that convey some definite semantics and,

only $100 would
n of "class-inclu-
s are included in
"; but when one
tween the classes
is an American,"

Figure 7 The meaning of an idiosyncratic annotation is not immediately obvious to anyone but the author.

ot change, i.e., for which
pring system (2), we first
. Then φ(r) = 0 tells us b
= ẍ = 0, we conclude tha

Figure 8 Highlighted atom without annotation.

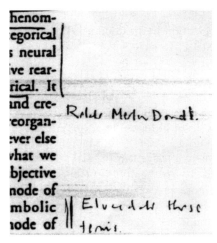

Figure 9 Compared to single, double ranges carry more emphasis.

sometimes, a degree associated with it. Fig. 9 shows two ranges, one of which is *single* and the other *double-line*. In certain cases, the single range may carry "this is important" semantics. The double-line range stresses this statement to make it "this is *very* important." A triple-line range would mean "very, very important" and so on. Sometimes the number of lines is replaced with the thickness of the range or underscore. A thicker selection corresponds to stressing or, literally, underscoring the statement contained in the atom.

Action markings are yet another form of implicit annotations. An indispensable tool of professional proofreaders, it associates actions with various forms of atom selections. For instance, a strike-out atom almost invariably corresponds to "deletion." A vertical line separating two characters often stands for "insert space" and so on.

For a moment let us return to explicit annotations and look at their forms. Fig. 3 shows an example of an explicit *text* annotation, which presents itself as a handwritten note. A different form is shown in Fig. 7. Here the annotation is represented with a *symbol*. Finally, the last but not the least form is pictorial when the reader creates an annotation by means of drawing.

Up to this moment we have discussed explicit annotations and implicit ones which are contained in atom markings. To complete this picture, we must also mention annotations that can be found implicit in links connecting clumps to notes. Suppose that an atom contains some statement, and the reader would like to add a conclusion that follows from it logically. At this point, it can be convenient to use a follows-from mathematical symbol in the form of an arrow to stress the causal relationship between the atom and the note. Thus, the arrow gets to carry semantics additional to its primary linking function.

Just like annotations, handwritten links can be either explicit or implicit. Examples of explicit links are shown in Figs. 3 and 10, where atoms are linked to their notes by means of an arrow. Readers tend to use explicit links

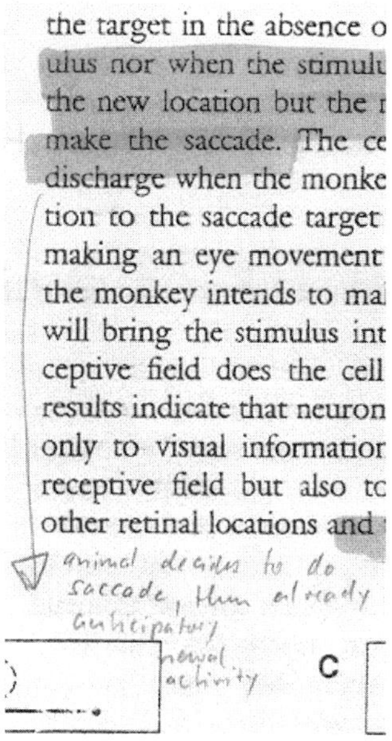

Figure 10 Arrow link connects atom with its annotation.

whenever there is an ambiguity as to which atom a particular note refers to. In Fig. 10, the arrow is necessary to show that the upper atom is the atom of interest and not the lower. Similarly, the arrow in Fig. 3 makes clear that the note refers to both of the formulas.

Links can take forms other than that of a connecting element. In certain cases, annotations refer to atoms by means of a certain symbolic label (e.g., a number). This is the form of choice whenever the annotation must be placed quite far from the atom or clump of atoms and using an arrow is impossible or would create unwanted clutter on the page. Color can also be used as a label to institute a connection; however, in the majority of cases, the link is implicit. The annotation is already connected by being *adjacent* to the clump of atoms. Fig. 5 shows an example of a remark being positioned next and parallel to the closed range. Such a placement tells us unambiguously that the two are directly related.

Distributed Clumps

We have briefly discussed handwritten annotations, their structure, and presentational forms. One important thing must be noted here. As a rule, the examples above deal with atoms and annotations located in the same document and even on the same page. If this restriction is lifted and notes are allowed to be distributed across many printed sources, the annotation clump transcends its locality and becomes *distributed*. Fig. 11 shows an example of this paradigm. In the figure, the note "Experiments from the two papers..." refers to two clumps of atoms in different papers. We can say that, when combined, these two local clumps result in a distributed clump or a set of clumps which share a common annotation.

In general, if we do not limit our scope with handwritten annotations and take into consideration hypertext and multimedia sources, one can hypothesize distributed clumps that cross document boundaries and include as well atoms of type video or sound and so on. As you may have noticed in Fig. 11, notes attached to a distributed clump refer to several papers and at the same time may reside in just one or even none of them. This is unlike handwritten annotations, where notes are almost invariably located right next to the clump of atoms. This is one reason why annotation technology treats annotations and atoms as individual entities. Our Annotation Database is an embodiment of this approach.

Annotation Databases

An Annotation Database (ADB) is a collection of information about the location and contents of atoms, annotations, and their dependencies in the form of links. Fig. 12 shows schematically an ADB divided into two major sections: the clump section and the annotation section. It contains all data about in what document and exactly where in that document a clump is located.

ADB also contains links, annotations, and, now optionally, their position in the document.

The ADB effectively takes clumps and annotations out of a document to be stored independently. This approach makes it possible to utilize the information in new ways. One such novel application is *annotation search*. When queried by a user, an ADB will respond with a collection of matching records. Having the information about location of atoms and annotations, the user can easily retrieve the original annotated document and go directly to the clump of interest. To make this paradigm more concrete, imagine ADB implemented as an online hypertext system. Querying such a database would amount to typing keywords of choice in the browser window. Once the request has been submitted and the ADB has returned a list of hits, going to the annotation within an online document would be as simple as clicking a link, as described in the next section about our Annotator system.

Consider the list of hits returned by an ADB. Generated dynamically, this list contains atoms and annotations and can be treated as a new annotation clump. This new clump can be used later on to assist writing a new

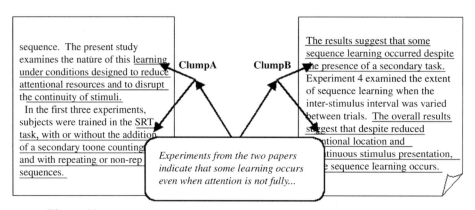

Figure 11 A distributed clump consists of atoms that reside in different documents.

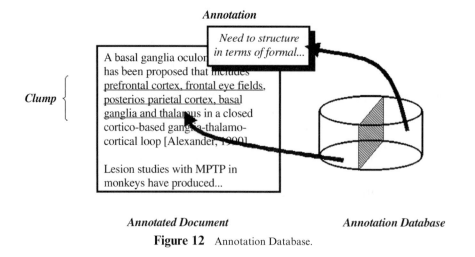

Figure 12 Annotation Database.

manuscript or in putting together a bibliography on a certain topic.

So far we have talked about Annotation Databases under the assumption that they are personal (i.e., individual for each user). What if parts of Annotation Databases were open for access to other people? In this case, ADB will become a shared resource. Access to shared annotations may come in handy when a team of people is working together, such as in an office environment or research laboratory. Opinions, solutions, thoughts, or evaluations—all can be shared concurrently between the members of such a group leading to increased productivity. It should be noted that annotations that are meant to be shared should always be unambiguous; otherwise, ADB users other than the owner may find interpreting the notes a very difficult task.

USCBP Databases: Integration of Independent Sources

Annotation technology research is a part of a greater effort supported by the USC Brain Project (USCBP). This effort is directed at development of a number of on-line databases for the neuroscience community. Consider, for instance, the problem of building databases for online models also known as Brain Models on the Web (BMW). A typical model consists of description, source code, interfaces, executables, and such and resides in the BMW database; however, having just these data for modeling is insufficient. First of all, the input and output data patterns and time series are indispensable elements. A good model should also refer to literature that supports or discusses relevant issues, assumptions made during development, and so on.

The USCBP has been developing databases to address these needs. The Time-Series Database (TSDB) has the capability to store and retrieve various patterns, electrode recordings, and images. On the other hand, the Summary Database (SDB) contains text summaries, notes, and logs. In this framework, AT is positioned to solve the important problem of integrating these databases. This capability for integration of independent information sources stems from the notion of distributed clump. With AT tools, the user has the freedom to impose arbitrary fine-grained structure on documents without actually modifying those documents. The functionality of annotations is also extended to include pointers, as in hypertext (see, for example, Inso, 1998). Taken together, these functions allow the models in BMW to be easily linked to supporting statements in documents and SDB summaries and to datasets in the TSDB.

5.3.4 Annotator

In this section we will introduce Annotator, a pilot implementation of the principles of annotation technology. Annotator is a distributed collaborative software system for making online annotations on the World Wide Web. Written completely in Java, it works with Netscape 4.x on any platform.

User Interface and Functionality

Working with Annotator is as easy as using Netscape Communicator to browse the Internet. By default, Annotator resides in the "disabled" mode. In this mode, all its features are disabled and the browsing is no different than if Annotator were not installed. To enable annotation features, the user must open a special URL. In response, the browser will prompt for a username and password to log in to an Annotation Database. (Some of these details of access to, and use of, Annotator will be described in the "Available Resources" section; here, we focus on the general functionality of Annotator.)

From this point on, all retrieved pages will show up along with annotations you or someone else created earlier. Fig. 13 shows an example of annotations embedded into a page. Atoms are shown in blue, and notes are shown in red. All atoms and annotations will also show up in another window called the Database Control Panel

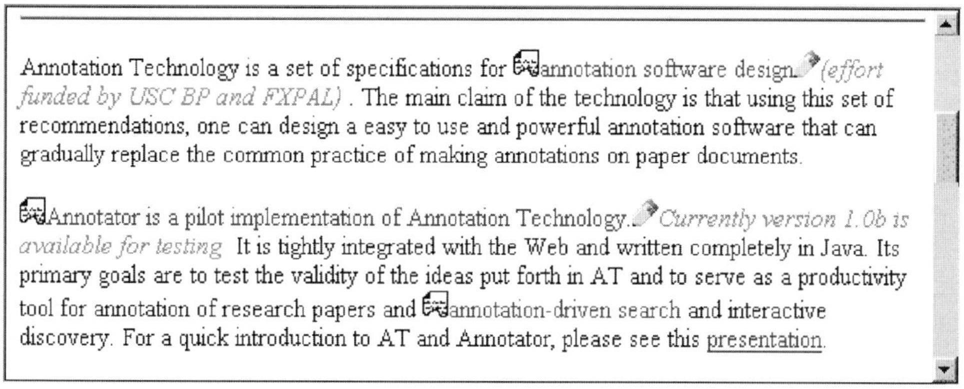

Figure 13 Viewing annotations in a browser window.

(DCP; see Fig. 14). The DCP displays annotations in a tree-like structured form, where branches correspond to links between atoms, clumps, and notes. The user can browse through such a tree by expanding or collapsing certain items. Double-clicking on an item causes the browser to open a page and scroll to the annotation of interest. Certain advanced features such as linking clumps and atoms by dragging and dropping the icons are in development. For a moment, let us return to Fig. 13, which shows a sample annotated page. On that page, the symbolic icons associated with atoms and notes work as miniature buttons. Clicking on one will make the DCP annotation tree change to reveal the chosen item.

The user can annotate a page by invoking the Netscape's HTML editor, Netscape Composer. Fig. 15 shows a screenshot of a page being annotated. From the editor's menu, the user commands a wide variety of annotation-related actions. For instance, to create an annotation, the user selects an atom of interest, clicks "Annotate," and types the note in the space allocated by the editor. A future version of the software will allow adding more atoms to a clump by making a selection and choosing "Add Atom." Long, out-of-line annotations called *memos* can be created in a manner similar to regular annotations. In the end, the user saves the work to the Annotation Database by clicking on the editor's "Publish Page" button.

One feature that is central and in some ways unique to the design of Annotator is the ability to search and summarize annotations. This is where the complexity and power of the Annotation Database comes to its full realization. To query ADB, the user must bring up the Database Control Panel, select the search mode, and type in a set of keywords. In return to this query, ADB will retrieve matching atoms and notes and display them in the tree form. By clicking on these items, the user now has direct access to the annotations in the original online documents. As an alternative, the user can select a number of retrieved items to create a distributed clump. This can be useful, for instance, to collect together annotations related to the topic of interest across various research papers. The user can also copy-and-paste such a clump in a word processor. In this case, Annotator will combine selected annotations and atoms into a readable text summary to be printed out or edited into some text.

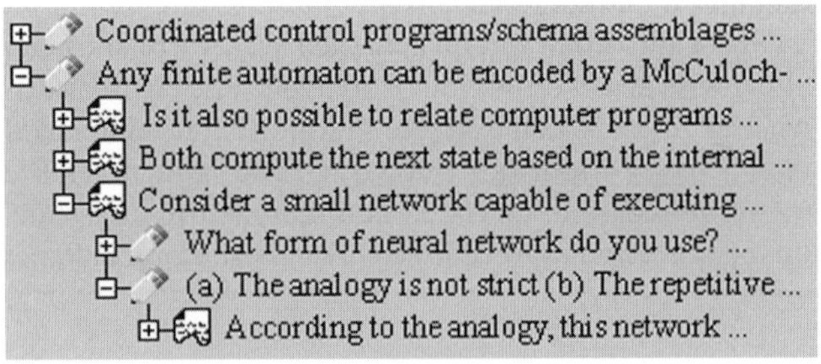

Implementation

Having discussed Annotator's functionality and user interface, let us take a look at the implementation details. Fig. 16 shows a diagram of the main system components. The proxy and Annotation Database comprise the system's core. Together, they are designed as shareable resources and can run on either a stand-alone server or on the user's PC in the single-user case. The browser's functionality is augmented by a plug-in package that must be installed before use in each workplace.

Consider a typical scenario where the browser issues a request to retrieve a hypertext document. In the disabled mode, the proxy acts as a repeater that forwards the request on to the server and passes back a response containing the page. In this mode, the user can browse the Web as if Annotator were not installed at all. Once the user switches the proxy over to the enabled mode by logging in, the pattern of communication changes. Every time the browser issues a request, the proxy will forward it on to the server and at the same time query the database for any annotations and atoms associated with the given URL. Once the server returns a document, the proxy checks the search results. If the database returns a non-empty set of records, the proxy will dynamically parse the document and insert the annotations and atoms at their proper locations. After this, the document is passed on to the browser for display. Note that the insertion is made in generating this display, not on its source document. This is a distinctive feature, for it

Figure 14 DCP annotation browser.

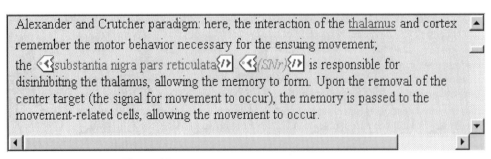

Figure 15 Editing annotations in Netscape Composer.

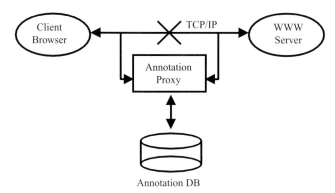

Figure 16 Annotation proxy queries ADB and dynamically inserts annotations into hypertext documents.

allows the user to make or view annotations on documents whose files are locked or otherwise unavailable to change. In general, downloading and storing any portions of document without permissions violates the copyright law. That gives one more reason why clumps must be annotated *in situ* rather than copied to the user's computer.

The Annotation Database is a custom-made Java-based database. To minimize the overhead due to data retrieval, it is designed for maximum real-time performance. The ADB schema contains several tables, three of which are most important. These are the atoms, annotations, and links tables. Along with the text contents and associated URL, the atoms and annotation data structures store information about their location within the documents called *anchors*. According to the schema, every annotation has one anchor assigned to it. Every atom has two anchors, one for the beginning of the selection and one for the ending.

A number of considerations are relevant to the implementation of anchors. There is no guarantee that an annotated page will never change. At any time, the owner of the page can edit it, and some pages are generated dynamically and may differ from one download to another. Hence, it is highly desirable that the system is robust with respect to small changes in hypertext pages and that the anchors in Annotator are implemented with this consideration in mind. To record a location in a page, Annotator finds a reasonably short but unique substring at that location. After that, the substring undergoes hashing and the hashcode is stored in the record along with the string's length. Anchors also store the first few characters of the search string. This is done in order to make finding the location during retrieval computationally more efficient.

The ADB is designed to be a shareable component. Because HTTP is a stateless protocol, the proxy must simulate a database session for each user from the moment users log in and up to the point when they log out. To this purpose the proxy maintains a table of sessions listing all database connections that are currently active. If any connection has not been active for a certain amount of time, the proxy will close it, effectively logging out that user after a period of inactivity.

The Annotation Database supports annotation searches. It takes a set of keywords as input and returns a collection of matching records. The search can be restricted to atoms or annotations only. Annotations are added by means of submitting the annotated document to the proxy. When the user clicks the "Publish" button in Netscape Composer, the editor uploads the page with annotations to the proxy. In its turn, the proxy intercepts the page, parses out any annotations and atoms contained there, computes anchor information, and saves all the data in the database along with the page's URL.

In this chapter, we have presented annotation technology, an integrative theory of annotations that offers a systematized set of recommendations for the design of advanced electronic annotation software. Based on an analysis of handwritten comments, AT proposes the annotation-clump framework that conveniently represents annotations as data structures. Annotations acquire new functionality during the transition from handwritten to electronic media. In relation to this fundamental process, AT introduces the notions of distributed clump, Annotation Database, and annotation search.

Available Resources

The paradigm described above has been put to the test with the development of the Annotator software system that is available for download from the USC Brain Project server at http://www-hbp.usc.edu/Projects/annotati.htm. The software consists of two packages, client and server. The client package must be downloaded and copied into the Netscape 4.x plug-in directory as explained in the accompanying manual. After downloading and unzipping, the server can run on any computer with Java Virtual Machine and network access. Note that the server can be used individually as well as shared. To log in to the Annotation Database, open URL http://annotation-proxy/login and log in as *admin* with password *admin*. To log out, go to http://annotation-proxy/logout.

Acknowledgments

NIMH grant MH 52194-03 and a grant from Fuji Xerox Palo Alto Laboratories supports this research. We thank Thomas McNeill, for support for the early stages of this work, and Abhitesh Kastuar and Shivani Aggarwal, for their work on programming Annotator.

References

Benz, H., Fischer, S., Mecklenburg, R. *et al.* (1997). DIANE: hypermedia documents in a distributed annotation environment. In *Proceedings of HIM '97*. Dortmund, Germany.

Inso (1998). *DynaText Professional Publishing System*, http://www.inso.com/dynatext.

Microsoft (1998). *Microsoft Word*. http://www.microsoft.com/office/word/default.htm.

Niles (1998). *EndNotes 3.0*. http://www.niles.com/.

Ovsiannikov, I., Arbib, M. A., and McNeill, T. H. (1999). Annotation technology. *Int. J. Human-Computer Studies*, 50.

Price, M., Schilit, B. N., and Golovchinsky, G. (1998). *XLibris: The Active Reading Machine*. ACM Press, Los Angeles, CA.

Third Voice (1999). *Third Voice*. http://www.thirdvoice.com.

CHAPTER 5.4

Management of Space in Hierarchical Storage Systems

Shahram Ghandeharizadeh, Douglas J. Ierardi, and Roger Zimmermann
University of Southern California Brain Project, University of Southern California, Los Angeles, California

Abstract

The past decade has witnessed a proliferation of repositories whose workload consists of queries that retrieve information. These repositories provide online access to vast amounts of data and serve as an integral component of many applications (e.g., library information systems, scientific applications, and the entertainment industry). Their storage subsystems are expected to be hierarchical, consisting of memory, magnetic disk drives, optical disk drives, and tape libraries. The database itself resides permanently on the tape. Objects are swapped onto either the magnetic or optical disk drives on demand and later deleted when the available space of a device is exhausted. This behavior will generally cause fragmentation of the disk space over a period of time, resulting in a non-contiguous layout of disk-resident objects. As a consequence, the disk is required to reposition its read head multiple times (incurring seek operations) whenever a resident object is retrieved. This may reduce the overall performance of the system, forcing the user to wait longer than necessary.

This study investigates four alternative techniques to manage the available space of mechanical devices in a hierarchical storage systems. Conceptually, these techniques can be categorized according to how they optimize several quantities, including: (1) the fragmentation of disk-resident objects, (2) the amount of wasted space, and (3) adaptation to the evolving access pattern of an application. For each of these alternative strategies, we identify the fundamental factors that impact the performance of the system and develop analytical models that quantify each factor. These models can be employed by a system designer to choose among competing strategies based on the physical characteristics of both the system and the target application.

5.4.1 Introduction

The Human Brain Project is a testament to a recent trend in the area of databases, namely, an increase in the number of repositories whose primary functionality is to disseminate information. These systems are expected to play a major role in library information systems; scientific applications, such as the Brookhaven protein repository (Bernstein *et al.*, 1977) and the human genome repository (Council, 1988); the entertainment industry, health-care information systems, knowledge-based systems, etc. These systems exhibit the following characteristics. First, they provide online access to a vast amount of data. Second, only a small subset of the data is accessed at a given point in time. Third, a major fraction of their workload consists of read-only queries. Fourth, objects managed by these systems are typically large and irregularly structured. Fifth, their applications consume the data at a high rate and almost always exhaust the available disk bandwidth. Hence, they face the traditional I/O bottleneck phenomenon.

As an example, consider the NeuroAnatomical Registration Viewer (NeuARt; see Chapter 4.3). It enables neuroscientists to browse, compare, and query the complex spatially distributed patterns of labels obtained from different experiments at their desktops through a network connection. Prior to NeuARt, scientists shared data by exchanging copies of Adobe Illustrator files that constitute the atlas and its associated imported layers representing data from different experiments. This sharing paradigm works well when a scientist is focused on data from two or three experiments. It becomes cumbersome when a scientist wishes to investigate patterns in the data over large numbers of experiments, requiring in excess of 30 to 40 layers in a single file. Assembling this data becomes labor intensive because a scientist must copy layers between files by manually scanning (opening and closing) several large Illustrator files and switching between them. From the start, we envisioned NeuARt as a flexbile tool that can be used as a viewer for mouse, human, etc. Thus, the size of underlying images may vary from several kilobytes to hundreds of megabytes (if not gigabytes), depending on whether the image is color or black and white, its resolution, and level of detail. For example, an uncompressed 2550 × 3300 pixel gray-scale image might be 8.4 megabytes in size. The same image in color would be 25.2 megabytes in size. A repository managing thousands of images might be hundreds of gigabytes in size, with only a small fraction of images being accessed frequently (e.g., those corresponding to experiments under investigation). Typically, an application retrieves an image in a sequential manner for display. The faster the image can be retrieved, the sooner it can be displayed (due to the availability of fast CPUs). This enhances system performance because a scientist waits for a shorter period of time when traversing between different images.

The large size of these databases has led to the use of hierarchical storage structures. This is motivated primarily by dollars and sense, as storing terabytes of data using random access memory (DRAM) would be very expensive. Moreover, it would be wasteful because only a small fraction of the data is referenced at any given instant in time (due to locality of references). A similar argument applies to other devices (e.g., magnetic disks). The most practical choice would be to employ a combination of fast and slow devices, where the system controls the placement of the data in order to hide the high latency of slow devices using fast devices.

Assume a hierarchical storage structure consisting of DRAM, magnetic disk drives, optical disks, and a tape library (Carey et al., 1993; see Fig. 1). As the different strata of the hierarchy are traversed starting with memory (stratum 0), both the density of the medium (the amount of data it can store) and its latency increase,

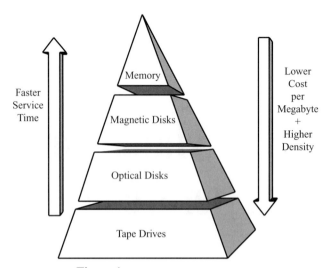

Figure 1 Hierarchical storage system.

while its cost per megabyte of storage decreases. At the time of this writing, in 1999, these costs vary from $1/megabyte of DRAM to $0.005/megabyte of disk storage (half a penny) to $0.002/megabyte of optical disk to less than $0.0002/megabyte of tape storage with robotic tape libraries. An application referencing an object that is disk resident observes both the average latency time and the delivery rate of a magnetic disk drive (which is superior to that of the tape library). An application would observe the best performance when its working set becomes resident at the highest level of the hierarchy: memory. However, in our assumed environment, the magnetic disk drives are the more likely staging area for this working set due to the large size of objects. Typically, memory would be used to stage a small fraction of an object for immediate processing and display. We define the working set (Denning, 1968) of an application as a collection of objects that are repeatedly referenced. For example, in a scientific endeavor using NeuARt, typically a few experiments specific to a region of a brain are accessed frequently. These data sets constitute the working set of a database system whose application is NeuARt.

In general, assuming that the storage structure consists of n strata, we assume that the database resides permanently on the last stratum $(n-1)$. For example, Fig. 1 shows a system with four strata in which the database resides on stratum 3. We term a file that consists of either one or more images an object. Objects are swapped in and out of a device at strata $i < n$, based on their expected future access patterns, with the objective of minimizing the frequency of access to the slower devices at higher strata. This objective minimizes the average latency time incurred by requests referencing objects.

At some point during the normal mode of operation, the storage capacity of the device at stratum i will be exhausted. Once an object o_x is referenced, the system may determine that the expected future reference to o_x is such that it should reside on a device at this stratum. In this case, other objects should be swapped out in order to allow o_x to become resident here; however, this migration of objects in and out of strata may cause the available space of the devices to become fragmented, resulting in the non-contiguous layout of its resident objects. Unlike DRAM, optical and magnetic disk drives and tape drives are mechanical devices. Storing an object non-contiguously would cause the device to reposition its read head when retrieving the object, reducing the overall performance of the device. However, when it is known that a collection of blocks will be retrieved sequentially, as in the applications considered here, then it is advantageous to store the file contiguously. To demonstrate the significance of this factor, Gray and Reuter (1993a) report that a typical magnetic disk supported by an adequate I/O subsystem can sustain a data rate of 24 to 40 Mbps as long as it is allowed to move its arm monotonously in one direction. With random block accesses scattered across the disk, at saturation point, one would observe a data rate of 3.2 Mbps from that same disk. (This analysis assumes a block size of 8 kilobytes and a service time of 20 msec to read a block.) This would diminishes the number of simultaneous scientists who can access a database at the same time unless each scientist would be willing to wait for a lengthy delay before the data are retrieved (termed *latency*).

Note that the degradation in system performance is incremental as the system swaps objects in and out of the space provided by a mechanical storage device. To illustrate, the Performance Results section shows that with a naive space management algorithm, the disk starts with a few seeks on behalf of each object and settles on an average of 70 seeks during its later stages of operation. In addition to an increase in the average number of seeks, the variance in this quantity also increases. Of course, the system may periodically reorganize data to ensure the contiguity of blocks that comprise an object. With systems that have down-time—i.e., become unavailable for some duration of time periodically—the reorganization procedure can be activated as an offline activity during this period; however, applications are increasingly intolerant of such down-time. For example, health-care information systems are expected to provide uninterrupted service 24 hours a day, year round. For systems of this sort, the reorganization procedure must be an online process. One may design an effective reorganization process based on the characteristics of the target application; however, it is likely to suffer from the following limitations:

1. The overhead of reorganization can be high if invoked frequently.
2. The reorganization process can respond only when it has detected an undesirable behavior, namely too many seeks. Consequently, the user may observe a lower performance than expected for a while before the reorganization process can remedy the situation.
3. The reorganization process will almost certainly fail in environments where the frequency of access to the objects changes in a manner that the identity of objects resident in a stratum changes frequently. For example, it may happen that, by the time the reorganization process groups the blocks of the object o_x together, the system has already elected to replace o_x with another object that is expected to have a higher number of future references.

One may design a space-management technique that ensures a contiguous layout of each object (e.g., REBATE; see Ghandeharizadeh and Ierardi, 1994). Generally speaking, there is a trade-off between the amount of contiguity guaranteed for the layout of each object on a device at stratum i and the amount of wasted space on that device. For example, a technique that ensures the contiguous layout of each object on a magnetic disk may waste a substantial amount of disk space. This trade-off might be worthwhile if the working set of the target application can become resident on the magnetic disks; otherwise, it would not be worthwhile if the penalty incurred due to an increasing number of references to slower devices at lower strata outweighs the benefits of eliminating disk seeks.

The contributions of this paper are twofold. First, it employs the design of the UNIX Fast File System (McKusick *et al.*, 1984), termed *Standard*, to describe the design of three new space-management policies: Dynamic, REBATE (Ghandeharizadeh and Ierardi, 1994), and EVEREST. While Standard packs objects onto the disk without ensuring the contiguity of each object, both Dynamic and REBATE strive to ensure the contiguous layout of each object. EVEREST, on the other hand, strikes a compromise between the two conflicting goals (contiguous layout vs. wasted space) by approximating a contiguous layout of each object. In a dynamic environment where the frequency of access to the objects evolves over time, the design of Standard, Dynamic, and REBATE benefits from a reorganization process that detects and eliminates an undesirable side effect:

1. Standard benefits because the reorganization process can ensure a contiguous layout of each object once the system has detected too many seeks per request.
2. Dynamic benefits because the reorganization process detects and eliminates its wasted space.
3. REBATE benefits because the reorganization process maximizes the utilization of space by detecting and re-allocating space occupied by the infrequently accessed objects.
4. EVEREST is a preventive technique that does not require a reorganization process.

Second, this chapter identifies the fundamental factors that impact the average service time of the system using alternative space-management policies and models them analytically. These models were verified using a simulation study. They quantify the amount of useful work (transfer of data) and wasteful work (seeks, preventive operations, reorganization, access to slower devices due to wasted space) attributed to a design. The models are independent of those strategies described in this paper and can be employed to evaluate other space-management techniques. Thus, they can be employed by a system designer to quantify the tradeoff of one technique relative to another with respect to a target application and hardware platform.

The rest of this chapter is organized as follows. Section 5.4.2 describes our target hardware platform. In Section 5.4.3, we describe the four alternative space management techniques using this platform. Section 5.4.4 demonstrates the trade-off associated with these techniques using a simulation study. In Section 5.4.5, we develop analytical models that quantify the factors that impact the average service time of a system with alternative strategies. Our conclusions are contained in Section 5.4.6.

5.4.2 Target Environment

In order to focus on alternative techniques to manage the space of a mechanical device, we make the following simplifying assumptions:

1. The environment consists of three strata: memory, disk, and tape library. The service time of retrieving an object from tape is significantly higher than that from magnetic disk.
2. The database resides permanently on the tape. The magnetic disk is used as a temporary staging area for the frequently accessed objects in order to minimize the number of references to the tape.
3. All devices are visible to the user via memory (see Fig. 2). The memory is used as a temporary staging area either to service a pending request or to transfer an object from tape to the magnetic disk drive.
4. The system accumulates statistics on the frequency of access (*heat*; see Copeland *et al.*, 1988) to the objects as it services requests for the users. It employs this past history to predict the future number of references to an object.
5. Each referenced object is retrieved sequentially and in its entirety.
6. Either all or none of an object is resident at a stratum; the system does not maintain a portion of an object resident on a stratum. This assumption is justified in the following paragraphs.

With the assumed architecture, the time required to read the object o_x from a device is a function of the size of o_x, the transfer rate of the device, and the number of seeks incurred when reading o_x. The time to perform a seek may include: (1) the time required for the read-head to travel to the appropriate location containing the referenced data, (2) rotational latency time, and (3) the time required to change the physical medium (when necessary). The number of accesses to a device on behalf of an object depends on the frequency of access to that object (its heat).

Once object o_x is referenced, if it is not disk resident and there is sufficient space to store o_x, then o_x is rendered disk resident. If the disk drive has insufficient space to store o_x, then the system must determine if it should delete one or more objects (victims) from this device in favor of o_x. Generally speaking, the following policy is employed for object replacement. The system determines a collection of least frequently accessed objects (say k of them) whose total size exceeds the size of o_x. If the total heat of these objects ($\sum_{j=1}^{k}$ heat(o_j)) is lower than heat(o_x), then the system deletes these k objects in favor of o_x. As described in Section 5.4.3, this general replacement policy cannot be enforced with all space-management techniques (in particular, REBATE and Dynamic). In those cases, we describe its necessary extensions.

An alternative to the assignment imposed by assumption 6 might be to "stripe" an object across the different strata such that each stratum performs its fair share of the overall imposed work when a request references this object. In the following paragraph, we describe this paradigm and its limitations that justify assumption 6.

With the striping paradigm, each object is striped into $n-1$ fragments with each fragment assigned to a device at stratum $i = 1, \ldots, n-1$ (no fragments are assigned to memory). In order to avoid the situation in which one device is waiting for another while requests wait in a queue, the system can choose appropriate sizes for different fragments of each object so that the service time of each device at stratum $i = 1, \ldots, n-1$ is almost identical. Every time the object is referenced, devices at all strata are activated, each for the same amount of time. Hence, all devices contribute an equal share to the work performed for each request. To illustrate, assume that the rate of data delivery is t for tape, and $4t$ for magnetic disk. Moreover, assume that this delivery rate is computed by considering the overhead of initial and sub-

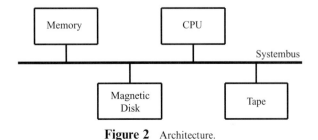

Figure 2 Architecture.

sequent seeks attributed to retrieval of an object from a device. With these parameters, this paradigm assigns $\frac{4}{5}$ of o_x to the magnetic disk and $\frac{1}{5}$ of o_x to the tape. Once o_x is referenced, both devices are activated simultaneously, with each completing its retrieval of the fragment of o_x at approximately the same time.

This paradigm suffers from the following limitations. First, for each object o_x, it requires the size of o_x's disk-resident fragment to be larger than the fragment that is resident on the tape, placing larger fragments on devices that have a smaller storage capacity. If all objects are required to have their disk-resident fragments physically present on the disk drive, then the amount of required disk storage would be larger than that of the tape, resulting in a high storage cost. One may reduce the amount of required disk storage (in order to reduce cost) by rendering a subset of objects tape-resident in their entirety. Once such an object (say, o_x) is referenced, the system employs the tape to retrieve o_x, without the participation of the magnetic disk. During this time, the system may service other requests by retrieving their disk-resident fragments. However, should these requests require access to tape-resident fragments of their referenced objects, then, in effect, the tape has fallen behind and becomes a bottleneck for the entire system; at some point, the memory (as a temporary staging area for these other objects) will be exhausted, and the disk will sit idle and wait for the tape to catch up.

A second limitation of this approach is its requirement that the different devices must be synchronized so that they complete servicing requests simultaneously. This involves a computation of the delivery rate of a device, perhaps in the presence of a variable number of seeks. This synchronization avoids the scenario in which one device waits for another in the presence of pending requests. Such synchrony is difficult to achieve even in an environment that consists of homogeneous devices, such as multiple magnetic disk drives (Gibson, 1992; Patterson et al., 1988). It becomes more challenging in a heterogeneous system, where each device exhibits its own unique physical characteristics. Due to these limitations, we elected to eliminate striping from further consideration for the remainder of this paper.

5.4.3 Four Alternative Space Management Techniques

This section presents four alternative space management techniques: Standard, Dynamic, EVEREST, and REBATE (see Table 1). Standard refers to the most common organization of disk space in current operating systems. We have elected to use the UNIX Fast File System (McKusick et al., 1984), termed UNIX FFS, to represent this class of models; hence, for the remainder of this paper, Standard = UNIX FFS. While Dynamic and EVEREST are two different algorithms, each can be viewed as an extension of the Standard model.

Table 1 Characteristics of Alternative Space-Management Techniques

	Contiguous Layout?	May Require Reorganization?	Wastes Space?
Standard	No	Yes	No
Dynamic	Yes	Yes	Yes
EVEREST	No	No	No
REBATE	Yes	Yes	Yes

REBATE, however, is a more radical departure that partitions the available disk space into regions, where each region manages a unique collection of similarly sized objects. (A region is not equivalent to a cylinder group, as described by UNIX FFS.)

Both Dynamic and REBATE ensure a contiguous layout of each object. Moreover, both techniques illustrate the benefits and difficulties involved in providing such a guarantee while maintaining sufficiently high utilization of the available space. The design of Dynamic, for example, demonstrates that a "smart" algorithm for ensuring contiguity of objects is both difficult to implement and computationally expensive to support. In addition, it wastes space. REBATE, on the other hand, attempts to simplify the problem by partitioning the available disk space into regions. Within each region, the space is partitioned into fixed-sized frames that are shared by the objects corresponding to that region. In general, however, the use of frames of fixed size will increase the amount of wasted space, and the partitioning of resources makes the technique sensitive to changes in the heat of objects. This sensitivity to changing heats motivates the introduction of a reorganization process that detects these changes and adjusts the amount of space allocated to each region and/or the sets of objects managed by each region.

EVEREST, on the other hand, does not ensure a contiguous layout of each object. Instead, it approximates a contiguous layout by representing an object as a small collection of chunks. Each chunk consists of a variable number of contiguous blocks; however, the number of chunks per object and the number of blocks per chunk are a fixed function of the size of an object and configuration parameters. Moreover, the number of chunks is small, bounded logarithmically in the object's size. In contrast to the other strategies, EVEREST is preventive (rather than detective) in its management of space fragmentation. Its advantages include: (1) ease of implementation, (2) a minimal amount of wasted space (comparable to the Standard, in this respect), and (3) no need for an auxiliary reorganization technique. Moreover, the basic parameters of the EVEREST scheme can serve to "tune" the performance of the system, in trading-off time spent in its preventive maintenance and the time attributed to seeks between chunks of resident objects. We describe each technique in turn, starting with Standard.

Standard

Traditionally, file systems have provided a device-independent storage service to their clients. They were not targeted to manage the available space of a hierarchical storage structure; however, they serve as an ideal foundation to describe the techniques proposed in this study. We use the Unix Fast File System (UNIX FFS) as a representative of the traditional file systems. We could not justify the use of Sprite-LFS in this role (and its detailed design) because it is an extended version of UNIX-FFS designed to enhance the performance of file system for small writes (Rosenblum and Ousterhout, 1992); conversely, our target environment assumes a workload consisting of large sequential reads and writes. Similarly, we have avoided file systems that support extent-based allocation (e.g., WiSS; see Chou *et al.*, 1985) because their design targets files that are allowed to grow and shrink dynamically; the objects in our assumed environment are static in size.

With UNIX-FFS, the size of a block for device i determines the unit of transfer between this device and the memory. With objects (files) that are retrieved in a sequential manner, the utilization of a device is enhanced with larger block sizes because the device spends more of its time transfering data (performing useful work) than repositioning its read head (wasteful, or at least potentially avoidable, work). For example, with UNIX FFS, which supports small files, the performance of a magnetic disk drive was improved by a factor of more than two by changing the block size of the file system from 512 to 1024 bytes (McKusick *et al.*, 1984). A disadvantage of using large block sizes is internal fragmentation of space allocated to a block: an object consists of several blocks with its last block remaining partially empty. UNIX FFS minimizes this waste of space as follows. It divides a single block into m fragments; the value of m is determined at system configuration time. Physically, a file is represented as 1 block and at most $m - 1$ fragments. Once a file grows to consist of m fragments, UNIX FFS restructures the space to form a contiguous block from these fragments.

The advantages of this technique include: (1) its simplicity, (2) its ready availability from the commercial arena, (3) its enhancement of the utilization of space by minimizing waste, and (4) its flexibility. It can employ the general replacement policy that was outlined in Section 5.4.2 to respond to changing patterns of access to the objects. A limitation of this technique, however, is its inability to ensure contiguous layout of the blocks of an object on the surface of a device. As the system continues operation and objects are swapped in and out, it will scatter the blocks of a newly materialized object across the surface of the device. This motivates the adoption of a reorganization procedure that will groups the blocks of each object together to ensure its contiguity. Section 5.4.1 sketched the drawbacks inherent in such a procedure.

Dynamic

The method that we term Dynamic is an extension of Standard that attempts to guarantee the contiguity of all disk-resident objects. Similar to Standard, the available disk space is partitioned into blocks of a fixed size; however, whenever an object is rendered disk-resident, Dynamic requires that the sequence of blocks allocated to that object be physically adjacent. The goal of the object-replacement criterion is similar to that described in Section 5.4.1, namely, to maximize the workload of the device, or the total heat contributed by the collection of disk-resident objects. However, the way that Dynamic strives to achieve this objective differs in the following way.

Let o_x be an object requiring b blocks, and assume that o_x is not disk resident. The replacement policy considers all possible contiguous placements of o_x on the disk. If there is some free region that contains b free blocks, then o_x can be made disk resident in this region, and the workload of the set of disk-resident objects increases. On the other hand, if no such free region exists, then it must be the case that every sequence of b contiguous blocks contains all or part of some other resident objects. To be specific, let us fix some sequence of b blocks. Assume that these blocks contain all or part of objects $o_i, ..., o_j$, together with zero or more free blocks. If Dynamic were to make o_x resident in these blocks, the disk-resident copies of these objects would be destroyed, in whole or in part. However, because we have assumed that no objects may reside partially on the disk, whenever a single block occupied by a resident object is overwritten this object is destroyed in its entirety. To determine how the workload might change if o_x is made resident in these blocks, we would like to quantify the amount of work contributed by the current configuration and compare that to the work expected from the proposed change. To do this we define work as follows.

Definition 1

If o_x is an object, then $work(o_x) = heat(o_x) \times size(o_x)$.

The definition captures the idea that part of the disk's workload that may be attributed to requests for object o_x is not merely a function of its heat, but also depends on the amount of time used by the disk to service these requests. This in turn depends upon the object's size. During any period of time during which the objects' heats remain fixed, one expects that this time will be proportional to $work(o_x)$, for each o_x that is actually disk-resident, if we neglect the initial seek for each access to o_x.

To illustrate why work, rather than heat alone, is required by Dynamic's replacement policy, consider the following example. An object o_x of heat $\frac{1}{10}$ requires 100 blocks to become disk resident. On the disk, there is a region of 100 contiguous blocks in which 90 are free and 10 are occupied by an object o_y of heat $\frac{1}{5}$. On the one hand, we can expect that object o_y will receive twice as many requests as object o_x. On the other hand, suppose that the time required to service a single request for o_y is t (neglecting the initial seek). Then each request for o_x requires $10t$ time. Based on these heats, we can expect that about one in ten requests will reference o_x and one in five will access o_y. So, over a sufficiently long sequence of requests, one expects that:

$$\frac{\text{time servicing requests for } o_x}{\text{timeservicing requests for } o_y} = \frac{\frac{1}{10} \times 10t}{\frac{1}{5}t} = \frac{work(o_x)}{work(o_y)} = \frac{5}{1}.$$

Materializing o_x in this region will thus increase the expected workload of the disk.

Dynamic's replacement policy may be stated succinctly as follows. On each request for an object (o_x) that is not disk resident, Dynamic considers all sequences of blocks where o_x may be placed. For each possible placement, it evaluates the expected change in the workload of the disk. If materialization of o_x will increase this quantity, Dynamic stores o_x in the region that maximizes the workload of the disk; otherwise, o_x is not materialized.

When Dynamic considers placing o_x in a sequence of b blocks, it first evaluates the work contributed by the current residents $(o_i,..., o_j)$ of those blocks. We define this quantity,

$$\sum_{k=i}^{j} work(o_k),$$

to be the work associated with these blocks. Rendering o_x disk resident by overwriting these objects would: 1) increase the workload of the disk by $work(o_x)$, and (2) reduce the workload by the current work associated with objects $o_i,..., o_j$. Hence, the expected change in the workload of the disk will be

$$work(o_x) - \sum_{k=i}^{j} work(o_k). \qquad (1)$$

If Equation 5.4.1 is positive, then Dynamic materializes o_x as it increases the workload of the disk.

The algorithm that Dynamic uses to determine when and where to materialize an object o_x is a straightforward scan of the disk—or rather a memory-resident data structure that records the layout of the current disk-resident population. To illustrate, assume that o_x needs 100 blocks to become disk resident. In order to maximize the device utilization, Dynamic must find the 100 contiguous blocks on the device that contribute the least to the device workload. Conceptually, this can be achieved by placing a *window* of 100 blocks at one end of the device and calculating the total workload of all objects that can be seen through this window. The window then slides down the length of the disk. Every time that the set of objects visible through this window changes, the visible workload is recalculated, and the overall minimum value m is recorded. After the entire disk is scanned, m is compared to $work(o_x)$, and if $work(o_x) > m$, then o_x is materialized in that sequence of blocks with associated workload m;. Otherwise, o_x is not materialized on the disk.

The actual calculation can be simplified somewhat by keeping an appropriate memory-resident image of the disk's organization. For this, we employ a list of intervals. Each interval corresponds either (1) to some sequence of blocks occupied by a single object, or (2) to a maximal contiguous sequence of free blocks. All intervals are annotated with their size and their resident object (when the blocks are not free). When a request is made for an object o_x requiring b blocks, Dynamic begins by gathering intervals from the head of the list until at least b blocks have been accumulated. Say this window consists of intervals $I_1,..., I_j$. The total workload of the objects represented among these intervals is recorded. Then, to slide the window to its next interval, the first interval I_1 is omitted, zero or more of the intervals $I_{j+1}, I_{j+2}, \ldots$ are added to the window, until it again contains at least b blocks. The process is repeated across the entire list, while retaining a pointer to the window of minimum workload. It is easy to see that the entire algorithm is linear in the number of disk-resident objects d, as the number of intervals (free and occupied) is no more than $2d + 1$, and each interval is added to and removed from the window at most once.

The advantage of this procedure is its ability to guarantee the contiguous layout of objects. In addition, similar to Standard, it always uses the most up-to-date heat information in making decisions concerning disk residency, so its disk configuration is adaptive and responds to changing access patterns. It uses the heat statistics to "pack" the hottest objects onto the disk. Colder objects are removed from the disk when it is found that hotter objects can increase the disk's workload, but each of these decisions must be made in a "greedy," local manner by considering objects as they are requested. The decisions are further constrained by the current organization of the disk, as Dynamic does not change the layout of those objects that remain disk resident. (More global reorganization of this sort may be effected by an auxiliary reorganization policy.)

Nevertheless, Dynamic can suffer from the following limitations. First, it will almost certainly waste disk space because it does not permit the discontinuous layout of an object. Smaller cold or free sequences of blocks can become temporarily unusable when sandwiched between two hot objects. In effect, the method is restricted in its later placement of data by the current layout, which in turn can evolve in an unpredictable manner. Moreover, Dynamic optimizes the workload of the disk by considering only a local (greedy) perspective; hence, it may perform wasteful work. For example, it may render an object o_x disk resident, only to overwrite it with a hotter object soon thereafter. Of course, as in the case of Standard, Dynamic can also be augmented with a reorganization scheme that attempts to optimize the layout of disk-resident objects from a more global perspective. Such a reorganization process would be subject to the same limitations as outlined in Section 5.4.1. Finally, we note that the algorithm discussed above (for determining whether an object should be materialized and where it should be placed), although linear in the number of disk-resident objects, may be time consuming when the number of objects is large. This may add a significant computational overhead to every request.

EVEREST

EVEREST is an extension of Standard designed to approximate a contiguous layout of each object on the disk drive. Its basic unit of allocation is a block, also termed *section,* of height 0. Within the EVEREST scheme, these blocks can be combined in a tree-like fashion to form larger, contiguous sections. As illustrated in Fig. 3, only sections of size(block) $\times$ B^i (for $i \geq 0$) are valid, where the base B is a system configuration parameter. If a section consists of B^i blocks, then i is said to be the height of the section. In general, B height i sections (physically adjacent) might be combined to construct a height $i + 1$ section.

To illustrate, the disk in Fig. 3 consists of 16 blocks. The system is configured with $B = 2$. Thus, the size of a section may vary from 1, 2, 4, 8, up to 16 blocks. In essence, a binary tree is imposed upon the sequence of blocks. The maximum height, given by $N = \lceil \log_B(\lfloor \frac{\text{Capacity}}{\text{size(block)}} \rfloor) \rceil$, is 4. With this organization imposed upon the device, sections of height $i \geq 0$ cannot start at just any block number, but only at offsets that are multiples of B^i. This restriction ensures that any section, with the exception of the one at height N, has a total of $B - 1$ adjacent *buddy* sections of the same size at all times. With the base 2 organization of Fig. 3, each block has one buddy. This property of the hierarchy of sections is used when objects are allocated, as described below.

ORGANIZATION AND MANAGEMENT OF THE FREE LIST

With EVEREST, a portion of the available disk space is allocated to objects. The remainder, should any exist, is free. The sections that constitute the available space are handled by a memory-resident *free list*. This free list is actually maintained as a sequence of lists, one for each section height. The information about an unused section of height i is enqueued in the list that handles sections of that height. In order to simplify object allocation, the following *bounded list length property* is always maintained:

PROPERTY 1

For each height $i = 0, \ldots, N$ at most $B - 1$ free sections of i are allowed.

Informally, the above property implies that whenever there exists sufficient free space at the free list of height i, EVEREST *must* compact these free sections into sections of a larger height.

ALLOCATION OF AN OBJECT

Property 1 allows for straightforward object materialization. The first step is to check whether the total number of blocks in all the sections on the free list is either greater than or equal to the number of blocks, denoted no-of-blocks(o_x), that the new object o_x requires. If this is not the case, one or more victim objects are elected and deleted. (The procedure for selecting a victim is the same as that described in Section 5.4.2. The deletion of a victim object is described further below.) Assuming at this point that there is enough free space available, o_x is divided into its corresponding sections according to the following scheme. First, the number $m = $ no-of-blocks(o_x) is converted to base B. For example, if $B = 2$, and no-of-blocks(o_x) $= 13_{10}$, then its binary representation is 1101_2. The full representation of such a converted number is $m = d_{j-1} \times B^{j-1} + \ldots + d_2 \times B^2 + d_1 \times b^1 + d_0 \times B^0$. In our example, the number 1101_2 can be written as $1 \times 2^3 + 1 \times 2^2 + 0 \times 2^1 + 1 \times 2^0$. In general, for every digit d_i that is nonzero, d_i sections are allocated from height i of the free list on behalf of o_x. In our example, o_x requires 1 section from height 0, no sections from height 1, 1 section from height 2, and 1 section from height 3.

For each object, the number k of contiguous pieces is equal to the number of ones in the binary representation of m, or with a general base $\sum_{i=0}^{j} d_i$ (where j is the total number of digits). Note that k is always bounded by $B \lceil \log_B m \rceil$. For any object, k defines the maximum num-

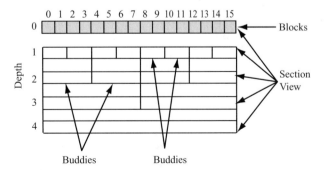

Figure 3 Physical division of disk space into blocks and the corresponding logical view of the sections with an example base of $B = 2$.

ber of disk seeks required to retrieve that object. (The minimum is 1 if all k sections are physically adjacent.) A complication arises when no section at the correct height exists. For example, suppose that a section of size B^i is required, but the smallest section larger than B^i on the free list is of size B^j ($j > i$). In this case, the section of size B^j can be split into B sections of size B^{j-1}. If $j - 1 = i$, then $B - 1$ of these are enqueued on the list of height i and the remainder is allocated. However, if $j - 1 > i$ then $B - 1$ of these sections are again enqueued at level $j - 1$, and the splitting procedure is repeated on the remaining section. It is easy to see that, whenever the total amount of free space on these lists is sufficient to accommodate the object, then for each section that the object occupies, there is always a section of the appropriate size, or larger, on the list. The splitting procedure sketched above will guarantee that the appropriate number of sections, each of the appropriate size, will be allocated and that Property 1 is never violated.

The design of EVEREST is related to the buddy system proposed in Knowlton (1965) and Lewis and Denenberg (1991) for an efficient main memory storage allocator (DRAM). The difference is that EVEREST satisfies a request for b blocks by allocating a number of sections such that their total number of blocks equals b. The storage allocator algorithm, on the other hand, will allocate *one* section that is rounded up to $2^{\lceil \lg b \rceil}$ blocks, resulting in fragmentation and motivating the need for either a reorganization process or a garbage collector (Gray and Reuter, 1993b).

DE-ALLOCATION OF AN OBJECT

When the system elects that an object must be materialized and there is insufficient free space, then one or more victims are removed from the disk. Reclaiming the space of a victim requires two steps for each of its sections. First, the section must be appended to the free list at the appropriate height. The second step is to ensure that Property 1 is not violated. Therefore, whenever a section is enqueued in the free list at height i and the number of sections at that height is equal to or greater than B, then B sections must be combined into one section at height $i + 1$. If the list at $i + 1$ now violates Property 1, then once again space must be compacted and moved to section $i + 2$. This procedure might be repeated several times. It terminates when the length of the list for a higher height is less than B.

Compaction of B free sections into a larger section is simple when the sections are all adjacent to each other; in this case, the combined space is already contiguous. Otherwise, the system might be forced to exchange one occupied section of an object with one on the free list in order to ensure contiguity of an appropriate sequence of B sections at the same height. The following algorithm achieves space contiguity among B free sections at height I:

1. Check if there are at least B sections for height i on the free list. If not, stop.
2. Select the first section (denoted s_j) and record its block number (i.e., the offset on the disk drive). The goal is to free $B - 1$ sections physically adjacent to s_j.
3. Calculate the block numbers of s_j's buddies. EVEREST's division of disk space guarantees the existence of $B - 1$ buddy sections physically adjacent to s_j.
4. For every buddy s_k, $k \leq 0 \leq B - 1$, $k \neq j$, if it exists on the free list then mark it.
5. Any of the s_k unmarked buddies currently store parts of other object(s). The space must be rearranged by swapping these s_j sections with those on the free list. Note that for every buddy section that should be freed there exists a section on the free list. After swapping space between every unmarked buddy section and a free list section, enough contiguous space has been acquired to create a section at height $i + 1$ of the free list.

To illustrate, consider the organization of space in Fig. 4a (see color insert). The initial set of disk-resident objects is $\{o_1, o_2, o_3\}$, and the system is configured with $B = 2$. In Fig. 4a, two sections are on the free list at height 0 and 1 (addresses 7 and 14, respectively), and o_3 is the victim object that is deleted. Once block 13 is placed on the free list in Fig. 4b (see color insert), the number of sections at height 0 is increased to B and it must be compacted according to Step 1. As sections 7 and 13 are not contiguous, section 13 is elected to be swapped with section 7's buddy, section 6 (Fig. 4c; see color insert). In Fig. 4d (see color insert), the data of section 6 are moved to section 13, and section 6 is now on the free list. The compaction of sections 6 and 7 results in a new section with address 6 at height 1 of the free list. Once again, a list of length two at height 1 violates Property 1 and blocks (4,5) are identified as the buddy of section 6 in Fig. 4e (see color insert). After moving the data in Figure 4f from blocks (4,5) to (14,15), another compaction is performed with the final state of the disk space emerging as in Fig. 4g (see color insert).

Once all sections of a de-allocated object are on the free list, the iterative algorithm above is run on each list, from the lowest to the highest height. The previous algorithm is somewhat simplified because it does not support the following scenario. A section at height i is not on the free list; however, it has been broken down to a lower height (say, $i - 1$) and not all subsections have been used. One of them is still on the free list at height $i - 1$. In these cases, the free list for height $i-1$ should be updated with care because those free sections have moved to new locations. In addition, note that the algorithm described above actually performs more work than is strictly necessary. A single section of a small height, for example, may end up being read and written several times as its section is combined into larger and larger sections. This can be eliminated in the following manner. The algorithm is first

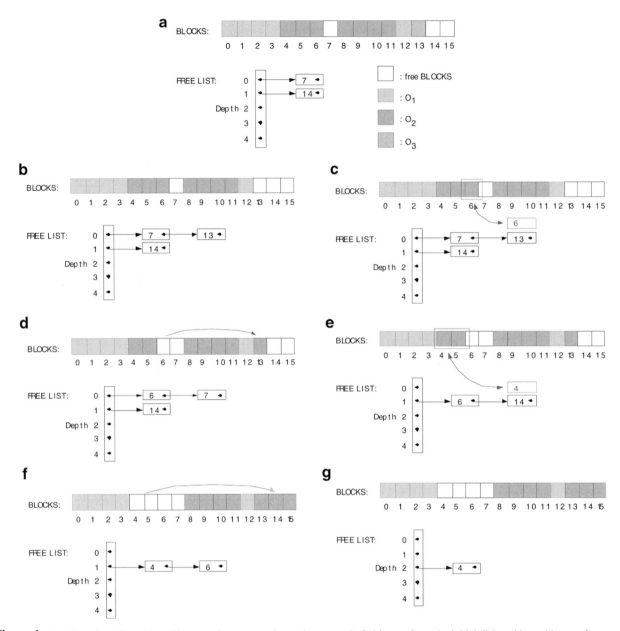

Figure 4 De-allocation of an object. The example sequence shows the removal of object o_3 from the initial disk resident object set $\{o_1, o_2, o_3\}$. Base two, $B = 2$. (a) Two sections are on the free list already (7 and 14) and object o_3 is de-allocated. (b) Sections 7 and 13 should be combined, however they are not contiguous. (c) The buddy of Section 7 is 6. Data must move from 6 to 13. (d) Sections 6 and 7 are contiguous and can be combined. (e) The buddy of Section 6 is 4. Data must move from (4, 5) to (14, 15). (f) Sections 4 and 6 are now adjacent and can be combined. (g) The final view of the disk and the free list after removal of o_3. (See color plates.)

performed "virtually"—that is, in main memory, as a compaction algorithm on the free lists. Once completed, the entire sequence of operations that have been performed determines the ultimate destination of each of the modified sections. These sections are then read and written directly to their final locations. One may observe that the total amount of data moved (read and then written) during any compaction operation is no more than $B - 1$ times the total amount of free space on the list. For example, when $B = 2$, in the worst case the number of bytes written due to preventive operations is no more than the number of bytes materialized, in an amortized sense. One may expect, however, this number to be smaller for a collection of objects of varying sizes.

The value of B impacts the frequency of preventive operations. If B is set to its minimum value (e.g., $B = 2$), then preventive operations would be invoked frequently because every time a new section is enqueued there would be a 50% chance for a height of the free list to consist of two sections (violating Property 1). Increasing the value of B would therefore "relax" the system because it reduces the probability that an insertion to the free list would violate Property 1. However, this would increase: (1) the number of seeks observed when retrieving an object, and (2) the expected number of bytes migrated per preventive operation. For example, at the extreme value of $B = n$ (where n is the total number of blocks), the organization of blocks will consist of two levels, and for all practical purposes EVEREST reduces to a variant of Standard.

The design of EVEREST suffers from the following two limitations. First, it incurs a fixed number of seeks (although few) when reading an object. Second, the overhead of its preventive operations may become significant if many objects are swapped in and out of the disk drive (this happens when the working set of an application cannot become resident on the disk drive). The primary advantage of the elaborate object de-allocation technique of EVEREST is that it avoids internal and external fragmentation of space as described for traditional buddy systems (see Gray and Reuter, 1993b).

IMPLEMENTATION CONSIDERATIONS

In an actual implementation of EVEREST, it might not be feasible to fix the number of blocks as an exact power of B. Rather, one would generally fix the block size of the file system in a manner dependent upon physical characteristics of both the device and the objects in the database. This is possible with some minor modifications to EVEREST. The most important implication of an arbitrary number of blocks is that some sections may not have the correct number of buddies ($B - 1$ of them); however, we can always move those sections to one end of the medium—for example, to the side with the highest block offsets. Then, instead of choosing the first section in Step 2 in the object de-allocation algorithm, one should choose the one with the lowest block number. This ensures that the sections towards the critical end of the disk—which might not have the correct number of buddies—are never used in both Steps 4 and 5 of the algorithm.

REBATE

REBATE (Ghandeharizadeh and Ierardi, 1994) partitions the available space of a device i into g regions ($G_1, G_2, \ldots, G_g$) by analyzing: (1) the storage capacity of the device, termed C_i, and (2) the size and frequency of access to each object in the database, termed size(o_x) and heat(o_x), respectively (Copeland et al., 1988). Each region G_j occupies a contiguous amount of space. The amount of space allocated to region G_j (termed space(G_j)) is determined such that the overall utilization of the space is maximized; that is, the probability of a byte from a region containing useful data—data most likely to be accessed in the future—is maximized and is approximately the same for all regions. Each region manages a set of unique objects (OBJ(G_j) = $\{o_1, o_2, \ldots, o_k\}$. The minimum and maximum size of an object managed by a region G_j (termed min(G_j) and max(G_j), respectively) are unique. The space allocated to region G_j is split into l_j fixed-size frames, where $l_j = \lfloor \frac{space(G_j)}{max(G_j)} \rfloor$. All objects whose sizes lie in the range from min(G_j) to max(G_j) are managed by region G_j and compete for its frames. REBATE wastes disk space when the size of a frame is larger than its occupying object. In order to minimize this waste, the regions are constructed so that the size of all objects in OBJ(G_j) is approximately the same. (Hence, REBATE attempts to minimize the value max(G_j)−min(G_j) for each region G_j.) If o_x maps to region G_j and does not currently occupy a frame of G_j, then the system compares work(o_x) with the other objects that currently occupy a frame of G_j. It replaces the object with the least imposed work (say, object o_y) only if work(o_y) < work(o_x). Otherwise, o_x does not become resident on this stratum. Further details are presented in Ghandeharizadeh and Ierardi (1994), which also provides an efficient dynamic programming algorithm for constructing optimal *region-based partitions* when accurate data on the heat of objects are available.

With REBATE, the system might be required to either construct new regions or re-allocate space among the existing regions for at least two reasons. First, a new object might be introduced whose size is larger than the size of objects that constitute the present database. In this case, none of the existing frames can accommodate this object, so a new region of larger frames must be introduced. Second, the access pattern to the objects might evolve in a manner that dictates the following: one or several of the current regions deserves more space than already allocated to it, while other regions deserve proportionally less space. Hence, the design of REBATE includes a reorganization technique that periodically proposes a new organization of the regions and renders it effective only if its expected improvement in the actual hit ratio observed by the device (that is, the effective utilization of its space) exceeds a preset threshold. This online reorganization procedure is described further in Ghandeharizadeh and Ierardi (1994).

REBATE may suffer from two limitations. First, in a system where the objects sizes are not naturally clustered into like-sized classes, REBATE may waste space. This under-utilization of available space in turn increases the frequency of access to the tertiary storage device (when compared to Standard, which packs objects on the disk drive without ensuring their contiguity). Yet, even in the case where such natural classes exist, determining a truly optimal partition of the device's space among regions is an *NP*-hard problem. REBATE's compromise—settling for region-based partitions—may in fact be suboptimal in certain worst-case scenarios (Ghandeharizadeh and Ierardi, 1994). Overall, whether REBATE outperforms Standard depends on a number of factors, including the amount of wasted space, the size of the working set of an application relative to the capacity of the device, the overhead attributed to performing seeks when retrieving an object, and the penalty incurred in accessing the device at the next stratum. (With our assumptions, the impact of last factor is significant.)

Second, REBATE partitions the available disk space among multiple regions, necessitating a reorganization process when deployed for a database where the frequency of access to its objects varies dynamically over time. This reorganization procedure is undesirable for several reasons. First, the overhead of reorganization can be expensive if it evaluates alternative layout of regions (by invoking the REBATE algorithm) too frequently. Second, the reorganization procedure can only respond after it has detected a lower hit ratio than is expected. Consequently, the user must observe a *higher* latency than expected for a while before the reorganization procedure can recognize and remedy the situation. Third, the reorganization procedure will almost certainly fail in environments where the frequency of access to the objects changes in a manner that forces a frequent reallocation of space among regions (a "ping-pong" effect, in which space is shuffled back and forth to follow the regions of hottest objects). When these frequencies change too often, an even worse situation arises where the reorganization procedure instantiates a new layout corresponding to heat values that have already changed. The system is thus trying to "predict the future;" yet, its only guide in this task is the statistical information that it can accumulate. When these quantities are unreliable, or show large and frequent variation, these predictive methods are bound to fail. In this circumstance, the use of an online reorganization method can itself cause further degradation in the system's performance.

5.4.4 Performance Evaluation

To quantify the performance trade-offs of Standard, Dynamic, EVEREST, and REBATE, we conducted a number of simulation studies. The simulation model evolved over a period of 12 months. During this period, we conducted many experiments and gained insights into: (1) the factors that impact the performance of the system with alternative space management techniques, (2) the experimental design of the simulator, and (3) the results that were important to present. Indeed, the design of EVEREST was introduced once we had understood the trade-offs associated with Standard, Dynamic, and REBATE.

Almost all the components of the simulator are straightforward, except for the Driver module that generates requests, with each request referencing an object. It is complicated because it employs several distributions to generate the requests pertaining to an arbitrary pattern of access to the objects. An arbitrary request generator was desirable for several reasons. First, it eliminated the possibility of bias towards a technique. Second, we believe that it models reality because the pattern of access to the objects is typically unknown in real applications. Third, using this model, it is straightforward to evaluate the accuracy of the statistical modules for estimating heats. We observed that the statistics module is fairly accurate.

The design of the Driver is based on the assumption that the heat of objects evolves gradually. For example, one may sample the distribution of access to the objects at two different points in time and observe that 80% of requests are directed to 20% of the objects for the first sample (an 80–20 access pattern) and a 90–10 access pattern for the second sample (90% of accesses are directed to 10% of the objects). Our assumption states that the heat evolved incrementally and that at some point it was more uniform than both 80–20 and 90–10 access patterns. The Driver models this paradigm by using a normal distribution to model each of 80–20 and 90–10 access patterns. Next, it migrates the heat of objects in 10 steps from 80–20 to 90–10. After the first interval, the distribution is more uniform than both 80–20 and 90–10. It is most uniform at the fifth interval. Starting with the sixth interval, the Driver starts its progress towards a 90–10 access pattern. By the tenth interval, the Driver is producing requests to the objects based on a 90–10 access pattern. (The details of this are provided ibelow.) We investigated simpler designs for generating requests (e.g., changing the distribution of access from 80–20 to 90–10 in one step) and observed no change in the final conclusions.

Early on, we employed the average service time observed with alternative space management techniques as the criterion to compare one strategy with another. This was a mistake because it hid the factors that impact the performance of the system with alternative techniques by associating weights to them. (These weights describe the physical characteristics of the devices in the hierarchy.) By focusing on these factors (instead of the average service time), we were able to develop analytical models that incorporate the physical parameters of a system to compute the average service time (see Section 5.4.5). These analytical models were validated using the simulation study (with less than 2% margin of error). Using these models, the system designer may choose the value of parameters corresponding to a target hardware platform to evaluate alternative techniques.

Simulation Model

The simulation model consists of four components: Driver, Space Management, Device Emulation, and Heat Statistics (see Fig. 5). The *Driver* module generates a synthetic workload by constructing a queue of object requests. The *Space Management* module realizes the different space-management algorithms and interfaces with the *Device Emulation* module. The Device Emulator models a magnetic disk drive with its seeks and transfer

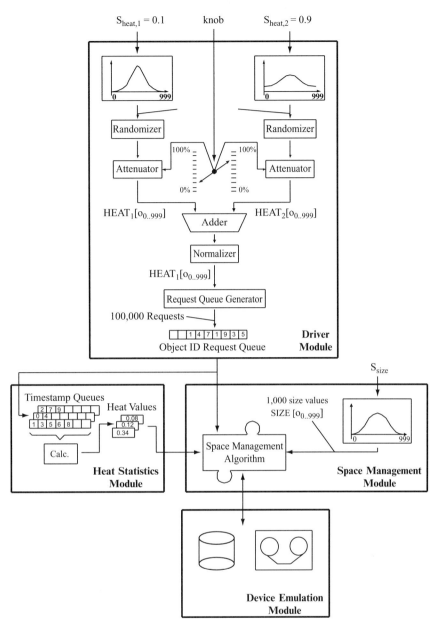

Figure 5 Block diagram of the simulation model.

times (see Ruemmler and Wilkes, 1994) and a simple tertiary device (i.e., a tape drive). Finally, the *Heat Statistics* module gathers information about the sequence of requests and compiles these data into a heat value for every object. The estimated heat value is then used by the Space Management module to decide which objects should be disk resident. The simulator was implemented using the C programming language.

The Driver module uses three input parameters to generate a sequence of object requests: two heat distributions ($\sigma_{heat,1}$ and $\sigma_{heat,2}$), and a *knob*. The knob determines the role of a given σ_{heat} in generating the requests. As illustrated in Fig. 5, when knob is equal to 0% for $\sigma_{heat,2}$, its value is 100% for $\sigma_{heat,1}$ (in this case, the value chosen by $\sigma_{heat,1}$ determines the final queue of requests).

To describe how requests are generated using a normal distribution, assume that knob is equal to 0% for $\sigma_{heat,2}$. In this case, object heats (or frequency of their appearance in the request queue) obey a normal distribution with a mean of zero and a standard deviation of $\sigma_{heat,1}$. Objects are viewed as uniformly distributed sample points in the interval $[-1,1]$. With a small value of $\sigma_{heat,1}$, the access pattern is skewed, and most accesses will be concentrated on a smaller subset of the objects. With larger values of $\sigma_{heat,1'}$ the heats of the objects become more uniform. As a rule of thumb, approximately 60% of the heat will be concentrated on a $\sigma_{heat'}$ fraction of all objects, and 90% of the heat on a $\sigma_{heat,1'}$ fraction. For example, when $\sigma_{heat,1'} = 0.1$, then approximately 98% of the heat is concentrated in 25% of the objects. As the value of $\sigma_{heat,1}$ increases, this ratio changes rapidly: At $\sigma_{heat,1'} = 0.2$, nearly 78% of the heat is concentrated in 25% of the objects, while at $\sigma_{heat,1} = 0.4$, less than 50% of the heat is concentrated in 25% of the objects. For values of $\sigma_{heat,1} \geq 1$, this distribution is nearly uniform. Objects are assigned heats in a purely random manner; there is no intentional correlations between the size and heat of objects.

The Driver uses two heat distributions to generate the final queue of requests. Both distributions are based on the same normal distribution with a mean of zero. The knob controls to what extent each of the two heat distributions is used in generating the request queue. When the value of a knob changes, the heat essentially migrates from one set of objects to another. At one end, 0% of the $\sigma_{heat,1}$ curve and 100% of the $\sigma_{heat,2}$-curve are in effect. At the other end, the percentages are reversed: 100% of the $\sigma_{heat,1}$-curve and 0% of the $\sigma_{heat,2}$-curve are used. For every object o_x, $heat_1(o_x)$ and $heat_2(o_x)$ are added and stored in the array $heat(o_{0...999})$. This array of results is further normalized such that $\sum_{j=0}^{999} heat(O_j) = 1$. Finally, the request queue is generated from the $heat(0_{0...;999})$ array.

The Space Management module services the requests that are generated by the Driver module. It has access to a synthetic database that consists of 1000 objects. A normal distribution of the sizes guarantees a fixed-average object size for all experiments (controlled by the input parameter σ_{size}). Each different space-management algorithm that implements Standard, Dynamic, EVEREST, and REBATE policies is a plug-in module.

The Device Emulation part of the simulator consists of data structures and routines to emulate a magnetic disk drive and a tertiary device. We employed the analytical models of Ruemmler and Wilkes (1994) to represent the seek operation, the transfer rate, and latency of a magnetic disk drive. The tertiary device is simplified and only its transfer rate is modeled. This module is also responsible for gathering the statistics that are used to compare the effectiveness of the different space-management policies (the number of seeks performed on behalf of an object, the average percentage of disk space that remains idle).

The Space Management module does not have access to any heat information that exists in the Driver and must learn about it by gathering statistics from the issued requests. The learning process is as follows. The module keeps a queue of time stamps for every object as well as an estimated heat value. All the queues are initially empty and the heat values are uniformly set to $\frac{1}{n}$, where n is the total number of objects. Upon the arrival of a request referencing object o_x, the current time is recorded in the queue of object o_x. Whenever the time stamp queue of object o_x becomes full, the heat value of that object is updated according to:

$$\text{heat}_{\text{new}}(o_x) = (1-c) \times \frac{1}{\frac{1}{K} \times \sum_{i=1}^{k-1}(t_{i+1} - t_i)} + c \times \text{heat}_{\text{old}}(o_x)$$

where K is the length of the time stamp queue (set to 50), c is a constant between 0 and 1 (set to 0.5), and t_x is one individual time stamp. After the update is completed, the queue of this object is flushed and new time stamps can be recorded. This approach is similar to the concept of the *Backward K-distance* used in the LRU-K algorithm (O'Neil et al., 1993). The two schemes differ in three ways. First, the heat estimates are not based on the interval between the first and the last time stamp in the queue but are averages over all the intervals. Second, the heat value of an object o_x is only updated when the time-stamp queue of o_x is full, therefore reducing overhead. And, third, the previous heat value $\text{heat}_{\text{old}}(o_x)$ is by a fraction of c taken into account when $\text{heat}_{\text{new}}(o_x)$ is calculated. The above measures balance the need for smoothing out short-term fluctuations in the access pattern and guaranteeing responsiveness to longer term trends.

Experimental Design

The two simulation model input parameters $\sigma_{heat,1}$ and $\sigma_{heat,2}$ are used to model how the heat of individual objects might change over time. The relevant parameters of the experiments are summarized in Table 2. The value of $\sigma_{heat,1}$ is always held constant at 0.1. The parameter $\sigma_{heat,2}$ is initially set to 0.17. The value of the knob is initialized to 100% of $\sigma_{heat,1}$. After 100,000 requests the value of knob is decremented by 10% (and therefore the ratio of requests corresponding to $\sigma_{heat,2}$ increased from 0% to 10% and that of $\sigma_{heat,1}$ to 90%). This process continues with the knob value decreasing by 10% after every 100,000 requests until its value reaches 0% (at this point, all requests correspond to the $\sigma_{heat,2}$ distribution). The heat represented by $\sigma_{heat,1} = 0.1$ is now re-distributed by invoking the randomization routine. At this point the value of knob starts to increase by 10% increments. Each time a new queue of requests is generated. This procedure is repeated many times. At extreme values of knob for $\sigma_{heat,1}$ (i.e., 0% and 100% for $\sigma_{heat,1}$), a random number generator is employed to ensure that the identity of frequently accessed objects changes, requiring the system to learn the identity of the frequently accessed objects each time.

The above experiment was repeated a total of 10 times, each time with a different $\sigma_{heat,2}$ parameter. The values used are listed in Table 2 and Fig. 6 illustrates the process.

Performance Results

Fig. 7 presents the number of seeks observed per request that finds the referenced object on the disk drive (a *disk hit*). With an empty disk, Standard lays the

Table 2 Simulation Parameters

Device Parameters	
Disk size	1 GB
Database (also tertiary) size	4 GB
Block size (where applicable)	4 kB
Object Parameters	
Number of objects	1000
Maximum object size	7.9 MB
Minimum object size	0.1 MB
Average object size	4.0 MB
Input Parameters	
σ_{size}	0.3
$\sigma_{heat,1}$	0.1
$\sigma_{heat,2}$	0.17, 0.2, 0.3, 0.4, 0.5, 0.6, 0.7, 0.8, 0.9, 1.0

referenced object contiguously. However, after a few iterations of knobs changing its value, each request observes on the average more than 60 seeks. Dynamic and REBATE ensure a contiguous layout and observe zero seeks per disk hit. As expected, due to its preventive style, EVEREST renders the number of seeks a constant (4.5 in this experiment; this number represents the total seeks required to both service a request observing a disk hit and the preventive operations performed by EVEREST).

Fig. 8 demonstrates the disadvantages of laying out an object contiguously with Dynamic and REBATE. Dynamic wastes 1 to 3% of the available disk space (explained in Dynamic section). REBATE wastes approximately 14% of the disk space due to internal fragmentation of a frame. Both Standard and EVEREST utilize the available space to its fullest potential. They do waste a small fraction of space (less that 0.1%) due to our assumption that an object should be resident in its entirety (no partial materialization of an object is allowed).

Fig. 9 quantifies the overhead attributed to the preventive characteristics of EVEREST. The number of preventive operations performed depends on how frequently the replacement policy is activated to locate and delete victim objects. To illustrate, the peaks in Fig. 9a correspond to the value of $\sigma_{heat,2}$ (knob = 100% for $\sigma_{heat,2}$). As described above (Table 2), the value of $\sigma_{heat,2}$ increases from 0.17 to 1. At $\sigma_{heat,2} = 1$, the distribution of access to the objects is fairly uniform, motivating the replacement policy to delete several objects from the disk in favor of the others. The amount of work (disk activity) performed per preventive operation depends on the degree of fragmentation of sections on the disk drive. Figs. 9b and 9c demonstrate the number of seeks incurred and the amount of migrated data attributed to a preventive operation. While there is significant variation, on the average a preventive operation requires 8 seeks and the migration of 1 MByte of data (note, this is 25% of the average requested object size). Once amortized across all the requests, this overhead becomes negligible (as illustrated by the number of incurred seeks in Fig. 7).

Fig. 9 demonstrates the following. First, the number of preventive operations should be a small fraction of the total number of requests serviced by a device. This clearly states the need for the existence of a working set. Otherwise, the number of objects replaced may become significant and, in turn, cause the overhead attributed to the preventive nature of EVEREST to dominate the average service time of the device. Second, the latency incurred by a request might be variable depending on: (1) whether a preventive operation is invoked, and (2) the amount of work performed by this preventive operation.

Finally, we compared the obtained results with the scenario where the system was allowed access to the queue of requests and could compute the heat of

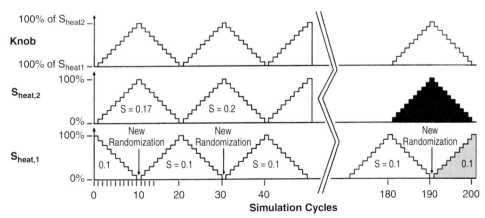

Figure 6 Values over time of three of the input parameters for the simulation experiments.

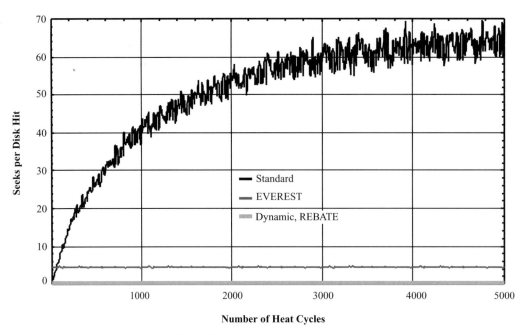

Figure 7 Number of seeks per disk hit.

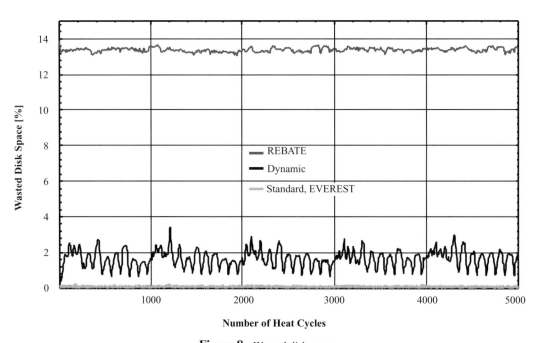

Figure 8 Wasted disk space.

the objects with 100% accuracy (as compared to employing the Heat Statistics Module to learn the heat information; see Simulation Model section for the details of this module). The obtained results were almost identical, demonstrating that: (1) the technique employed by the Heat Statistics Module has no impact on the obtained results, and (2) the employed technique to compute the heat statistics is effective in our experimental design.

5.4.5 Analytical Models

In this section, we develop analytical models that approximate the average service time of a system based on: (1) its physical characteristics, and (2) the fundamental factors that impact the performance of the system with the alternative space management techniques. These abstract models are useful because a system designer may manipulate the value of their parameters

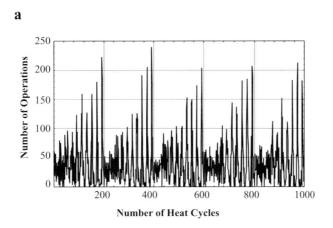

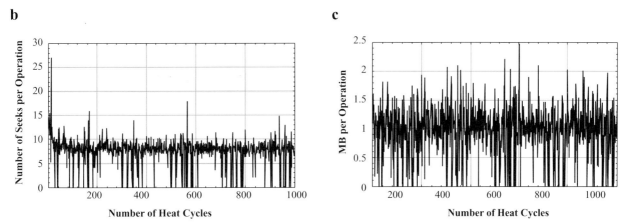

Figure 9 Overhead attributed to the preventive characteristic of EVEREST. (a) The number of preventative operations. (b) The number of seeks per preventative operation. Each migration of a section requires two seeks (1 read + 1 write). (c) Mbytes of data migrated per preventive operation.

to understand the benefits of one strategy as compared to another. They have been validated using the experimental simulation model.

The performance of the system with a space-management strategy is impacted by the following factors:

1. Average number of seeks incurred when reading an object (F) and the average time to perform a seek (S_{Seek})
2. Number of preventive operations performed (P) and the average time to perform one such operation (S_{Prev})
3. Number of bytes reorganized (U) and the number of seeks attributed to the reorganization procedure (E)
4. The amount of wasted space (W) and its expected hit ratio.

We analyze the average service time of the system with a given strategy to service a fixed number of requests and quantify what fraction of this service time is attributed to each of these factors (Fig. 10). One or more of these factors might be nonexistent for a strategy. For example, REBATE incurs the overhead of neither preventive operations nor the seeks attributed to retrieval of an object. In this case, the value of appropriate parameters will be zero (F and P for REBATE), enabling the model to eliminate the impact of these factors. (Refer to Table 1 for a list of factors attributed to the different space-management techniques described in this paper.) We assume that the system has accumulated the statistics shown in Table 3. We describe each factor and its corresponding analytical model in turn.

The portion of average service time attributed to seeks incurred when reading an object is

$$F \times S_{Seek} \qquad (5.4.2)$$

where F is the average number of seeks performed on behalf of a retrieval from disk, and S_{Seek} is the average service time to perform a seek. These statistics can be gathered as the system services requests.

The portion of average service time attributed to preventive operations is defined as:

$$P \times S_{Prev} \qquad (5.4.3)$$

P defines the average number of preventive operations per disk hit, and S_{Prev} is the average service time to perform a preventive operation.

The amount of time attributed to disk transfer time is a function of the average number of bytes retrieved from

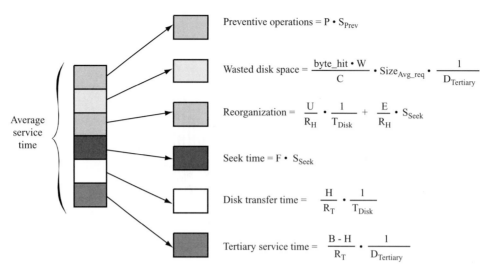

Figure 10 Components of average service time for a single queue of requests.

Table 3 List of Parameters Used by the Analytical Models

Term	Definition
C	Storage capacity of the magnetic disk
R_T	Total number of requests issued during a fixed period of time
R_H	Total number of requests that observe a disk hit during a fixed period of time
$Size_{Avg\_req}$	Average number of bytes retrieved per request
B	Total number of bytes retrieved by R_T requests
H	Total number of bytes found on the disk by R_T requests
P	Average number of preventive operations per disk hit
W	Total number of bytes wasted by a strategy
U	Total number of bytes reorganized (read + write)
E	Total number of seeks attributed to the reorganization procedure
F	Average number of seeks per disk hit
$D_{Tertiary}$	Delivery rate of tertiary storage device (incorporates the seek time of the device)
T_{Disk}	Transfer rate of the disk drive
$byte\_hit$	Fraction of a byte that observed a hit, $byte\_hit = H/B$
S_{Seek}	Average service time for a seek
S_{Prev}	Average service time for a preventive operation

the disk per request ($\frac{H}{R_T}$) and the transfer rate of the disk drive (T_{Disk}):

$$\frac{H}{R_T} \times \frac{1}{T_{Disk}} \quad (5.4.4)$$

Similarly, the amount of time spent transfering data from tertiary is a function of the average number of bytes retrieved from the tertiary per request ($\frac{B-H}{R_T}$) and the delivery rate of the tertiary storage device:

$$\frac{B-H}{R_T} \times \frac{1}{D_{Tertiary}} \quad (5.4.5)$$

The delivery rate of tertiary storage device incorporates the average number of seeks incurred by this device per request, and the overhead of such seeks.

A technique that employs a reorganization process, reads and writes a fixed number of bytes (U) causing the device to incur a fixed number of seek operations (E). This overhead averaged across all requests (R_H) is:

$$\frac{U}{R_H} \times \frac{1}{T_{Disk}} + \frac{E}{R_H} \times S_{Seek} \quad (5.4.6)$$

A technique such as REBATE may waste disk space; however, its impact might be negligible if the wasted space is not expected to have a high hit ratio. Assume the existence of a unit that defines what fraction of each byte on the disk should observe a hit, termed *byte-hit ratio* (its details are presented in the following paragraphs). The wasted space reduces the change in *byte_hit* by a fixed margin:

$$\frac{byte\_hit \times W}{C}.$$

his causes a fixed number of bytes of the average request to be retrieved from the tertiary storage device ($\frac{byte_hit \times W}{C} \times Size_{Avg\_reg}$), and the overhead of reading these bytes can be quantified as:

$$\frac{byte\_hit \times W}{C} \times Size_{Avg\_reg} \times \frac{1}{D_{Tertiary}} \quad (5.4.7)$$

The byte-hit ratio is a function of the size of both the working set of the system and the storage capacity of the disk drive. When the size of the working set of an application is larger than the storage capacity of the disk, every byte becomes valuable because it minimizes the number of bytes retrieved from the tertiary storage device. In this case, byte-hit ratio is defined as $\frac{H}{B}$. When the working set is smaller than the storage capacity of the disk, the probability of a wasted byte observing a hit is a function of the database size, the amount of wasted space, and the pattern of access to the objects. For example, if one assumes that references are randomly distributed across the objects (bytes) that constitute the remainder of database (except for those that are part of the working set), then *byte_hit* might be defined as:

$$\frac{1}{\text{size}(DB) - W}.$$

We verified the analytical models using the simulator. This was achieved as follows. The simulation was invoked for a period of time in order to accumulate the value of parameters outlined in Table 3. Next, the average service time of the system as computed by the simulator was compared with that of the analytical model. In almost all cases, there was a perfect match. The highest observed margin of error was less than 2%. It is important to note that these models should be extended with queuing times in the presence of both multiple users and multiple disk drives (this is a future research direction).

5.4.6 Conclusions

In this chapter, we have studied alternatives in space management for large repositories of objects that are generally retrieved sequentially and in their entirety. These repositories might be found in various multimedia applications (such as video-on-demand servers) and in numerous scientific applications (such as the Brookhaven and Cambridge database of molecular structures). To isolate those factors that contribute significantly to the performance of such a system, we have sampled the space of storage-management policies for a fixed hierarchical architecture. The simulation results, and their analyses, permit one to isolate the trade-offs inherent in various designs: trade-offs between wasted time (seeking) and wasted space; between local, greedy techniques for optimizing a device's workload (as in Standard or Dynamic) and those that impose a more global order on the medium (REBATE or EVEREST); and between detective and preventive strategies for adapting to a changing workload. However, a complete evaluation of these trade-offs is dependent on both the physical characteristics of the system and the target application. For example, the impact of wasted space and wasted time upon the actual workload of the device depends critically on its seek time and bandwidth, block size and average object size, and the size of the resident working set relative to the capacity of the device. Whether a system should impose a global order on its device, and effectively partition its space, may depend upon the changeability of the working set and the expected or observed variance in the heats of objects. Similarly, the policy adopted to organize and reorganize space depends upon characteristics of the tasks for which it is deployed: whether a working set exists, how quickly it changes, and how predictable its evolution is.

While we believe that this study is complete in its treatment of issues that arise with design of strategies to manage the space of a mechanical device, it raises two related research topics that deserve further investigation. First, implementation details of Dynamic, REBATE, and EVEREST are lacking and require further consideration should a system designer elect to employ one of these strategies in a file system. In particular, we intend to investigate the crash-recovery component of these strategies (i.e., it enables the device to recover to a consistent state after a power failure). Second, the management of objects across the different strata of a hierarchical storage structure requires further analysis. In particular, a management technique should decide if it is worthwhile to allow multiple copies of an object with one copy residing at a different stratum of the hierarchy (e.g., one copy on the magnetic disk and a second on the optical disk; see Fig. 1).

References

Bernstein, F., Koetzle, T., Williams, G., Mayer, E., Bryce, M., Rodgers, J., Kennard, O., Himanuchi, T., and Tasumi, M. (1977). The protein databank, a computer based archival file for macromolecular structures. *J. Mol. Biol.*, **112(2)**, 535–542.

Berson, S., Ghandeharizadeh, S., Muntz, R., and Ju, X. (1994). Staggered striping in multimedia information systems, In *Proceedings of the ACM SIGMOD International Conference on Management of Data*.

Carey, M., Haas, L., and Livny, M. (1993). Tapes hold data, too, challenges of tuples on tertiary storage. In *Proceedings of the ACM SIGMOD International Conference on Management of Data*, pp. 413–417.

Chen, H., and Little, T. (1993). Physical storage organizations for time-dependent multimedia data. In *Proceedings of the Foundations of Data Organization and Algorithms (FODO). Conference* (October).

Chou, H. T., DeWitt, D., Katz, R., and Klug, T. (1985). Design and implementation of the Wisconsin Storage System. *Software Practices and Experience* **15**, 10.

Copeland, G., Alexander, W., Boughter, E., and Keller, T. (1988). Data placement in Bubba. In *Proceedings of the ACM SIGMOD International Conference on Management of Data,* pp. 100–110.

Council, N. R. (1988). Mapping and sequencing the human genome. In *Committee on the Human Genome Board on Basic Biology*, National Academy Press.

Denning, P. J. (1968). The working set model for program behavior. *Commun. ACM* **11(5)**, 323–333.

Gall, D. L. (1991). MPEG, a video compression standard for multimedia applications. *Commun. ACM*.

Ghandeharizadeh, S., and Ierardi, D. (1994). Management of disk space with REBATE. In *Proceedings of the Third International Conference on Information and Knowledge Management (CIKM)*.

Gibson, G. A. (1992). *Redundant Disk Arrays*, Reliabls Secondary Storage, MIT Press, Cambridge, MA.

Gray, J., and Reuter, A. (1993a). *Transaction Processing*, Concepts and Techniques, Morgan Kaufmann, San Mateo, CA, chap. 13, pp. 670–671.

Gray, J., and Reuter, A. (1993b). *Transaction Processing*, Concepts and Techniques, Morgan Kaufmann, San Mateo, CA, chap. 13, pp. 682–684.

Knowlton, K. C. (1965). A fast storage allocator. *Commun. ACM* **8(10)**, 623–625.

Lewis, H. R., and Denenberg, L. (1991). *Data Structures and Their Algorithms* Harper Collins, New York, chap. 10, pp. 367–372.

McKusick, M., Joy, W., Leffler, S., and Fabry, R. (1984). A Fast File System for UNIX. *ACM Trans. Computer Syst.,*

Ng, R., and Yang, J. (1994). Maximizing Buffer and disk utilizations for news on demand. In *Proceedings of the International Conference on Very Large Databases*, September.

O'Neil, E. J., O'Neil, P. E., and Weikum, G. (1993). The LRU-K page replacement algorithm for database disk buffering. In *Proceedings of the ACM SIGMOD International Conference on Management of Data,* pp. 413–417.

Patterson, D., Gibson, G., and Katz, R. (1988). A case for redundant arrays of inexpensive disks (RAID). In *Proceedings of the ACM SIGMOD International Conference on Management of Data,* May.

Rosenblum, M., and Ousterhout, J. (1992). The design and implementation of a log-structured file system. *Trans. Computer Syst.* **10(1)**, 26–52.

Ruemmler, C., and Wilkes, J. (1994). An introduction to disk drive modeling. *IEEE Computer*.

Tobagi, F., Pang, J., Baird, R., and Gang, M. (1993). Streaming RAID-A disk array management system for video files. In *Proceedings of the First ACM Conference on Multimedia*, August.

PART 6

Summary Databases and Model Repositories

CHAPTER 6.1

Summary Databases and Model Repositories

Michael A. Arbib[1] and Amanda Bischoff-Grethe[2]

[1] University of Southern California Brain Project, University of Southern California, Los Angeles, California
[2] Center for Cognitive Neuroscience, Dartmouth College, Hanover, New Hampshire

Abstract

A *Summary Database* is like a review article but is structured as entries in a database rather than as one narrative, storing such high-level data as assertions, summaries, hypotheses, tables, and figures that encapsulate the "state of knowledge" in a particular domain. A *Model Repository* is a database that not only provides access to computational models but also links each model to the Empirical and Summary Databases to provide evidence for hypotheses in the model or to data to test predictions from simulation runs made with the model. This chapter provides a conceptual framework for databases of these types, pointing to specific instances of such databases—one implicit in the structure of our Brain Models on the Web (BMW) Model Repository as well as the NeuroScholar and NeuroHomology databases—described in later chapters and setting an agenda for future research and development.

6.1.1 The Database Typology and the NeuroInformatics Workbench

In Chapter 1.1, we introduced a fourfold typology of types of databases of relevance to Neuroinformatics: Article Repositories, Repositories of Empirical Data, Summary Databases, and Model Repositories. In this section, we briefly review what the earlier chapters of this volume have shown concerning our approach to providing tools for the construction and use of such databases. In particular, we will see the relationship of the key components of our NeuroInformatics Workbench—NeuroCore, NeuARt, NSLJ, and Annotator—to this typology. The rest of the chapter will provide a conceptual framework for the specific Summary Databases (see Chapters 6.3 and 6.4) and Model Repository (Chapter 6.2) that we have created to date, emphasizing not only our current status but also the long-term goals that have informed our work.

Article Repositories

These are in part the domain of commercial publishers and scientific societies, but can also take the form of resources constructed for access by some community over the Web or as a repository for technical reports for some organization. As such, we can leave the vigorous development of the tools for creating and using such repositories to others, but the USCBP has developed a simple prototype Model of Online Publishing (Chapter 5.3) to show how to increase the utility of online journals, etc. by offering new ways to link them to Repositories of Empirical Data and personal databases.

Repositories of Empirical Data

Our approach to Repositories of Empirical Data has emphasized the notion of a protocol for each experiment for which data are stored. The *protocol* provides information on the hypotheses being tested, the experimental methods used, etc. We have developed NeuroCore (Part 3) as our basic design for such databases, providing a general data schema which neuroscientists can extend readily to provide a tailored data structure that is still easy to understand by investigators from other laboratories. NeuroCore is enriched by the NeuroCore Time-Series Datablade (Appendix A2) to support storage of neuroscience time-series data. We further extended the utility of such repositories by developing NeuARt (Chapter 4.3) and NeuroSlicer (Chapter 4.4) to support the registration of data against standard brain atlases, with implications for database structuring and data retrieval.

Summary Databases

A *Summary Database* is like a review article but is structured as entries in a database rather than as one narrative, storing such high-level data as assertions, summaries, hypotheses, tables, and figures that encapsulate the "state of knowledge" in a particular domain. Our three USCBP Summary Databases, one implicit in the structure of our Brain Models on the Web (BMW) Model Repository, as well as the NeuroScholar[1] and NeuroHomology databases, will be described in subsequent chapters. We have already presented Annotator, our annotation technology (Chapter 5.4) for building a database of annotations on documents and databases scattered throughout the Web. The key observation is that a Summary Database can be seen as a form of annotation database with each summary serving as an annotation on all the clumps (selected items) that it summarizes. However, our work on BMW, NeuroScholar, the NeuroHomology database, and Annotator to date have proceeded in parallel and are only just now reaching the stage where we can begin to implement the conceptual integration that has been a foundation for our work.

Model Repositories

We envision a *Model Repository* as a database that not only provides access to computational models but also links each model to Empirical and Summary Databases to provide evidence for hypotheses in the model or to data to test predictions from simulation runs made with the model. Chapter 6.3 presents the current status of our Model Repository BMW (Brain Models on the Web). At present, BMW primarily includes models written in various versions of our Neural Simulation Language NSL (Chapter 2.2); future versions will reach all the way from models of brain imaging (Chapter 2.4) through NSL models to detailed compartmental models (NEURON and GENESIS) down to the finest levels of macromolecular structure (as in some of the EONS modeling of Chapter 2.3). We have already noted (Chapter 2.1) the utility of providing the implementation of a model with simulation interfaces which mimic experimental protocols so that operations on the model capture the manipulations the experiment might have made on corresponding portions of the nervous system.

6.1.2 An Overall Perspective

Our aim is to create an environment in which modelers and experimentalists can work together to increase our understanding of nervous systems. Fig. 1 is based upon the following points:

1. Laboratory data mean little unless they are stored in the database with information about the protocol used to obtain them.
2. Much of our access to such data will be indirect, via empirical generalizations and summary tables and figures.
3. Such summaries should be linked to laboratory data from which they are derived and/or journal articles in which they have been published.
4. Tools are needed to summarize laboratory data for publication.
5. Tools are needed to contrast and compare experimental results, generalizations, and the predictions of models of neural activity.

When working in neuroscience, a multitude of questions may occur regarding the data: "Is there a projection in the rat from the substantia nigra pars compacta to the subthalamic nucleus? How strong is this projection? Has this projection been modeled?" These and other questions can lead to time-consuming searches through published articles retrieved via such services as PubMed or Current Contents. First, there is the problem of knowing the appropriate keywords to retrieve the data relevant to the search; second, each article must be examined to determine if it does, indeed, cover the topic in question; and, finally, the article must be read more closely to find the sentence, paragraph, or figure that adequately answers the question. Although the answer is known for now, what about several weeks or months later when the material is needed again? Much of the same process is repeated, only this time the search is through a

---

[1]The NeuroScholar project was initiated as part of the USC Brain Project and is now being further developed by Gully Burns as a separate but federated research project.

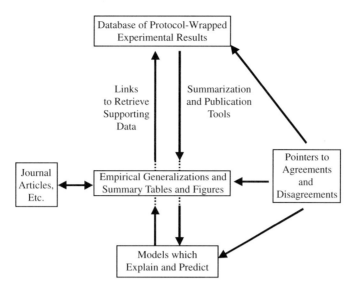

Figure 1 A functional perspective on the integration of diverse databases in the USC Brain Project framework. The figure stresses the task of relating experimental data to empirical generalizations and the assumptions and predictions of computational models of neural function.

pile of papers on one's desk instead of through a search engine, culminating in the search for the paragraph that may or may not have been marked to indicate an important point. These summary statements that support our data can occur again and again and can necessitate a document search again and again, leading to time lost and often frustration when the answer does not immediately appear. Using a database to store this material, then, may be the answer to faster research and better organization of the data most important to one's work. The reader will note here a resonance with the general issues discussed in our introduction to annotation databases (Chapter 5.4).

A Summary Database (SDB), then, is a repository for summary data, documents, and annotations. A summary is a brief statement or description that summarizes a more complex body of text (or graphics or tables or graphs, etc.). For example, several different documents may give a precise description of a methodology used to determine connectivity between two regions and the results obtained by applying it. The summary data, then, would state that the connection does (or does not) exist, linked to the precise description from the source data, or "clumps" of the original document. In this fashion, one may search for generalities (i.e., does region A project to region B) and them examine more closely the source to help form a personal opinion on the summarized data.

Data stored within the SDB may relate to various aspects of neuroscience. The most obvious type is neuroanatomical data describing a cell group, receptor, channel, neurochemical, and so on. Neuroanatomical data, though, may be biological or simulated. A model stored within BMW (see Chapter 6.2 for details) may describe relations among several cell groups. For example, a model of the basal ganglia may state that a population of neurons, called put_inh, project to modeled external globus pallidus (GPe) neurons. When one does a search of summaries, then, one may choose to search for biological data, simulated data, or both. Storing these kinds of data together—whether in a single database or the databases of a well-designed federation—makes it easy to search for support (or inconsistencies). For example, a model of the basal ganglia in monkey may include a projection from the substantia nigra pars compacta (SNc) to the subthalamic nucleus (STN). Biologically, this projection is not known to exist in monkey; however, it does exist in the rat.

A summary statement may be a simple statement, such as region A projects to region B, or may contain more detailed information, such as the description of a cell type's dendritic tree or the pattern of connectivity between two regions. These data may serve further as support for a homology (Chapter 6.4). For example, a homology based upon relative position will provide the details regarding the relative position as a summary; another homology based upon connectivity will provide details regarding the afferents and/or efferents to the region and the pattern of connectivity. Alternatively, the user may view the originating text(s) supporting a summary statement.

In the SDBs we have developed in USCBP to date, we have stored the supporting data, called "*clumps*," for summaries directly within our databases. These clumps have been primarily portions of text from various documents, usually published research articles, but one can include figures, extracts from tables, etc. In Chapter 5.4 on annotations (a summary may be viewed as an annotation on the clumps that it summarizes), we have presented our software for creating and using Extended URLs to locate clumps from documents on the Web, not only retrieving the document but also highlighting the clump of interest. It is a challenge to extend the Extended URL methodology to characterize other types of clumps, such as subtables of a database or extracts from graphics, videos, or simulation runs. When using this technology, only the Extended URL, rather than the clump itself, need be stored in the Summary Database. This is important either when the clump is larger than a "fair use" portion of copyrighted material or when the clump is a very large object (e.g., recordings from multiple electrodes, video clips, etc.). The latter consideration reminds us of the mass storage issues discussed in Chapter 5.5.

Another general issue for SDBs is that neuroscience data do not have the unequivocal nature of entries such as, say, "Mr. X has seat Y on flight W on date D" in an airline reservation database. A seemingly clear statement such as "Region A connects to Region B" may depend on the protocol used, variability between animals within and across species, the skill of the experimenter, and the specific interpretation of how best to characterize the two regions. Thus, a summary (like a review article) is itself the result of both the choice of clumps to summarize and of the collator's choices in synthesizing what may be apparently discordant material. Both Chapters 6.3 (NeuroScholar) and 6.4 (NeuroHomology) pay attention to these issues. Each summary should contain a field naming the collator; each summary should come not only with copies of, or pointers to, the clumps that are summarized, but also—where possible—with a rationale for the summary and a confidence level (between 0 and 1) for the assertion that the summary provides. Moving beyond the current state of our art, we envisage the following.

Summaries may be either unrefereed or refereed. In any case, summaries will be annotated with new data of relevance, critiques, annotations, etc. New search tools (such as those planned for our annotation technology, Chapter 5.4) will make it easy to use an existing summary to anchor a search for related data. The pace of research means that summaries will need constant updating—even if their general conclusions remain correct, new data may markedly change the confidence levels for the assertions they contain. When a critical mass of such changes has accumulated, or when a synthesis of related

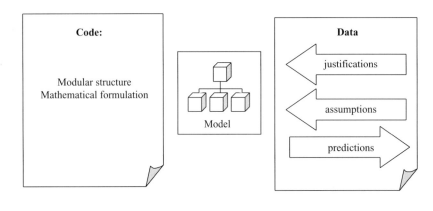

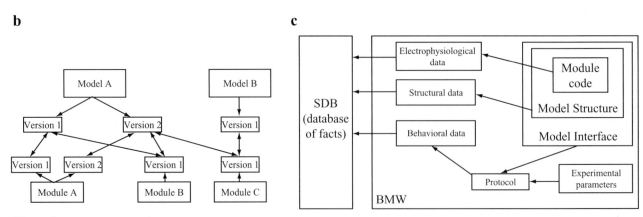

Figure 2 (a) Overview of two views of a model. (b) More detailed schematic of the modular structure of the model, emphasizing that a model, and modules, may have different versions; shows how links to a Summary Database may be used to test and/or justify assumptions and predictions of the model. (c) Emphasizes the other dimension of a Model Repository, documenting the relation of the model to empirical data and (although this is not made explicit in the diagram) other models (see also Fig. 1).

summaries is undertaken, then a new summary can be posted which can be seen as a new version of the one it replaces. As in all cases of versioning, there will be the editorial decision regarding refereed databases as to when a new version becomes the "official" version, in which case the old versions should still be available for (rare) consultation, but perhaps in an archive based on a slower but less expensive mass storage medium (*cf.* the issues discussed in Chapter 5.5).

6.1.3 General Considerations on Model Repositories

Fig. 2 presents our general perspective on Model Repositories. On the one hand (part b), the modeler will want access to the model in all its detail, ranging from viewing the modular structure at various levels of detail all the way down to the code. The repository should also allow access to different versions of each model and model. As emphasized in Chapter 2.2, we have developed the Schematic Capture System (SCS) to ease the creation and viewing of the modular structure of modules written in NSL. It should be a straightforward extension to enable the SCS to display any and all modular models. The greater challenge (one to be shared with other groups beyond USCBP) will be to ensure the *interoperability* of models developed using different simulators, "wrapping" modules so they can work together, thus allowing the creation of new versions of old models, and new models as well. A current project is to develop SCS so that it not only allows both the modeler and experimentalist to view the components of a model and their connections, with links to the supporting NSL code, but will also allow one to follow links to summary data that test or justify the model assumptions expressed in the SCS graphics.

We also show in Fig. 2 the SDB as a database separate from the Model Repository (in this case, BMW); however, in the current prototype of BMW, the SDB is a subset of the BMW database, rather than a separate database. In the remainder of this chapter, we briefly review the structure of the SDB contained within the current implementation of BMW. We then introduce key features of our other two current USCBP Summary Databases, NeuroScholar and NeuroHomology. Further details of these three SDBs are provided in Chapters 6.2, 6.3, and 6.4, respectively.

6.1.4 Brain Models on the Web

Brain Models on the Web (BMW) is a Model Repository which allows users to run experiments, build new models, and keep track of versions within a database. SDB and BMW are constructed to work together and take advantage of material that each database stores. Fig. 3 describes the relationship between the two data

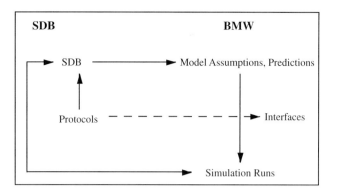

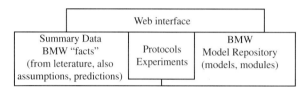

Figure 3 A diagram showing the relationship between SDB and BMW. Summary statements within SDB can be used to form model assumptions and predictions. The completed model and its simulation runs can be documented via SDB to allow researchers to access both biological and simulated data on the same topic. In the current implementation of BMW, the Summary Database is embedded within BMW as a set of facts that can be related to modules, rather than being a separate database or federation of Summary Databases.

bases. A researcher creating a model within BMW will rely upon SDB for data to support his model assumptions and test predictions and will tie these summaries to the appropriate areas of the model. When the model is complete, he may choose to store several simulation runs of the different neural populations represented within the model. These simulations may also be documented within SDB. Similarly, SDB may hold information on the model's protocols and interface. Another researcher working in SDB, then, may perform a search on modeling data and be returned information regarding the model in BMW and its supporting data. He may then decide to run the model himself, perhaps trying slightly different protocols, to validate that the neural population within the model does indeed match expectations based upon available literature. This researcher may choose to take this one step further and review side-by-side comparisons of the simulated model and real biological data stored within NeuroCore.

As already noted, the current BMW contains an embedded Summary Database (Fig. 3); details are provided in Chapter 6.2. Here, we briefly note the way in which clumps are handled in the SDB of the current BMW. The entry for a clump contains the text of the clump (as distinct from Extended URLs pointing to clumps; see Chapter 5.4), together with the reference to the document that contains the clump and a set of keywords. Fig. 4 shows a clump retrieved by a search on the keyword "prism." The current version does not contain explicit summaries; rather, it stores relations between

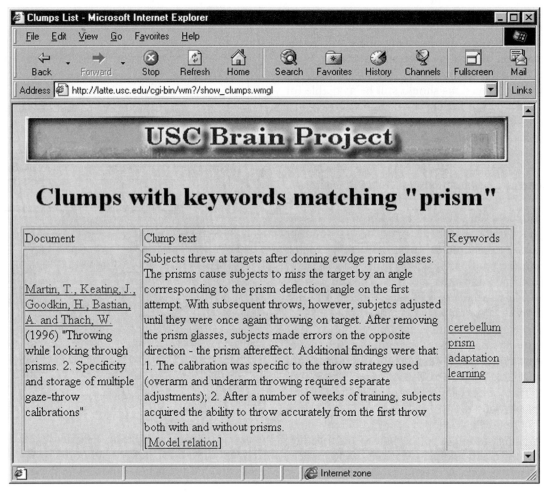

Figure 4 The current version of BMW contains an SDB which contains clumps. The entry for the clump contains the text of the clump (as distinct from Extended URLs pointing to clumps; see Chapter 5.4) together with the reference to the document that contains the clump and a set of keywords. Here we see a clump retrieved by a search on the keyword "prism."

clumps expressing empirical data and aspects of the model—whether model assumptions or simulation results—stored in BMW. Thus, a search can reveal a summary implicit in all these links. For example, given a clump, we can search for modules and models to which it is related. Fig. 5 gives an example relating a clump to the "Dart" model via the relation "Replicates behavior." Fig. 6 shows some of the clumps retrieved by querying which clumps are related to a given model. Note that the response is related to the hierarchical structure of the model—we see clumps related to the top-level description of the model as well as model behavior; we also see clumps related to the constituent modules of the model. At present, the set of such relations is relatively limited, and the search for such relations is still conducted by hand.

6.1.5 NeuroScholar

NeuroScholar is a Summary Database, that is, a knowledge-base management system for neuroscientific information taken from the literature. The key issue that drives the design of this SDB is to allow users to compile, interpret, and annotate low-level, textual descriptions of neuroanatomical data from the literature (Chapter 6.3). NeuroScholar is founded on the following conceptual entities (Fig. 7):

1. *Atoms:* These are raw text and figures that can be found in the literature (we call these clumps elsewhere).

2. *Primitives:* These are classifications of atoms so that the information they are concerned with can be structured in a useful way. Here, we differentiate between "knowledge types" representing qualitative differences between data based on data from a paper's abstract or results section as author interpretations, and descriptions. The properties of a given primitive returns subprimitives, allowing the construction of complex data entities that closely correspond to neuroscientific concepts found in the literature.

3. *Relations:* These are used to store rules that compare and link different primitives. This is especially useful

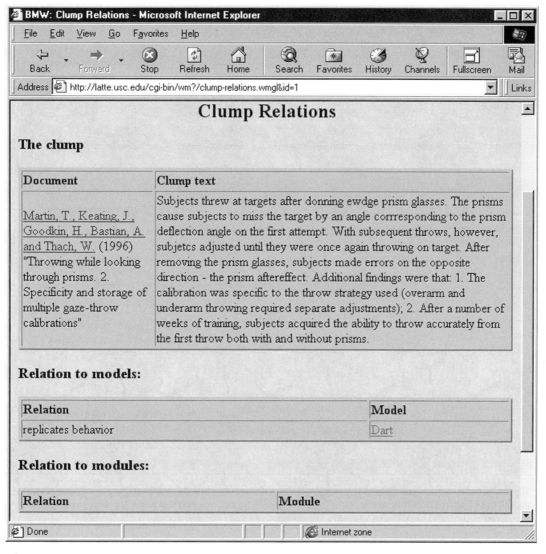

Figure 5 Given a clump, we can search for modules and models to which it is related. Here we see an example relating the clump to the "Dart" model via the relation "Replicates behavior." At present, the set of such relations is relatively limited, and the search for such relations is still conducted by hand.

when linking primitives that are derived from different papers.

4. *Annotations:* NeuroScholar allows users to make two sorts of annotations—comments, which are just remarks that users wish to attach to atoms, publications, primitives, or relations, and justifications, which are required to substantiate any relation or primitives that are classified as user interpretations.

Fig. 8 shows a number of examples of NeuroScholar Primitives: *brainVolumes* are simply volumes of brain tissue possessing geometrical properties; *blackBoxConnections* represent neural connections and have two subsidiary object: the *somata brainVolume* denotes the region where the connection originates and the *terminalField brainVolume* denotes the region where the connection terminates.

In Fig. 8b, we add more advanced NeuroScholar primitives: *neuronPopulations* are populations of cells, grouped on the basis of a stated criterion; *dendrites, somata, axons,* and *terminalField* objects represent shared properties of the constituent neurons of *neuronPopulation* objects. The properties of these subprimitives could be defined to accommodate a range of different properties from different papers (such as the presence of labeled mRNA, the morphology of the somata, or firing characteristics of neurons).

We intend to combine NeuARt, our atlas-based user interface for a NeuroCore database that handles neuroanatomical data, and NeuroScholar so that specified brainVolumes can be linked to volumes embedded in the relevant brain atlas (*cf.* Chapters 4.3 and 4.4 for the relevant context for this ongoing work). This means that the framework for published data can be used to structure and interpret experimental data while also linking the geometrical properties of the of published data to an atlas. This exemplifies the spatial query manager concept

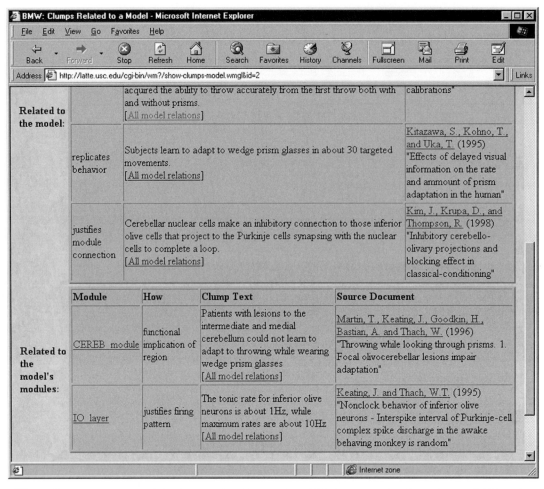

Figure 6 Here we see some of the clumps retrieved by querying which clumps are related to a given model. Note that the response is related to the hierarchical structure of the model; we see clumps related to the top-level description of the model as well as model behavior; we also see clumps related to the constituent modules of the model.

of NeuARt—that relevant neuroscience data can be based on geometrical queries to an atlas, as well as on text-based and other forms of query.

Chapter 6.3 provides much more information on NeuroScholar, including a critical discussion of the evaluation data used to form summaries and an indication of the data analysis tools that have been used to mine interesting conclusions from connectivity data. NeuroScholar's weighted scheme for "believability" in connectivity provides a useful model for evaluating data summaries for other areas of neuroscience.

6.1.6 NeuroHomology

The NeuroHomology database (Chapter 6.4) is a knowledge-based Summary Database that is designed to aid the search for homologies between brain structures for human/monkey, human/rat, and human/rodent species. To support this, the database contains searchable entries for brain structures and connections as interpreted from the literature, with quantification of the degrees of confidence of connections and staining techniques that are used. As in NeuroScholar (Chapter 6.3), we also propose a method of computing an overall confidence level for a number of entries related to a single connection, and based on it the user can evaluate the reliability of the searched connection as reflected in the literature.

We have identified eight criteria to define a homology between two brain structures: cell morphology, relative position, cytoarchitecture, chemoarchitecture, myeloarchitecture, afferent and efferent connections, and function. A *"homology inference engine"* lets one retrieve all data relevant to a possible homology between regions in different species and compute a degree of confidence that a homology does indeed exist.

The homologies part of the NeuroHomology database can handle data about brain structures from any species for which data are available. On the other hand, due to the fact that the homologies are seen as a tool for computational neuroscientists and we focus on models of brain structures from humans, monkeys, and rodents, the database contains human/rat, human/monkey, and

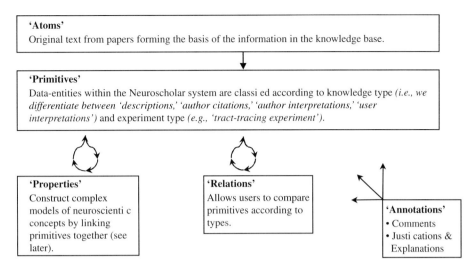

Figure 7 NeuroScholar uses the term "atom" where we use the term "clump" but offers useful typologies not yet attempted in other work on Summary Databases in the USC Brain Project.

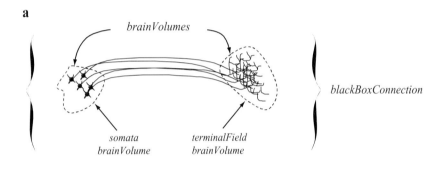

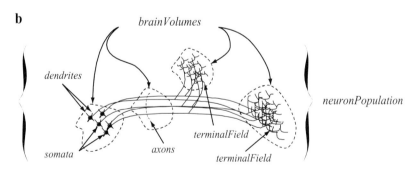

Figure 8 Examples of NeuroScholar primitives, with increasing detail in (a) as compared to (b).

monkey/rat homologies. The search of the NeuroHomology database can be made by abbreviations of brain structures. The user can enter the abbreviations for two brain structure from two different species (Fig. 9). The result of a search will retrieve all the homologies that are found in the database with regard to the searched pair of brain structures. The user can inspect all the details of any retrieved entry by clicking on the cited reference. The result will be eight different tables containing information about each homology criterion. Associated with each entry is a short description of the homology, which is inserted by the collator. The user can enter and inspect the additional annotations that are attached to each investigated entry. In this way, the user can evaluate the relative importance of common patterns of connectivity in evaluating the degree of homology. A second reason for showing the common afferent and efferent connections, as reflected from the inserted data in the database, is related to the functionality of brain areas. Two brain structures that share a common pattern of connectivity are likely to have the same position in the hierarchy of processing of information in the central

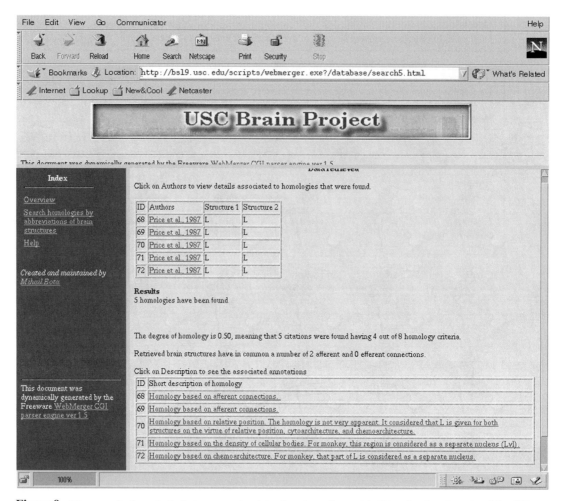

Figure 9 The query for homologies between monkey lateral nucleus of amygdala (L) and rodent L retrieved five different entries. The homology criteria that are fulfilled are afferent connections, relative position, cytoarchitecture, and chemoarchitecture (staining for AchE).

nervous system and have common functions. Moreover, the computational neuroscientists that use the NeuroHomology database can evaluate the degree of reliability to model neural systems from one species by using data from other species.

6.1.7 Future Plans

Although at present they are separate structures, we envision that the NeuroHomology database will be merged with the SDB currently embedded within BMW to form a generic SDB enriched with a "homology inference engine", as well as a number of features developed within the NeuroScholar project. We also plan to make full use of our annotation technology in the creation of SDBs and their linkage to BMW. On the more formal side, we are developing a set of key properties and relations to better relate neural structures to the model structures that represent them, enriching the anatomical examples coded in NeuroScholar with the rich set of constructs contained within the models already embedded in BMW. As noted, we are extending SCS from the representation of NSL models to the integration of models with empirical data and hope to engage the Neuroinformatics community in solving the interoperability problem for developing models, versioning, and linking them to empirical data in a modeling environment that contains a variety of simulators. Finally, much more needs to be done in developing criteria for comparing model simulations and predictions with empirical data, with special attention given to parameter identification techniques that will determine which settings of model parameters best match model performance to a given suite of test data.

Acknowledgments

This work was supported in part by the Human Brain Project (with funding from NIMH, NASA, and NIDA) under the P20 Program Project Grant HBP: 5–P20–52194 for work on "Neural Plasticity: Data and Computational Structures" (M. A. Arbib, Director). We thank Jacob Spoelstra, Gully A. P. C. Burns, and Mihai Bota for their work which enriched this chapter and which is reported in more detail in Chapters 6.2, 6.3, and 6.4, respectively.

Brain Models on the Web and the Need for Summary Data

Amanda Bischoff-Grethe,[1] Jacob Spoelstra,[2] and Michael A. Arbib[3]

[1] *Center for Cognitive Neuroscience, Dartmouth College Hanover, New Hampshire*
[2] *HNC Software, Inc., San Diego, California*
[3] *University of Southern California Brain Project, University of Southern California, Los Angeles, California*

Abstract

Computational models of the brain have become an increasingly common way in which to study brain function. These models vary in design from the study of cable properties of a single neuron to a large-scale network of neurons for modeling behavior or overall function. These models may be hard-coded in a programming language, such as C, or may be developed in one of many programming environments meant for model design (e.g., NSL, Neuron, GENESIS, and others). However, the model's main point (or the hypotheses behind its construction) may be difficult to ascertain. A user who is able to read the code might be able to determine whether a connection or property exists, but may not know the reason for that property unless it has been documented in the code or a README file. Alternatively, the user may have to refer to other documentation (such as a journal publication) which often does not contain enough detail at the code level. Also, models of similar functions may be written in different programming environments and rely upon different hypotheses. Used in conjunction with a summary database, the Brain Models on the Web database is one way to bring these models together in a common environment for study by both modelers and experimenters alike.

6.2.1 Storing Brain Models

Brain Models on the Web (BMW) is a development, research, and publications environment for models. The main goal of the BMW database is to act as an online repository for brain models and to facilitate the publishing of links between models and the underlying data. The idea is to allow designers not only to explore the structure and code of a model, but also to see the data underlying each assumption or design decision. A user is then able to search the database (or federated databases of summary and empirical data) for new data that either support or contradict the existing "facts" and from there develop a new model to take into account the new evidence. This new model version can be partly or completely based upon code from the previous model, or it may link together two separate models in a brand new way. Multiple versions can be stored within the database, a useful tool when comparing alternative methods. Alternatively, experimenters can start at the data end (which might be an experimental protocol) and from there find models that either have used the data in their design or replicate/contradict that particular data item. These models might produce testable predictions which could lead the way to more experiments. Finally, BMW serves as a repository for "works in progress" for users creating and conducting experiments on models. It relies upon a Summary Database (SDB) for the summary data which justifies a given model, as well as the storage of assertions made concerning a model. We will describe the relationship between the SDB and BMW in more detail in Section 6.2.2. BMW thus makes it easy to find models, to store data summaries culled from the literature and to keep track of versions as a model evolves.

A model may be characterized by its *conformancy* with BMW standards as belonging to one of five different classes: work-in-progress (WIP), poster, bronze, silver, and gold. WIP models are kept private to the individual or group working on them. These are models that are in development and are not available for public viewing. Once the model is completed, it may be submitted as a poster. A poster model is stored in BMW and is available for public viewing but has not yet been reviewed. A committee or peer-reviewed journal publication would review the submitted poster's scientific merit and grant it full status using a descriptor—bronze, silver, or gold—to indicate that the poster had been reviewed and satisfied certain criteria. Bronze, silver, and gold by no means indicate the quality of the model but are labels for keeping track of how well the model is documented and linked to other databases within the BMW framework. A bronze model simply involves code stored at an ftp site (or within the database for retrieval) with a simple README document describing the model. Silver implies that the user has put some effort into "BMWfying" (pronounced "beamifying") the model; for example, for NSL models, the modules have been constructed in a form acceptable to database storage, and there is some limited documentation on the model. Gold is a fully BMWfied model. All the code has been referenced appropriately within module tables, and full documentation, including summaries and annotations, exist within the SDB. Any documentation must conform to one or more of the following formats: a plain text file, HTML, or PDF.

6.2.2 The BMW-SDB Relationship

The "An Overall Perspective" section in Chapter 6.1 has already made clear the importance of summary databases in general and noted the three summary databases developed to date by the USC Brain Project: the Summary Database developed as part of BMW (referred to as SDB throughout this chapter); NeuroScholar (see Chapter 6.3); and the NeuroHomology database (see Chapter 6.4). Summary data summarize a more complex body of texts (and/or graphics, tables, graphs, simulation results, etc.). The summary may also be linked to excerpts from the source data which we refer to as *clumps*. Because an increasing number of journals are available online, we expect to take advantage of our

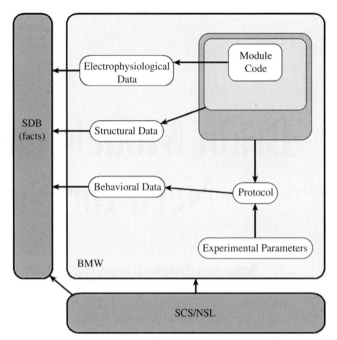

Figure 1 Relationship of the SDB summary database and the remainder of BMW. A model is composed of modules and is used via a model interface which captures aspects of an empirical protocol. These may be supported or tested by links to data stored in the SDB. SCS/NSL is the Schematic Capture System (SCS) developed for NSL. SCS can be used to view data and NSL models and modules schematically in terms of the connections and the structure that comprise them.

annotation technology (Chapter 5.4) to link summary statements to the clumps supporting the summary as found within the online document itself.

Fig. 1 shows the relation of the SDB (the specific summary database developed as part of the BMW effort, focused on data linked to modeling efforts) to the remainder of BMW. The box labeled "BMW" shows that a model is composed of modules and is used via a model interface which captures aspects of an empirical protocol. (See Chapters 1.1 and 2.1 for more on this methodology.) Thus, the design of a module and the way in which modules are combined to form a model may be supported or tested by links to structural, electrophysiological, and other data stored in the SDB, while simulation experiments may yield predictions that can either be posted in the SDB or tested by comparison with summary data in the SDB. At present, BMW-SDB is a relatively small database, and in this chapter we will emphasize the explicit links entered into the database by the modeler. As the database grows, we expect much modeling work to involve searches for data that support or undermine hypotheses used in the model and for empirical results that can test or guide simulation experiments to be conducted with the model. The bottom box refers to the Schematic Capture System (SCS) we have developed for NSL (see Chapter 2.2 for details), which can be used to view data and models schematically in terms of diagrams expressing the way in which models and modules are formed as connections of the modules that comprise them. We describe the proposed relationship between BMW and SCS in the "Links to SCS" section below.

The data within the SDB function as a method for documenting the model code; they provide justifications for choices made by the modeler when designing the model. This encourages the modeler to be exacting in determining parameters. A parameter value may be chosen either because the value may be documented by the literature summarized in the SDB, or because the value was found by "parameter identification," whether analytical or using multiple simulation runs, which determine what parameter values allow the model to exhibit behavior or other empirical constraints documented in the SDB. To date, most of our work with the SDB is based on previously published documents. This is limiting to some extent, as researchers do not publish all their data, nor do they necessarily provide precise documentation as to the methodology used in generating the data. It is our hope that as the SDB and Repositories of Experimental Data (REDs) are more widely used, researchers will use REDs to store their data and protocols, and the SDB will hold summaries of the RED data.

6.2.3 Reviewing a Model: The Dart Model of Prism Adaptation

Given the broad-brush overview of BMW-SDB provided by Fig. 1, we will now use a specific example of reviewing a model to introduce a number of features embodied in the current system. With the understanding that this provides, we will then present the tables that constitute the actual database design. These include tables for *persons, models, modules, documents, clumps,* and *experiments*. We then present in detail, with exemplary screen dumps, instructions on how to access the database, including material on logging into the database, browsing the database, searching for models and data, and inserting models and data.

Suppose that we have arrived (by means described in later sections) at the page showing Model Versions for the Dart model (Fig. 2). Let us assume we are interested in looking at version 1.1.1 more closely. Clicking on its hyperlink we now begin to reach the guts of the model (Fig. 3). This particular version of the model has been given a gold conformancy status. This means that the model has been constructed to conform with BMW standards; it relies upon the database for storage of not only the model itself, but of its justifications as well. For this particular version, we can explore several options: we can choose to work with the executable model; we can review a tutorial to understand the model; we can view the data associated with the model; we can view only the top-level module; or we can choose to look at all modules which are contained within the model.

By clicking on the Toplevel Module hyperlink, we can explore the hierarchy of this module. Fig. 4 displays the module implementation for the top level of the Dart model, namely, DART_top. Notice that the module itself has a version. This version number may be different from the version number of the model. This may occur for various reasons: the model is new but relies upon previously written modules; the module itself has been updated; or other modules have been updated, representing a different version of the model. For a given module, we are given a list of the associated files; these typically are the code that handles that module's behavior. We can also see a list of models that use this module. In this case, versions 1.1.1 and 2.1.1 of the Dart model currently use the Toplevel module.

Some models will allow users to execute them from the database. As an example, we show the results of a trial run of the Dart model during basic adaptation (Fig. 5; see color insert). "The Dart" model hypothesizes a role for the intermediate cerebellum during prism adaptation. In experiments by Martin *et al.* (1996a,b), both normal subjects and cerebellar patients were required to throw at a target after donning 30° wedge prism glasses. The prisms caused the subjects to initially miss the target by an angle proportional to the prism deflection angle. With subsequent throws, however, normal subjects adjusted until they were once again throwing on target. In contrast, cerebellar patients did not adapt and continued to miss the target. After removing the glasses, the prism gaze-throw calibration remained and subjects made corresponding errors in the opposite direction.

Fig. 5 (see color insert) shows the result of a simulation experiment using a normal subject. Throws are shown in chronological order on the x-axis, while the y-axis indicates horizontal error (the distance by which the target was missed). The simulation allows for two throw strategies (overhand and underhand) and two conditions (prisms-on and prisms-off). In this particular experiment, the model starts with 10 warm-up throws (5 overhand and 5 underhand) before donning the prism glasses. The first throw while prisms are worn (red squares) misses the target proportional to the prism angle, but the model improves with subsequent throws until it once again hits the target. When the prisms are removed (blue squares), however, we observe an after-effect proportional to the adjustment angle, and the model has to re-adjust its normal throwing, in agreement with the results reported by Martin *et al.* (1996b). The graphical interface allows the user to choose a different experimental protocol (for instance, to observe the effect of the adaptation on underhand throwing) and also to adjust both the experimental and model parameters.

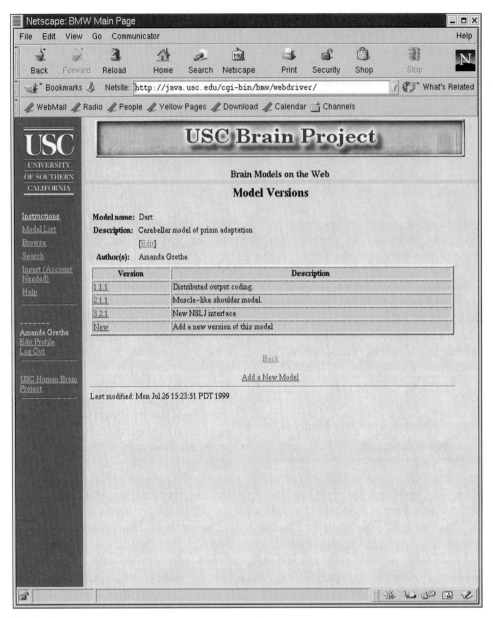

Figure 2 An example of viewing the Brain Models on the Web database of the USC Brain Project. Here, the Dart model has been selected, and information regarding the model versions are presented on the screen. The Dart model has three separate versions. Versioning may be based upon the addition of new material, such as the addition of a shoulder model as seen in version 2.1.1. Clicking the hyperlink related to the model version displays a new page with a list of available information associated with that particular model version.

6.2.4 Database Design

The database was implemented in SQL and is hosted on a Sun workstation running the Informix DBMS. This was done to enhance its ability to connect with other aspects of the NeuroInformatics Workbench and to allow for a consistency across databases contained within the Workbench. There are five main groups of data that may be stored within BMW/SDB:

1. *Persons:* The Person group is fairly self-explanatory; this group contains tables that define persons involved with the database. These include users of the database, modelers, authors of articles, and reviewers.
2. *Models:* The Model group describes tables that relate to models. A model contains descriptive information regarding it, but it also links together the smaller aspects, or modules, which describe it.
3. *Modules:* The Module group contains tables that define modules. These are the building blocks which, when linked together, comprise a model.
4. *Documents and clumps:* Documents are the articles used to support or refute the research. Clumps contain pieces of information that can be used to support (or refute) a model and/or module. These clumps can

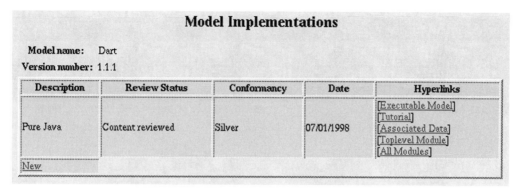

Figure 3 This panel from a BMW screen lists information regarding a model version along with links to allow the user to run the model, examine a tutorial, review the associated data, or view the modules that were used to create the model.

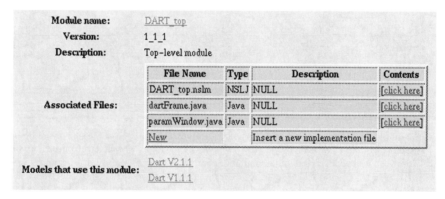

Figure 4 An example of a module and its associated files. The module was implemented in NSL 3.0 (Java implementation) and hence has several files written in Java and in NSL that are available for viewing. Models using this module are listed at the bottom of the screen.

contain data gleaned from journal articles or experiments or can be hypotheses that the model author puts forth within the model. (As noted earlier, this material constitutes the current SDB embedded within BMW).

5. *Experiments:* The Experiment group contains tables that describe the experiments the model is capable of performing. For example, the Dominey and Arbib (1992) saccade model is able to perform several different kinds of saccade experiments, including simple saccades, memory saccades, and double saccades. The experiments are defined through protocols. Note that a single protocol interface may be used to carry out many experiments.

In the following sections, we will briefly describe these tables and their attributes used to represent these data and provide a graphical representation of their relationships. To better understand the relationships between the tables, we have defined different graphic entities to represent different table types (Fig. 6). We briefly recapitulate the description from the "Overview of Table Descriptions" section of Chapter 3.2.

BMW uses the following table definitions as part of its methodology:

1. *Core table:* A core table allows the storage of information related to a specific concept or entity that is part of the core framework. For example, a person is a basic entity which is a core table describing persons stored within the database.

2. *Set backing table:* A set backing table is used to define the mappings from multiple entries to multiple entries (m-n). As an example, a model may contain many modules. A particular module may be used within many models. A set backing table would therefore be used to describe the mapping between models and modules.

3. *Dictionary table:* These are core tables that describe basic necessary tables used for looking up information. For example, we have previously described levels of conformancy. These would be placed within a dictionary table for reference by other tables within BMW. Dictionary tables are also a way to encourage users to use common nomenclature, but new terms can be added as needed (e.g., platinum as a model conformancy level).

4. *Extendible parent:* An extendible parent table allows individual users to tailor the database to their own needs. For example, a person working with saccadic

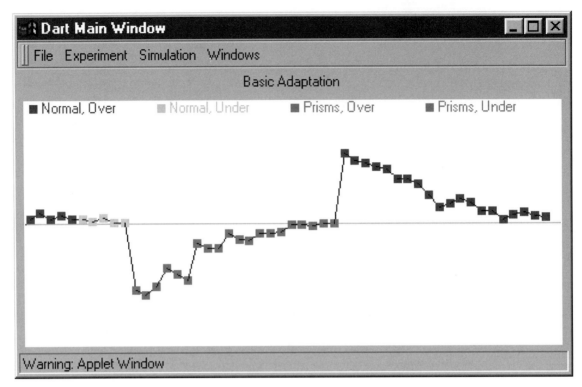

Figure 5 Example of output from the Dart model. The graph demonstrates the effects of wedge prisms on underhand and overhand throwing, along with the adaptation of the user's ability to hit the target when a dart is thrown. In this simulation, the model initially throws both overhand and underhand under normal conditions and is fairly accurate. When prisms are donned, the model must adapt to the related change between sight and movement. Once the model has reached an appropriate level of accuracy, the prisms are removed. The model makes errors in the opposite direction before re-adapting to normal conditions. (see color plates.)

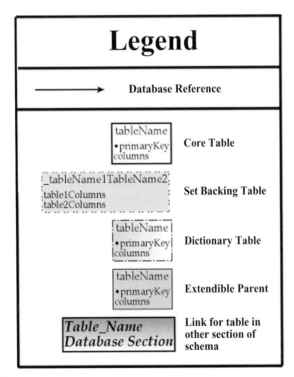

Figure 6 The legend used in describing the database tables and the relationships as represented within BMW.

models may wish to establish a protocol specific to saccadic models. Another modeler may work only with arm-reaching protocols.

As seen within NeuroCore (Chapter 3.2), we have created a table *ids*. This table is used for the generation of unique IDs needed by the database for the storage of entries. In particular, it is used with the *model*, *modelVersion*, *module*, and *moduleVersion* tables to simplify establishing certain relationships. For example, clumps may be linked to the above tables easily by relying upon the *ids* table.

Person

Within the database, there are several different kinds of persons. Consider first a document used as documentation for a model. This document will have one or more authors. If the document is a book chapter, we will also want to store information about the book's editors. Saving the author (or editor) as a person within the database allows us to search the database for other documents this author (or editor) may have written. Another kind of person within BMW is a reviewer, an individual who may be called upon to review models submitted to BMW. We therefore created a table, *person*, within our database (Fig. 7):

```
users
• uniqueID
  lastName
  firstName
  userName
  password
  email
```

```
person
• uniqueID
  lastName
  middleName
  firstName
  email
  address
  affiliation
```

Figure 7 The tables describing people associated with the database. The *person* table describes people stored within the database (such as authors of articles), while the *users* table represents users of the BMW database.

person: A core table that stores information about model and module authors and reviewers. This information includes the person's name, email, current address, and affiliation.

We also need to keep track of the users within the database. Which users have permission to enter data vs. run models? Is the user permitted to create, store, and alter models within the database? Is the user working in conjunction with other users on a model? Because users are treated differently than persons, we created a table, *user*, for describing them:

users: A core table that stores information about the authorized users of the database. These are people who may make changes to the database, as opposed to the general public who may just browse the entries. This information is limited to the user's name, email address, and password.

Models and Modules

A model can be organized in many ways. A model is a packaged, executable program that may be used to run experiments. A module is a logical segment of code that has clearly defined inputs, outputs, and behavior and as such is sufficiently decoupled from the model that it could be re-used in other models. Models are built using modules, which in turn may consist of submodules linked in a hierarchical manner. For example, a model of just the basal ganglia can consist of several modules (Fig. 1 of Chapter 1.1). These modules represent the different regions of the basal ganglia (putamen, globus pallidus, substantia nigra). In turn, the module globus pallidus may be made up of submodules (e.g., its internal, GPi, and external, GPe, segments). This method of organization makes the code not only easier to read, but also easier to plug into other models in development. A motor control model may require the putamen, whereas a saccade model requires the caudate as basal ganglia input, but both models may contain the same representation for substantia nigra pars compacta (SNc) because that brain region is known to project to both putamen and caudate.

Several versions of a model can exist. An early version of a basal ganglia model may only have the direct pathway projection (from caudate to substantia nigra pars reticulata, SNr), but a later version may add the indirect pathway (from caudate to SNr by way of a projection to subthalamic nucleus, STN, and GPe). An important point is that modules and models are versioned independently. This is to allow for independent upgrades or changes to the items without necessarily changing everything. Obviously, a model using a new version of a module must itself be an upgrade to a new model version; the reverse, however, need not be true. A specific model implementation uses a specific version of a given module, but other versions of the model might not use that particular module or use a different version thereof. Also, a module version could be used in any number of models. A module representing the caudate, for example, may be used in several different models of saccadic eye movements. This may occur when a module captures the features that a user desires without the need for modification.

Taken together, tables representing a model include (Fig. 8):

model: This is the highest-level description of a model and is defined as a core table. As such, it contains the most basic information regarding a model: the name of the model (name), a description of the model (what does the model represent?), a reference to information about the owner/creator of the model (ownerID), its creation date (created), and the date when it was last modified (modified). It is to this base table that all other model-related tables must be linked. Because this is the top-level table for describing a model and the modules it contains, the information stored here does not change for subsequent revisions of the model. For example, the Dominey and Arbib model of saccade generation may have multiple versions in which new brain regions are added, but the basic model hypothesis and protocol are unchanged. The only field that changes for this table over time is the modified field.

modelVersion: This table identifies a specific version of a model. A specific version need not be created by the person who originally built the model; we use this table to document who is responsible for the current version via the ownerID field. A model version also has a creation date (created) and last modified date (modified) and a note field for recording information specific to the version. Typically, new versions of models consist of three numbers separated by periods (e.g., 3.2.1). These numbers represent the major, release, and revision information of the model. A version is thus described by the major, release, and revision fields.

modelOSSpecific: Here we store information about a specific version of a model for a specific operating system or implementation. A specific model version may be implemented in various ways (e.g., C code, a simulation program such as NSL, etc.) but accomplishes the same

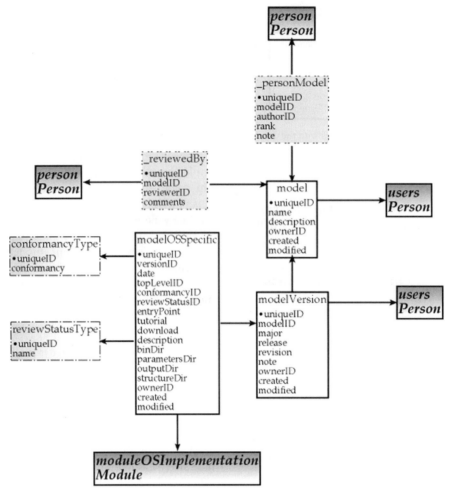

Figure 8 The tables describing a model within BMW. A *model* contains basic information which is expanded upon by the *modelVersion* and *modelOSSpecific* tables. These tables are useful in describing multiple versions of the model, which may vary due to model detail, the implementation (due to operating system or programming language used), and other features.

task and adheres to the same principles and hypotheses. It is not, therefore, a new version of the model but rather a different implementation. Some programming languages are not standardized across operating systems; therefore, alternate versions of the same model code are needed. This table also provides us with the model's version number (versionID), its conformancy (conformancyID) status (is the model a work-in-progress, poster, etc.?), its review (reviewStatusID) status (has the model been reviewed?), and numerous files associated with the model. We also have the owner of this version (ownerID) defined within this level. While a model may initially be created by one person, others are free to modify it and create new versions; we therefore need a method of documenting the person(s) responsible for each version, as well as a description as to how this model is different from other versions. The data files to which this table refers (e.g., tutorial, binDir, parametersDir, etc.) contain documentation and code related to the model and can include the locations of various data files with which the model interacts (e.g., parameters and outputs), a descrip-

tion of the model (which may include information as to why this version is different from another version), and so on.

reviewStatusType: This dictionary table stores all the options for defining the review status of a model. Examples are "not reviewed," "admin review," etc. All work-in-progress and models waiting for review models are therefore of the status "not reviewed;" posters may be either "not reviewed" or "admin review" (meaning they are in the review process); and bronze, silver, and gold models are "reviewed."

_reviewedBy: This is a set backing table used to define the many-to-many relationship between multiple reviewers (reviewerID) and models (modelID). Set backing tables are ways of defining relationships between entries found within separate tables. This set backing table relates models and persons (or reviewers). Thus, a model may be reviewed by many reviewers; correspondingly, a reviewer may review many models. A review may also include comments by reviewers.

conformancyType: This is a dictionary table that stores the options for defining model conformance to BMW requirements. Currently, this includes WIP (work-in-progress), poster, bronze, silver, and gold.

_personModel: This is a set backing table used to encode the many-to-many relationship between authors (authorID) and models (modelID). Thus, a model may have many authors, and an author may own many models through the use of this table.

Fig. 9 illustrates the tables involved in the representation of a module:

module: This is the highest-level description of a module and is represented with a core table. This module consists of a name, a description, its creation date (created), its last modified date, and the module's owner (ownerID). The libraryID field allows us to define a module as belonging to a particular library of routines. Similar to the *model* table, it contains the generic information for describing a module, regardless of the number of versions or alternate implementations of the module.

moduleVersion: This core table identifies a specific version of a module and resembles the *modelVersion* table; it contains fields describing the version (major, release, and revision), a note field, the creator/owner of the module (ownerID), the creation date (created), and the modify date (modified).

library: Modules can be grouped in libraries. This core table stores the names and descriptions of a library. A library is represented by a name and a description of what the library represents.

_modelModules. A set backing table that associates models (modelID) and modules (moduleID), this table allows us to define many-to-many relationships between models and modules. One of the key features of BMW is the reusability of modeling code. With this table, we can establish relationships allowing different models to use the same module, and a single model can contain multiple modules.

implementation: This core table stores the file information that forms part of the implementation of a specific module version. The implementation gives us a better idea as to the module's version (moduleVersionID), what kind of file it is (Java, NSL, GENESIS, etc.) using the fileTypeID field, and the location of the module's file (theFile). It also describes the owner/creator of the implementation (ownerID), as different implementations of the same module may be programmed by different people. Finally, it also gives the creation (created) and last modified dates (modified).

_personImplementation: This is a set backing table that identifies the author (authorID) of a file (fileID) that forms part of an implementation. This table allows an implementation to be created (or modified) by several people, as well as allowing a single person to create/modify multiple implementations (of any module for which they have permission).

fileType: This is a dictionary table that identifies file types (C, Java, Makefiles, etc.).

The BMW database has currently been constructed so as to store hyperlinks pointing to a particular model, but the model's code is not physically stored within the database. Obviously, these hyperlinks may point to a model created using any computer language or modeling package. The model tables do have a way of representing non-NSL models, which may not be built in a modular fashion as are those written in NSL. The *modelOSSpecific* and *implementation* tables are constructed with the non-NSL model in mind. Each contains rows that are related to storing executables, parameter files, output files, etc., as well as the file type. With the use of SCS, only NSL models will currently be capable of executing within BMW. In the near future, we will be adding an applet field associated with the particular instance of a model (*modelOSSpecific*). Any model that contains a Java

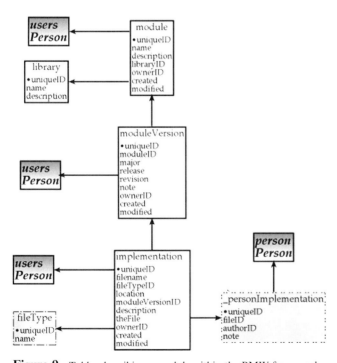

Figure 9 Tables describing a module within the BMW framework. A *module* is the basic building block for a model. The *moduleVersion* and *implementation* tables handle module versions and implementations in a similar fashion to that seen with the model-related tables.

applet, such as the NSL 3.0 models, will be executable by selecting this field within the BMW Internet interface.

Documents and Clumps

SDB data are stored as "clumps" (which at present—but not in the future—take the form of pieces of text) with a reference to the source document. For each clump, a number of keywords may be stored to facilitate searching. Clumps can be linked to models or modules using a standardized descriptor (e.g., "replicates data," "justifies connection," etc.). This allows two-way searching; for each clump, a list of models and modules that are linked to it can be generated. Similarly, for each model or module, a list of clumps that describe data related to it can be displayed.

Below we briefly describe the tables commonly associated with documents (Fig. 10):

document: This core table stores bibliographic information about clump sources. The current table description is based upon LaTeX format. It is used to store all allowed types of documents (see *documentType*, below) within the same table. It also includes the name of the owner of the document (ownerID), the date the document was created (created), and the last modified date (modified). Normally, a document entry would not need modification unless it contained an error.

documentType: This dictionary table defines the different kinds of documents available within BMW/SDB. These documents include articles, booklets, chapters, conference articles, manuals, theses, technical reports, and unpublished documents.

clump. This core table stores a piece of textual data taken from a reference. It contains the identifier that links it to the originating document (documentID), the text of the clump (theText), the document's URL (if the document is available online), the clump's owner (ownerID, which need not be the same as the document owner but is simply the person who inserted the clump into the database), its creation date (created), and its last modified date (modified). As with the *document* table, modification would occur only if there were errors (such as typographical) which occur within the clump.

keyword: This dictionary table defines the list of all keywords used with clumps. This allows searches of clumps by keyword rather than by searching the text of each clump. For example, "Patients with lesions to the intermediate and medial cerebellum could not learn to adapt to throwing while wearing wedge prism glasses" would have the keywords cerebellum, learning, and prism adaptation. Users can add as many keywords as they like to a given clump.

_keywordClumps: This set backing table encodes many-to-many relationship between keywords (keywordID) and clumps (clumpID). Thus, a single clump may have multiple keywords, and a single keyword may be associated with many clumps.

Clumps form the basis for data summaries. Presently, clumps are only in text form, but future implementations will involve a wide variety of different data types (e.g., figures, tables, etc.). For models striving to achieve silver or gold status, clumps are essential. They can be used in numerous ways: to characterize the relationships of connections and parameters within a model/module, to document hypotheses, to reference known experimental data which the model is meant to replicate, or to support or refute assumptions made by the model. On the most basic level, a clump contains one or more statements about a topic and links to the reference from which it was paraphrased. We can improve upon this by adding information regarding the relevancy of the clump to the model/module. We have created the following tables to allow us to link with the clump table defined within the SDB (Fig. 11):

_clumpModel: This set backing table associates clumps (clumpID) with either models or modules (atomID). Models/modules may have many clumps of supporting information; conversely, a clump may support several different models/modules.

atomTypes: This dictionary table defines the type of atom, or table type, for tuples in the relevant table. Because *model*, *modelVersion*, *module*, and *moduleVersion* all rely upon a "global" unique ID (generated by the *ids* table), we need a way of determining to what object the clump is attached. This table contains the terms "model," "module," and "moduleversion" so that we know to which item type the clump is related. Without this information, the database would not be able to

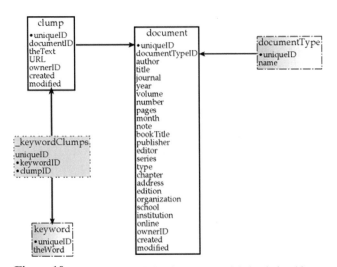

Figure 10 The tables describing documents and their relationship to clumps within BMW. A *document* may be one of several different kinds of documents as defined by *documentType*. A *clump* is a portion of text taken from the document. It may have one or more *keywords* associated with it.

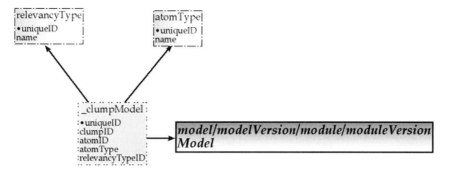

Figure 11 The tables describing the relationships between clumps (found within the SDB database) and objects within BMW. The *atomType* table allows us to define the different object types within BMW (model, module, etc.) that may be linked to a *clump*.

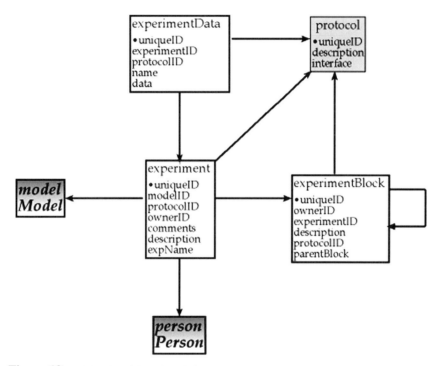

Figure 12 Tables describing the relationship between experiments and models within the database. An *experiment* defines the experiment to be conducted. An *experimentBlock* may define one or more blocks within an experiment (such as seen in learning experiments). A *protocol* may be defined for an experiment, in addition to each *experimentBlock*. Finally, the results of the simulation are stored within *experimentResults*.

follow a clump to a given model/module that uses it for reference (be it support or contradiction).

relevancyType: This dictionary table stores all the strings that are used to describe the relation between a clump and an atom (module/model). Examples include "justifies connection," "replicates results," and "empirically defined parameter."

Experiment

Finally, a model can be used with many experiments. For example, the Dominey and Arbib saccade model can perform simple, memory, and double saccades with only a change to its input parameters. These experiments may include protocols, experiment blocks, and experimental results associated with the experiment (Fig. 12):

experiment: This stores information to execute an experiment using a particular model. Typically, an experiment is defined by the model for which the experiment is conducted (modelID), the creator of the experiment (ownerID), the protocol used for the experiment (protocolID), and the name of the experiment (expName). Sometimes an experiment involves multiple blocks, or steps, such as is seen in a learning experiment. To accommodate this, we have included a field (expBlock) linking to the table responsible for experiment blocks, *experimentBlock*.

experimentBlock: Experiment blocks may be defined in experiments involving multiple parts. For example, the Dart model involves several blocks: one each for learning to throw overhand and underhand without prisms, blocks for learning to throw overhand or underhand while wearing prisms, and the final blocks during which the model performs overhand or underhand throws without prisms. This table stores the name of the experiment block creator (ownerID), the protocol used (protocolID), and a description of the block (description). Because there are typically several blocks, performed in a sequence, associated with an experiment, the field parentBlock allows us to form a linked list of blocks, with each block also providing a reference to the experiment (experimentID) to which it belongs. Finally, the data associated with each block must be stored. In the Dart model, associated data may refer to the presence or absence of prisms or the target location, among other parameters.

protocol: This links an experiment to a model protocol. Because a protocol may apply to both experiments and experiment blocks, a protocol may be linked with either data type.

experimentResults: This stores results of the conducted experiment/protocol. The links to *experiment* and *protocol* allow us to see which protocol and which experiment generated the results under scrutiny.

6.2.5 Accessing the Database

There are numerous ways to explore BMW and its functionality. We will begin by browsing through the data in order to provide an understanding of how the different tables relate. We will follow this with an example of searching for a particular model and, finally, how to insert information into the database.

Logging into the Database

Upon accessing BMW (http://www-hbp.usc.edu/Projects/bmw.htm), the visitor is first greeted with a brief summary description of the purpose of the database. From this html page, a link is provided to the database. On the left side of the window is a menu used for site navigation. This menu is present for all screens within BMW; at any time the user may click on Home Page and return to this page. There are several items listed within the menu: Model List, for listing the models within BMW; Browse, for browsing the database for models, modules, references, and clumps; Search, for searching the database; Insert, for inserting new information into BMW; Help, a document describing BMW, its tables, and how to use it; and Log In. The screen also provides information to the user about login status and information about the creators and maintainers of BMW. In order to log into BMW, the user clicks the hyperlink titled Log In.

The Log In screen is straightforward. A returning user need only enter a username and a password and then press Enter to access BMW. New users are permitted to request an account using the bottom portion of the screen. The user is prompted for name, email address, and a preferred username. Upon pressing Enter, the user is informed that the request has been sent to USCBP and will be processed shortly. Once users have acquired an account, they can add models to BMW. This account is BMW specific; while the USCBP will have a general account and password, allowing a user to browse all the databases existing within USCBP, the BMW account allows the user further access into BMW. That is, any general user may perform searches or browse the material stored, but only those with a BMW-specific account will be permitted to enter data into the database.

Browsing the Database

Instead of searching for a specific model, module, clump, or reference, users have the option of browsing all the entries in the database. By clicking the hyperlink, Browse, a screen listing the different kinds of data available for browsing is presented. Currently, users can browse models, modules, references, and clumps. We provide a brief description of each in the sections below.

BROWSING MODELS

Fig. 13 is an example list of models stored within the database. This list provides the name of the model, the primary author of each model, and a brief description of the model. Because it is expected that browsing for models will be a common goal of the user, this page may also be reached directly via the Model List hyperlink in the menu. The *model* table (Fig. 8) provides the display with the name, description, and first author of the model. Let's assume that the user is interested in viewing the Dart model. By clicking on the Dart hyperlink, the user is presented with the Web page discussed earlier (Fig. 2) where we see that the Dart model has three different versions: the original version (1.1.1), a version that has added a shoulder model (2.1.1), and a third model with a new interface written in NSL 3.0 (3.2.1) implemented in Java. (Incidentally, this illustrates that BMW is not restricted to storing only models in NSL 3.0.) To display this information, BMW relies not only upon the *model* table for the model name (name) and author (ownerID), but also the *modelVersion* and *modelOSSpecific* tables to describe the different versions of the model (major, release, revision, etc.).

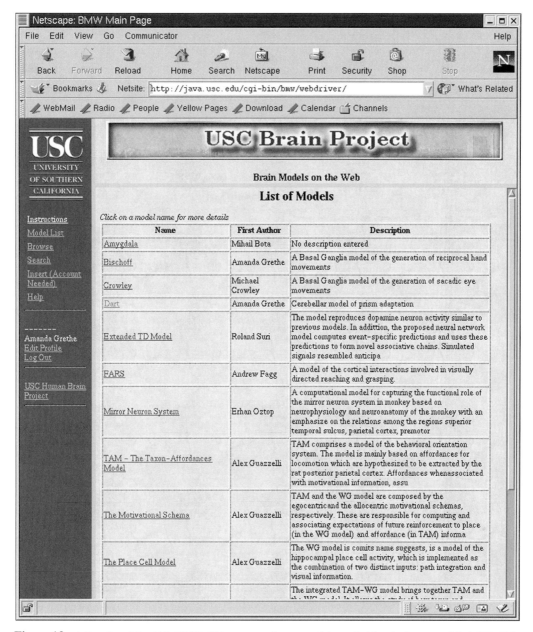

Figure 13 A list of models within BMW. This page may be accessed through either the Browse or Model List links in the menu. The models list relies upon the *model* table to display the name, first author, and description of models stored within BMW. Selecting the hyperlinked model name will take the user to Web pages providing more details about the chosen model.

BROWSING MODULES

To browse existing modules within BMW, the user chooses the Module option from the Browse Page. This will display a new page (Fig. 14) containing a tabular list of all the modules currently stored within BMW. The information displayed includes the name of the module and a brief description, obtained from the *module* table (Fig. 9). For further information regarding a module, the user clicks the hyperlink for that module. For example, clicking the DART_top hyperlink brings up a new page showing a list of versions that exist for this module and a hyperlink for viewing clumps related to the module. The information regarding module versions is stored within the *moduleVersion* table. Using the _clumpModel_ set backing table, we are able to find which clumps (stored in the *clump* table) are associated with the module.

BROWSING REFERENCES

Browsing references is useful should the user wish to see what references already exist in the database without doing a search on keywords or authors. Choosing the All References hyperlink on the Browse page brings up a new page listing all the references currently in BMW. Reference information (found in the *document* table; Fig. 10) displayed includes the name of the author(s)

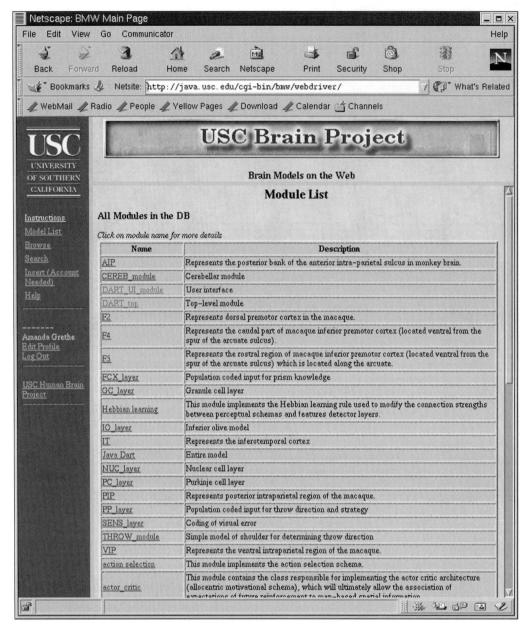

Figure 14 Browsing the modules list. Here, the different modules and their descriptions are presented when the user chooses to browse modules via the Browse hyperlink. The information displayed originates from the *module* table. Selecting a module name will display a new Web page with more details regarding the module.

(author), the year of publication (year), the title (title), the journal (journal), the volume (volume), and the page number (pages) of the reference (Fig. 15). Users can then view a reference more closely by clicking on its hyperlink. The user is told the type of reference (in this case, a journal article) as well as the publication information, provided by the *documentType* and *document* tables. Additionally, the user is also shown a list of clumps related to the article. The *clump* table contains a field titled documentID; it provides the identity of the document from which a clump was obtained. This field is used to list all clumps in the database associated with a particular document. A reference need not have clumps in order to exist within BMW; it merely needs to be entered by an authorized user.

BROWSING CLUMPS

Finally, users can browse clumps within the database by clicking the All Clumps hyperlink on the Browse page. This brings up a page listing all the clumps that reside within BMW, as described by the *clump* table (Fig. 10). The user is given the name of the source document (via the *document* table) and a title for the clump and is also given the opportunity to view the clump in closer detail by clicking on its title hyperlink. Details of a clump (Fig. 16) include its source reference (documentID), the

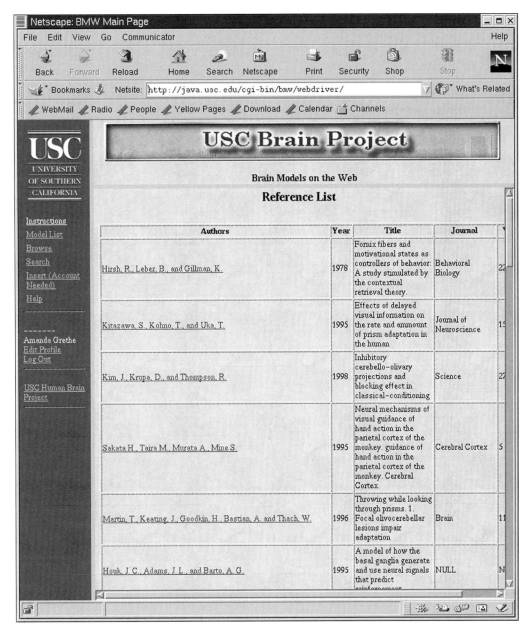

Figure 15 Browsing the reference list. The user is presented with the following reference information (found in the *document* table): the authors, year, title, journal (or other document as appropriate), volume, and the page number of the reference. Selecting the author hyperlink will create a new Web page with further information about the selected document, such as clumps related to the document.

Figure 16 Details of a clump. The user is presented with the author(s), year, and title of the article from which the clump was taken. The compete text of the clump is also presented, along with the keywords associated with the clump and a hyperlink to the list of models and modules that use it.

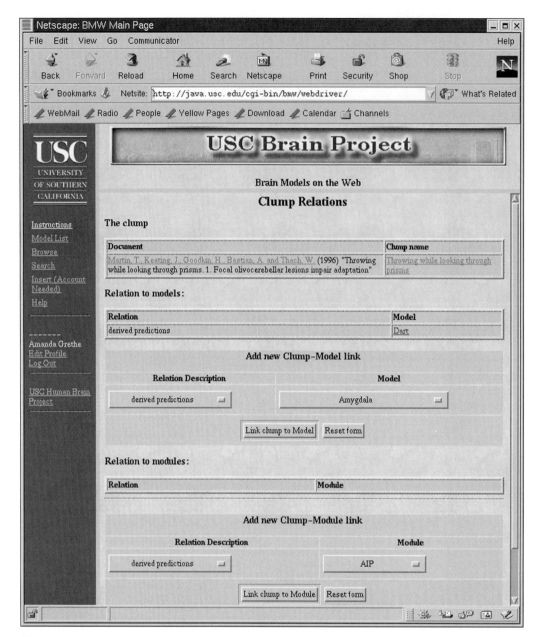

Figure 17 This clump can also be used to illustrate the precise relationship between the text and the model(s) that used it. Here, the Dart model is shown to derive predictions from the selected clump of text.

complete text of the clump (theText), keywords associated with the clump (via the _keywordClumps set backing table), and a link to the models and modules linked to the clump. The latter link exists both on the Clump List and the Clump Details pages. Clicking on this link displays a screen describing which models are associated with the chosen clump (Fig. 17). This is determined by the _clumpModel set backing table; given a uniqueID of a clump, it provides a list of identifiers for models which use the specified clump. In the example shown, the clump related to prism throwing is associated to the Dart model and was used to derive predictions. The pull-down menus below allow the user to add links between clumps and models and between clumps and modules.

Searching for Models and Data

Users are able to search the database for three different kinds of data: models, clumps, and documents. Upon clicking the Search link within the menu, the user is presented with the entry page for search capabilities. The different searches are presented below.

SEARCHING FOR MODELS

To search for models from the Search Page, the user chooses the Models hyperlink. This brings up a search screen related to models. The user is given the option of searching for either a text string within a model's title or for a specific model author. Depending upon the search chosen, different fields will be checked for the search

Click on a model name for more details		
Name	First Author	Description
Dart	Amanda Grethe	Cerebellar model of prism adaptation

Figure 18 The result of a search for a model containing the word "Dart" in its title. In this example, only one model satisfies the search criteria.

term; a text string search will check the name field of the *model* table, while an author search will rely upon the ownerID field of the *model* table (Fig. 8). In our example, the user has chosen to search for all models containing the word "Dart" in the title. The search is initiated when the user clicks the Word Search button, and the next screen displays the search results (Fig. 18). In this case, there is only one model containing "Dart" in the title.

SEARCHING FOR CLUMPS

A user is able to do searches for clumps in three ways: search for a text string contained within a clump, search by a keyword substring, or search by choosing the keyword from a pull-down list of keywords. A text string search within a clump will involve searching the Text field within the *clump* table. This can be very time consuming, but it is useful if the user is looking for something that may not exist as a keyword. Keyword searches are quicker, whether using a predefined list or a user-specified substring. Keywords available within the database are stored within the *keyword* table (Fig. 10). The *_keywordClumps* set backing table is searched for the entered keyword (keywordID) and links to the *clump* table (via clumpID) to return a list of clumps associated with the given keyword. For example, if the user chose to search by the keyword "prism adaptation" and then clicked the Find button, the *_keywordClumps* table would be used to determine a list of clumps related to prism adaptation. Upon submitting this search, the user is presented with a new screen listing clumps that contain it. This list includes the source document, a portion of the clump of text, and all keywords associated with that clump. The user can continue searching on different keywords by clicking on the desired hyperlinks on the right-hand side of the screen. Clicking on the clump text will bring the user to the screen describing the clump as previously seen in Fig. 16.

SEARCHING FOR DOCUMENTS

Finally, users can search for specific documents within BMW. By choosing the Documents hyperlink on the Search page, the user is presented with a search screen related to references. Users can currently perform a search by author or by a text substring. For example, a search for all articles within BMW that are authored by "Thach" will use the author field of the *document* table to bring up a new page listing the articles co-authored by Thach that the database currently contains. Clicking on the hyperlink for a given reference brings up a new display giving document details and listing related clumps. Searching by a text substring will involve searching the title field within the document table.

Inserting Models and Data

Before data can be entered into the database, the user must first log in (see above for details on the login process). To begin inserting data of any kind within the database, the user selects the Insert hyperlink from the menu. A page will appear that details the different kinds of data that may be entered into BMW (Fig. 19). Currently, users may choose to insert references, persons, models, and modules.

INSERTING REFERENCES

From the Insert New Data page, the user selects the Reference hyperlink. The database allows the entry of a large number of classes of bibliographic references, modeled after the format used by LaTeX. The supported reference types include articles, booklets, chapters, conference articles, manuals, theses, technical reports, and unpublished documents (as defined within the *documentType* table; see Fig. 10). In the current example, the user has chosen to enter a journal article. An entry screen appropriate for this reference type appears (Fig. 20). The On-line field is for an optional URL to an online version of the reference. Entering it without the "http://" prefix (e.g., www.name.org/here/this.html) would be a valid entry. The rest of the fields should be self-explanatory; they include the author, title, year, journal name, volume, number, and pages for the article, along with any notes the user may wish to enter regarding that

Insert New Data			
Reference	Person	Model	Module
Add a new bibliographic entry to insert clumps from.	Add a new model author.	Start inserting a new model in the database.	Start inserting a new module (part of a model implementation) in the database.

Figure 19 The start screen, which enables users to insert new data of various kinds. Currently, insertion of reference, person, model, or module is supported. In order to insert one of these data types, the user must be logged in and must select the appropriately titled hyperlink.

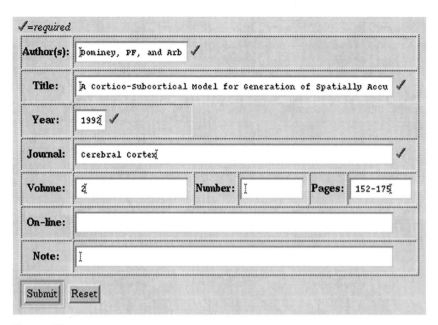

Figure 20 The entry page for storing a new article reference within BMW. At a minimum, the user is required to enter the author(s), the title, the year of publication, and the journal name for insertion within the *document* table.

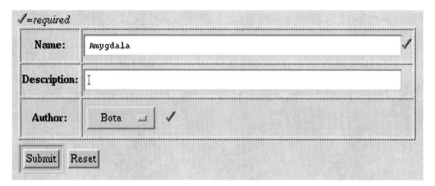

Figure 21 An example of a model being entered into BMW. The minimum requirements are a name for the model and the name of the first author. Model authors are available from a pull-down list of persons stored within the database. The data are then stored within the *model* table.

article. Once the user presses the Submit button, a message appears telling the user if the insert was successful or if there is information missing from the entry. The entered document is added to the *document* table within BMW.

Inserting Persons

Although a model is referenced with its first, principal, author, there may be multiple authors involved with that project. We have therefore provided a method for adding new people to the database who can later be attached to the model. Note that the person is not the same as the BMW users list. From the Insert New Data page, the user chooses Person. This will bring up a page for entering a new person. The minimum required information about a new person includes the first and last name and email address. The user can additionally add the address and the affiliation of the new person. Entered information will be stored within the *person* table.

Inserting Models

From the Insert New Data page, the Model hyperlink leads to the pages used for inserting a model and its modules into BMW. The minimum requirements for the *model* table are a name and the first author (Fig. 21). Note that the Author entry is obtained by selecting from a menu, based upon the names stored within the person table. If the principal author's name does not appear in the menu, it has to be inserted into the database first (see "Inserting Persons"). When the information is entered, the user presses the submit button. The model name will now be entered into the database.

Once a model exists, the user needs to establish the initial version of the model. After a model is entered, the user is presented with a page that allows entry of versions (Fig. 22). After pressing the hyperlink New, the user can begin entering a new model version. When presented with an Insert Version screen (Fig. 23), the user may enter a version number and add notes detailing how the

Figure 22 The screen for inserting new versions of a model. The model name, description, and author (found in the *model* table) are displayed. Previously defined versions of the model may be listed in the table. To insert a new version, the user selects the New hyperlink.

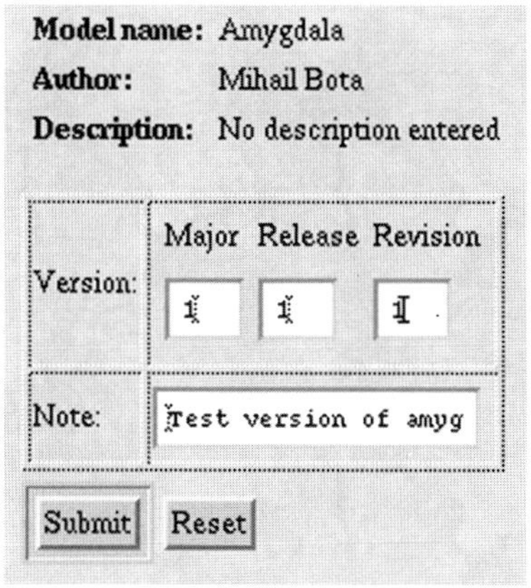

Figure 23 An example of entering a new version for a model. A version is defined by three fields: major, release, and revision. The user can change the version number and add notes describing any difference between this and previous versions of the model.

model is different from previous versions (should they exist). This is stored within the *modelVersion* table, including the modelID that was created when the model was inserted into BMW. Once the version information has been submitted, the user can begin entering model implementations.

The implementation of a model can refer not only to the modules comprising a model, but also to other aspects, such as the machine, the operating system, the compiler, and the graphics environment. We can add new implementations of a specific model version as the need arises. After clicking the New hyperlink while looking at a model version, the user is then presented a page for entering implementation information (Fig. 24) based upon the fields within the *modelOSSpecific* table. The user is given the option of entering a description, the review status, its conformancy (defined in the *conformancyType* table), and the locations of the executable and tutorials on the model. In order to add the executable code and the tutorial, a URL must be provided. BMW will store the URL, allowing later users to run the model or read the tutorial. Note that the executable and the tutorials are not explicitly stored within BMW.

After the model implementation has been established, the user can now enter modules. From a model's

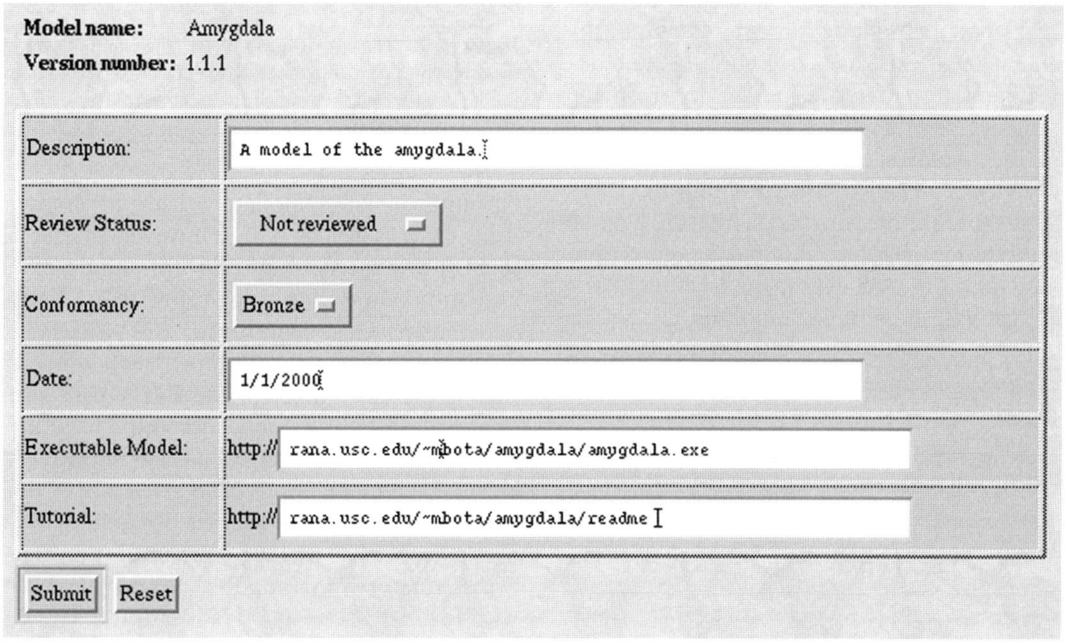

Figure 24 The user is invited to enter pertinent information regarding the implementation of the model. This can include the locations of the executable version of the model and a tutorial explaining how to use the model. This information is stored within the *modelOSSpecific* table.

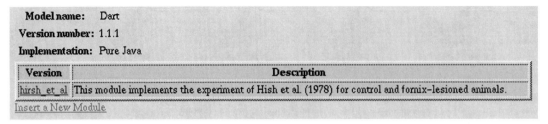

Figure 25 The module list screen. Modules previously defined for a model will be listed within the table. Clicking the Insert a New Module hyperlink allows the user to create new modules associated with the model.

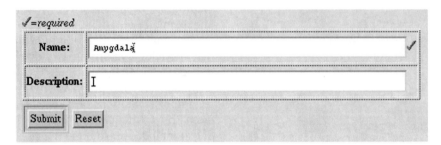

Figure 26 The new module entry screen. The user is required to name a newly defined module and is able to enter a brief description. The entered information is stored within the *module* table.

implementation (Fig. 25), the user may click on the Insert a New Module hyperlink to bring up a screen for entering modules (Fig. 26). The module insertion screens follow a logic similar to that seen for inserting a model: create the module (using the *module* table); create a version of the module (using the *moduleVersion* table); and, finally, define the module's implementation (using the *implementation* table).

INSERTING CLUMPS

Clumps are short summaries of information contained in a reference. Before one can enter a clump, the source document must exist as a reference in the database. New Clumps are entered using the form on the page displaying details of a reference; therefore, the user must first find the document and then click on the author of the clump. All documents are listed within the *document* table, and those people capable of entering clumps are listed within the *person* table. Clump text is entered in the text area (Fig. 27). A URL may be entered for referral to a page with more information on this server in the user's personal space. The default path is set to "~bmw/contrib/users/[user login]/" in case users would prefer to upload the pages to their user spaces on our server. In this case, users must use ftp to post the pages at the specified locations.

6.2.6 Future Plans

Links to SCS

Most of the features of BMW-SDB can be exploited for any model that has a modular structure. BMW

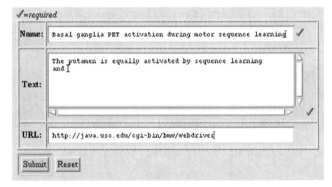

Figure 27 The clump entry screen. The clump is given a name, and the text is quoted within the text area provided. The user may also enter the URL at which the originating document may be found. The data are stored within the *clump* table.

currently includes models written in modular fashion using Java as well as NSL. However, the current version of SCS is written specifically for the development and execution of models written using NSL 3.0 (Chapter 2.2) and thus provides features for BMW-SDB that are not available for models developed with other simulators. SCS is therefore useful if the BMW user wishes to run or modify a NSL model. To execute a NSL model, the user must first download the SCS/NSL software onto her client machine and start SCS. When SCS downloads a model, it will be saved as a set of module files within the user's account to reduce communication and database access time. A NSL module typically contains one file with the extensions .nsl (NSL scripting language), .mod (NSL modules), and .sif (schematic and icon file). If it is intended for execution on the NSL 3.0 system implemented in Java, then it will also include a .class file, as well. If it is intended for execution on the NSL 3.0 system implemented in C++, it will also include a .o and .exe file. The current implementation of SCS provides access to a library of models and modules and tools for model development. Work currently under way will extend SCS so that, should the model also contain summaries (i.e., the model is silver or gold), documentation from the SDB will be viewable via the database browser within as well as (the case documented in this chapter) outside of SCS/NSL.

After making modifications to an existing model, the user may also choose to upload a model to BMW via SCS. If the model has been changed, it would need to go though the review process again. To make major updates known, the user must choose to submit these updates to the poster session. When the SCS/NSL session ends, the user will be asked if the current work should be returned to BMW as a new version. If the user elects to do so, the files will be stored in the appropriate manner. SCS/NSL will also have the capability to submit documentation to the database; alternatively, users may work with BMW using their favorite Internet browsers. In the examples throughout this chapter, we have worked with BMW via the Web interface.

Links to Experimental Neuroscience Data

The NeuroCore schema (Chapter 3.2) describes a way to organize large datasets from the neuroscience community. These diverse datasets provide the details needed by a computational model if it is to capture the behavioral phenomenon in question. Currently, BMW relies upon summary statements that describe the experimental data; it also relies upon published results for a graphical comparison of the modeled and experimental data. One significant method of extending BMW is to allow it to link to the experimental data stored within NeuroCore. This would serve several purposes: (1) it would allow for a direct comparison between experimental and modeling data; (2) it would directly provide the data necessary to determine the model design; (3) model results for particular simulations could be stored within the NeuroCore framework; and (4) it would further serve as model documentation (e.g., a simulated neuron matches parameters from a neuron studied *in vivo* and stored within NeuroCore). This enhanced functionality would lead to a more powerful tool in the development and testing of computational models and would serve to better inform the neuroscience community on the linkages between simulated and experimental datasets. The tables in BMW that represent experiments and results have yet to be implemented. Once complete, these tables will be used in comparisons between NeuroCore data and simulated data.

Links to Synthetic PET

Synthetic PET, as described in Chapter 2.4, is a method of calculating PET images from a computational model. These images may later be compared with human PET data for later model refinement or can assist in the development of new experimental ideas. Synthetic PET relies upon the overall synaptic activity generated by the model's connections. At this time there is no direct interface between BMW and Synthetic PET; rather, Synthetic PET is a standalone applet that accesses the BMW database for its synaptic data. We would like BMW and Synthetic PET to be more fully integrated, such that a model being viewed within BMW can automatically launch a Synthetic PET comparison. With plans to include neuroimaging data within NeuroCore, users would be able to use Synthetic PET as yet another way to compare simulated and experimental data.

References

Dominey, P. F., and Arbib, M. A. (1992). A cortico-subcortical model for generation of spatially accurate sequential saccades. *Cerebral Cortex* **2**, 153–175.

Martin, T. A., Keating, J. G., Goodkin, H. P., Bastian, A. J., and Thach, W. T. (1996a). Throwing while looking through prisms. 1. Focal olivocerebellar lesions impair adaptation. *Brain.* **4**, 1183–1198.

Martin, T. A., Keating, J. G., Goodkin, H. P., Bastian, A. J., and Thach, W. T. (1996b). Throwing while looking through prisms. 2. Specificity and storage of multiple gaze-throw calibrations. *Brain* **4**, 1199–1211.

CHAPTER 6.3

Knowledge Mechanics and the Neuroscholar Project: A New Approach to Neuroscientific Theory

Gully A. P. C. Burns
University of Southern California Brain Project, University of Southern California, Los Angeles, California

6.3.1 An Introduction to Knowledge Mechanics

In any scientific discipline, the role of the published literature is to provide an "intellectual environment" for a domain of knowledge where new ideas are presented, contradictions and controversy may be formally aired, and experimental results and their interpreted meanings are archived for subsequent retrieval by other workers. Individual scientists interact with this environment by writing and reading papers. Recent technological developments have accelerated the speed at which this interaction may take place, in terms of the delivery of published information to scientists and of their submissions to the literature, but at present few tools have been developed to help them manage and understand this ever-increasing body of information.

The neuroscience literature has intrinsic complications due to the fact that its subject matter is both broad and deep. It involves many different scientific subdisciplines ranging from animal behavior and psychology through cellular anatomy and physiology to studies of molecular biophysics and biochemistry. Each of these subdisciplines involves different types of data and different conceptual approaches. The scholarly basis of a given field usually involves an enormous amount of in-depth information. For example, there are over 500 different brain regions in the rat (Swanson, 1998), each one of which may be made up of approximately five cell groups, which may, in turn, project to ten other cell groups. This means that the number of connections defining the circuitry between cell groups is probably of the order of 25,000 different macroconnections (see Chapter 4.1). No single individual, no matter how great a scholar, can hope to incorporate anything more than a subset of the available information into his or her thinking without some form of computational support.

A trend of research began in the early part of the 1990s that was concerned with the problem of summarizing the contents of large numbers of papers in a computational format so the summaries could then be analyzed with mathematical methods. These studies were concerned with the global organization of neuroanatomical circuits in the brain and used formal approaches to store the interpretations of the researchers who were collating data from the literature (Burns, 1997; Felleman and van Essen, 1991; Scannell *et al.*, 1995, 1999; Stephan *et al.* 2000; Young, 1993). The practical experience of performing these studies highlighted the main practical obstacles of this approach. First, the process of summarizing the literature is extremely time consuming and requires expert knowledge. Second, no formal methods exist for making summaries forwardly compatible with subsequent work by other collators. Finally, these studies were limited to the neuroanatomical tract-tracing literature and were based on a data model that was based on a simplification of neurobiological concepts present in the primary literature (i.e., descriptions of neuronal connections did not permit axon collaterals to be represented).

The NeuroScholar project represents a progression of this technology into a new phase by introducing key concepts from the field of informatics. Consider "data" as the lowest level of known facts; "information" as data that have been sorted, analyzed, and interpreted; and "knowledge" as information that has been placed in context of other known information (Blum, 1986). If we apply these concepts to the contents of a given research publication, each individual experimental observation or fact may be represented as "data," the results and methods sections can be considered to be a conglomeration of data and therefore comprise the publication's "information," and the introduction and conclusions comprise the publication's "knowledge." Within the system described here, we will also consider the interpretations of the readers of the paper to be knowledge that may be potentially informative. Additionally, there is no reason to assume *a priori* that different collators will interpret the literature in the same way; therefore, the practicalities of solving the stated problem require a large-scale, multi-user, knowledge-base management system that can represent and contrast multiple interpretations of neuroanatomical data present in the literature.

By defining a rigorous computational approach to implementing these concepts, we seek to define a utilitarian theoretical framework for published neuroscientific data, information, and knowledge as a computational tool. Not only should this framework capture and represent the relevant information, but it should also synthesize it into predictive theories that can be used both by experimentalists and theorists. Throughout this chapter, we will refer to this framework as "Knowledge Mechanics," alluding to the way in which "Quantum Mechanics" or "Statistical Mechanics" defines a useful theoretical framework for certain physical systems. The long-term objective of NeuroScholar as a system that implements a knowledge mechanical paradigm is to define a "computational intellectual environment" for neuroscientists that they can use to manipulate published information from the literature in the same way that an application programming interface (API) allows a software engineer to manipulate programming data. Although the work to date on the NeuroScholar project has emphasized the representation and manipulation of neural connectivity data, the knowledge mechanics framework has been designed and implemented as a general software solution. This will be discussed in more detail in the section concerned with software design.

This chapter will seek to describe how Knowledge Mechanics could achieve these goals and will examine the philosophical basis of the concept of "theory" in neuroscience and how a knowledge mechanical approach may address key issues that would otherwise be impossible to challenge. A discussion about the software requirements of NeuroScholar is followed by a description of the main design features of our data model with worked examples.

6.3.2 Concept of "Theory" in Neuroscience

The definitions of the word "theory" that are appropriate to scientific application of the word are listed below (Merriam-Webster, 2000):

1. "The analysis of a set of facts in their relation to one another"
2. "The general or abstract principles of a body of fact, a science, or an art"
3. "A plausible or scientifically acceptable general principle or body of principles offered to explain phenomena (e.g., wave theory)"
4. "A body of theorems presenting a concise systematic view of a subject (e.g., theory of equations)"

These definitions rely on the concept that a theory is some form of explanation of a set of data and among the listed definitions there is a progression of reliability and rigidity as one descends the list. This graded measure of reliability can be defined in terms of the synonyms "hypothesis," "theory," and "law." A hypothesis does not provide anything more than a tentative explanation of a given phenomenon, whereas a theory implies a more tightly constrained explanation backed up by more supporting evidence. A law suggests a carefully defined predictive statement to which there would be few, if any, exceptions (Merriam-Webster, 2000).

The goal of scientists should ultimately be to progress from hypotheses to laws in their explanations of nature's phenomena. These explanations must be made in a way that is appropriate to the state of our knowledge of the subject in question so that rigidly defined laws are only applied in situations where enough experimental evidence can be produced to support them. If the practical goal of scientific theory is to explain the mechanisms that give rise to existing data and make predictions that can be tested experimentally, then this project's objective should be to provide users with a way of making testable predictions for experiments.

Within the discipline of neuroscience, we postulate that hypotheses are put forward in the conclusions sections of papers as possible explanations of specific experimental results. When considered *en masse*, these hypotheses represent the theoretical framework of the subject and are often expressed in review articles or textbooks. Different hypotheses often contradict each other, and the process of constructing a coherent theoretical "story" (as these things are often described) from this mosaic of different ideas is difficult. We impose the following practical guidelines for theories derived from the literature under the knowledge mechanics framework:

1. The theory should take as much as possible of the data pertaining to a system, behavior, or phenomenon into account.
2. The theory's limitations, including any supporting assumptions, should be stated clearly so that when the theory encounters anomalous data the assumptions may be tested to see if an alternative assumption could accommodate the anomalous data.
3. The degree of abstraction of the theory should not simplify the essential details of existing data. Unfortunately, much of the data in neuroscience are so complex that some simplification is necessary and judgment must be employed to decide which aspects of the data are "essential" and which are not.
4. Any specialized terminology should be defined unambiguously. This is important, as commonly used words (such as the word "connection," for example) may have more than one meaning. Some aspects of nomenclature carry very different meanings depending on the source, meaning that often it is essential to qualify the use of an expression according to a cited definition of a third party.
5. The logical argument of the theory should be self-contained, transparent, and fully explained, so nothing need be taken on faith from the theory's originator.
6. Theories should provide predictions that may be tested in addition to sensible explanations of existing data. This not only provides a validation of the theory but also gives the theory value in terms of its usefulness to the scientific process.

Although these requirements might seem somewhat obvious and based on straightforward common sense, the task of actually implementing them is extremely difficult for the following reasons.

1. The literature is too huge for single individuals to search exhaustively.
2. The act of linking highly heterogeneous and disparate facts together to form theories is very difficult to achieve computationally, and the vast majority of neuroscientists use interpretative reasoning rather than mathematics to solve their problems.
3. Some of the information in the literature is inaccurate.
4. Neuroscientific data are often qualitative and are not measured according to standard units.
5. Functional explanations of neural phenomena are often vague and are usually stated with some qualifying doubt. This is because many of the fundamental questions concerning these explanations have simply not been answered so that many important concepts cannot be defined precisely (e.g., "neural information," "signals," "pathway," "system," "organization").

Neuroscientists often express their ideas, concepts, and models in the form of written prose. This could be considered the lowest common denominator in terms of communicating precise ideas about data. Consequently, the way in which information is transferred between papers is dependent on the language used to express the information which can lead to problems of semantics, vocabulary, and nomenclature; for example, the problems surrounding neuroanatomical nomenclature can be extremely restrictive and confusing (Swanson, 1998).

The remainder of the chapter will be involved with how the NeuroScholar system seeks to solve these problems.

6.3.3 High-Level Software Requirements and Fundamental Design Concepts of the NeuroScholar System

The general issues raised in the last section may be addressed more fruitfully if we consider a specific example: previous work in building databases for summarizing neuroanatomical connection data from the literature (Burns, 1997; Burns and Young, 2000) highlighted several design principles that should be considered carefully in the design of the next generation of solutions to this problem. These issues are listed below:

1. The task of collating information from experimental neuroscientific papers into a machine-readable format is extremely time consuming and should be performed by expert scientists in the field. Even then, there will be differences in interpretation between different collators.
2. It is not sufficient to read a paper, to classify and summarize it, and then only record the summary. It is essential to record the reasoning that led collators to their chosen classification so that subsequent users transparently understand the basis for the chosen summary. Thus, an important design feature for any computational classification of the literature is that interpretations of data should be easy to follow, either by providing the data that underlie the interpretation or by explicitly explaining the interpretation.
3. The way that data are selected as "reliable" or rejected as "unreliable" should be made explicit.
4. The way in which the data are represented computationally should be as conceptually close to the intended representation of the publication's author as possible. I feel that we have to place trust in the experimentalists producing the data and to stay as faithful as possible to the way they see their data.

5. Translating between nomenclatures is of fundamental importance and may be formally addressed by encoding a computational method of translation (Stephan et al., 2000b).

The translation of these design principles into a real application is an exercise in software engineering, requiring at least two separate and clearly written components: a list of the requirements of the software and a description of the software's design. This section will discuss the fundamental, general aspects of the requirements and design, and more detail will be provided in following sections.

At the time of this writing, the NeuroScholar system is a work in progress. A small-scale demonstration is implemented on the World Wide Web and may be accessed via http://neuroscholar.usc.edu/. This chapter is a presentation of software requirements and design principles with the understanding that very few of the concepts presented here have been implemented as an online service at the present time.

Software Requirements

In order to appreciate the design requirements of the system, consider the following worked example. Let us imagine the scene at a neuroscientific laboratory meeting in the future where three users of the NeuroScholar system are discussing aspects of neuronal circuitry. The discussion concerns how two regions of the brain interact with each other, and there is some disagreement. For illustrative purposes, let's call the users Alice, Bob, and Colin.

Alice is an expert on the connections of one of the regions in question (region A). Similarly, Bob is an expert on the second region (region B), and Colin is an expert on the electrophysiology of both regions under certain behavioral paradigms. All have used NeuroScholar to structure their consideration of the literature. Alice and Bob concur about the majority of pathways between the two regions but disagree about some of the connections. Colin knows much less than his two colleagues about the connections but would like to use their expertise to plan his next experiment. In this situation, the NeuroScholar system would allow them to perform the following tasks:

1. If Alice describes all the outputs of region A and Bob describes the inputs of region B, their accounts could be joined by searching for overlap between the two schemes. NeuroScholar could identify any causes of disagreement or differences of interpretation. Alice and Bob could then reconcile any controversy by looking at the reasons why each of them chose the interpretations that they did by examining the annotations they made when reading and re-reading the original papers and by refreshing their understanding of important points in primary literature. Thus, they could then put forward a unified scheme of connections in which they both believe and which describes the pathways from region A to region B.
2. Alice and Bob might then search the global NeuroScholar database for any information that they might have left out of their individual descriptions. If they disagree with interpretations of other users, then they can comment on the other users' interpretations and make their own.
3. They could analyze this scheme, either intuitively or with multivariate statistics, to look for hitherto-unknown high-level organizational principles.
4. They can make this scheme or high-level interpretations available to Colin, who would add his accounts of the electrophysiological properties of the regions in question. It would be immediately apparent to Colin where gaps in the literature revealed lack of data. He could attempt to synthesize his electrophysiological scheme into the scheme already provided by the anatomists or point out where the anatomist's ideas did not mesh with the electrophysiological data.

This example is only illustrative (descriptions of how some of these software requirements may be fulfilled will follow in later sections of this chapter), but it describes elements of the practical requirements made on the NeuroScholar system. More technically, the functionality of the system is based on the concept that users may build a computational representation of their knowledge, which they may then use in several different ways. This functionality is broken down into its component parts below.

BUILDING THE REPRESENTATION

The process of building a representation of users' knowledge in NeuroScholar is accomplished by placing a computational structure on the data that adequately captures the concepts of neuroscience theory. This is equivalent to devising a suitable ontology or data model for it (an ontology may be described as "descriptions of the domain knowledge of some field;" see van Heijst et al., 1997). This ontology must be able to capture as much relevant information as possible, including low-level data from experimental results and methods, high-level interpretations from within the literature, and comments, observations, judgments, and interpretations from individual users of the system. The ontology should be well defined enough to be able to formalize any common aspects of papers from different experimental protocols, but it must also be extensible to allow non-standard, original data to be included.

The adage that "knowledge is power" is especially true in the context of informatics. NeuroScholar is directly concerned with the storage and manipulation of users' expert opinions concerning their colleagues' published work. Thus, the contents of the knowledge base carry political significance, and care must be taken concerning the security of the system. Users may choose to publish their knowledge model within the system and to restrict access to specified users or everyone, or to maintain absolute privacy, sharing data with no one.

Adjusting the Representation

Just as an individual's interpretations will change over time, individual users' knowledge models will need to be updated, adjusted, and maybe even completely reinvented. In order to keep track of all changes made to the database, detailed access logs will be maintained. Users will be encouraged to timestamp "versions" of their knowledge representation based on their particular outlook at a particular time.

Querying the Representation

The retrieval of knowledge from the system may rely on the structured, interconnected nature of the ontology to give the user the capability to query data, information, or knowledge in a combinatorial way. At present, no plans are being implemented to translate language-based queries into the system, so users would require a certain degree of familiarity with the data model in order to construct their queries appropriately. However, a well-designed user interface will provide much of the desired functionality.

Users will be able to tailor their questions in terms of the type of experiments in which they are interested (e.g., "show me all the available injection sites from tract-tracing studies using horseradish peroxidase") or in terms of a specific interpretation (e.g., "show me all the inputs to the supramammillary nucleus in the rat") or even in terms of the interpretations of data (e.g., "show me M. A. Arbib's interpretations of the connections reported in Swanson et al., 1997"). Queries will be made combining data from different experiments (e.g., "show me the areas in the brain that contain neurons that project to the lateral hypothalamic area and also express ranatensin" or "what is the distribution of all peptides that are expressed in the brain of the rat when it is performing hunting behavior?").

Validating the Representation

The system may automatically search for contradictions between users or publications in order to highlight areas of controversy. When a specific contradiction is found, the system informs relevant users and allows them to make a judgment about the contradicting options. In this circumstance, users will be required to explain any judgments they make in order to make the logic behind their decision as transparent as possible for other users.

Analyzing the Representation

In some types of experiment it will be possible to perform "interpretive analyses" where information from the results section of the paper may be interpreted by the system according to logical parameters that would be stipulated as part of the definition of the experiment type. This will be discussed in more detail in the "Software Design" section of this chapter.

Users' knowledge representation could be compiled into a global summary that represents their understanding of a specific phenomenon or system. This summary could provide the input to secondary analyses either for visualization (as was the case for the neuroanatomical connection databases that were the forerunners of NeuroScholar; see Young, 1992), or to investigate organizational properties of the data. Techniques such as non-metric multidimensional analysis (NMDS), non-parametric cluster analysis (MCLUS), and optimal set analysis (OSA) have been successfully applied to similar problems in the past. Analyses such as these may help expert users interpret their data to make experimental predictions. Scannell et al. (1995) used NMDS to analyze a database of connections to successfully predict the presence of plaid-pattern-sensitive cells in the anterior ectosylvian sulcus of the cat. They subsequently identified these cells electrophysiologically as predicted (Scannell et al., 1996). It must be stressed that these approaches are simply tools to aid the intuitive powers of neuroscientists and that we are not claiming that these approaches will circumvent the role of human reasoning in neuroscience.

Statistical approaches may also be used to generate accurate summaries by searching for optimal solutions that satisfy the many well-defined constraints that exist in the data. For example, consider a situation where the act of comparing data that originate from two different experiments is unreliable, but comparisons between data points from within the same experiment are reliable. In cases such as these, it would be possible to compile a set of meta-data comprising all the relevant constraints that could then be analyzed globally to produce global constraints for the whole system. Methods like these were used to calculate finely graded connection weights for the rat visual system (Burns et al., 1996). This was accomplished on the assumption that the density of labeling in tract-tracing studies is correlated with the anatomical strength of the connection, but different tracer chemicals have different sensitivities. Thus, comparisons of labeling density within the same experiment (or tentatively between experiments that use the same technique) reflected differences in connection strength that could not be inferred from comparisons between experiments. This general approach was also used to convert between parcellation schemes in macaque monkey (Stephan et al., 2000b).

Inter-representation Communication: Importing and Exporting

The apparently simple task of collating information from the literature into some form of repository is too time consuming for a single individual to tackle.

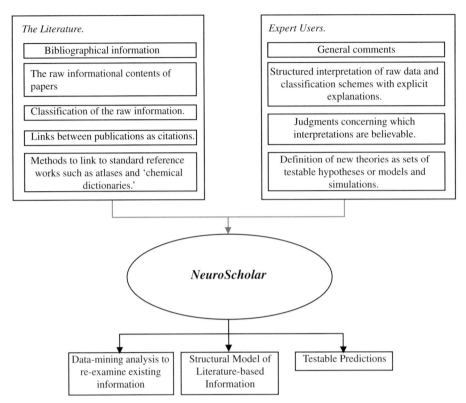

Figure 1 High-level inputs and outputs of NeuroScholar.

Consequently, NeuroScholar has been designed to be a multi-user system in order to distribute the workload of data entry effectively. This will mean that data entered by any given user may be made available to other users through conventional database security methods.

HIGH-LEVEL INPUTS AND OUTPUTS

The high-level functionality of the system may also be described in terms of data flow, illustrated in Fig. 1.

Software Design

The data structure of our ontology for published information is probably the most important single feature of the system. We implement a hierarchical scheme for data structures in NeuroScholar which we here propose as a possible conceptual framework for "Knowledge Mechanics" as an example of a knowledge management system to fulfill our software requirements. This scheme is based on relational database design and is illustrated in Fig. 2. The highest level structures are referred to as *views*, which are made up of *primitives*. The different types of *primitives* are illustrated below in Fig. 3 and are themselves made up of *tables* (which, in turn, are made up of *tableColumns*). *Tables* and *tableColumns* are defined in terms of the computational implementation of the system and will not be considered in detail here.

Views are equivalent to the high-level representations of primitives and their links to other primitives. For

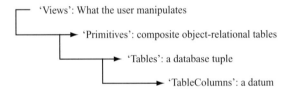

Figure 2 Global hierarchy of data structures in NeuroScholar.

example, if a journal article has three authors, a view that represents that article contains a *journalArticle* primitive and three *person* primitives. The types of primitive that we define within the knowledge mechanics paradigm are *publications*, *fragment*, *objects*, *properties*, *relations*, and *annotations* (where the *annotations* have been separated into three types: *comments*, *justifications*, and *judgements*). Each primitive is responsible for a different function within in the system. The associations between primitives are shown in Fig. 3 as a class diagram using the universal modeling language notation (UML; see Rational Software Corp., 1997). The use of UML in database design provides a powerful, versatile way of communicating schema design (see Muller, 1999, for an excellent introduction to this methodology).

We describe the following *primitives* within our system: *publications*, *fragments*, *objects*, *properties*, *relations*, and *annotations*. There are two subtypes of objects: *interpretationObjects* and *citationObjects*. Annotations are themselves an abstract class (i.e., every annotation must belong to one of three subtypes: *comments*, *justifications*,

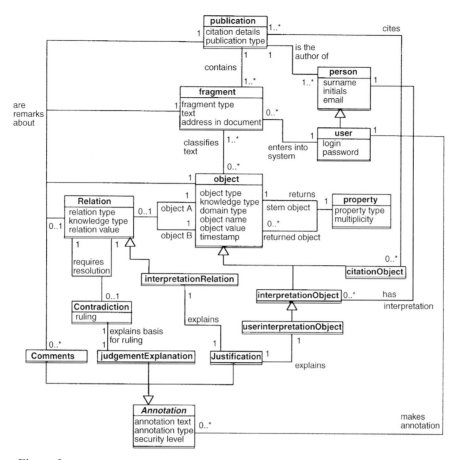

Figure 3 The ontology of the Knowledge Mechanics core expressed in a simplified universal modeling language (UML) class diagram (see text for details).

and *judgmentExplanations*). These inheritance relationships are represented as triangular-headed arrows ("generalization associations" in the UML) in Fig. 3. In the interest of presenting a concise figure, we abbreviated the schema in Fig. 3 by omitting the mechanism of how types are constrained in order to ensure that fragments, objects, their properties and relations, and annotations fit together in a highly structured way. UML representations of methods have also been omitted from the diagram.

PUBLICATIONS THEMSELVES REPRESENTING THE AUTHOR'S OWN WORDS

One of the problems encountered in designing a database system for neural connectivity was that, when I returned to my interpretations after some time had passed, I could no longer remember the original reasoning I had used to generate them. I then had to reread the original text to reconstruct my argument which led me to the realization that it was necessary to store both the individual phrases that were responsible for a given interpretation and their precise location in the text in the database itself. It should be noted that this requirement has legal implications, as it may contravene aspects of copyright law that control access to the content of commercial publications. This is a vital design feature and could be performed by storing links to published material in online journals, if necessary. *Publication* primitives contain the details of the source of the information being represented in the system, corresponding to instructions on how to obtain a copy of the source from outside the system. *Fragment* primitives contain a single datum of information as it appears within the original source material.

CONSTRUCTING COMPLEX OBJECTS FOR NEUROSCIENCE

The role that is played by *object* primitives in NeuroScholar is symbolized by their central position in the schema shown in Fig. 3. We use *objects* as the representations of neuroscientific concepts, and we define the interrelationships between these concepts by defining properties and relations for them.

Object primitives contain the classification of a concept and are classified according to "domain type" and "knowledge type." Domain types are based on a classification of the subject under consideration; in NeuroScholar, this is defined by the experimental method being used to obtain the results that support whatever classification is being represented by the object in question.

For example, this means that we differentiate between objects defined from "tract-tracing studies" and "electrophysiological studies." Knowledge types are based on informatics-based considerations of the classification being used for the object so that we differentiate between an object that is based on primary descriptions of experimental results and an object that is based on a citation of another author's interpretations of the results. Knowledge types in use in the current design of the system are *author citation* (which requires links to a *publication* primitive), *interpretation* (with subtypes *author interpretation* and *user interpretation*), *description*, and *calculation* (the representation of interpretive analyses described above). Domain types in use in the current design of the system are *anatomical experiments*, *core objects*, *physiological recording experiment*, *physiological stimulation experiment*, *tract-tracing experiment*, and *tract-tracing method experiment*.

Within the framework of this system, *properties* and *relations* are *primitives* that operate on *objects*. A given *property* acts on an *object* to return a second *object*, thus enabling the systems' designers to utilize an object-oriented-like approach to classifying information and knowledge. The cardinality of these relations (i.e., how many *objects* should be returned by a given *property* when acting on a given *object* and how many *objects* return a given *object* when acted on by a given *property*) may be defined explicitly, allowing the construction of large numbers of interconnected, versatile, complex data constructs (for a full description of object-oriented design, see Pressman, 1992). *Relations* may be used to attach typed values to pairs of *objects*. This enables users to compare and link *objects* according to structured rule sets that represent characteristics of the information or knowledge in use. Thus, users may describe potential overlap between parcellation schemes (Stephan *et al.*, 2000b), compare the strength of neuroanatomical connections (Burns *et al.*, 1996), highlight contradictions between different classifications of the same data, or define any number of rule-based observations concerning two *objects*.

Fig. 4 illustrates the most important composite object in NeuroScholar: a so-called *neuronPopulation (interpretation, core)* object. This defines a generalized population of neurons and is made up of *somata(interpretation, core)*, *dendrites(interpretation, core)*, *axons (interpretation, core)*, and *terminalField(interpretation, core)* subobjects which contain relevant data describing the characteristics of those components. A type of object of importance in this scheme is the *brainVolume(interpretation, core) object*. These *objects* are simply regions of brain tissue that are defined in terms of a specific publication's parcellation scheme, providing the geometrical substrate for the data in the system. According to this treatment, brain atlases are considered to be "just another publication" with a parcellation scheme expressed as *brainVolume(interpretation, core) objects* to which we link other publications' *brainVolume* objects with relations (see later). Each subobject of any given *neuronPopulation(interpretation, core)* objects may be linked to a given *brainVolume(interpretation, core)* object in order to express its location.

This representation of neuroscientific data is very general and consequently may be used as a template for more specialized representations. We will illustrate different forms of *neuronPopulation* objects that are specific to individual domain-type/knowledge-type pairs in the next section. Additionally, many aspects of the structure of the NeuroScholar model are very similar to those encountered in the design of the USC Brain Project's federated scheme for repositories of experimental data (NeuroCore; see Part 3). In order to make the most of this similarity, we converted the data model of the NeuroCore federation of databases into our object-oriented system in order to make use of NeuroCore's widespread applicability and other desirable design features.

BUILDING RULE SETS WITH RELATIONS

Objects define separate things that may be compared. In the case of neuroscientific data, a good example of this derives from the process of converting the parcellation scheme of one paper to that of another. Stephan and Kötter mathematically formalize this process using set theory so that data can be computationally translated between different cortical maps (a procedure referred to as the Objective Relational Transform, or ORT; see Stephan *et al.*, 2000b). They use one of four rules to describe how the representation of a cortical area in

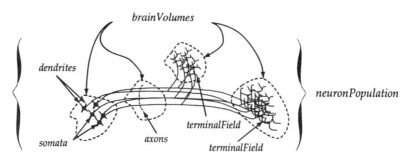

Figure 4 The general *neuronPopulation(interpretation, core)* object (see text for description).

one map relates to a similar representation in another map. Given two areas (A and B) in two maps (1 and 2), the rules would take the following forms:

1. "Area A in map 1 *is equivalent to* area B in map 2."
2. "Area A in map 1 *is enclosed by* area B in map 2."
3. "Area A in map 1 *encloses* area B in map 2."
4. "Area A in map 1 *overlaps* area B in map 2."

Stephan and Kötter develop this framework further by using an algorithmic approach to automatically translate data between maps or to define hybrid maps that use a mixture of nomenclature from different maps.

We use these spatial relationships in NeuroScholar to compare *brainVolume* objects from different parcellation schemes and to describe how data from a given paper may be related to the parcellation scheme in an atlas. Authors often describe a specific location in a paper with an alphanumeric abbreviation denoting the region of interest, and the process of linking data from different papers together is based on linking the regions of interest in the papers to standard brain structures defined in an atlas. For example, Allen and Cechetto performed a combined electrophysiological and tract-tracing study in 1993 where they stimulated sites in the lateral hypothalamus with injections of excitatory amino acids and then performed tract-tracing experiments by injecting tracer into stereotaxically equivalent regions in the contralateral hypothalamus (Allen and Cechetto, 1993). In one case, the injection site of a tract-tracing study is described as being contained within a region referred to as the "tuberal part of the lateral hypothalamic area (LHAt)" by the authors. When the accompanying figure of the injection site is viewed and delineated on a standard brain atlas plate (Swanson, 1998), the region in question appears to span two atlas regions: the zona incerta ("ZI") and the lateral hypothalamic area ("LHA; see Fig. 5). We can capture this detail by generating relations and use these relations to translate any description of the properties of neurons within that region to a standardized parcellation scheme. It would be possible to allow users to trace the exact position of their interpretation of the *brainVolume* on the atlas through the use of a drawing tool for storage and subsequent retrieval (such as the NeuARt data browser; see Chapter 4.3).

In this case, these rules can be widened to accommodate other NeuroScholar objects as well. The equivalent relations of "*is equivalent to,*" "*is enclosed by,*" "*encloses,*" and "*overlaps*" when applied to pairs of *neuron Populations* are "*is equivalent to,*" "*is a subset of,*" "*is a superset of,*" and "*overlaps.*" These relations could be used to link *neuronPopulation* objects in different parts of the knowledge base.

MAKING IT MAKE SENSE TO USERS: THE USE OF ANNOTATIONS

Annotation primitives represent personalized human interactions between users and the contents of the system. They are designed to be read subsequently by other users of the system. There are three types of *annotation primitive,* and each one fulfills a different role. *Comments* may be attached to any *primitive* in order to express the user's opinion concerning that *primitive*. *Justifications* must be attached to *objects* that are based on a *user's interpretation* and simply explain the logic that supports the definition of the *object*. *Judgments* must be entered in order to support a chosen stance where a user selects one *relation* over another when contradictions occur.

6.3.4 Neuroscholar in Detail

In order to describe more detailed aspects of the system, it may be useful to describe how we would represent

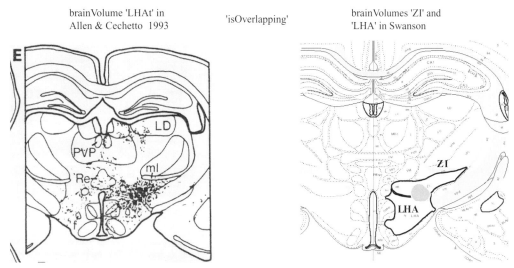

Figure 5 An example of a spatial relation.

data from a type of specific experiment in detail and then extrapolate to show how this approach could be generalized to other experiments. In this section, we discuss NeuroScholar's representation of tract-tracing experiments from a conceptual perspective.

A Short Introduction to Tract-Tracing Experiments

All tract-tracing experiments are based on the same basic design. An experimental animal is anesthetized and small amounts of tracer chemicals are injected into the region of the brain under investigation. The tracer is taken up by the parts of neurons that extend into the injection site, typically at axonal terminals or dendritic arbors. Then the tracer is transported along the cell's axon by intracellular transport mechanisms, either anterogradely from the cell body to the axonal terminals or retrogradely in the opposite direction, or both. Finally, the animal is killed, and its brain is sectioned and stained to reveal a pattern that shows the distribution of transported label (Blackstad et al., 1981).

The end product of a given tract-tracing experiment is a three-dimensional map of tracer concentration at the time of the animal's death. Because tracer is originally applied in a localized injection to a small region of the brain, labeled regions of the brain must be connected to the injection site. The data are usually presented as photographs or camera lucida drawings of sections which show the distribution of label throughout the brain (for example, see Fig. 4 of Sesack et al., 1989).

Studies are classified into "retrograde" and "anterograde," based on the type of axonal transport mechanism mediating the tracer's movement (Vallee and Bloom, 1991). Tracers that have been classified as "retrograde" are assumed to be taken up by axon terminals and then transported along the axons to the soma of neurons projecting to the injection site. In contrast, tracers that have been classified as "anterograde" are assumed to be taken up by the soma and dendrites of neurons and then transported to label the axonal terminals of that cell.

Unfortunately, interpreting labeling patterns is not as simple as described above. The categorization of tract-tracing techniques as "anterograde" and "retrograde" is a simplification. In reality, the uptake and transport processes of the cells can give rise to different patterns of labeling in different circumstances. The sensitivity of the tracer can be described as the ratio of the amount injected to the amount transported. This property determines how much tracer is used in a given injection and, therefore, the size of a given injection site. Tracer sensitivity and other tracer properties are largely determined by the uptake and transport mechanisms of tracer by dendrites, soma, axons, or terminals of cells at the injection site.

This often gives rise to "fibers of passage" labeling, where axons passing through the injection site have taken up tracer and transported it to terminals or soma. Some tracer chemicals can be transported transynaptically, so it cannot be assumed that neuroanatomical tracers only label cells that are directly connected to the injection site (Cliffer and Giesler, 1988; Dado et al., 1990; Sawchenko and Swanson, 1981). If tracer can be taken up by the dendrites of cells that lie on the periphery of the injection site, it may be possible for anterograde label to be transported to the terminals of cells whose soma lie outside the injection site. Transport of some tracers can also occur in both anterograde and retrograde directions. In some cases, tracer could conceivably be taken up by axonal terminals, transported retrogradely, and then transported anterogradely down a collateral branch of the neuron (see Fig. 3 of Warr et al., 1981). When attempting to minimize spurious results and uninterpretable labeling patterns, neuroanatomists cross-reference their studies so that both "anterograde" and "retrograde" methods are applied to the problem in order to corroborate the presence of connections between structures.

Readers of neuroanatomical papers have to trust the papers' descriptions of injection sites and the absence of transynaptic labeling or labeling from damaged and undamaged fibers of passage. Some methods are more prone to error than others and may be treated with some element of suspicion (Sawchenko and Swanson, 1981). If the injection site's boundaries do not coincide exactly with those of the nucleus or area being studied, the labeling produced in a single experiment can only be regarded as a subsample of the total population of cells involved. It would be necessary to take this subsampling factor into account when performing quantitative studies where the number of cells projecting to or from any one region might be counted. The opposite problem is evident when attempting to label the connections of a small structure. If the technique being used produces large injection sites, then the zone of active uptake will extend beyond the boundaries of the nucleus under study and produce false-positive labeling from neighboring regions.

Some techniques are more sensitive than others, labeling cells that others miss (Wan et al., 1982). This may imply that the cells labeled in these experiments are usually an underestimation of the total population of cells with active uptake zones in the injection site. One would normally assume that this subsampling does not suffer from systematic errors, but it is possible that some classes of cells may be more strongly labeled than others; for example, cells with very extensive terminal arbors might tend to take up more retrograde tracer than neurons with small trees.

There are many different tract-tracing techniques available to the experimental neuroanatomist, but they do not, in general, produce results that conform simply to an ideal "retrograde" or "anterograde" model. The most commonly used methods include the autoradiographic technique using tritiated amino acids, horseradish peroxidase, horseradish peroxidase conjugated to wheat germ agglutinin, the most common fluorescent dyes, and *Phaseolus vulgaris* leuco-agglutinin. These methods have been extensively studied and reviewed by Bolam (1992), Heimer and Robards (1981), and Heimer and Zaborszky (1989); the reader is referred to these standard neuroanatomical texts for more information.

We do not consider lesion studies or electrical stimulation studies to be tract-tracing methods, as neither method involves injection of an actively transported substance. The earliest techniques for understanding connections were based on introducing damage to the brain structures under study and then using the Fink-Heimer silver-staining procedure to label damaged axons and terminals and indicate the course of projections from the damaged area (see description in Blackstad *et al.*, 1981). However, the nature of the lesion can affect the data results. Aspiration and radiofrequency lesions suffer from the problem that fibers passing through the area where the lesion is delivered also degenerate. Cytochemical lesions avoid this problem by selectively damaging cell bodies and leaving fibers of passage unaffected. Electrical stimulation experiments involve placing a stimulating electrode in one area and a recording electrode in another and then measuring responses at the recording site after passing current through the stimulating electrode. This technique is used relatively rarely (but see Finnerty and Jefferys, 1993).

Representing the Logical Structure of Tract-Tracing Injections in NeuroScholar

Based on the characteristics of tracers described in the last section and a working knowledge of the tract-tracing literature, we devised an object scheme for experimental data that is schematically illustrated for an ideal anterograde tracer in Fig. 6. A *neuronPopulation(description, tract-tracing experiment)* object that is based on a *description* of data from a tract-tracing experiment is characterized by the following list of subobjects:

1. An *injectionSite(description, tract-tracing experiment)* subobject (this has a *brainVolume(description, core)* sub-subobject denoting the enclosing region that contains it)
2. An array of different *labeling(description, tract-tracing experiment)* subobjects, each of which could correspond to different types of labeling in different regions (so that each member of the array has a *brainVolume(description, core)* sub-subobject)
3. The *neuronPopulation(description, tract-tracing experiment)* object linked to an *experimentalProtocol(description, tract-tracing experiment)* object describing all aspects of how the experiment was performed

If a different protocol is used, then the same subobjects are used to represent the data (i.e., an *injectionSite* object, an array of *labeling* objects, and an *experimentalProtocol* object), but the meaning of this information

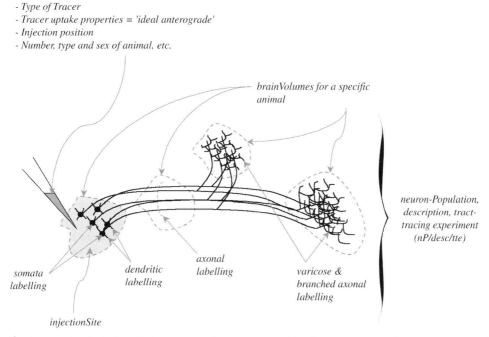

Figure 6 A *neuronPopulation(description, tract-tracing experiment)* object from an anterograde tract-tracing experiment.

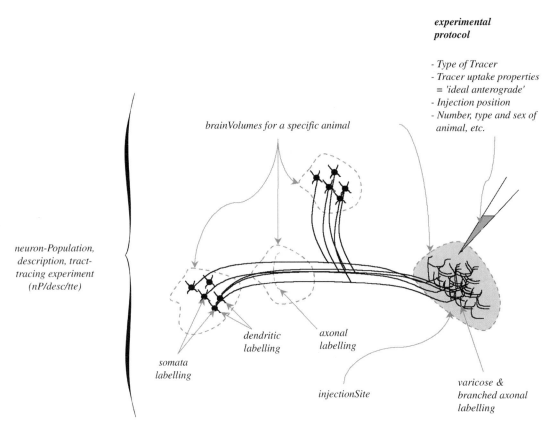

Figure 7 A *neuronPopulation(description)* object from a retrograde tract-tracing experiment.

may be completely different. A case illustrating the use of an ideal retrograde tracer is illustrated schematically in Fig. 7. Sadly, all retrograde tracers suffer from problems of uptake by fibers of passage (meaning that the tracer chemical is taken up by axons that are simply passing through the injection site without forming synapses), so a simple classification of tracer methods into ideal anterograde or retrograde types is unsatisfactory.

Instead, for each tracer chemical, we define an array of four values that denotes the strength of tracer uptake in the *dendrites, somata, axons,* and *terminalField* of a *neuronPopulation(interpretation, core)* object (see above). Thus, if we assume that PHAL tracers have ideal anterograde properties, we define the *tracerProperties* array of PHAL to be [*dendriteUptake* = "strong," *somataUptake* = "strong," *axonsUptake* = "none," *terminalFieldUptake* = "none"]. By the same token, we might describe the *tracerProperties* array of Fluoro-Gold to be [*dendriteUptake* = "none," *somataUptake* = "none," *axonsUptake* = "weak," *terminalFieldUptake* = "strong"]. At present, we define these arrays as designers of the system, but they could be derived from experimental data by deriving these arrays from NeuroScholar representations of the neuroanatomical experiments that define and test the methodology (e.g., Gerfen and Sawchenko, 1984; Wessendorf, 1991). This illustrates a very important aspect of the knowledge mechanics design

strategy: to represent every logical stage of the reasoning process fully transparent and driven by experimental data (as much as possible). Practically, this may be defined within NeuroScholar by requiring that certain types of analysis require the results of other analyses as input data.

The results and conclusions of tract-tracing papers are most often concerned with "connections," the concept that one specified brain region contains neurons that project to and terminate in another region. This is a very commonly used concept, and we represent it in NeuroScholar by defining a new class of objects referred to as *blackBoxConnections(author interpretation, tract-tracing experiment)*. These objects have three subobjects: two *brainVolume* subobjects denoting the regions of origin and termination of the connection and the *connectionStrength* object taking an ordinal value to denote the strength of the connection (i.e., "zero," "weak," "moderate," "strong"). A schematic illustration of this design is shown in Fig. 8.

The schematic descriptions we have laid out are simple illustrations of the data model being implemented. This is illustrated as an Entity-Relationship diagram in Fig. 9, which shows how the NeuroCore scheme was used a basis for the object-object relationships within the broader scheme of NeuroScholar's framework for knowledge mechanics (see Fig. 3).

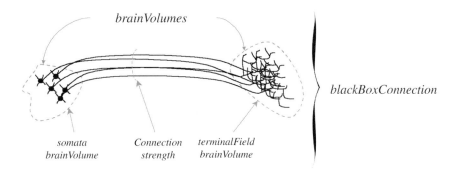

Figure 8 A *blackBoxConnection(author interpretation)* object from a retrograde tract-tracing experiment.

Broadly speaking, the object model may be broken into the following parts: (1) experimental meta-data (see the *expMetaData* objects in Fig. 9), made up of *interpretation* objects that capture a paper's knowledge; (2) research data (see the *researchData* objects in Fig. 9), which are made up of *description* objects from a paper's results section; and (3) *description* objects, which capture the experimental structure (see *experimentCore* objects in Fig. 9), the protocol and experimental manipulations (see *protocol* and *expManip* objects in Fig. 9), and details of the experimental subjects being used (see *researchGroup* and *researchSubject* objects in Fig. 9). There are two sets of "add-on" objects: *brainVolume(interpretation, core)* objects, which capture details of the parcellation scheme of the paper, and the *chemicalDictionary* objects, which capture details of the chemical properties of the tracer chemicals used.

Interpretive analysis procedures are represented in Fig. 9 as a dashed box superimposed on an arrow that leads from the children of the *researchData* object to the children of the *expMetaData* object. These procedures apply queries to the research data contained in a paper and then apply rules derived from the *tracerProperties* (see above) of the tracer chemical used in the experimen to generate *neuronPopulation(calculation, core)* and *blackBoxConnection(calculation, tract-tracing expt)* objects. These objects can be used to augment a user's interpretation of the data in a paper, especially where an author's own interpretations may be incorrect (if authors assumed that early tracer methods did not suffer from the fibers of passage problem, as was often the case).

More Relations in NeuroScholar

The so-called "strength" of neuroanatomical connections is often used to prioritize how different connections influence the global organization of a system (e.g., Scannell *et al.*, 1995), but the task of analyzing the interrelationships between descriptions of connections taken from different studies is difficult to approach from a formal perspective. Most neuroanatomical papers offer qualitative descriptions of connection strength rather than quantitative measurements. In the rat, quantitative data do exist for a limited number of connections that have been particularly well studied (Linden and Perry, 1983; Martin, 1986) but does not exist for the vast majority of connection reports. For example, the number of retinal ganglion cells that project to the superior colliculus is of the order 105 (Linden and Perry, 1983), and the weakest retinal efferents may only involve a few fibers, such as the retinal projection to the inferior colliculus (see Fig. 5 in Itaya and van Hoesen, 1982). In addition to this, the sensitivity of different neuroanatomical methods varies over at least two orders of magnitude (Ter Horst *et al.*, 1984; Trojanowski, 1983; Wan *et al.*, 1982). Furthermore, the density of label produced in any neuroanatomical experiment is dependent on the concentration and volume of tracer injected (Behzadi *et al.*, 1990).

A formal approach to generate finely graded connection strengths has been described that uses the basic premise that comparisons between labeling patterns representing connections within one experiment are very likely to be accurate, and comparisons between labeling patterns representing connections within one paper are quite likely to be accurate (Burns, *et al.*, 1996). Large numbers of these comparisons can be evaluated as a set of constraints for an optimization engine in order to calculate ordinal schemes of connections that best fit the data. Burns *et al.* used the MBOLTZMANN program to perform this process (Burns *et al.*, 1996; Hilgetag *et al.*, 1995). These comparisons can be built up within NeuroScholar as a set of ordinal relations ("is greater than," "is less than," etc.) between the *connectionStrength* subobjects of *blackBoxConnection* objects.

This process is a general model for a class of analyses, where the knowledge base can be queried to provide a large set of constraints between objects and then this set can be analyzed with optimization approaches such as MBOLTZMANN or optimal set analysis (OSA; see Hilgetag *et al.*, 2000) to give an accurate appraisal of the organization of the system within the bounds of the constraints defined by the literature.

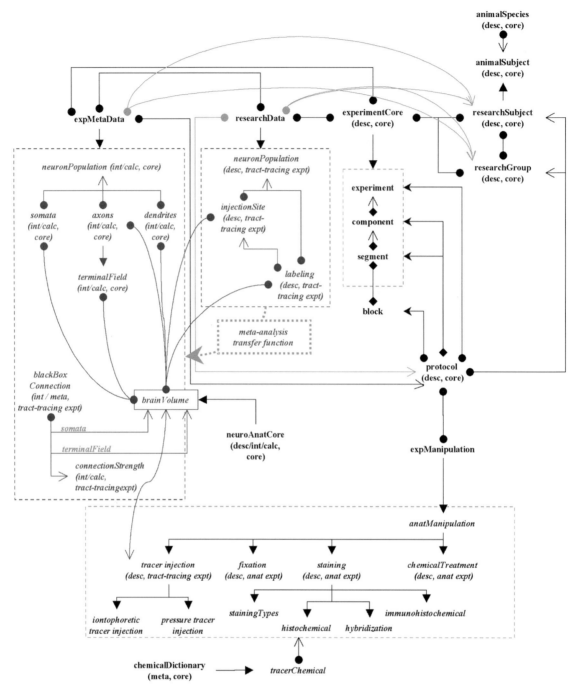

Figure 9 An entity-relationship diagram of the NeuroScholar representation of the NeuroCore database design. Arrows denote relationships where "──▶" represents a parent-child inheritance relationship and all other double-headed arrows denote an association. Multiplicity types of these associations are expressed as different types of arrowheads: Arrows ending in "──▶" represent a multiplicity of "0–1"; "──▷" represent a multiplicity of "1"; "──◆" represent a multiplicity of "0–n;" and "──●" represent a multiplicity of "1–n."

Extending NeuroScholar to Other Data

NeuroScholar may be used as a knowledge mechanics framework for information from non-tract-tracing data from other neuroscientific experiments. Here, we describe electrophysiological or *in situ* hybridization data in the same illustrative framework as we used for tract-tracing experiments. As with tract-tracing studies, the key object in these studies will be the *neuronPopulation* object. In electrophysiological experiments, the *neuronPopulation* will be defined by the position of the recording electrode and will represent a "physiologically defined cell class," whereas in neurochemical experiments the *neuronPopulation* will be defined by the presence of labeling across the whole brain. These experiments may be broken into four parts:

1. *Preparation:* Training, lesions, conditioning, etc. (anything that an experimentalist might do to an animal before stimulating it within the experiment; see below)
2. *Stimulation:* A specific physiologically defined stimulus (such as an airpuff blown onto the cornea to elicit an eyeblink, an audible tone, or an injection of electrical current or a pharmacological agent)
3. *Cellular dynamics:* How a group of cells is functionally active electrophysiologically or neurochemically
4. *Behavior:* What the animal is doing

The causal relationships between these different parts of the system are shown in Fig. 10. These relationships have no significant relationship concerning the way that the information is handled in the database itself but may strongly influence the way that the information is handled within the system. If the four parts are related according to the flow diagram in Fig. 10, then several types of experiment may be defined according to Table 1.

Linkages between *neuronPopulations* from different experiments must be built on the basis of the actual criteria that exist within the literature. If the constituent *brainVolumes* of each of the *neuronPopulations* are located in the same region, then spatial relations can be used to link the data. If researchers have explicitly described linked phenomena within the paper (such as with double-labeling studies or by combining electrophysiological recording with cell-filling, for example) then the *neuronPopulation* objects themselves may be linked using set-theoretical relations.

6.3.5 The Significance of Knowledge Mechanics

The first issue of the journal *Trends in NeuroSciences* began with the phrase, "Even the most active neuro-

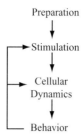

Figure 10 Flow diagram representing the causal interactions between parts of a physiological experiment in NeuroScholar.

scientist spends more working hours on reading, reviewing, and writing scientific reports that on direct experimental effort" (Bloom, 1978). Eighteen years later, Floyd Bloom, the author of that quote, wrote in the same journal how he anticipated the need for knowledge-management systems concerned with making sense of published neuroscientific data (Bloom, 1995). Bloom and Young devised one of the earliest effective informatics solutions to the problem of representing as much of the breadth of neuroscientific data as possible, and their database (called Brain Browser) still provides a valuable reference work today (Bloom *et al.*, 1989).

Bloom's foresight in both recognizing the scope of the problem of applying informatics solutions to analyzing the literature and his conceptualization of the software required to tackle the problem cannot be faulted. In fact, many of the design features of NeuroScholar attempt to replicate some of the functionality of Brain Browser; however, we view the definition of NeuroScholar as an implementation of a theoretical approach called "knowledge mechanics" that is new and innovative. As this chapter has described, this approach employs simple, extendable data models made up of a relatively small number of data structures (so-called *fragments*, *objects*, *properties*, *relations*, and *annotations*) according to tested

Table 1

Code	Experiment Type	Description
C	Cellular Dynamics	Passively recording activity from an animal
SC	Stimulation–cellular dynamics	Recording from an animal under a specific stimulus paradigm
PSC	Preparation—stimulation—cellular dynamics	Recording from an animal under a specific stimulus when the animal has been "primed" in some way
PSCB	Preparation—stimulation—cellular dynamics—behavior	Recording from a behaving animal under a specific stimulus when the animal has been "primed" in some way
PCB	Preparation—cellular dynamics—behavior	Recording from a behaving animal when the animal has been "primed" in some way
PSB	Preparation—stimulation—behavior	Observing an animal"s behavior under a specific stimulus when the animal has been primed (e.g., lesion experiments)
B	Behavior	Simple observation of animal's behavior

software engineering principles to define a pragmatic ontology. Given that the user base of this software consists of scientists who have an inherent interest in understanding how their data, information, and knowledge are handled, we present the underlying design of this approach as a theoretical representation for neuroscientific knowledge so that the complex details of our computational representation are explicitly stated.

Finally, we believe that this approach can only be effective if it is incorporated into the process of publication itself and is used by a large number of scientists. Thus, the significance of the approach will depend heavily on its usefulness and predictive power as a theoretical approach and this will be affected by design features of the system's user interface and the quality of technology supporting the software. For appraisal of this and other aspects of our work, see the NeuroScholar Website at http://neuroscholar.usc.edu/.

References

Allen, G. V., and Cechetto, D. F. (1993). Functional and anatomical organization of cardiovascular pressor and depressor sites in the lateral hypothalamic area. II. Ascending projections. *J. Comp. Neurol.*, **330(3)**, 421–438.

Behzadi, G., Kalen, P., Parvopassu, F., and Wiklund, L. (1990). Afferents to the median raphe nucleus of the rat, retrograde cholera toxin and wheat germ conjugated horseradish peroxidase tracing, and selective D-[3H]aspartate labelling of possible excitatory amino acid inputs. *Neuroscience* **37(1)**, 77–100.

Blackstad, T. W., Heimer, L., and Mugaini, E. (1981). General approaches and laboratory procedures. In *Neuroanatomical Tract Tracing Techniques* (L. Heimer and M. Robads, Eds.). Plenum Press, New York.

Bloom, F. (1995). Neuroscience-knowledge management, slow change so far. *Trends Neurosci.* **18(2)**, 48–49.

Bloom, F. (1978). New solutions for science communication problems needed now. *Trends Neurosci.* **1(1)**, 1.

Bloom, F., Young, W., and Kim, Y. (1989). *Brain Browser.* Academic Press, New York.

Blum, B. (1986). *Clinical Information Systems.* Springer, New York.

Bolam, J. (1992). *Experimental Neuroanatomy, A Practical Approach.* Oxford University Press, New York.

Burns, G. A. P. C. (1997). Neural connectivity in the rat, theory, methods and applications. In *Physiology*, Oxford University Press, New York, p. 481.

Burns, G. A. P. C., and Young, M. P. (2000). Analysis of the connectional organisation of neural systems associated with the hippocampus in rats. *Phil. Trans. Roy. Soc. London B,* Biol. Sci. **255(1393)**, 55–70.

Burns, G. A. P. C., O'Neill, M. A. and Young, M. P. (1996). Calculating finely-graded ordinal weights for neural connections from neuroanatomical data from different anatomical studies. In *Computational Neuroscience.* Trends in Research, J. Bower, Boston, MA.

Cliffer, K. D., and Giesler, G. J., Jr. (1988). PHA-L can be transported anterogradely through fibers of passage. *Brain Res.* **458(1)**, 185–191.

Dado, R. J., Burstein, R., Cliffer, K. D., and Giesler, G. J., Jr. (1990). Evidence that Fluoro-Gold can be transported avidly through fibers of passage. *Brain Res.* **533(2)**, 329–333.

Felleman, D. J., and van Essen, D. C. (1991). Distributed hierarchical processing in the primate cerebral cortex. *Cerebral Cortex* **1(1)**, 1–47.

Finnerty, G. T., and Jefferys, J. G. (1993). Functional connectivity from CA3 to the ipsilateral and contralateral CA1 in the rat dorsal hippocampus. *Neuroscience* **56(1)**, 101–108.

Gerfen, C. R., and Sawchenko, P. E. (1984). An anterograde neuroanatomical tracing method that shows detailed morphology of neurons, their axons and terminals, immunohistochemical localization of an axonally transported plant lectin. *Phaseolus vulgaris* leukoagglutinin (PHA-L). *Brain Res.* **290**, 219–238.

Heimer, L., and Robards, M. J., Eds. (1981). *Neuroanatomical Tract-Tracing Techniques.* Plenum Press, New York.

Heimer, L., and Zaborszky, L. Eds. (1989). *Neuroanatomical Tract-Tracing Methods.* Vol. 2. *Recent Progress.* Plenum Press, New York.

Hilgetag, C., Burns, G., O'Neill, M., Scannell, J., and Young, M. (2000). Anatomical connectivity defines the organization of clusters of cortical areas in the macaque monkey and the cat. *Phil. Trans. R. Soc. London B* **335(1393)**, 92–110.

Hilgetag, C. C., O'Neil, M. A., Scannell, J. W., and Young, M. P. (1995). A novel network classifier and its application, optimal hierarchical orderings of the cat visual system from anatomical data. In *Genetic Algorithms in Engineering Systems,* Innovations and Applications.

Itaya, S. K., and van Hoesen, G. W. (1982). Retinal innervation of the inferior colliculus in rat and monkey. *Brain Res.* **233(1)**, 45–52.

Linden, R., and Perry, V. H. (1983). Massive retinotectal projection in rats. *Brain Res.* **272(1)**, 145–149.

Martin, P. R. (1986). The projection of different retinal ganglion cell classes to the dorsal lateral geniculate nucleus in the hooded rat. *Exp. Brain Res.* **62(1)**, 77–88.

Merriam Webster (2000). *WWWebster Dictionary* http://www.m-w.com/netdict.htm.

Muller, R. J. (1999). *Database Design for Smarties,* Using UML for Data Modeling. Morgan Kaufmann, San Francisco, CA.

Pressman, R. (1992). *Software Engineering, A Practitioner's Approach.* McGraw-Hill, New York.

Rational Software Corp. (1997). *UML Semantics* Version 1.1. Rational Software Corp., Santa Clara, CA, www.rational.com.

Sawchenko, P. E., and Swanson, L. W. (1981). A method for tracing biochemically defined pathways in the central nervous system using combined fluorescence retrograde transport and immunohistochemical techniques. *Brain Res.* **210(1–2)**, 31–51.

Scannell, J. W., Burns, G. A. P. C., Hilgetag, C. C., O'Neil, M. A., and Young, M. P. (1999). The connectional organization of the corticothalamic system of the cat. *Cerebral Cortex* **9**, 277–299.

Scannell, J. W., Sengpiel, F., Benson, P. J., Tovée, M. J., Blakemore, C., and Young, M. P. (1996). Visual motion processing in anterior ectosylvian sulcus of the cat. *J. Neurophysiol* **76(2)**, 895–907.

Scannell, J. W., Blakemore, C., and Young, M. P. (1995). Analysis of connectivity in the cat cerebral cortex. *J. Neurosci.* **15(2)**, 1463–1483.

Sesack, S. R., Deutch, A. Y., Roth, R. H., and Bunney, B. S. (1989). Topographical organization of the efferent projections of the medial prefrontal cortex in the rat, an anterograde tract-tracing study with *Phaseolus vulgaris* leuco-agglutinin. *J. Comp. Neurol.* **290(2)**, 213–242.

Stephan, K. E., Hilgetag,, C. C., Burns, G. A. P. C., O'Neill, M. A., Young, M. P. and Kötter, R. (2000a). Computational analysis of functional connectivity between areas of primate cerebral cortex. *Phil. Trans. R. Soc. London B* **355(1393)**, 111–126.

Stephan, K. E., Zilles, K., and Kötter, R. (2000b). Coordinate-independent mapping of structural and functional data by objective relational transformation (ORT). *Phil. Trans. R. Soc. London B* **335(2393)**, 37–54.

Swanson, L. W. (1998). *Brain Maps, Structure of the Rat Brain.* Elsevier Science, Amsterdam.

Ter Horst, G. J., Groenewegen, H. J., Karst, H. and Luiten, P. G. M. (1984). *Phaseolus vulgaris* leuco-agglutinin immunohistochemistry, a comparison between autoradiographic and lectin tracing of neuronal efferents. *Brain Res.* **307(1–2)**, 379–383.

Trojanowski, J. Q. (1983). Native and derivatized lectins for *in vivo* studies of neuronal connectivity and neuronal cell biology. *J. Neurosci. Methods* **9**.

Vallee, R. B., and Bloom, G. S. (1991). Mechanisms of fast and slow axonal transport. *Annu. Rev. Neurosci.* **14**, 59–92.

van Heijst, G., Schrieber, A., and Wielinga, B. (1997). Using Explicit Ontologies in KBS development. *Int. J. Human-Computer Studies* **45**, 183–292.

Wan, X. C. S., Trojanowski, J. Q., and Gonatas, J. O., (1982). The dynamics of uptake, transport and clearance of horseradish-peroxidase (Hrp). conjugates of cholera toxin (CtHrp). and wheat-germ agglutinin (WgHrp). *J. Neuropathol. Exp. Neurol.* **41(3)**, 350.

Warr, A., Olmos, J. D., and Heimer, L. (1981). Horseradish peroxidase, the basic procedure. In *Neuroanatomical Tract Tracing Techniques* (Heimer, L., and Robads, M., Eds.). Plenum Press, New York, pp. 207–262.

Wessendorf, M. W. (1991). Fluoro-Gold, composition, and mechanism of uptake. *Brain Res.* **553(1)**, 135–148.

Young, M. P. (1993). The organization of neural systems in the primate cerebral cortex. *Proc. R. Soc. London B,* Biol Sci. **252(1333)**, 13–8.

Young, M. P. (1992). Objective analysis of the topological organization of the primate cortical visual system [see comments]. *Nature* **358(6382)**, 152–155.

CHAPTER 6.4

The NeuroHomology Database

Mihail Bota and Michael A. Arbib
*University of Southern California Brain Project, University of Southern California,
Los Angeles, California*

6.4.1 Introduction: The Definition of the Concept of Homology in Neurobiology

The concept of homology is central in comparative biology. It expresses the existence of typical and specific correspondences between parts of members of natural groups of living organisms (Nieuwenhuys, 1998). The term was first introduced by Owen, who defined a *homolog* as "the same organ in different animals under every variety of form and function" (Butler and Hodos, 1996). This definition was given before Darwin's theory of evolution, thus the modern concept of homology was changed by evolutionary biology and genetics (Butler and Hodos, 1996). Accordingly, the concept of homology was defined in relation to "continuity of information," inheritance of features from a common ancestry, or phyletic continuity.

When discussing homologies at the level of the nervous system, one has to distinguish three levels of organization (Striedter, 1999):

1. Hierarchy of cellular aggregates, composed of major brain regions, brain nuclei, and nuclear subdivisions
2. Hierarchy of cell types, including major types and subtypes
3. Hierarchy of molecules grouped into families and superfamilies.

Even though specific homologies between two species can be identified at each level of organization of the nervous system, it is not necessarily true that a homology at a level will transcend it or imply a homology that is specific for another level. As an example, Reiner (1991) considers that the dorsal cortex of reptiles is homologous to mammalian neocortex, but the reptilian neocortex does not have cells specific to the mammalian neocortex. Due to the great complexity of the task of finding homologies between two species at each level of organization of the nervous system, we will mainly discuss in the following those that are characteristic of the hierarchy of cellular aggregates in adult organisms.

The process of definition of homology at the level of brain structures is not a direct one; rather, it implies a process of inference from distinct clusters of attributes. Thus, the concept of *degree* of homology is more appropriate to use when discussing homologies of brain structures across species. If one wants to define and evaluate the degree of homology between two brain structures from two species, then it has to correlate with what makes a given brain structure distinguishable from other structures. To define a neural structure, neuroscientists use numerous attributes, including gross morphology, relative location, cytoarchitecture, types of cell responses to different ways of stimulation, and function (Bowden and Martin, 1997).

In this sense, Bowden and Martin (1997), using a unified dictionary for human and monkey brain structures (*Nomina Anatomica*), established homologies between monkey and rat brain structures based on their morphology (gross and Nissl stain appearance). Bowden and Martin consider that at least 84% of the primary landmark structures of the macaque brain have morphologic homologs in the rat, excluding the cortex. The status of another 8% of brain structures is unclear, so it is possible that as many as 92% of macaque brain structures have morphologic homologs in rat. At the cortical level the rats lacking the cortical sulci, have no morphological equivalents of about 25 cortical gyri of the macaque cortex. At the subcortical level, the structures of the macaque brain that do not have morphological homologs are the level of subdivisions of the thalamus (mainly the pulvinar nucleus), the striatum, and the lateral ventricle. This way of defining homologies between rat and monkey brain structures is incomplete. Areas that are morphologically homologous are not necessarily homologous from the functional point of view, or areas that are not homologous according to their morphology can be homologous according to other criteria.

A more complete way of defining homologies between two brain structures has to take into account as many attributes as possible. In this sense, following Butler and Hodos (1996) and Nieuwenhuys et al. (1998), we identified eight criteria that can make a brain area distinguishable and can be used to evaluate the degree of homology between two brain structures. These eight criteria are the morphology of cells within a brain structure, the relative position, the cytoarchitecture, chemoarchitecture (neurotransmitters and enzymes that are found within a brain structure), myeloarchitecture, afferent and efferent connections (hodology), and function. Accordingly, we take into account all these identified attributes when evaluating the degree of homology of a pair of brain structures from two species.

The first criterion, that of *morphology of neurons*, refers to the hierarchy of cell types within the nervous system. As an example, the structure of pyramidal neurons of cerebellum is a constant feature across tetrapods (Striedter, 1999). The principal type of cell in the lateral, basal, and accessory basal nuclei is a "pyramidal" neuron (Price *et al.*, 1987). This structure was found to be common for basolateral nuclei across species. In cat, these cells are called P cell, while in rat and mouse, about 70% percent of the basolateral nuclei are made of pyramidal cells. A similar type of cell was recognized in the opossum amygdala. Moreover, the principal cell type of the human basolateral amygdala appears to be a pyramid-like cell (Price *et al.*, 1987).

The criterion of *relative position* refers to the position of a brain structure, relative to other brain nuclei or to brain landmarks. As examples of homologies based on relative position we consider the periamygdaloid cortex which occupies the ventral surface of the amygdala in rat, cat, and monkey (Price *et al.*, 1987) and the retrosplenial cortex in the monkey and human which includes areas 29 and 30 and is buried in the callosal sulcus (Vogt, 1993).

The criterion of *cytoarchitecture* refers to the appearance of brain structures in Nissl staining. In this sense,

Krieg's parietal Area 1 in rat cytoarchitectonically resembles Area 1 in humans, due to the fact that cells in layers IV and VI are noticeably sparse (Krieg, 1947). In rat and rabbit, the anterior cingulate cortex is agranular, and the posterior cingulate cortex has a dysgranular component, with layers II and III poorly differentiated, a thin layer IV, and a prominent layer V (Vogt, 1993). A poor differentiation of the anterior cingulate cortex is characteristic for monkeys and humans, too (Vogt, 1993).

The criterion of *chemoarchitecture* of brain structures refers to specific neurotransmitters and enzymes that are found within these structures. One example of homology based on chemoarchitecture is the case of the accessory basal nucleus of amygdala in rat and cat and the basomedial nucleus in monkey. Even though this nucleus of amygdala has different sizes and appearances across species, it lightly stains for acetylcholine esterase (AChE) in monkey, cat, and rat (Price *et al.*, 1987). The same situation is found for the magnocellular portion of the basal nucleus of amygdala; it is that part of the basal nucleus of the amygdala that stains most intensely for AchE in rats, cats, and monkeys.

The method of staining of myelin of axons is used not only for identification of specific areas, but also as a criterion of homology. Accordingly, the distinction between the lateral intraparietal area (LIP) and the ventral intraparietal area (VIP) can be made on the basis of difference in myelin staining. VIP appears to be highly myelinated, while the myelin stain in LIP is light (Colby *et al.*, 1988). Examples of homologies established on the basis of *myeloarchitectonics* include area DM that appears to be a visual area in all non-human primates (Krubitzer and Kaas, 1993) and the cerebellum in avians and mammals (Feirabend and Voogd, 1986).

The homology criteria of *common afferent and efferent connections* have a particular importance. This is due to the fact that the pattern of connectivity of a brain structure can be directly related to the function of that brain structure, and the pattern of afferences and efferences, respectively, establish the position of that brain structure in the hierarchy of processing information by the brain. In many cases, a pair of structures from two different species can be remotely related in terms of relative position and cytoarchitecture but can share a common pattern of connections. This is the case of the rat and monkey prefrontal cortices. The existence of rodent prefrontal cortex is still under debate, but if one follows the definition of the prefrontal cortex as that cortex that receives input from the mediodorsal nucleus of thalamus, then several prefrontal cortices can be identified in the rat (Kolb, 1990). Based on the fact that MD projects in primate brains mainly to the prefrontal cortices, one can assume that the prefrontal cortices in rat and monkey are homologous with respect to the input from MD.

Another case of defining and finding homologies of a brain structure is that of the rodent posterior parietal cortex (PPC). In rodents, PPC is distinguished from the adjacent areas on the basis of thalamic inputs: it receives inputs exclusively from the laterodorsal (LD) and lateral posterior (LP) thalamic nuclei (Chandler *et al.*, 1992; Corwin and Reep, 1998). Regarding the putative homology between the rodent PPC and the monkey PPC, one can note that the monkey PPC receives thalamic connections from the pulvinar nucleus and from LP. Corwin and Reep consider that the existence of a pulvinar-LP complex is recognized across mammalian species that lack a pulvinar, and it is likely that the LP is a homologous structure of the pulvinar nucleus. Corwin and Reep conclude that the monkey and rat PPC have a common thalamic input. The patterns of connections for rat and monkey PPC bring more information in favor of the existence of homology between these two areas. In both species, the PPC has extensive connections with the prefrontal cortices (Corwin and Reep, 1998); therefore, one can assume the existence of homology between the rat and monkey posterior parietal cortices.

The homology criterion of *function* is not considered to be a proper one by all schools of comparative neuroanatomy. As an example, de Beer defines the homology between two organs on the basis of their characteristics (what they are) and not what they do (Nieuwenhuys *et al.*, 1998). On the other hand, Campbell and Hodos (1970) include the physiological data and behavioral changes resulting from stimulations and lesions as information that has to be used in establishing the homologies between two brain structures. The homology criterion of function depends in fact on the above-discussed criteria. Thus, if two brain structures have common cell types, present common chemo-and cytoarchitectonical characteristics, and common connectivity patterns, then one should expect that those two brain structures have the same function or related functions. Accordingly, we follow the definition given by Campbell and Hodos and consider the function as a homology criterion. A typical example is the primary visual area (area 17). In each major branch of mammalian species, area 17 can be delimited on the basis of myelo-(heavy myelination) and cytoarchitecture (the presence of a granular layer IV), the presence of a single and systematic visuotopic map, a well-defined pattern of subcortical afferents, small receptive fields relative to the extrastriate areas, and the presence of many orientation-selective neurons with simple receptive fields; therefore, one should expect that area 17 has the same function across species.

The discussion of the homology criteria that can be established between pairs of brain structures across species indicates that two brain structures are homologous within a certain degree. Discussion about whether two brain structures are homologous should take into account the constellation of attributes that define those. Also, the existence of homology at one level does not necessarily imply the existence of homology at other levels; therefore, the discussion of whether two brain structures are homologous should be focused on the degree of homology. The degree of homology can take a maximal value, when all the homology criteria are fulfilled, and a minimal value, when none of the homology criteria is fulfilled.

6.4.2 Theory of Degrees of Homology

The degree of homology is calculated on the basis of the number of fulfilled homology criteria for each retrieved reference and of the number of retrieved references. As described earlier, we have taken into account eight criteria to define a homology between two brain structures in different species. Due to the fact that the literature on the homologies at the level of neural systems is not in agreement when stating the hierarchy of importance of the criteria that are used to establish a homology between two brain structures, we have considered that all criteria have the same weight. This can be seen as a limitation of the formalism proposed, but it is a result of the heterogeneity found in the literature dedicated to comparative neuroanatomy.

We have inserted data found in literature that explicitly states homologies between brain structures for humans, monkeys, and rats. That is, if a reference states a homology between two brain structures for a pair of species, then the criteria that are fulfilled are recorded in the database. In order to compute the degree of homology between two brain structures, we have assigned an index for each considered criterion. Thus, for any given pair of brain structures from two different species, the indexes for those criteria that are fulfilled in each inserted reference will take value 1; otherwise, the value will be zero. The query for homologies for two brain structures from two different species will retrieve all the references in the database that refer to the searched pair. Accordingly, the query will retrieve all the associated indexes to the homology criteria and will evaluate the number of indexes that have value 1. If two references have a number of common criteria that are fulfilled, then the indexes do not change their values. The degree of homology is then calculated as the sum of all indexes retrieved from all references related to the searched pair of brain structures and divided by the total number of criteria of homology (8). The degree of homology increases only if the number of fulfilled criteria is increased in all searched papers and remains constant if the same criteria are found in any number of entries. Thus, the database tries to maximize the degree of homology for any pair of brain structures that are inserted in it, regardless of the number of papers that cite that pair as being homologous.

We have chosen this way of computing the degree of homology making the following assumption: the process of establishing homologies between brain structures from different species is an incremental one. That is, with the development of neuroanatomical and neurophysiological techniques, more homology criteria can be investigated and more accurate answers can be given to the question of whether two brain structures fulfill a given set of criteria. Another reason for choosing the above-mentioned method of computing the degree of homology is related to the redundancy found in the literature. One reference can establish that a number of homology criteria are fulfilled and also can cite other sources for extending the number of criteria or in support of the findings described in the reference.

6.4.3 The NeuroHomology Database: Description

The NeuroHomology Database is designed in Microsoft® Access and uses the WebMerger CGI parser engine as a Web interface. NeuroHomology is a knowledge-based summary database and contains three interconnected modules: *Brain Structures, Connections*, and *Homologies*. In Fig. 1 we show the modules and relationships that are contained in the database. Descriptions of each module of the database will be provided in later sections of this chapter. The conventions that are used to express relationships between entities and objects are as following:

1. A unidirectional connection denotes a 1-to-*n* relationship. This is the case of the relationship between Brain

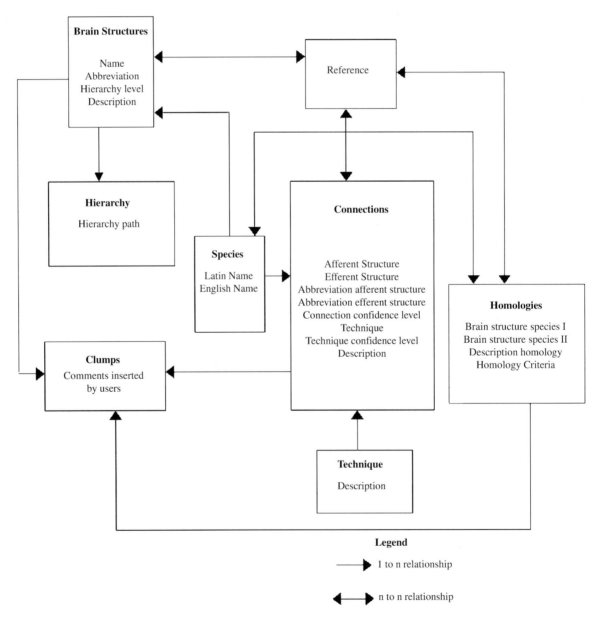

Figure 1 Schematic of the NeuroHomology Database structure.

Structures and the Hierarchy module. One brain structure (one entry in the database) has a single hierarchy path, but one hierarchy path can be common to a number of brain structures that have the same degree of specialization.

2. A bidirectional connection denotes an *n*-to-*n* relationship. This is the case of the relationship between references and any of the three entities. One reference can refer to many brain structures, and a brain structure can be defined in a number of different papers.

The parts of the NeuroHomology Database—Brain Structures, Connectivity Issues, and Homologies—can be accessed independently. We have designed the Web interface in independent parts to answer to queries from a larger category of users. In this way, a user who wants to find if there is any homology between structures X and Y from two different species can also find the definitions of structures X and Y according to different sources, as well as the afferents and efferents of these two structures.

6.4.4 The NeuroHomology Database: Brain Structures

The Brain Structures part of the NeuroHomology Database has brain structures as objects, as found in the literature. The description of a brain structure is taken as it is from the inserted papers. When searching the neuroanatomical literature, one can find brain structures defined on the basis of the topological or topographical position or on the basis of chemoarchitecture or myeloarchitecture. Usually, those brain structures that cannot be distinguished from the neighbors on the basis of gross appearance are defined according to other criteria. As can be seen in Fig. 2, PPC is defined in three sources on different grounds: morphological appearance and relative position. The description for each entered brain structure is the minimal one that can be found in the paper that describes it. To ensure a proper definition of a brain area, we seek to insert in the database those research papers that define brain structures according to at least one criterion. Each brain structure is captured in a hierarchy of brain superstructures. In this way, the hierarchy path that is assigned to a brain nucleus shows the successive specializations of brain superstructures that lead to that brain nucleus. The hierarchy path for each brain structure is established on the basis of the paper that describes it. Generally, a hierarchy path contains volumetric superstructures up to the level where a superstructure is defined by other criteria.

As an example, the lateral intraparietal area (LIP) in the macaque brain is defined by Felleman and van Essen (1991) on a myeloarchitectonical basis. The level of hierarchy for LIP is 5, so the hierarchy path is "Forebrain/Telencephalon/Cerebral Cortex/Parietal Lobe/Intraparietal Sulcus." Thus, area LIP has the hierarchy path determined only on volumetric structures, but this structure is defined on the basis of the criterion of myeloarchitecture. A similar situation is that of the lateral intraparietal area-ventral part (LIP-v). Felleman and van Essen (1991) define this region from LIP as being the target of the input from area MT. Accordingly, the

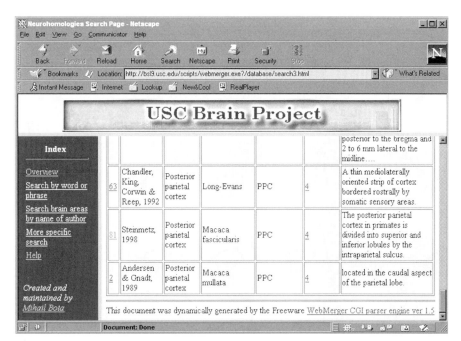

Figure 2 The result of a query using the abbreviation of a brain structure. The searched abbreviation was PPC. The user can compare definitions of brain areas with the same denomination, but in different species.

hierarchy level of this area is 6, and the hierarchy path is "Forebrain/Telencephalon/Cerebral Cortex/Parietal Lobe/Intraparietal Sulcus/LIP."

A possible source of confusion may arise from the use of different names for the same structure and from definitions of brain structures that are defined on different criteria and partially overlap. In this situation, the user has the possibility of searching by superstructure. The search by superstructure will return all the brain structures that exist in the database and are under it. The user then can compare the criteria that were used to define structures that are identical in terms of localization in the brain but have different names or areas that partially overlap. This is the situation for area 7a defined by Cavada and Goldman-Rakic (1989a,b) and area PG defined by Pandya and Seltzer (1982). Both sources describe the same part of the macaque parietal lobe but use different names. Whenever possible, we have used the Latin name for species (genus, or strain). If the Latin name is not available (as in the case of different strains of rats), we have used the English name.

The search of the Brain Structures can be extended or made specific. The brain structures that can be found in the database are regardless of species; in this way, the user is able to compare definitions of brain structures that have the same name in different species. Fig. 2 gives comparative definitions of brain structures in *Macaca fascicularis*, *Macaca mulatta*, and the Long-Evans rat. One type of extended search can be made using a word or phrase from the description of entered brain structures.

Another possible extended search can be made by the name of an author of an article in the database. This type of query will retrieve the names of authors that partially or totally match the searched string, together with the names, abbreviations, and hierarchies of brain areas that have been investigated by them and the descriptions of brain areas that are associated with the retrieved brain structures.

The search of brain structures can be more specific, by using any combination of three search possibilities: by abbreviations of brain structures, by superstructure, and by species. The page that contains these search options is shown in Fig. 3. The user is offered seven different combinations for searching the Brain Structures part of the NeuroHomology Database. By clicking on only one of the three checkboxes, the user can perform a more general search. The search will be narrowed if two out of three or all three checkboxes are used. In the case of search by species, a more general search can be done if the name of the family of species (i.e., "Macaca") or the generic name for a strain (i.e., "rat") is searched for. This type of search will return all records that have those species under the same genus. As an example, searching by the species "Macaca" will result in all the entries that contain as species *Macaca fascicularis*, *Macaca fuscata*, *Macaca nemestrina*, *Macaca mulatta*, etc. In the same way, if one searches by the species "rat," then the result will include all records that contain the strains Albino rat, Sprague-Dawley rat, Long-Evans rat, etc. A narrower search can be performed if strings such as

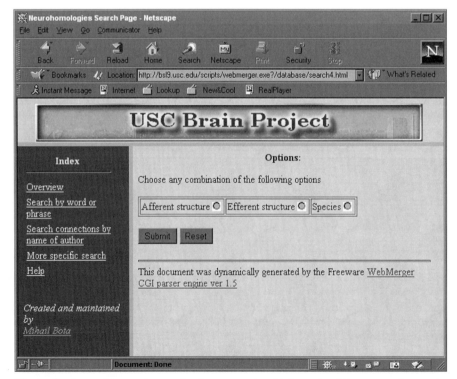

Figure 3 Annotations can be associated to any retrieved information. The user can inspect the associated annotations that were entered previously.

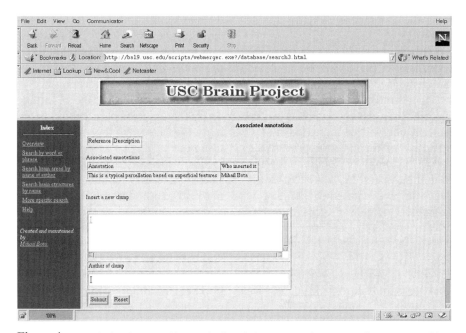

Figure 4 The window for a specific search of Brain Structures. The user can choose any combination of the three options for searching: by abbreviation of brain structure, name of superstructure, and species.

"nemestrina," or "Albino" are entered in the query. This type of search will result only in entries that belong to *Macaca nemestrina* or Albino rat, respectively. The most specific search is made if all three possibilities are chosen. In this case, the result will be restricted to a given species and will contain a single brain structure with at least one definition. In other words, the query will return those structures that have the same abbreviations, are under the same superstructure, and can be found in a single species. That is, a single structure will be retrieved having at least one definition from at least a single source.

In this way, the user can begin with a broad search of brain structures by using a word or phrase from the descriptions of brain structures or by author, choosing the appropriate option from the menu of the search page, or the user can narrow subsequent searches in order to inspect brain regions for a given species as specializations of a specific superstructure. Another specific feature of the NeuroHomology Database is associating new annotations (comments) with any retrieved entry (Fig. 4). By clicking on the ID that is associated with each retrieved entry, the user can inspect the annotations entered by previous users, and this function also allows the user to add new comments to the selected entry.

6.4.5 NeuroHomology Database: Connectivity Issues

The Connections part of the NeuroHomology Database has brain structures related by connections as objects. A typical result of a search of Connections contains the sources that discuss the retrieved connection; the species (strain); the afferent, efferent, and abbreviations of structures as found in the reference; the type of connection (excitatory/inhibitory or in terms of neurotransmitters); the technique that was used; the connection confidence level; the technique confidence level; and a description of the connection as found in the reference. The field "Technique" contains the specific technique that was used (e.g., "injection with tritiated amino-acids" or "PHAL"). By clicking on this field, the user has the option of inspecting the procedure that was followed and its description in the inserted paper. The evaluation of the connection confidence level was inspired by the confidence level used in the NeuroScholar Database (Burns, 1997). We considered five possible integer values for the connection confidence level: the maximum value is 3 and the minimum is -1. The assignment of a specific value for connection confidence level is the result of interpretation of the data from an article by the collator.

The value of -1 is used for an assertion such as "contrary to previous studies, we did not find the connection between area X and Y in the inserted article." A value of 0 is considered when no connection was found between two brain structures. A value of 1 is found when expressions such as "light connection," "light labeling," "sparse labeling," or "few cells were labeled" are found in the description of the connection. We consider that a connection has a confidence level of 2 when in the description of the connections statements such as "moderate staining" or "moderate labeling" are found. Finally, a value of 3 is assigned to those connections that are described as having "dense labeling" or "dense staining" attributes or the connection is "strong" or "dense."

Due to the fact that a number of sources describe connections as one relative to the other, we assigned values for connection confidence level as follows. If the confidence level for a connection was already assigned, then the confidence level for a connection that is discussed relative to the first one will be increased by 1 if the connection is "stronger" or the injection produced a "stronger" or "heavier" "labeling" or "staining." Conversely, a connection that is discussed relative to another will have a confidence level decreased by 1 if it is described as having "lighter" or "fewer" cells "labeled" or "stained."

We also assigned a confidence level for the technique that was used to investigate a given connection. The confidence levels can take values between 1 and 3. In order to evaluate the technique confidence level, we investigated the advantages and limitations for each commonly used technique for labeling brain connections. In this sense, we have taken into account the following tract-tracing techniques: injections with tritiated amino acids, horseradish peroxidase (HRP) alone or conjugated to wheat-germ agglutinin (WGA) or cholera toxin (CBT), fluorescent dyes, combinations of retrograde fluorescent markers and antigens, and *Phaseolus vulgaris* leuco-agglutinin (PHAL). The limitations of tract-tracing techniques that were taken into account when evaluating the technique confidence level were the mechanism of incorporation of chemicals by the neurons, the trans-synaptic labeling, the labeling of damaged or intact fibers of passage, and the difficulty of evaluating the number of labeled cells (Gerfen and Sawchenko, 1984; Llewellyn-Smith *et al.*, 1992; Sawchenko and Swanson, 1981; Sawchenko *et al.*, 1990; Skirboll *et al.*, 1989; Smith, 1992).

Injections with tritiated amino acids can lead to transsynaptic labeling; the degree of staining is time dependent, and the degree of opaque grains on the sensitive film can be influenced by the background radioactivity. Moreover, the axonal trajectories are difficult to trace, and the axonal and terminal labeling fields are difficult to discriminate (Gerfen and Sawchenko, 1984). Therefore, we assigned a value of 1 to this technique. The HRP technique is successfully used for both retrograde and anterograde labeling and does not lead to trans-synaptic labeling but is taken up non-specifically by the axons, dendrites, and cell bodies of neurons (Llewellyn-Smith *et al.*, 1992). Moreover, HRP is taken up by fibers of passage (Gerfen and Sawchenko, 1984). Accordingly, we assigned the value of 1.5 to the technique.

In the case of combination of HRP with WGA or CBT, the binding process is more specific. These specific uptake mechanisms make WGA-HRP and WGA-CBT more sensitive retrograde tracers than free HRP. On the other hand, just as for HRP, these techniques are prone to label fibers of passage (Gerfen and Sawchenko, 1984). Thus, we assigned the value of 1.75 for each of these techniques. One improvement to the WGA-HRP and WGA-CBT techniques is that of adding gold to these tracers; the cut or damaged axons do not take up these tracers, although they are taken up by the fibers of passage and axon terminals. Accordingly, we assigned the value of 2 to this tract-tracing technique.

One widely used group of techniques for retrograde staining of neural connections employs such fluorescent markers as rhodamine beads, fast blue, Fluoro-Gold, and propidium iodide (Kuypers *et al.*, 1979; Skirboll *et al.*, 1989). Each of these markers has its own advantages and limitations. As an example, fast blue and Fluoro-Gold are taken up by fibers of passage, and these dyes are not very useful for studying neurons with small projection fields; staining with rhodamine beads has disadvantages regarding evaluating the number of cells that are labeled (Skirboll *et al.*, 1989). For all these techniques and the combinations of those with antigens we assigned the value of 2.5; however, if the source mentions that no fibers of passage were labeled, and the number of labeled neurons could be evaluated, then the confidence level increases to 3.

The last discussed tract-tracing technique is that of PHAL. The principal advantages of this anterograde labeling technique include the clarity and completeness of labeling of neurons at the site of injection and of their axons; the tracer is not taken up by fibers of passage, and the results of injections into pairs of adjacent cell groups show that the anterogradely filled fibers and the terminal fields arise strictly among cell groups in which the labeled cells are seen (Gerfen and Sawchenko, 1984; Smith, 1992). Moreover, the fibers that are morphologically distinct, the collateral branches, and terminal specializations can be easily visualized (Gerfen and Sawchenko, 1984). For all these advantages, we assigned the value of 3 to PHAL. Also, reports of using combinations of PHAL as an anterograde staining technique with retrograde tracers that state that no fibers of passage were labeled will have the confidence level of 3. In conclusion, by evaluating the relative advantages and limitations of each widely used tract-tracing technique, we have offered a partial quantification of the reliability of results obtained using each technique.

The search of the Connections part of the NeuroHomology database is analogous to that of the Brain Structures. As in the case of Brain Structures, the search can be broad, or it can have different degrees of specificity. The broad search can be made by a word or phrase from the description of a connection or by the name of an author. Also, as for Brain Structures, the user can insert comments to any retrieved entries. A search can be narrowed by using a combination of three additional possibilities: afferent structure, efferent structure, and species (Fig. 5).

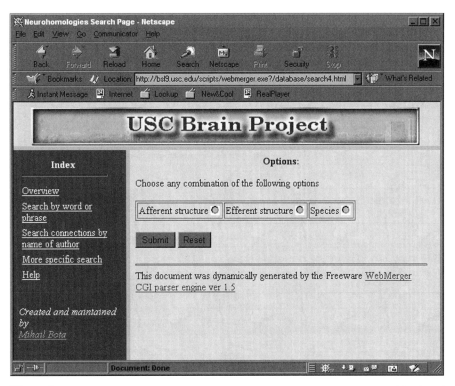

Figure 5 The window for specific search of Connections. The user can choose any combination of options for searching by afferent structure, efferent structure, and species.

A third degree of confidence, that of combined confidence level, is computed for the narrowest query: search by afferent structure and by efferent structure and by species. For each entry that has a pair of structures, the searched afferent and efferent brain nuclei, and for the searched species, the connection confidence level and technique confidence levels are calculated. The combined confidence level for each retrieved entry is calculated as following:

$$CC = CCL^*TCL/\max(TCL) \qquad (1)$$

where CC is the combined confidence level, CCL is the connection confidence level, and TCL is the technique confidence level. The value of $\max(TCL)$ is 3. In this way, the strength of the connection is weighted with the relative confidence of the technique. The minimum value of CC is -1, and the maximum value is 3. For a single retrieved entry, a value of CC smaller than that of CCL is an indication of the fact that the retrieved connection could have another confidence level if a more reliable technique is used. If there is more than one entry for a given connection, then an overall confidence level is computed as the average of combined confidence levels for all retrieved entries. The overall confidence level shows how a connection is reflected in a number of papers, taken together. The overall confidence level should be interpreted in correlation with overall confidence levels for each individual paper. As an example, if, in one paper, a connection is described as being "strong" and the technique that was used was injection with tritiated amino acids, then the connection confidence level will be 3 and the technique confidence level will be 1, with the combined confidence level being 1. If, in a second paper, the same connection is described as being "sparse" and the technique was PHAL, then the second confidence level will be 1 and the technique confidence level will be 3, with the second combined confidence level being 1 again. The overall confidence level for both entries will be 1, indicating the fact that the connection is possibly "sparse" and the results obtained by using the PHAL technique are more reliable. Therefore, one can conclude that the degree of reliability of the strength of a specific connection is increased not only by the specific technique that was used, but also by the number of entries found in the database that are related to it. On the other hand, an entry that states that there is no connection between two structures, contrary to the previous studies, will decrease the overall confidence level. A special case would be that of two entries, both with the same technique confidence level but one with a connection confidence level of 1 and one with -1. The overall confidence level in this case would be equal to zero, meaning that there is no connection between those two structures. This result can be an indication that further studies might be needed to elucidate the existence of that connection or that more data should be entered in the database.

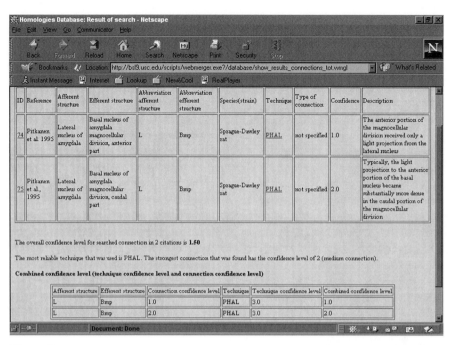

Figure 6 The result of a search of connections between the lateral nucleus of amygdala and basal nucleus of amygdala magnocellular division. For each search by afferent and efferent structures and by species, the combined confidence levels and the overall confidence level are calculated (see text for details).

Another example is shown in Fig. 6, which shows two entries for connections from the rat lateral nucleus of amygdala (L) to the basal nucleus of amygdala, magnocellular division (Bmp), as found in Pitkanen *et al.*, (1995), who follow the terminology and parcellation defined by Reep *et al.*, (1987) for amygdala. The authors dissociate between the anterior and caudal parts of Bmp, based on the density of afferents from L, but do not provide distinct abbreviations for these two parts. Thus, the anterior part of Bmp receives a light projection from L (the confidence level has the value 1), while the caudal part receives a more dense projection (we assigned a value of 2 to this connection). Both connections have been traced using PHAL (technique confidence level 3). As can be seen in Fig. 6, the connection confidence level is identical with the combined confidence level; therefore, the reliability of the connections is maximal. On the other hand, the overall confidence level for both entries is 1.5, suggesting that the connection between L and Bmp as an individual nucleus is between a light and a medium one. This result can be interpreted as follows: if one considers the Bmp as an individual nucleus, then the connection from L to Bmp is between light and medium. On the other hand, this interpretation is a simplistic one, and the topography of projections from L to Bmp, as suggested by Pitkanen *et al.*, is lost. In this situation, based on the afferents from L, one can propose a further dissociation of Bmp into an anterior and caudal part, seen as distinct substructures. This suggestion can be seen as a prediction for future experiments to establish the possible subparcellation of Bmp according to different criteria; therefore, by providing the above-described confidence levels, we offer a way for evaluating different results of tract-tracing experiments. By using the overall confidence level for a number of entries, we offer an overall view of the connections between brain structures as reflected in the literature. Moreover, as we saw above, we can offer predictions for future experiments to elucidate the existence of connections between two brain structures that retrieve two contradictory entries with the same confidence level or the existence of brain nuclei as distinct structures on the basis of their connectivity patterns.

6.4.6 The NeuroHomology Database: Homologies

As described earlier, we have identified eight criteria to define a homology between two brain structures: cell morphology, relative position, cytoarchitecture, chemoarchitecture, myeloarchitecture, afferent and efferent connections, and function. We use these eight criteria to evaluate a degree of homology between two brain nuclei. The Homologies part of the NeuroHomology database can handle data about brain structures from any species; however, due to the fact that the Homologies are seen as a tool for computational neuroscientists and we focus on models of brain structures for humans, monkeys, and rodents, the database contains human/rat, human/monkey, and monkey/rat homologies.

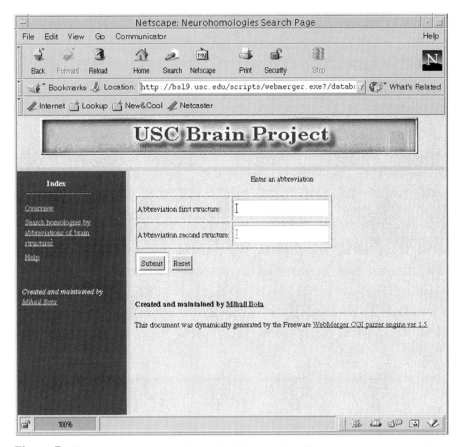

Figure 7 The window for a specific search of homologies. The user can search homologies between brain structures in different species by abbreviations of structures.

The search of the Homologies database can be made by abbreviations of brain structures that are considered to be homologous (Fig. 7). The user can enter in the fields "Abbreviation first structure" and "Abbreviation second structure" the abbreviations for two brain structures from two different species. The result of a search will retrieve all the homologies that are retrieved by the homology inference engine of the database that was described in a previous section, with regard to the searched pair of brain structures. An example of a typical search is shown in Fig. 8. The situation shown in Fig. 8 is the result of a search for homologies between the monkey frontal eye fields (Fef) and the rodent medial agranular cortex (PrCm). The abbreviations of structures are entered as found in the entered sources. The user can inspect all the details of any retrieved entry by clicking on the cited reference. The result will be eight different tables containing information about each homology criterion. Associated with each entry is a short description of the homology, which is inserted by the collator. The user can enter and inspect additional annotations that are attached to each investigated entry.

In the case of homology between Fef and PrCm, the degree of homology is 0.13, while in the case of homology between the rodent L and monkey L (Fig. 9), the degree of homology is 0.5, with four criteria (afferent connections, relative position, cytoarchitecture, and chemoarchitecture) being fulfilled in five retrieved entries. Each entry that is shown in Fig. 9 refers to a different citation of fulfilled criteria of homology.

The user can investigate each entry by clicking on the "Reference" field. A typical example is shown in Fig. 10. The example shown in Fig. 10 involves the homologies between monkey PPC and rat PPC. This page shows the details of the homologies found in the citation Reep *et al.* (1994). The user can further investigate the details of the afferents and efferents by clicking on the abbreviations of efferent and/or afferent brain structures.

Separately, we calculate the common afferent and efferent connections for each homologous pair of brain structures (as shown in Fig. 9). In this way, the user can evaluate the relative importance of common patterns of connectivity in evaluating the degree of homology. A second reason for showing the common afferent and efferent connections, as reflected by the data in the database, is related to the functionality of brain areas. Two brain structures that share a common pattern of connectivity are likely to have the same position in the hierarchy of processing of information in the central nervous system and to have common functions. Moreover, the computational neuroscientists who use the NeuroHomology Database can evaluate the degree of reliability to model

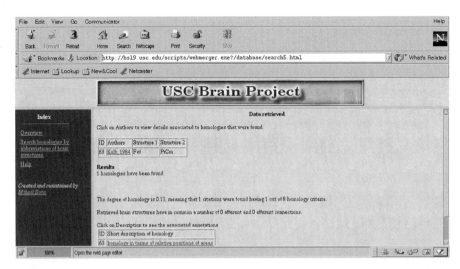

Figure 8 The result of a search of homologies between primate Fef and rodent PrCm (see text for details).

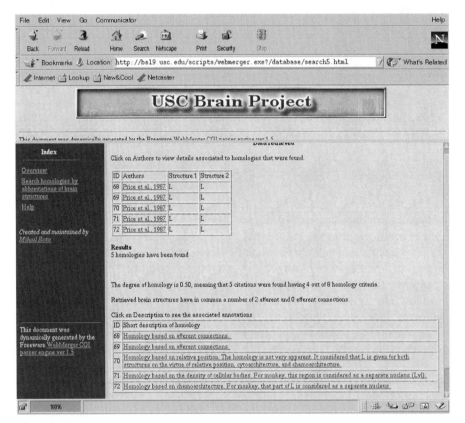

Figure 9 A query for homologies between monkey lateral nucleus of amygdala (L) and rodent L retrieved five different entries. The homology criteria that are fulfilled are afferent connections, relative position, cytoarchitecture, and chemoarchitecture (staining for AchE).

neural systems from one species by using data from other species.

6.4.7 Conclusion and Future Development

The NeuroHomology Database is a knowledge-based summary database that contains homologies between brain structures for human/monkey, human/rat, and human/rodent species. The database can be used for searching brain structures and connections as interpreted from the literature. We have proposed a way to quantify the degrees of confidence of connections and staining techniques that are used. We also have proposed a method for computing the overall confidence level for a number of entries related to a single connection; based on it, the user can evaluate the reliability of the searched connection as reflected in the literature.

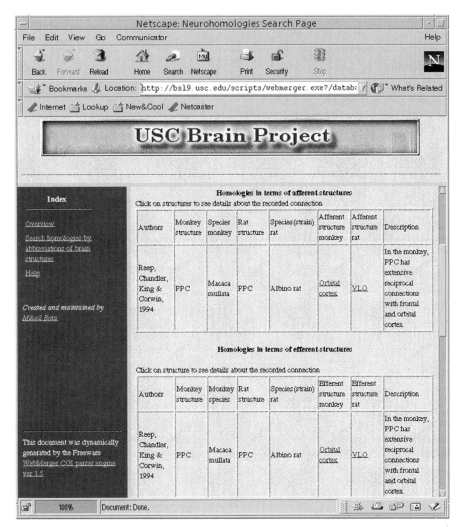

Figure 10 The details of an entry retrieved by the search for homologies between *Macaca mulatta* PPC and Albino rat PPC.

Related to homologies between brain structures, we propose that the homology between two structures from different species can be better described by the degree of homology computed by taking into account the criteria described earlier. In this way, we can offer a quantitative measure of the homology between two brain structures. The NeuroHomology Database is the first attempt to systematize the literature on homologies between brain structures. Therefore, the usefulness of the NeuroHomology Database can be threefold: (1) as a tool for systematization of existing information about homologies at the level of the central nervous system that could be of special interest for comparative neuroanatomists; (2) as a tool for neuroscientists who study a brain structure in a given species and want to compare their findings with those that are characteristic of a possibly homologous structure from another species; and (3) for computational neuroscientists that can evaluate the degree of reliability to model brain structures from one species by using experimental data or characteristics of brain structures from other species. The future development of the NeuroHomology Database will be focused on two main directions: (1) further refinement of the existent criteria of homology and insertion of new criteria, and (2) extension of the NeuroHomology Database with electrophysiological information and linking it to the BMW database.

Related to the first direction of development, the criteria of relative position and cytoarchitecture will be refined. The relative position will be considered in both topological and topographical terms. The criterion of cytoarchitecture will make the dissociation between the layers of the isocortex and allocortex on one hand, and the appearance of subcortical nuclei. Related to this direction of development is the issue of assigning specific weights for each criterion taken into account. As stated in a previous part of this chapter, we consider in the current version of the NeuroHomology Database that all criteria have the same importance in establishing a degree of homology, each having the same weight. A more accurate and realistic way of calculating the degree of homology for a pair of brain structures from two species would be to assign specific weights for each considered criterion. In the first part of this chapter, we restricted our discussion to the homology of brain structures at the level of hierarchy of cellular aggregates. Future development of the NeuroHomology Database will be also be in the direction of establishing homologies at the levels of hierarchies of brain cells and their subtypes and molecules. To do so, special work will be devoted to add the homologies at the level of gene expression. The homology criterion will include in this way the ontogenetic similarities in different species.

The second line of development, which is related to the first one, is the incorporation of electrophysiological data at the cellular level and the connection of the NeuroHomology Database with the BMW database. In this way, the Brain Structures part of the NeuroHomology Database will describe more accurately a brain structure, and a link with the BMW database will be possible. The user will be able to inspect both the neuroanatomical and neurophysiological properties of a brain structure in a given species and the computational models/modules that are associated with that structure or model some features of it.

References

Bowden, D. M, and Martin R. F. (1997). A digital Rosetta stone for primate brain terminology. In *Handbook of Chemical Neuroanatomy*. Vol. 13. *The Primate Nervous System, Part I* (Bloom, F. E., Bjorklund, A., and Hokfelt, T., Eds.), pp. 1–37.

Bowden, D. M., and Martin, R. F. (1995). *NeuroNames Brain Hierarchy* 2(1), 63–83.

Burns, G. A. P. C (1997). Neural Connectivity of the Rat, Theory, Methods, and Applications, Ph.D. thesis. Oxford University, London.

Butler, A. B., and Hodos, W. (1996). *Comparative Vertebrate Neuroanatomy*, Evolution and Adaptation, Wiley-Liss, New York, pp. 7–13.

Campbell, C. B., and Hodos, W. (1970). The concept of homology and the evolution of the nervous system. *Brain Behav. Evol.* 3(5), 353–67.

Cavada, C., and Goldman-Rakic, P. S. (1989a). Posterior parietal cortex in rhesus monkey. I. Parcellation of areas based on distinctive limbic and sensory corticocortical connections. *J. Comp. Neurol.* 287(4), 393–421.

Cavada, C., and Goldman-Rakic, P. S. (1989b). Posterior parietal cortex in rhesus monkey. I. Evidence for segregated corticocortical networks linking sensory and limbic areas with the frontal lobe. *J. Comp. Neurol.* 287(4), 422–445.

Chandler, H. C., King, V., Corwin, J. V., and Reep, R. L. (1992). Thalamocortical connections of rat posterior parietal cortex. *Neurosci. Lett.* 143(1–2), 237–242.

Colby, C. L., Gattass, R., Olson, C. R., and Gross, C. G. (1988). Topographical organization of cortical afferents to extrastriate visual area PO in the macaque, a dual tracer study. *J. Comp. Neurol.* 269(3), 392–413.

Corwin, J. V., and Reep, R. L. (1998). Rodent posterior parietal cortex as a component of a cortical network mediating directed spatial attention. *Psychobiology* 26(2), 87–102.

Feirabend, H. K., and Voogd, J. (1986). Myeloarchitecture of the cerebellum of the chicken (*Gallus domesticus*), an atlas of the compartmental subdivision of the cerebellar white matter. *J. Comp. Neurol.* 251(1), 44–66.

Felleman, D. J., and van Essen, D. C., (1991). Distributed hierarchical processing in the primate cerebral cortex. *Cerebral Cortex* 1(1), 1–47.

Gerfen, C. R., and Sawchenko, P. E. (1984). An anterograde neuroanatomical tracing method that shows the detailed morphology of neurons, their axons and terminals, immunohistochemical techniques. *Brain Res.* 210(1–2), 31–51.

Gerfen, C. R., Sawchenko, P. E., and Carlsen, J. (1989). The PHA-L anterograde axonal tracing method. In *Neuroanatomical Tract-Tracing Methods*. 2. *Recent Progress*. (Heimer, L., and Zaborsky, L., Eds.), Plenum Press, New York, pp. 19–46.

Kolb, B. (1990). Posterior parietal and temporal association cortex. In *The Cerebral Cortex of the Rat*. (Kolb, B., and Tees, R. C., Eds.). The MIT Press, Cambridge, MA, pp. 459–471.

Krieg, W. J. S. (1947). Connections of the cerebral cortex. I. The albino rat, a topography of the cortical areas. *J. Comp. Neurol.* 84, 221–275.

Krubitzer, L. A., and Kaas, J. H. J. (1993). The dorsomedial visual area of owl monkeys, connections, myeloarchitecture, and homologies in other primates. *J. Comp. Neurol.* 334(4), 497–528.

Kuypers, H. G., Bentivoglio, M., van der Kooy, D., and Catsman-Berrevoets, C. E. (1979). Retrograde transport of bisbenzimide and propidium iodide through axons to their parent cell bodies. *Neurosci. Lett.* 12(1), 1–7.

Llewellyn-Smith, I. J., Pilowsky, P., and Minson, J. B. (1992). Retrograde tracers for light and electron microscopy. In *Experimental Neuroanatomy*, A Practical Approach (Bolam, J. P., Ed.). Oxford University Press, New York, pp. 31–60.

Nieuwenhuys, R, ten Donkelaar, H. C., and Nicholson, C. (1998). *The Central Nervous System of Vertebrates*. Springer, New York, pp. 273–326.

Pachen, A. L. (1999). Homology, history of a concept. In *Homology* (Bock, G. R., and Gail, C., Eds.). John Wiley & Sons, London, pp. 5–18.

Pandya, D. N., and Seltzer, B. (1982). Intrinsic connections and architectonics of posterior parietal cortex in the rhesus monkey. *J. Comp. Neurol.* 204(2), 196–210.

Pitkanen, A., Stefanacci, L., Farb, C. R., Go, G. G., LeDoux, J. E., and Amaral, D. G. (1995). Intrinsic connections of the rat amygdaloid complex, projections originating in the lateral nucleus. *J. Comp. Neurol.* 356(2), 288–310.

Price, J. L., Russchen, F. T., and Amaral, D. G. (1987). The limbic region. II. The amygdaloid complex. In *Handbook of Chemical Neuroanatomy*. Vol. 5. *Integrated Systems in the CNS, Part I*, Hypothalamus, Hippocampus, Amygdala, Retina (Bjorklund, A., Hokfelt, T., and Swanson L. W., Eds.). Elsevier, New York, pp. 279–388.

Reep, R. L., Chandler, H. C., King, V., and Corwin, J. V. (1994). Rat posterior parietal cortex, topography of corticocortical and thalamic connections. *Exp. Brain Res.* 100(1), 67–84.

Reiner, A. (1991a). A comparison of neurotransmitter-specific and neuropeptide-specific neuronal cell types present in the dorsal cortex in turtles with those present in the isocortex in mammals, implications for the evolution of isocortex. *Brain Behav. Evol.* 38(2–3), 53–91.

Reiner, A. (1991b). Levels of organization and the evolution of isocortex. *Trends Neurosci.* 19(3), 89–91.

Sawchenko, P. E., Cunningham, E. T., Jr., Mortrud, M. T., Pfeiffer, S. W., and Gerfen, S. W. (1990). *Phaseoulus vulgaris* leucoagglutinin anterograde axonal transport technique. In *Methods in Neurosciences*. Vol. 3. Academic Press, New York, pp. 247–260.

Sawchenko, P. E., and Swanson, L. W. (1981). A method for tracing biochemically defined pathways in the central nervous system using combined fluorescence retrograde transport and localization of an axonally transported plant lectin. *Phaseolus vulgaris* leucoagglutinin (PHA-L). *Brain Res.* 290(2), 219–38.

Skirboll, L. R., Thor, K., Helke, C., Hokfelt, T, Robertson, B., and Long, R. (1989). Use of retrograde fluorescent tracers in combination with immunohistochemical methods, In *Neuroanatomical Tract-Tracing Methods*. 2. *Recent Progress*. (Heimer, L., and Zaborsky, L., Eds.) Plenum Press, New York, pp. 5–18.

Smith, Y. (1992). Anterograde tracing with PHA-L and biocytin at the electron microscopic level. In *Experimental Neuroanatomy*, A Practical Approach. (Bolam, J. P., Ed.). Oxford University Press, New York, pp. 61–80.

Striedter, G. F. (1999). Homology in the nervous system, of characters, embryology, and levels of analysis. In *Homology* (Bock, G. R., and Gail, C., Eds.). John Wiley & Sons, London, pp. 158–170.

Vogt, B. A (1993). Structural organization of cingulate cortex, areas, neurons, and somatodendritic transmitter receptors. In *Neurobiology of Cingulate Cortex and Limbic Thalamus*, A Comprehensive Handbook (Vogt, B. A., and Gabriel, M., Eds.). Birkhauser, Boston, MA, pp. 19–69.

Wiley E. O. (1981). *Phylogenetics*, The Theory and Practice of Phylogenetic Systematics. John Wiley & Sons, London.

APPENDICES

APPENDIX A1

Introduction to Informix

Jonas Mureika and Edriss N. Merchant
*University of Southern California Brain Project,
University of Southern California, Los Angeles, California*

Informix SQL Tutorial: A Practical Example

As a practical example of SQL in Informix, we will review the steps to create a portion of the NeuroCore database (see Chapter 3.1 and 3.2 for a complete description of the NeuroCore database). A sub-structure of the contact portion of the database will be featured. Specifically, our example database will be used to store contact information for researchers in the lab. In order to construct the best possible (and least confusing) database schema of this type, the following questions should be addressed: *who* are our contacts, *where* do they work, and *how* does one contact them?

This exemplar database will demonstrate the following versatile features of an Informix database: referential integrity, set backing tables, and inheritance. These elements will be explicitly defined when they are introduced later in the text. For a quick-reference guide to the syntax used, please see Appendix B1.

The first step in the creation of our database is to create the database. This will initialize the database and create the necessary internal information needed by the database server. To create this database, which we will name *contactDB*, we can execute the following SQL command:

CREATE DATABASE contactDB;

The database will contain a set of relational (inheritance-based) tables that will store the information necessary to contact the persons stored in this database. Each table is defined as containing certain columns where one can store information related to the concept defined by the table. An entry into the table is referred to as a *tuple*. Relations between tables can be defined so that columns in a table can reference values stored in columns residing in other tables. The structure of a database is known as the *schema*. The schema for our exemplar database can be seen in Fig. 1.

In defining the schema for our contact database, the first objects we must define are the contacts we wish to store. In our case, we will store the information concerning both persons and the labs to which they belong. Inheritance is immediately evident in this structure. In our example, the tables, "lab" and "person" are children of "contact." Both labs and persons are contacts and should inherit all the properties of a contact. For example, all contacts in the database will have a phone number. Instead of defining this at each individual table, we can define this relation for a parent-table (i.e., "contact") which will then be passed down to its children, namely "person" and "lab." However, we must now define how persons and labs are related. A set backing table is a special table that allows a many-to-many relationship between other tables to be defined. For example, many persons may be associated with a lab and, similarly, many labs may be associated with a person. The set backing table holds the identification elements (relations) to link the two tables. The "_labPerson" table is such a set backing table that identifies which individuals in the database belong to which lab. Once we have defined the contacts we wish to store in the database, we need to define how we want to contact them. In our example, we will store telephone contact information. For each contact, we can define various types of phone numbers (e.g., office, home, mobile). To accomplish this, one can define a "phone" table, which can contain the individual tuples that define the phone contact information for each contact. However, in order to classify the types of phone information stored in the database, we will need to define a list of allowable phone

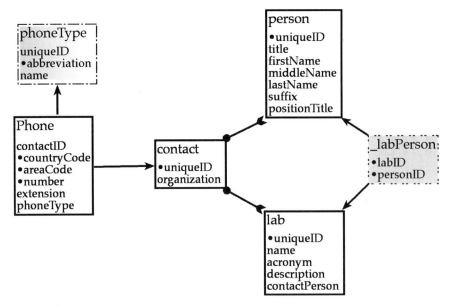

Figure 1 Database schema for the SQL tutorial. Each table is represented as a single entity, with the relations depicted as arrows. Inheritance (parent-child relations) are depicted as arrows with round heads.

types. The "phoneType" table is a dictionary table that is referred to by the "phone" table. Each tuple stored in the "phone" table must reference a valid entry in the "phone" type table.

The first step in implementing the database for our contact information is to create a parent table for the contacts we want to maintain, namely "contact." All types of contacts (labs, people) are stored in tables that are children of the "contact" table. This allows us to associate various entities and relations with all contacts (e.g., phone numbers). Because a row is actually a datatype, it facilitates matters to manipulate the tables in terms of row datatypes created by the syntax:

```
CREATE ROW TYPE contact_t
(
    uniqueID      INT8          NOT NULL,
    organization  VARCHAR(128)
);
CREATE TABLE contact OF TYPE contact_t
(
    PRIMARY KEY (uniqueID)
);
```

These statements create a new row type that defines what a contact is and then creates the actual table to store that information. In order to reference tuples within the database, one must define a collection of columns that contain unique values. This allows any item within the table to be uniquely referred to. A collection of such columns is termed a *primary key* and is defined in the "create table" statement. The primary key for all tables within the contact hierarchy is the uniqueID. We next create a table called "lab" to store the names of the locations/institutes at which the persons work:

```
CREATE ROW TYPE lab_t
(
    name          VARCHAR(128)  NOT NULL,
    acronym       VARCHAR(32),
    description   LVARCHAR,
    contactPerson INT8
)
under contact_t;

CREATE TABLE lab OF TYPE lab_t
(
    UNIQUE (uniqueID)
    UNIQUE (name, organization)
)
under contact;
```

Notice that, because the lab is a sub-table of contact (it inherits information from the contact super-table), the syntax of the statement has changed slightly. For a sub-table we need to define *under* which table the new information will reside in the inheritance hierarchy. Each lab must have a unique name (the keyword UNIQUE ensures a one-to-one correspondence between the pair [name, organization] and "lab"). That is, there cannot be two labs with the same name in the same organization.

The table that describes the actual person will be denoted as "person":

```
CREATE ROW TYPE person_t
(
    title          VARCHAR(16),
    firstName      VARCHAR(64) NOT NULL,
    middleName     VARCHAR(64),
    lastName       VARCHAR(64) NOT NULL,
    suffix         VARCHAR(16),
    positionTitle  VARCHAR(128)
)
UNDER contact_t;

CREATE TABLE person OF TYPE person_t
(
    UNIQUE (uniqueID)
)
UNDER contact;
```

The major stipulations for this table are that the person must have a first and last name; all other entries can be blank (NULL). To further develop the structure of the database, we need to create a set backing table to relate persons and labs. In this case, the table "_labPerson" (and associated datatype "_labPerson_t") relates the entries in the tables "lab" and "person". This implements a many-to-many relationship between the entries in the lab table and the entries in the person table. The syntax for such a table is

```
CREATE ROW TYPE _labPerson_t
(
    labID     INT8    NOT NULL,
    personID  INT8    NOT NULL
);

CREATE TABLE _labPerson OF TYPE
    _labPerson_t
(
PRIMARY KEY (labID, personID),
FOREIGN KEY (labID) REFERENCES lab
    (uniqueID),
FOREIGN KEY (personID) REFERENCES person
    (uniqueID)
);
```

The table consists entirely of primary keys. Each tuple stored in this table contains an ID for a person and the lab they belong to. A set backing table can be used as a way of joining tables in one-to-many or many-to-many relationships. In other words, we can store information on people who belong to a single lab, people who belong to multiple labs, and labs containing a single person or multiple people. We have now implemented the information necessary to define a contact.

In our example, we chose the telephone as the best method of contacting an individual; however, to store the phone data we need to know what type of phone is being stored (e.g., cellular, home, office). To accomplish this, a dictionary table was implemented:

```
CREATE ROW TYPE phoneType_t
(
    abbreviation  VARCHAR(16) NOT NULL,
    name          VARCHAR(64) NOT NULL
);

CREATE TABLE phoneType OF TYPE
    phoneType_t
(
UNIQUE (name),
PRIMARY KEY (abbreviation)
);
```

In this case, the new datatype "phoneType_t" has been assigned to contain two strings of variable length (16 and 64 characters, respectively). Note the stipulation NOT NULL; this is a check to ensure that the entries *do* contain some value. The keyword UNIQUE acts as a check to the database contents and structure. Unique instructs the database that the name variable in the table "phoneType" cannot be assigned the same value more than once. That is, no phone type should have exactly the same name (cellular, beeper, etc.).

Because the table "phoneType" only describes the type of phone number being stored, we need to create another table with more explicit contents in which to store the actual phone contact data. This will be the table "phone," which has the associated datatype "phone_t":

```
CREATE ROW TYPE phone_t
(
    contactID    INT8         NOT NULL,
    countryCode  VARCHAR(8)   NOT NULL,
    areaCode     VARCHAR(8)   NOT NULL,
    number       VARCHAR(32)  NOT NULL,
    extension    VARCHAR(8),
    phoneType    VARCHAR(16)  NOT NULL
);

CREATE TABLE phone OF TYPE phone_t
(
PRIMARY KEY (countryCode, areaCode, number),
FOREIGN KEY (phoneType) REFERENCES
    phoneType (abbreviation),
FOREIGN KEY (contactID) REFERENCES
    contact (uniqueID)
);
```

Here, the variable "phoneType" should not be confused with the table of the same name. Note, however, that in the table creation syntax the varchar "phoneType" introduced in the definition of the datatype "phone_t" is assigned as a FOREIGN KEY that REFERENCES "abbreviation" in the table phoneType (likewise for "contactID" and "uniqueID"). This is an example of *referential integrity*. The value of the VARCHAR(16) "phoneType" is the same as the value of "abbreviation" in the "phoneType" table. They are referentially linked. The foreign key is the same value as the primary key it references. Ideally, such primary/foreign key links can propagate down through the database and create an intricate and delicate web of joined tables. It would be easy to destroy such a fragile structure if the primary key were to be (accidentally) deleted. Luckily, this is usually not permitted in a database.

Another of the versatile features of SQL is the ability to create indices associated with a data entry. Creating an index speeds up queries to the database by providing the query processor with additional information to search a specific table. To demonstrate this feature, we index parts of the datatype "person:"

CREATE INDEX person_asc_ix ON person (title ASC, firstName ASC, middleName ASC, lastName ASC, suffix ASC) USING btree;

CREATE INDEX userInfo_asc_ix ON person (username ASC) USING btree;

This is the general form of the syntax used to create an index. An index is a structure of pointers that point to rows of data in a table. An index optimizes the performance of database queries by ordering rows to make access faster. The end result of the index is that it provides a secondary access method to the data. Creating indices on data can help speed search and retrieval times for arbitrarily large databases. Indices can list their contents in ascending order (ASC) or descending order (DESC), and one can index the same data using both methods.

We have now finished the implementation of our contact database. The next step in the implementation is to actually store data in the database. To illustrate this process, we will add two individuals to the database who work in the same laboratory. First, we add the laboratory and then the individuals:

INSERT INTO lab (uniqueID, name, organization, acronym, description) values (1, 'Our Example Lab', 'Our University', 'OEL', 'A lab for our example database');

Notice that we only insert data into the lab table and not into the lab and contact table, even though the organization column is defined in the contact table. We must then add our two individuals:

INSERT INTO person (uniqueID, firstName, middleName, lastName) values (2, 'John', 'Q', 'Public');

INSERT INTO person (uniqueID, firstName, middleName, lastName) values (3, 'Jane', NULL, 'Doe');

As we inserted data into the person table, notice that we did not define all the values contained within the person table. We defined only the values we were required to and had values for. Also notice that Jane does not have a middle name, so we have declared that value to be NULL, containing no value. Once we have our people and labs defined, we will need to define which lab our people belong to. In our example, John does not belong to a lab, but Jane does. To define this, we can insert the following values into the "_labPerson" table:

INSERT INTO _labPerson values (1,3);

Notice that we did not define the actual columns we were inserting data into. Because we did not select specific columns in the insert statement, we must provide data for all columns defined in the table and list them in the order in which they were defined. In our example above, we have defined that the person with the uniqueID = 3 belongs to the lab with uniqueID = 1. Now that we have defined our contacts, we must define the contact information:

INSERT INTO phoneType values ('office', 'Office Number');
INSERT INTO phone values (2, '001', '213–555', '1212', 'office');

Here we have inserted a new phone type, namely a person's office phone. We are now able to insert an office phone number into the phone table. If we had not defined a phone type, we would not be able to insert any values into the phone table, as each phone number must be of a particular type (this policy is enforced through the use of referential integrity defined by the foreign key – -primary key structure). We have entered a phone number for John. Notice that the contactID refers to John's uniqueID in the contact hierarchy. Now that we have entered some contact information, we can now perform queries to extract information from the database. To list all persons in the database we can perform the simple query:

SELECT * FROM person;

Here we have defined that we wish to select all (*) columns from the persons table. This will return information for John and Jane. A more complex query would allow us to select John's phone numbers:

SELECT person.firstName, phone.areaCode, phone.number FROM
phone,
person
WHERE
        phone.contactID == person.uniqueID
        and person.firstName == 'John';

Here we are selecting for all tuples in the phone table where the contact ID is equal to the ID of persons with a firstName of John. Notice that we are only returning the area code and phone number from the records stored in the phone table. It is hoped that the previous section has demonstrated various features of a typical database and how a user would design, implement, and query such a database.

For more information on SQL and Informix's implementation of SQL, users can visit the documentation available at the Informix website http://www.informix.com.

APPENDIX A2

NeuroCore TimeSeries Datablade

Jonas Mureika and Edriss N. Merchant
*University of Southern California Brain Project,
University of Southern California, Los Angeles, California*

Introduction

To be able to store neurophysiological and behavioral data (for more information on the experiments and data being stored within NeuroCore, please see Chapter 3.1) in a usable format within the database, a new data type had to be created. In an object-relational database, this new type can be accessed and manipulated by the database just as for the database's inherent types (e.g., integer, float, varchar). Being able to directly access and manipulate this new time series data types allows users to perform queries and perform analysis on the data directly and not just on metadata that has been computed at some prior time.

Internal Structure and Description

The NeuroCore TimeSeries Datatype is constructed by using Informix's opaque data type. The opaque data type is a fundamental data type that a user can build upon (a fundamental data type is atomic; it cannot be broken into smaller pieces, and it can serve as the building block for other data types). When one defines an opaque data type, the data type system of the database server is extended. The new opaque data type can be used in the same way as any built-in data type that the database server provides. To define the opaque data type to the database server, one must provide the following information:

- A data structure that serves as the internal storage of the opaque data type
- Support functions that allow the database server to interact with this internal structure
- Optional additional routines that can be called by other support functions or by end users to operate on the opaque data type

The NeuroCore TimeSeries Datatype is able to store data in both binned (data collected at regular time intervals) and time-stamped (data collected at irregular time intervals) formats. The general structure of the data type (see Fig. 1 for an example of one-dimensional binned data and Fig. 2 for an example of two-dimensional time-stamped data) contains a header, data record, and optional time record. The header itself stores vital information regarding the structure of the data being stored. The data record contains the actual data being stored, whereas the time record contains the collection time points if the data were collected at irregular time intervals (it is NULL otherwise). The structure of the NeuroCore TimeSeries Datatype can also be extended to the N-dimensional case (e.g., for the storage of three-dimensional coordinates recorded from a moving object).

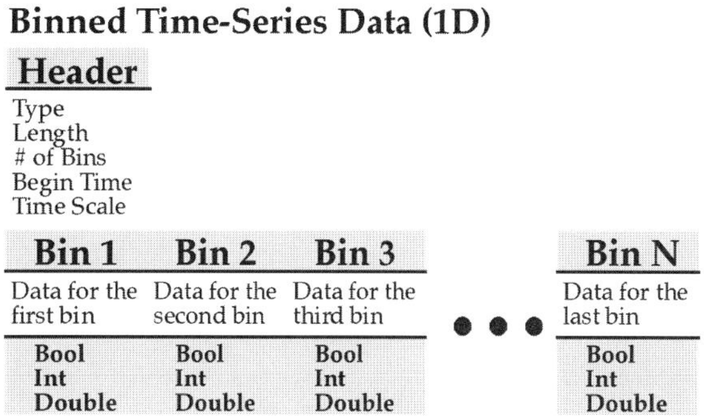

Binned Time-Series Data (1D)

Header
Type
Length
# of Bins
Begin Time
Time Scale

Bin 1	Bin 2	Bin 3	•••	Bin N
Data for the first bin	Data for the second bin	Data for the third bin		Data for the last bin
Bool Int Double	Bool Int Double	Bool Int Double		Bool Int Double

Figure 1 General structure of binned time-series data (one-dimensional case). For the binned data, only the data values themselves are stored. The time interval of the sampling is stored in the header.

Timestamped Time-Series Data (2D)

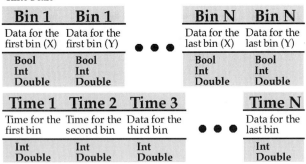

Figure 2 Internal format of time-stamped time-series data (two-dimensional case). For the time-stamped data, the data values are stored along with a time stamp.

Support Functions

The support functions of an opaque data type tell the database server how to interact with the internal structure of the opaque type. Because an opaque type is encapsulated, the database server has no *a priori* knowledge of its internal structure. It is the support functions, which one writes, that interact with this internal structure. The database server calls these support functions to interact with the data type. The table found on the next page summarizes the support functions one needs to define for an opaque data type.

Additional SQL-Invoked Routines

Operator Functions

An operator function is a user-defined function that has a corresponding operator symbol. For an operator function to operate on a newly defined opaque data type, one must write a version of the function for the new data type. When one writes a new version of an operator function, the following rules must be adhered to:

- The name of the operator function must match the name on the server; however, the name is not case sensitive -the plus() function is the same as the Plus() function.
- The operator function must handle the correct number of parameters.
- The operator function must return the correct data type, where appropriate.

Operator functions are divided into three classes: arithmetic (e.g., the plus() operator, which corresponds to +); text (e.g., the concat() operator, which corresponds to ||); relational (e.g., the lessthan() operator which corresponds to <).

Aggregate and Built-In Functions

A built-in function is a predefined function (e.g., exp() and sqrt()) that Informix^{TM} provides for use in an SQL expression. An aggregate function (e.g., min or max) returns one value for a set of queried rows. Informix supports these functions on built-in data types. To have these functions supported on an opaque data type, one must write a new version of the function following these rules:

- The name of the function must match the name on the server; however, the name is not case sensitive -the abs() function is the same as the Abs() function.
- The function must be one that can be overridden.
- The function must handle the correct number of parameters, and these parameters must be of the correct data type.
- The function must return the correct data type, where appropriate.

End-User Routines

The server allows one to define SQL-invoked functions or procedures that the end user can use in SQL expressions and queries. These end-user routines provide additional functionality that a user might need to work with the newly defined data type. Examples of the end-user routines include the following:

- Functions that return a particular value in the opaque data type: Because the opaque type is encapsulated, an end-user function is the only way users can access fields of the internal structure.
- Casting functions: Several of the support functions serve as casting functions between basic data types that the database server uses. Additional casting functions may be written between the opaque type and other data types (built-in, complex, or opaque) of the database.
- Functions or procedures that perform common operations on the opaque data type: If an operation or task is performed often on the opaque type, an end-user routine to perform this task may be written.

Comparing Data

The server supports the following conditions on an opaque type in the conditional clause of SQL statements:

- The IS and IS NOT operators
- The IN operator, if the equal() function has been defined

Function	Purpose
input	Converts the opaque-type data from its external representation to its internal representation
output	Converts the opaque-type data from its internal representation to its external representation
receive	Converts the opaque-type data from its internal representation on the client computer to its internal representation on the server computer
send	Converts the opaque-type data from its internal representation on the server computer to its internal representation on the client computer
import	Performs any tasks required to process an opaque type when a bulk copy imports the opaque type in its external representation
export	Performs any tasks required to process an opaque type when a bulk copy exports the opaque type in its external representation
importbinary	Performs any tasks required to process an opaque type when a bulk copy imports the opaque type in its external representation
exportbinary	Performs any tasks needed to process an opaque type when a bulk copy exports the opaque type in its external representation
assign()	Performs any tasks required to process an opaque type before storing it to disk
destroy()	Performs any tasks required to process an opaque type before the database server removes a row that contains the type
compare()	Compares two values of opaque-type data during a sort

- The BETWEEN operator, if the compare() function has been defined
- Relational operators, if the corresponding operator functions has been defined
- Opaque-type sorting, if the compare() function has been defined

Discussion

The NeuroCore TimeSeries Datatype has been developed to store the complex time-series data acquired during many neurophysiological and behavioral experiments. For other neuroscience database applications, unique data types can be constructed to support the non-standard data that are an integral part of the experiments one wishes to store. For example, a database for neuro-imaging data (see Chapter 3.1 for a discussion of this type of data) must store the three-dimensional images acquired as a part of these experiments. Defining a new opaque data type enables the user to work with complex data as easily as one can work with the standard data types found in relational databases. Furthermore, by allowing a user to access and manipulate the actual data within a non-standard data type, complex queries and data-mining procedures can be developed. For more information on the NeuroCore TimeSeries Datatype, please visit the USCBP website.

APPENDIX A3

USCBP Development Team

The individuals listed in this appendix have all been a part of the USCBP development team. The work detailed in this volume would not have been possible without their contributions.

Principle Investigators

Michael A. Arbib (Director)
Michel Baudry
Theodore W. Berger
Peter Danzig
Shahram Ghandeharizadeh
Scott Grafton

Richard Leahy
Dennis McLeod
Tom McNeill
Larry Swanson
Richard F. Thompson

Core and Administrative Personnel

Amanda Alexander
Luigi Manna
Edriss Merchant

Jonas Mureika
Charles Neerdaels
Paulina Tagle

Staff, Research Associates, Post Doctoral Fellows, Research Assistants, Graduate Students, and Undergraduate Students

Goksel Aslan	Ben Cipriano	Alex Guazzelli
Shaowen Bao	Jeff Cogen	Ning Jiang
Martha Berg	Fernando Corbacho	Nripjeet Josan
Prameela Bhat	Michael Crowley	Jonghyun Kahng
Karan Bhatia	Sunil Dalal	George Kardaras
Xiaoning Bi	Ali Dashti	Dongho Kim
Amanda Bischoff	Alper Kamil Demir	Jeansok Kim
Mihail Bota	Lars Eggert	David King
Jean-Marie Bouteiller	Ragy Eleish	Mohammad Kolahdouzan
Gully Burns	Chetan Gandhi	Karthik Krishnan
Gilbert Case	David Gellerman	Shih Hao Li
Victor Chan	Gary Gitelson	Kevin Liao
Martin Chian	Grant Goodale	Wen-Hsiang Liao
Siu-Ching Ching	Jeffrey Grethe	Jim-Shih Liaw
Choi Choi	Tao Gu	Salvador Marmol

Alok Mishra
Aune Moro
Wael Musleh
Denny Naigowit
Sarveshwar Nigam
Ilya Ovsiannikov
Erhan Oztop
Danjie Pan
Salim Patel
Gorica Petrovich
Yuri Pryandkin
Tejas Rajkotia
Wendy Riley

Mehmet Saglam
Lalith Saldanha
William Scaringe
Weifeng Shi
Ying Shu
Rabi Simantov
Lori Simcox
Donna Simmons
Tayan Siu
Jacob Spoelstra
James Stone
Ashit Tandon
Judith Thompson

Richard H. Thompson
Bijan Timsari
Georges Tocco
JoAnne Tracy
Faheem Uddin
Karen Varilla
Alak Vasa
Zhuo Wang
Gregory Wehrer
Xiaping Xie
Elliot Yan
Roger Zimmerman

APPENDIX B1

Informix SQL Quick Reference

Jonas Mureika and Edriss N. Merchant

*University of Southern California Brain Project,
University of Southern California, Los Angeles, California*

Introduction

This appendix provides a quick reference for the Informix SQL language. This appendix is intended to be a guide to various SQL commands referenced within this book. For more information on Informix™ products and the SQL language implemented as part of the Informix Dynamic Server, one can visit the Informix website at http://www.informix.com.

SQL Statements

CREATE DATABASE

Syntax: CREATE DATABASE db_name
  [with (options)]

Description: The CREATE DATABASE statement creates a new database of the name specified by db_name. The name of the database must be unique on the installation and the name cannot be a keyword.

CREATE FUNCTION

Syntax: CREATE FUNCTION func_name[param_list]
  returns type_name
  [with (modifier [, modifier ...])]
  as {sql_statement | sql_statement_list} |
  external name file_path_spec [(function_name)]
  language {C | c} [[not] variant]

Description: The CREATE FUNCTION statement registers a new function in the database. The function body may be written in SQL or C.

CREATE TABLE

Syntax: CREATE TABLE table_name
  [with (options)]
  [of new type type_name]
  ({column_description | like table_name}
  [{, column_description ... |, like table name ...}]
  [table_constraints_list])
  [under table_name [,table_name ...]]
  [archive]

Description: The CREATE TABLE statement creates a table in the database. The table_name of the table being created must be unique among all table, view, and index names within the user's current schema.

CREATE ROW TYPE

Syntax: CREATE ROW TYPE type_name
  ([internallength = {integer_constant | variable},]
  [input = func_name,]
  [output = func_name]
  [, modifierb [, modifier ...]])
  [under type_name [, type_name ...]]

Description: The CREATE ROW TYPE statement creates a new type in the database. The type_name is the name of the type being created. The type_name must be unique within the schema in which it was created.

CREATE INDEX

Syntax: CREATE INDEX index_name_ix ON table_name ([row_name {ASC | DESC},...]) USING
  [index_method];

Description: The CREATE INDEX statement assigns an indexing marker to the data element (row_name) in the table table_name. Indices can be ascending (ASC) or descending (DESC), and only one of each type may be assigned to a given row. The argument index_type describes how the index is to be structured for search/retrieval purposes and may assume the values 'B_TREE,' 'R_TREE.'

DELETE

Syntax: DELETE from {table_name | only
  (table_name)}
  [using suggestion_list]
  [where search_condition]

Description: The DELETE statement removes rows from a table.

DROP

Syntax: DROP {database db_name | table table_name | type type_desc | func_name (param_list)}

Description: The DROP statement destroys the specified database and removes the specified table, type, and function from the database.

INSERT

Syntax: INSERT into table_name
    [(col_name [, col_name ...])]
    [using suggestion_list]
    {values (expression [, expression ...]) | SELECT_stmt|

Description: The INSERT statement adds rows to a table.

RETURN

Syntax: RETURN expression

Description: The RETURN statement retrieves the results from a function or query. If RETURN is used on a function, it returns a scalar value if that is what the function returns; otherwise, it returns the name of a cursor containing all the returned values.

SELECT

Syntax: SELECT [unique | distinct] SELECT_list
    from from_list
    [using suggestion_list]
    [where search_condition]
    [group by expression, [, expression]]
    [having search_condition]
    [order by order_spec]
    [union [all] SELECT_statement]

Description: The SELECT statement retrieves the data specified in the SELECT_list from the tables, views, or versions specified in the from_list.

APPENDIX C1

USC Brain Project Research Personnel

Name	Degree(s)	Role on Project
Administration		
Michael A. Arbib	Ph.D.	Project Director
Paulina Tagle	B.S.	Administrative Support
Luigi Manna	M.S.	Systems Administration
Tao Gu	PhD cand	RA – Systems Adm
Karthik Krishnan	M.S.	RA – Systems Admin
Thanawut Naigowit	B.S.	RA – Webmaster
Modeling		
Michael A. Arbib	Ph.D.	PI
Amanda Bischoff	PhD cand	Research Asst (RA)
Mihai Bota	PhD cand	RA
Fernando Corbacho	PhD cand	RA
Michael Crowley	PhD cand	RA
Jeffrey Sean Grethe	PhD cand	RA
Alex Guazzelli	PhD cand	RA
Jacob Spoelstra	PhD cand	RA
EONS		
Jim-Shih Liaw	Ph.D.	USCBP Faculty
Ying Shu	PhD cand	RA
Neural Simulation Language and Programming Support		
Amanda Alexander	M.S.	Systems Programmer
Prameela Bhat	B.S.	RA Asst programmer
Karan Bhatia	M.S.	RA Asst programmer
Siu-Ching Ching	M.S.	RA Asst programmer
Alper Kamil Demir	M.S.	RA Asst programmer
Lars Eggert	M.S.	RA Asst programmer
Ragy Eleish	M.S.	RA Asst programmer
Chetan Gandhi	M.S.	RA Asst programmer
David Gellerman	B.S.	RA
Grant Goodale	B.S.	RA Asst programmer
Nripjeet Josan	M.S.	RA Asst programmer
George Kardaras	M.S.	RA Asst programmer
Sarveshwar Nigam	M.S.	RA Asst programmer
Danjie Pan	M.S.	RA Asst programmer
Lalith Saldanha	M.S.	RA Asst programmer
Tayan Siu	M.S.	RA Asst programmer
Neuro Core		
Charles Neerdaels	B.S.	Systems Programmer
Jonas Mureika	B.S.	Systems Programmer
Edriss Merchant	B.S.	Systems Programmer
Jeffrey Sean Grethe	PhD cand	RA
Neurochemistry, Synaptic Plasticity, and Registration		
Michel Baudry	Ph.D.	Co-PI
Richard Leahy	Ph.D.	USCBP Faculty
Rabi Simantov	Ph.D.	Collaborator/Visiting Professor
Georges Tocco	Ph.D.	Research Associate
Xiaoning Bi	Ph.D. cand	RA
Jean-Marie Bouteiller	Ph.D. cand	RA
Wael Musleh	Ph.D. cand	RA
Salim Patel	PhD cand	RA
Bijan Timsari	Ph.D. cand	RA
Data Management		
Peter Danzig	Ph.D.	Co-PI
Shahram Ghandeharizadeh	Ph.D.	Co-PI
Dennis McLeod	Ph.D.	Co-PI
Cyrus Shahabi	Ph.D.	USCBP Faculty
Goksel Aslan	Ph.D. cand	RA
Jonghyun Kahng	Ph.D. cand	RA
Dongho Kim	M.S.	RA
Shih Hao Li	M.S.	RA
Wen-Hsiang Liao	PhD cand	RA
Ilya Ovsiannikov	PhD cand	RA
Weifeng Shi	M.S.	RA
Lori Simcox	B.S.	RA
Elliot Yan	M.S.	RA
Roger Zimmerman	PhD cand	RA
Hippocampus		
Theodore Berger	Ph.D.	Co-PI

Continues

Continued

Name	Degree(s)	Role on Project
Victor Chan	M.S.	RA
Choi Choi	M.S.	RA
Mehmet Saglam	M.S.	RA
William Scaringe	M.S.	RA
Xiaping Xie	M.D.	Research Associate
Basal Ganglia		
Tom McNeill	Ph.D.	Co-PI
Michael A. Arbib	Ph.D.	PI
Amanda Bischoff	PhD cand	Research Asst (RA)
Jeff Cogen	PhD cand	Grad Student
Michael Crowley	PhD cand	RA
Cerebellum/Time Series		
Richard Thompson	Ph.D.	Co-PI
Judith Thompson	M.S.	Research Associate
Shaowen Bao	PhD cand	RA
Martha Berg	PhD cand	RA
Gilbert Case	PhD cand	RA
Lu Chen	PhD cand	RA
Jeffrey Sean Grethe	PhD cand	RA

Name	Degree(s)	Role on Project
Stephanie Hauge	PhD cand	RA
Jeansok Kim	PhD cand	Research Associate
David King	PhD cand	RA
Aune Moro	M.S.	Technician
JoAnne Tracy	PhD cand	RA
Brain Maps and Atlases		
Larry Swanson	Ph.D.	Co-PI
Gully Burns	Ph.D.	Research Associate – Neuroscholar
Ali Dashti	PhD cand	RA
Gorica Petrovich	PhD cand	RA
Donna Simmons	Ph.D.	Research Associate
James Stone	PhD cand	RA
Brain Imaging/Synthetic PET		
Scott Grafton	M.D.	Co-PI
Michael A. Arbib	Ph.D.	PI
Amanda Bischoff	PhD cand	Research Asst (RA)
Yuri Pryadkin	PhD cand	Research Asst (RA)

APPENDIX C2

Doctoral Theses from the USC Brain Project (May 1997–August 2000)

Michael Crowley (May 1997): "Modeling Saccadic Motor Control: Normal Function, Sensory Remapping and Basal Ganglia Dysfunction"

Fernando Corbacho (August 1997): "Schema Based Learning: Towards a Theory of Organization for Adaptive Autonomous Agents"

Jonghyun Khang (December 1997): "Mediation of Information Sharing in Cooperative Federated Database Systems: Ontologies, Mediations and Data Mining"

Goksel Aslan (May 1998): "Semantic Heterogeneity Resolution in Federated Databases by Meta-Data Implementation and Stepwise Evolution"

Amanda Bischoff (May 1998): "Modeling the Basal Ganglia in the Control of Arm Movements"

Jacob Spoelstra (July 1999): "Cerebellar Learning of Internal Models for Reaching and Grasping: Adaptive Control in the Presence of Delays"

Ali Dashti (August 1999): Data Placement Techniques for Hierarchical Multimedia Storage Systems.

Alex Guazzelli (August 1999): "Integrating Motivation, Spatial Knowledge, and Response Behavior in a Model of Rodent Navigation"

Jeffrey Sean Grethe (May 2000): "Neuroinformatics and the Cerebellum: Towards and Understanding of the Cerebellar Microzone and its contribution to the Well-timed Classically Conditioned Eyeblink Response"

Bijan Timsari (May 2000): "Geometrical Modeling and Analysis of Cortical Surface: An Approach to Finding Flat Maps of the Human Brain"

Khan, Latifur (August 2000): "Ontology-Based Customization for Multimedia Information Retrieval"

Index

A

Access to databases, 7–8
Active objects, 34
Affordance cells, 63
Affordances, 46
Annotations
　atoms, 256–257, 292
　clumps, 20–21, 257, 259–260
　database, 260–261, 263
　description of, 7–8, 22–23
　explicit, 258–259
　handwritten, 257–259
　idiomatic, 258
　implicit, 258
　links, 259
　memos, 262
　NeuroScholar creation of, 293
　pictorial, 258–259
　prevalence of, 255
　ranges, 258–259
　search for, 260
　software technology for, 255–256
　symbol presentation of, 259
Annotator
　advantages of, 22
　description of, 22, 27, 261
　functionality of, 261–262
　implementation, 262–263
　resources, 263
　user interface, 261–262
Applets, 4
Applications, 4
Arm movements, basal ganglia's role in, 50–53
Article repositories, 5–6, 20, 287
Atlas, of brain
　cell groups, 171
　cryosection imaging-based, 206
　expandable database uses, 176–177
　experimental data transferred to, 173–174
　expression of, 170–171
　fiber tracts, 171
　future of, 176–177
　magnetic resonance imaging-based, 206
　multi-modality, 206
　slice-based sampling, 169–171
　Swanson, 17, 179–180, 184–185, 209
　typical type of, 171–173
Atlas-based databases
　challenges for, 19
　description of, 17–18
　NeuARt, 17–18
　neurochemical data
　　database design, 219–221
　　description of, 217
　　NeuroCore repository incorporation of, 222–225
　　synaptic neurotransmission, 217–218
Atoms, 256–257, 292
atomTypes, 306–307
Attributes
　definition of, 35
　inheritance of, 36–37
　in object-based database model, 35–36

B

Basal ganglia
　Crowley-Arbib model of, 73
　description of, 9, 43
　in saccade control, 48–50
　in sequential arm movements, 50–53
Brain. *see also specific anatomy*
　flatmaps of, 176
　organization of, 168–169
　slice-based sampling of, 169–171
　structures of
　　chemoarchitecture, 338
　　cryoarchitecture, 338
　　homology of, 339
　　modules as, 45

Brain (*continued*)
    time-series variation, 186
    three-dimensional computer graphics model of, 174–176
    weight of, 168
Brain atlases
    brain-slice images registered against, 186
    cell groups, 171
    cryosection imaging-based, 206
    expandable database uses, 176–177
    experimental data transferred to, 173–174
    expression of, 170–171
    fiber tracts, 171
    future of, 176–177
    magnetic resonance imaging-based, 206
    multi-modality, 206
    slice-based sampling, 169–171
    Swanson, 17, 179–180, 184–185, 209
    typical type of, 171–173
Brain maps
    experimental methods, 169
    features of, 167–168
    interactive, 171–173
    linear rescaling, 173–174
Brain models on the web
    components of, 291–292
    database creation, 261
    definition of, 297
    description of, 11–12, 22, 237, 291
    future directions for, 316–317
    goal of, 297
    NeuroCore links, 317
    principles of, 25–26
    schematic capture system links, 316–317
    standards, 298
    storing of, 297–298
    summary database embedded in
        browsing of, 308–313
        clumps
            browsing of, 310–312
            description of, 306
            inserting of, 316
            searching for, 313
        description of, 27, 261, 291–292, 298–299
        documents
            description of, 306
            searching for, 313
        experiments, 307–308
        logging into, 308
        models
            description of, 303–305
            inserting of, 314–316
            searching for, 312–313
        modules
            browsing of, 309
            description of, 305–306
        overview of, 300–301
        persons
            description of, 302–303
            inserting of, 314
        references
            browsing of, 309–311
            inserting of, 313–314

        synthetic positron emission tomography links, 317
        tables, 301–302
Brain theory, 43–44
Broca's area, 44
Buffering, of modules, 77

C

Central nervous system, 168
Cerebellum
    in classical conditioning
        description of, 56–61
        time-series database for studying, 120–125
    description of, 9, 43
    excitatory cells, 57–58
    functions of, 53
    inhibitory cells, 57–58
    microcomplexes of, 53, 58–61
    mossy fiber/parallel fiber inputs, 53, 57–59, 61
    motor control and coordination models, 53–56
    in prism adaptation, 54, 56
    schematic representation of interactions, 55
Circuit-tracing methods, 169
Classes
    definition of, 35
    in object-based database model, 35–36
Classical conditioning
    cerebellum's role in, 56–61, 120
    description of, 56–57
    time-series database designed for data
        gathering of, 121–122
        inputting of, 124–125
    description of, 120
    experimental preparation, 120–121
    NeuroCore database extensions, 122–124
Classificational object-based data model, 237–238, 247–338
_clumpModel, 306
Clumps
    browsing of, 310–312
    description of, 20–21, 257, 289–290
    distributed, 259–260
    function of, 306
    linking of, 306–307
    retrieval of, 293–294
Cognitive map
    description of, 61–62
    locale system, 62
    taxon system, 62
Collaboratory databases, 7
Composite concepts, 248–250
Conceptual intersection, 248
Conceptual negation, 248
Conceptual union, 248
Concurrency, of neural networks, 74
Conditioned stimulus, 14
conformacyType, 305
Connection masks, 78
Cooperative federated databases
    environmental characteristics of, 247
    heterogeneity in
        common ontology uses, 244
        data model, 242–244

INDEX 373

description of, 242
semantic, 242–244
information flow patterns in, 248
information frameworks, 241
information repositories, 241
information sharing, 241–242
schematic representation of, 242
Core tables, 137, 301
Coriolis forces, 54

D

Dart model, of prism adaptation, 299, 302
Data
  access to, 7–8
  accessibility of, 117
  annotated, 7–8, 22–23
  heterogeneous, 4, 131–132
  mining of, 132
  types of, 5–7
Data model mapping, 220
Data schema, 6
Data type
  description of, 37
  opaque, 37–38
Database
  access to, 7–8
  annotation, 260–261, 263
  article repositories, 5–6
  atlas-based
    challenges for, 19
    description of, 17–18
    NeuARt, 17–18
    neurochemical data
      database design, 219–221
      description of, 217
      NeuroCore repository incorporation of, 222–225
      synaptic neurotransmission, 217–218
  collaboratory, 7
  definition of, 29
  designing of
    schema complexity addressed in, 118–119
    user community considerations, 119–120
  developments, 30–31
  extending of, 132
  linking of, 5
  management system, 29–30
  NeuroCore. see NeuroCore database
  neuroimaging data
    data, 129–131
    description of, 129
  object-based
    classification of objects, 246
    description of, 4, 30
  object-relational, 4, 30
  personal laboratory, 7
  public refereed, 7
  relational
    classification of objects, 246
    description of, 4, 30–32
  structures stored in, 7
  summary. see Summary database
  time-series. see Time-series database
Database model
  definition of, 30
  object-based
    attributes, 35–37
    characteristics of, 34
    classes, 35–36
    components of, 35
    description of, 33–34
  object-relational, 37–39
  relational, 31–32
  schema complexity addressed, 118–119
DataMunch
  description of, 160
  development of, 161
  future applications, 160
  operating principles of, 160
DIANE, 256
Dictionary tables, 137, 301
documentType, 306
Domain integrity, 31
Dominey-Arbib model, 50
Dynamic classificational ontology
  base ontology, 232, 247–249
  classificational object-based data model, 237–238, 247–338
  components of, 232
  definition of, 241
  derived ontology, 232, 247, 249–250
  description of, 21, 247
  information sharing mediators
    description of, 241, 250–251
    discovery, 252–253
    export, 251–252
  principles of, 232
  prototype, 254
Dynamic synapses
  axon terminal, 97
  description of, 95–96
  leaky-integrator equation, 96
  learning algorithm, 96–97
  mathematical expressions of, 96
  speech recognition case study, 97

E

Elementary objects of the nervous system
  components of, 13
  description of, 12, 91–92, 100
  object library, 92–97, 100
  object-oriented design of, 91, 100
  principles of, 12–13, 91
  protocol-based simulations
    advantages of, 100
    description of, 98
    long-term potentiation hypotheses studied using, 98–99
  synapse modeling
    description of, 93–95
    dynamic synapse neural network
      axon terminal, 97
      description of, 95–96
      leaky-integrator equation, 96

Elementary objects of the nervous system (*continued*)
    learning algorithm, 96–97
    mathematical expressions of, 96
    speech recognition case study, 97
Embedded objects, 162
EndNotes, 256
Entity integrity, 31
EVEREST
  description of, 267–269
  free list, 272
  implementation of, 275
  limitations of, 275
  objects
    allocation of, 272–273
    de-allocation of, 273–275
  preventive characteristic of, 281
Excitatory amino acid transporters, 223
Experimental protocol
  data-gathering procedures and, 118
  description of, 5, 11, 46
experimentBlock, 308
Explicit semantic primitives, 34
Extendible parent tables, 137–138, 301–302
External globus pallidus, 50

F

Fagg-Arbib-Rizzolatti-Sakata model, of parietal-premotor interactions in control of grasping, 46, 107–111
Federated databases
  cooperative
    environmental characteristics of, 247
    heterogeneity in
      common ontology uses, 244
      data model, 242–244
      description of, 242
      semantic, 242–244
    information flow patterns in, 248
    information frameworks, 241
    information repositories, 241
    information sharing, challenges associated with, 241–242
    schematic representation of, 242
  definition of, 231
  description of, 5, 19–20, 231–232
  goal of, 231
  information discovery, 232
  information sharing
    caching method of, 235
    copy method of, 235
    direct link method of, 235
    export determinations, 236–237
    import determinations, 236–237
    import/export based sharing paradigm, 233–234
    primitives/tools for, 236–237
    remote-query execution method of, 236
    time-bounded copy of, 236
  information sharing phases, 38–39, 231
  NeuroCore provisions, 136
  neuroscience, 237–238
  overview of, 38–39
  schematic representation of, 232
  semantic heterogeneity resolution, 232–233
  system-level interconnection, 233–234
Flatmaps, 176

G

GENESIS modeling system, 5, 9
Geographical information system
  coverages, 182–183, 185–186
  definition of, 180
  description of, 179–180, 186–187
  layers, 181–182
  raster data
    coverages, 185–186
    description of, 180, 185
    spatial relationships, 185
  rectification, 183–184, 186
  registration, 183–184, 186
  spatial data, 180–181
  spatial relationships, 181
  vector data, 182
Glutamate
  receptors
    description of, 218
    regulation of, 218–219
  transporters of
    description of, 222–223
    excitatory amino acid transporters, 223
    kainic acid regulation, 224–225
Ground control point rectification/registration, 184

H

Heterogeneous data
  description of, 4
  handling considerations for, 131–132
Hierarchical storage systems
  capacity of, 266–267
  components of, 266
  description of, 265–266
  space management
    analytical model of, 279–283
    description of, 267
    techniques for
      Dynamic, 269–272
      performance evaluations of, 276–279
      REBATE, 269, 275–276, 282
      simulation studies, 276–279
      Standard, 269
  target environment, 268–269
Hippocampus
  description of, 9, 43
  in spatially guided navigation
    description of, 61–62
    locale system, 62
    parietal cortex interactions, 62–64
    taxon system, 62
    taxon-affordance model, 64–66
    world graph model, 64–66
  in trace conditioning, 57
Hodgkin-Huxley model, 72
Holism, 44
Homology. *see also* NeuroHomology database

criteria, 338–339, 349–350
degree of, 339–340
description of, 337
myeloarchitectonics, 338
organization levels, 337
in rat and monkey brains, 338
Hybridization histochemical localization, 169

I

Immunohistochemical localization, 169
Individual object identity, 34
Inferotemporal cortex, 48
Information discovery, in federated databases, 232
Information retrieval, 245
Information unit, 29
Informix, 355–358, 363–364
Interfaces
  commercial grade, 132
  definition of, 7
  extension, for NeuroCore database
    AnatData, 194
    areas for, 147–149
    description of, 136–137
    NeuARt system, 191–192
    time-series databases
      classical conditioning, 122–124
      long-term potentiation, 127–129
  NeuARt system
    Active Set Manager, 198–199
    data access, 194
    description of, 192–193
    Display Manager, 195–197
    Inquiry Manager, 200
    Level Manager, 199–200
    modules, 193–195
    Query Manager, 197
    Results Manager, 198–199
    schematic representation of, 195
    Spatial Index Manager, 197
    spatial queries, 197–198
    Viewer Manager, 200
Internet resource directory, 245–246

J

JADE. *see* Java applet for data entry
Java, 10, 71–72
Java applet for data entry, 124–125, 156–157

K

_keywordClumps, 306
Knowledge mechanics, 319–320, 333–334

L

Landmarks, 210
Leaky integrator model, 44, 72, 96
Leaky-integrator equation, 96
Lightweight personal database system description of, 238
  schematic representation of, 238

Linear rescaling, 173–174
Long-term depression, glutamatergic synapses in, 218
Long-term potentiation
  calpain protease effects, 222
  description of, 55
  glutamate transporters in, 223
  glutamatergic synapses in, 218
  simulation protocols for studying, 98–99
  stages of, 125
  time-series database created for
    data, 126–127
    description of, 125–126
    experimental preparation, 126
    extensions for NeuroCore, 127–129

M

Maps
  brain
    experimental methods, 169
    features of, 167–168
    interactive, 171–173
    linear rescaling, 173–174
  features of, 167–168
Maximum selector model, 85–89
Membrane potential
  definition of, 8, 44, 78
  time evolution of, 8, 44
Memos, 262
MLC cells
  description of, 62
  in spatially guided navigation, 63–64
Model. *see also specific model*
  compartmental, 9
  definition of, 45
  description of, 7
  methodological issues, 9–10
  system, 9
  types of, 9
Model for on-line publications
  description of, 22, 151
  prototype, 161
Model repository
  characteristics of, 291
  definition of, 287
  description of, 6–7, 288
Modeling
  definition of, 4
  experimentation and, integration of, 44–45
  of neural networks, 72
  of synapses, 78
_modelModules, 305
modelOSSpecific, 303–304
modelVersion, 303
Modularity
  description of, 72
  of neural networks, 72–73
  object-oriented programming, 74
Module
  as brain structures, 45
  as neural network, 77–79
  as schemas, 45–46

Module (*continued*)
  definition of, 7
  libraries, 305
Module assemblages, 76
moduleVersion, 305
Moment matching, 205
Motor pattern generators, 53–54
Multilevel simulations, 13

N

Navigational system
  hippocampal role in
    description of, 61–62
    locale system, 62
    parietal cortex interactions, 62–64
    taxon system, 62
  taxon-affordance model, 64–66
  world graph model, 64–66
NeuARt system
  description of, 17–19, 26, 189, 237
  development of, 189–190
  function of, 190–191, 266
  hierarchical storage system, 266
  neuroanatomical extensions for NeuroCore database, 191–192
  overview of, 18
  schematic representation of, 191
  three-dimensional registration integrated with, 227
  user interfaces
    Active Set Manager, 198–199
    data access, 194
    description of, 192–193
    Display Manager, 195–197
    Inquiry Manager, 200
    Level Manager, 199–200
    modules, 193–195
    Query Manager, 197
    Results Manager, 198–199
    schematic representation of, 195
    Spatial Index Manager, 197
    spatial queries, 197–198
    Viewer Manager, 200
Neural networks
  artificial
    description of, 43–44
    levels of detail in, 8–9
  concurrency of, 74
  definition of, 8
  dynamic synapse
    axon terminal, 97
    description of, 95–96
    leaky-integrator equation, 96
    learning algorithm, 96–97
    mathematical expressions of, 96
    speech recognition case study, 97
  modeling of, 72
  modularity of, 72–73
  modules as, 77–79
  simulation of, 72
  speech recognition case study, 97
Neural simulation language system
  applications of, 26–27
  brain models on the web, 11–12
  computational model, 75
  description of, 4, 26–27, 71
  environments, 71–72
  Executive window, 81–82
  history of, 71
  Java, 71–72
  legacy architecture, 84
  main system, 83
  models
    description of, 74–75
    maximum selector, 85–89
    simulating of, 85
  modules
    buffering, 77
    connectivity, 76–77
    description of, 74–75
    file types, 83
    hierarchical, 75–76
    interconnections, 75–77
    library structures, 83–85
    neural network use of, 77–79
    scheduling, 77
  overview of, 10, 67
  resources, 89–90
  schematic capture system, 10–11, 80–83, 85
  simulation, 79–80, 83, 85
  subsystems, 83
  threshold functions, 78
Neural summaries on the web, 237
Neuro Slicer
  description of, 208–209
  experimental data matched to atlas, 210–211
  landmarks, 210–211
  results obtained, 211–213
  surface atlas reconstruction, 209
  warping of experimental image, 211
NeuroChem
  architecture of, 221
  development of, 225–227
  three-dimensional registration and NeuARt integrated with, 227
Neurochemical data, atlas-based databases of
  database design, 219–221
  description of, 217
  NeuroCore repository incorporation of, 222–225
  synaptic neurotransmission, 217–218
NeuroCore database
  access levels, 140
  architecture of, 15–16
  brain models on the web linked to, 317
  classical conditioning database extensions for, 122–124
  components of, 135–137
  contributions of, 131
  core database schema
    categories, 138
    for contact information, 140–141
    description of, 136
    protocols, 142–145
    reference sections, 145–147
    for scientific experiment, 141–145

top-level, 138–140
database browser
  components of, 158–159
  description of, 157–158
  future applications, 159
  illustration of, 158
  implementation of, 159
  specialized query result field, 158–159
DataMunch
  description of, 160
  development of, 161
  future applications, 160
  operating principles of, 160
data-sharing levels, 139
description of, 15, 26, 237
design principles of, 135
extension interface
  AnatData, 194
  areas for, 147–149
  description of, 136–137
  NeuARt system, 191–192
  time-series databases
    classical conditioning, 122–124
    long-term potentiation, 127–129
federation interface, 136
Java applet for data entry, 156–157
long-term potentiation database extensions, 127–129
neurochemical data incorporated into, 222–225
NeuroScholar representation of, 332
objects in, 137
on-line interface, 151–152
schema browser
  description of, 153
  evolution of, 153
  future work for, 154, 156
  illustration of, 155
  implementation of, 154
  operating principles of, 154
schematic representation of, 155
tables
  of Anat-RED, 192
  core, 137
  dictionary, 137
  extendible parent, 137–138
  extension, 137–138
  set backing, 137
time-series datablade, 137, 149, 359–361
Neurohomology database
  brain structures, 341–343
  connectivity, 343–346
  definition of, 294
  description of, 24–25, 340–341, 348
  homologies, 294–296, 346–349
  modules, 340
  schematic representation of, 24, 340
  tract-tracing techniques, 344
Neuroinformatics, defined, 3
NEURON modeling system, 5, 77
Neurons
  anatomy of, 168
  firing rate of, 78
  morphology of, 338

NeuroScholar
  annotations using, 293
  description of, 23–24, 237, 292, 320
  design concepts of, 321–322
  NeuroCore database design represented in, 332
  non-tract-tracing data applications of, 332–333
  primitives, 292–293, 295, 324–326
  principles of, 292–293
  relations in, 331
  software
    design considerations, 324–327
    requirements of, 322–324
  spatial relationships, 327
  summary database, 23–24
  tract-tracing experiments
    description of, 328–329
    logical structure of, 329–331
  users' knowledge representation
    adjusting of, 323
    analyzing of, 323
    building of, 322–323
    querying of, 323
    validating of, 323
Neuroscience
  federated databases, 237–238
  hypotheses posited, 320–321
  theory concept in, 320–321
N-Methyl-D-aspartate, 125, 218
Nonlinear warping, 174
Non-metric multidimensional analysis, 323
Nonparametric cluster analysis, 323

O

Object uniformity, 34
Object-based database model
  attributes, 35–37
  characteristics of, 34
  classes, 35–36
  classificational, 237–238, 247–338
  components of, 35
  description of, 33–34
Object-based databases, 4
Object-oriented programming, 74, 92–93
Object-relational database, 4
Object-relational database model, 37–39
Objects
  active, 34
  classification of, 246
  embedded, 162
Ontology
  base, 232, 247–249
  common
    classifications, 246–247
    database interoperation, 245
    description of, 244
    information retrieval, 245
    information sharing issues regarding, 244
    Internet resource directory, 245–246
  definition of, 21, 244
  derived, 232, 247, 249–250
  dynamic classificational

Ontology (*continued*)
   base ontology, 232, 247–249
   classificational object-based data model, 237–238, 247–338
   components of, 232
   definition of, 241
   derived ontology, 232, 247, 249–250
   description of, 21, 247
   information sharing mediators
      description of, 241, 250–251
      discovery, 252–253
      export, 251–252
   principles of, 232
   prototype, 254
Optimal set analysis, 323

P

Parietal cortex
   hippocampus and, interactions between, 62–64
   premotor cortex and, in control of grasping
      description of, 43, 46
      F5 selection, 46, 48, 108, 110
      Fagg-Arbib-Rizzolatti-Sakata model, 46, 107–111
      synthetic positron emission tomography findings, 48, 107–111
Personal laboratory databases, 7
_personImplementation, 305
_personModel, 305
Phaseolus vulgaris leuco-agglutinin, 189–190
Point pairs matching, 205
Positron emission tomography
   neurophysiology considerations, 104
   operating principles of, 103–104
   synthetic
      activation differences, 111–112
      advantages of, 111–112
      definition of, 104
      description of, 13–14
      human positron emission tomography findings compared with, 109–111
      localization, 104
      modeling activation, 104–105
      parietal-premotor cortex interactions in control of grasping, 48, 107–111
      principles of, 104
      saccadic eye movement studies using, 105–107
   three-dimensional models of brain structure, 174–176
Prefrontal cortex, 74
Premotor-parietal cortex, in control of grasping. *see* Parietal cortex
Primitives
   definition of, 292
   explicit semantic, 34
   NeuroScholar, 292–293, 295
Prism adaptation
   cerebellar role in, 56
   Dart model of, 299, 302
   description of, 56
Protocol viewer, 162
Protocol-based simulations
   description of, 98

long-term potentiation hypotheses studied using, 98–99
Public refereed databases, 7
Publishing, model for on-line, 22
Purkinje cells
   climbing fibers, 55
   firing rates of, 58–60
   input for, 55

R

Random access memory, 266–267
REBATE, 267, 275–276, 282
Rectification, 183–184
Refereed databases, 7
Referential integrity, 31
Registration, of autoradiographs
   classification criteria, 204
   description of, 203–204
   intersubject
      brain atlases, 206
      curve matching, 207
      definition of, 206
      elastic approach, 208
      intensity-based strategies, 207–208
      model-based strategies, 206–207
      point pairs matching, 207
      surface matching, 207
      viscous fluid approach, 208
   intrasubject
      curve matching, 205
      description of, 204–205
      intensity-based strategies, 205–206
      point pairs matching, 205
      surface matching, 205
   NeuARt integration, 227
   transformation method, 203–204
Relational databases, 4, 30–32
relevancyType, 307
Repositories
   description of, 117–118, 265–266
   empirical data, 6, 20, 288, 299
   model, 6–7
   space management of
      analytical model of, 279–283
      description of, 267
      techniques for
         Dynamic, 269–272
         performance evaluations of, 276–279
         REBATE, 269, 275–276, 282
         simulation studies, 276–279
         Standard, 269
_reviewedBy, 304
reviewStatusType, 304
Rubbet sheeting operation, 184

S

Saccades, reflex control of
   basal ganglia's role in, 9, 48–50
   cerebral cortex's role, 49
   description of, 10, 105–106
   model of, 51

synthetic positron emission tomography studies of, 105–107
Scheduling, of modules, 77
Schema
    data model for, 118–119
    definition of, 30
    modules as, 45–46
Schematic capture system, 10–11, 80–82, 85, 291, 316–317
Semantic heterogeneity resolution, 232–233
Set backing tables, 137, 301
Simulation
    definition of, 7, 72
    multilevel, 13
    of neural networks, 72
    in neural simulation language system, 79–80, 83, 85
    protocol-based
        description of, 98
        long-term potentiation hypotheses studied using, 98–99
    space management techniques for hierarchical storage systems, 276–279
Simulation interface, 46
Spatial Index Manager, 197
Speech recognition, dynamic synapses neural modeling study of, 97
SQL, 32–33, 151–152, 355–358, 363–364
SQM, 197–198
Submodules, 76
Subthalamic nucleus, 50
Summary database
    in Brain models on the web
        browsing of, 308–313
        clumps
            browsing of, 310–312
            description of, 306
            inserting of, 316
            searching for, 313
        description of, 27, 261, 291–292, 298–299
        documents
            description of, 306
            searching for, 313
        experiments, 307–308
        logging into, 308
        models
            description of, 303–305
            inserting of, 314–316
            searching for, 312–313
        modules
            browsing of, 309
            description of, 305–306
        overview of, 300–301
        persons
            description of, 302–303
            inserting of, 314
        references
            browsing of, 309–311
            inserting of, 313–314
    data stored in, 289
    definition of, 287, 289
    description of, 6, 20–21, 261, 288
    overview of, 288–289
    refereed, 290
Summary statement, 289
Superior colliculus, 48

Swanson atlas, 17, 179–180, 184–185, 209
Synapse
    dynamic
        axon terminal, 97
        description of, 95–96
        leaky-integrator equation, 96
        learning algorithm, 96–97
        mathematical expressions of, 96
        speech recognition case study, 97
    in long-term depression, 218
    in long-term potentiation, 218
    modeling of
        description of, 78
        elementary objects of the nervous system, 93–95
        simulations, 94–95
    neurotransmission methods, 217–218
    plasticity of, 218–219
    schematic representation of, 93
Synthetic positron emission tomography
    activation differences, 111–112
    advantages of, 111–112
    brain models on the web linked to, 317
    definition of, 104
    description of, 13–14
    human positron emission tomography findings compared with, 109–111
    localization, 104
    modeling activation, 104–105
    parietal-premotor cortex interactions in control of grasping, 48, 107–111
    principles of, 104
    saccadic eye movement studies using, 105–107

T

Tables
    of Anat-RED, 192
    core, 137, 301
    dictionary, 137, 301
    extendible parent, 137–138, 301–302
    extension, 137–138
    set backing, 137, 301
Taxon-affordance model, 64–66
Taxon-affordance model–world graph model, 66
Three-dimensional registration, of two-dimensional cryosection models
    classification criteria, 204
    description of, 203–204
    intersubject
        brain atlases, 206
        curve matching, 207
        definition of, 206
        elastic approach, 208
        intensity-based strategies, 207–208
        model-based strategies, 206–207
        point pairs matching, 207
        surface matching, 207
        viscous fluid approach, 208
    intrasubject
        curve matching, 205
        description of, 204–205
        intensity-based strategies, 205–206

Three-dimensional registration, of two-dimensional cryosection models (*continued*)
    point pairs matching, 205
    surface matching, 205
  NeuARt integration, 227
  transformation method, 203–204
Time series data
  description of, 14–15
  historical types of, 14
Time series data protocol, 5
Time-series database
  capabilities of, 261
  classical conditioning
    data
      gathering of, 121–122
      inputting of, 124–125
    description of, 120
    experimental preparation, 120–121
    extensions for NeuroCore, 122–124
  description of, 261
  long-term potentiation
    data, 126–127
    description of, 125–126
    experimental preparation, 126
    extensions for NeuroCore, 127–129
Trace conditioning
  definition of, 57
  hippocampal role in, 57

U

Unconditioned response, 14
Unconditioned stimulus, 14
UNIX Fast File System, 269–270

W

World graph model, 64–66
World Wide Web
  brain models on. *see* Brain models on the web
  classification of, 246–247
  database federation, 5
  description of, 3–4
  graphical databases, 174
  resource-sharing tools, 246
  textual databases, 174